17th European Conference on Mixing (MIXING 17)

AA005036

Porto, Portugal
2-5 July 2023

ISBN: 978-1-7138-9473-5
eISBN: 978-1-7138-9472-8

Printed from e-media with permission by:

Curran Associates, Inc.
57 Morehouse Lane
Red Hook, NY 12571

Some format issues inherent in the e-media version may also appear in this print version.

Printed with permission by Curran Associates, Inc. (2024)

For permission requests, please contact MIXING 17
at the address below.

MIXING 17
Departamento de Engenharia Química
Faculdade de Engenharia da Universidade do Porto
Rua do Dr. Roberto Frias
4200-465 Porto
Portugal

Phone: 351 22 041 3628

organization@mixing17.eu

Additional copies of this publication are available from:

Curran Associates, Inc.
57 Morehouse Lane
Red Hook, NY 12571 USA
Phone: 845-758-0400
Fax: 845-758-2633
Email: curran@proceedings.com
Web: www.proceedings.com

TABLE OF CONTENTS

CF1 – CFD

B2 – BIO TECHNOLOGY

MT1 – MIXING TECHNOLOGY AND QUANTIFICATION

ER1 – EMULSION AND RHEOLOGY

PP1 – PRODUCTS

MT2 – MIXERS AND MIXING TECHNOLOGY

MF1 – MULTIPHASE MIXING

POSTER PRESENTATIONS

MIXING 17

17th European Conference on Mixing

2nd – 5th July 2023
Porto, Portugal

BOOK OF ABSTRACTS

http://mixing17.eu

Sponsored by

SRE CM
LABORATORY OF SEPARATION AND REACTION ENGINEERING
LABORATORY OF CATALYSIS AND MATERIALS

NOV

M-STAR

MICE molds & injected components engineering

U.PORTO
FEUP FACULDADE DE ENGENHARIA
UNIVERSIDADE DO PORTO

EFCE
European Federation of Chemical Engineering
Europäische Föderation für Chemie-Ingenieur-Wesen
Fédération Européenne du Génie Chimique

Preface

Mixing is at the heart of most chemical processes, and it is generally the cornerstone when moving from lab to industrial processes. The widespread of CFD, and advanced fluid mechanics experimental techniques led to great evolutions on this field of chemical engineering over the last decades. The European Conferences on Mixing have been a privileged forum attracting academics and industrialists from all over the world to discuss progress on mixing technology, research, and applications in the process industries.

The 17th European Conference on Mixing, Mixing 17, was held in Porto, Portugal, from July 2 to July 5, 2023. This successful conference series is organized by the European Federation of Chemical Engineering – Working Party on Mixing; it has been running since 1974. Mixing 17 ran under the unifying theme of ***mixing as a solution for fundamental problems of modern societies***. An exciting scientific program was developed around eight main scientific topics:

- Mixing in Turbulent Flows
- Mixing in Laminar Flows
- Mixing in Multiphase Flows
- Mixing and Complex Rheology
- Mixing Reactive Systems and Product Development
- Mixing Equipment and Technology
- Mixing Modelling, Simulation and Theory
- Mixing in Biotechnology Applications

This conference was previously held in several venues, reaching very high scientific levels, enabling all participants to keep up with state-of-the-art advances in Mixing and Chemical Engineering.

1 – 1974, Cambridge, United Kingdom
2 – 1976, Cambridge, United Kingdom
3 – 1979, York, United Kingdom
4 – 1982, Noordwijkerhout, Netherlands
5 – 1985, Würzburg, Germany
6 – 1988, Pavia, Italy
7 – 1991, Brugge, Belgium
8 – 1994, Cambridge, United Kingdom
9 – 1997, Paris, France
10 – 2000, Delft, Netherlands
11 – 2003, Bamberg, Germany
12 – 2006, Bologna, Italy
13 – 2009, London, United Kingdom
14 – 2012, Warsaw, Poland
15 – 2015, St. Petersburg, Russia
16 – 2018, Toulouse, France
17 – 2023, Porto, Portugal

Committees

Conference Chair:
José Carlos Lopes, CoLab Net4CO2, Portugal

Scientific Committee:
Joelle Aubin (Chair), Institut National Polytechnique de Toulouse, France
Alain Liné, Institut National des Sciences Appliquées de Toulouse, France
Andrea Ducci, University College London, United Kingdom
Cláudio Fonte, University of Manchester, United Kingdom
Daniela Dzhonova, Institute of Chemical Engineering - Bulgarian Academy of Sciences, Bulgaria
David Fletcher, University of Sydney, Australia
Federico Alberini, University of Bologna, Italy
Fouad Azizi, American University of Beirut, Lebanon
Frédéric Augier, IFP Energies nouvelles, France
George Papadakis, Imperial College London, United Kingdom
Giuseppina Montante, University of Bologna, Italy
Gul Ozcan Taskin, Loughborough University, United Kingdom
Gustavo Padron, BHR Group, United Kingdom
Harry van den Akker, University of Limerick, Ireland
Inci Ayranci, Middle East Technical University, Turkey
Jérôme Morchain, Institut National des Sciences Appliquées de Toulouse, France
Lena Hohl, Technical University of Berlin, Germany
Mark Simmons, University of Birmingham, United Kingdom
Matthias Kraume, Technical University of Berlin, Germany
Mounir Bouafi, Solvay, France
Nicolas Perret, Solvay, France
Pascal Pitiot, AmarilisPPT, France
Piero Armenante, New Jersey Institute of Technology, United States
Pierrette Guichardon, Ecole Centrale Marseille, France
Rajarathinam Parthasarathy, Royal Melbourne Institute of Technology, Australia
Richard Grenville, Philadelphia Mixing Solutions, United States
Tom Rodgers, University of Manchester, United Kingdom
Tomas Jirout, Czech Technical University, Czech Republic
Ville Alopaeus, Aalto University, Finland
Yoshiyuki Komoda, Kobe University, Japan
Zhenming Gao, Beijing University of Chemical Technology, China

Organizing Committee (LSRE-LCM, Faculty of Engineering, University of Porto, Portugal):
Ricardo Santos
Madalena Dias
Margarida Brito
Isabel Fernandes
Isabel Barbosa
Sofia Brandão
Maria Carolina Duarte
Paulo Marrocos

Contact:
E-mail: organization@mixing17.eu
Address:
Mixing 17, Departamento de Engenharia Química
Faculdade de Engenharia da Universidade do Porto, Rua do Dr. Roberto Frias, 4200-465 Porto, Portugal

Keynote Speakers

Harry Van den Akker

Professor, Fluid Mechanics, University of Limerick, Ireland

"A bird's eye view on 50 years of mixing"

Professor Harry E.A. Van den Akker, with a PhD degree from Eindhoven University of Technology, started in 1977 as a research engineer at the Royal/Dutch Shell Laboratories, Amsterdam (KSLA) and spent a year (1984/85) at Shell Oil Westhollow Research Center in Houston, TX. In 1988, he was appointed Full Professor of Transport Phenomena at Delft University of Technology and Director of the 'Kramers Laboratorium voor Fysische Technologie'.

When in 2002 this Laboratorium merged with other groups to form the Department of Multi-Scale Physics, he became Head of Department. He retained this position until in 2012 this Department was dissolved. Since 2013, he has been the Bernal Professor of Fluid Mechanics at the University of Limerick, Republic of Ireland, while retaining a part-time affiliation with 'Delft'. In all those years, he has been supervising 40 PhD students.

Harry Van den Akker organised European Mixing X in 2000 in Delft, was Visiting Professor at King's College, London, UK (2000-2003) and spent a short sabbatical at Princeton University in 2012. He was appointed the 2018 Burgers Visiting Professor of Fluid Mechanics at the University of Maryland and the 2021 'William R. Kenan Jr.' Visiting Professor for Distinguished Teaching at Princeton University.

Harry received the 2011 Master Teacher Award for Excellence at Delft University. In 2013, his 1999 paper (with Dr Jos Derksen) in AIChE Journal on lattice Boltzmann Large Eddy Simulations of a stirred vessel was selected by the North American Mixing Forum (NAMF) as one of the 21 most influential contributions to the field of mixing. In 2015, he was awarded the 2015 BHR Group Lifetime Recognition Award in Mixing by the EFCE 'Working Party on Mixing'.

Martina Micheletti
Professor, Bioprocess Fluid Dynamics, University College London, UK
"Mixing and scaling challenges in novel biopharmaceutical applications"

Martina is a Professor of Bioprocess Fluid Dynamics in the department of Biochemical Engineering at University College London. She has a first degree in Chemical Engineering from University of Pisa and a PhD in Mechanical Engineering from King's College London and has been at UCL since 2004. Martina leads research in mixing and fluid dynamics applied to bioprocesses and has applied her novel techniques and rigorous approach to obtain fundamental understanding of single-use bioreactor technologies and in increasingly important sectors like cell and gene therapy products, continuous production of monoclonal antibodies and vaccine manufacture. She has received several industrial awards and led collaborations with companies like Pfizer, MSD, AstraZeneca, among others. Since 2018 she has been Director of the UCL-Oxford Future Research Vaccine Manufacturing Hub (Vax-Hub) with Professor Dame Sarah Gilbert, aimed at developing lost-cost vaccines for low- and middle-income countries and whose work has been instrumental to the rapid development of the viral vectored vaccine against SARS-CoV-2. Martina is currently the Director of the recently awarded EPSRC Manufacturing Hub for a sustainable future (2023-2030) aimed at development of vaccines for pandemic preparedness.

Irina Ramos

Director, Bioprocess Technologies & Engineering, AstraZeneca, USA

"Mixing and processing in Pharma and vaccine industry"

Irina Ramos is a Director of downstream continuous manufacturing, in the Bioprocess Engineering and Technology group at AstraZeneca in Gaithersburg, MD, USA. She has been a key driver of the technology development and implementation strategy for continuous biomanufacturing. Her team works to increase fundamental understanding of the various downstream unit operations of drug substance manufacturing. They manage the design, testing and qualification of process equipment and automation on the systems to integrate them.

Irina holds a BS/MS in Chemical Engineering from University of Porto (Portugal) and a PhD in Chemical & Biochemical Engineering from University of Maryland, Baltimore County (UMBC) (USA). She has been involved with the National Institute for Innovation in Manufacturing Biopharmaceuticals (NIIMBL) Process Intensification program by leading the Control Strategy workstream since 2020.

For the past 14 years, Irina's work in the pharmaceutical industry has spanned from early to late-stage process development and validation, process scalability, CMC and technology transfer to both clinical and commercial manufacturing. In 2020-22 Irina lead a Technology Transfer team for the AstraZeneca's COVID-19 vaccine manufacturing process, as part of a global cross-functional team that successfully provided a substantial proportion of global COVID-19 vaccine supply at low cost (170 countries).

Irina has been teaching a class on Downstream Process Development, as part of the Professional Masters in Biotechnology from UMBC, since 2015. She is passionate about mentorship and developing resources to improve the communication of science.

Bento Amaral
Oenologist, Wine Tasting Teacher, Faculty of Biotechnology of Universidade Católica Portuguesa, Portugal

"Winemaking & Mixing"

Bento Amaral is a Professor in Oenology post-graduation at the Faculty of Biotechnology of Universidade Católica Portuguesa, where he graduated in Food Engineering. During his degree, he did an internship in the Faculty of Oenology in Bordeaux, where he realised his passion for the science dedicated to the study of wine and winemaking. Bento became a specialist in Portuguese and International wine and a leading expert in Port wine, summing up 23 years of experience while working at the Port and Douro Wines Institute (IVDP), first as head of the tasting committee and then as technical director.

Besides wine, the sea is Bento's other great passion. After the accident at the beach in 1994, it was precisely the sea that allowed him to live some of his biggest dreams: he was the World Disabled Sailing Champion in 2005 and was placed 9th in the Beijing Paralympic Games in 2008.

Recently, he launched a project, Humanwinety, that aims to promote and implement inclusion. It intends, through education, to be an active means of training minorities who continue to suffer from exclusion for their integration into the wine and tourism labour market.

Bento was awarded in 2019 the title of Officer of the Order of Infante D. Henriques by the President of the Republic and in 2014 with the title Chevalier dans l'Ordre du Mérite Agricole by the French Government. In 2020, he was nominated for "Person of the Year" by Wine Enthusiast.

MIXING IN UNBAFFLED TANKS: A REVIEW What We Know and What We Need to Know. Dedicated to the memory of Professor Alvin Nienow

Nienow Session.1

Arthur William Etchells III[1], *Robert P. Hesketh*[1]. *(1) Rowan University, Glassboro, NJ USA, Etchells@aol.com Hesketh@rowan.edu*

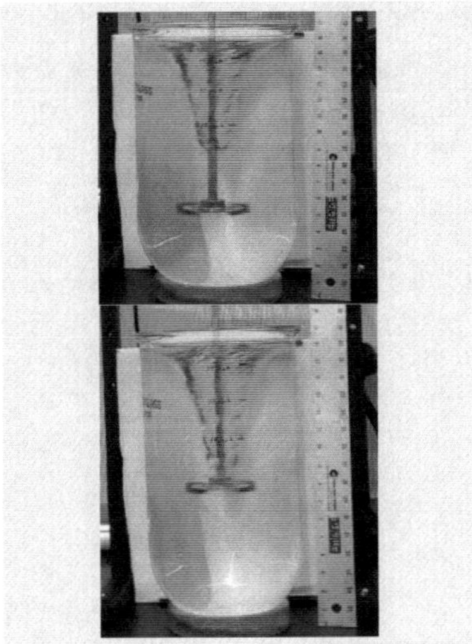

A common method of mixing in turbulent flow is a vertical cylindrical tank with baffles and a vertical center mounted shaft with an axial impeller. This is a well studied and characterized configuration and popular in industry. The alternate configuration without baffles is less popular with little characterization and less popular in industry though often seen. In this paper we will review what is known about several important phenomena in stirred tanks both baffled and unbaffled. We will review correlations for vortex depth, power and blend time for unbaffled tanks

The purpose of baffles is to turn a rotational motion caused by the agitator into an axial motion more favorable to several process results such as blending, solids suspension and distribution. They also increase the power input range which can be useful for area generation process results such as deagglomeration and secondary phase breakup and interphase mass transfer.

Correlations for the key variables are reviewed for unbaffled tanks and compared with those for baffled tanks. The most noticeable correlation missing is for unbaffled blend time over a wide range of Reynolds numbers.

Introduction

There are several reasons for industrial vessels not having baffles: perceived difficulties in cleaning, concerns of stagnation at higher viscosities, materials of construction such as glass and baffles not being part of the original design and change in viscosity with processing.

Additionally there is a general belief that baffles increase blend time at high viscosities compared to unbaffled tank by causing stagnation zones. Limited data indicate that blending takes four times longer with baffles than without baffles all else being equal. It is then worth the problems at low viscosities to get the benefits believed to be achieved at higher viscosities. Little data exists to validate this belief.

The second part of this work will look at blend time with unbaffled tanks in more detail than previously over a wide range of Reynolds number.

Power

A plot of Power number versus Reynolds number for the HE-3 impeller in baffled and unbaffled tanks is shown in Figure 1. This figure shows a region at Reynolds numbers below 10 where the Power number is inversely proportional to Reynolds numbers and the flow is presumed laminar. Between a Reynolds number of 10 and 1000, the slope gradually changes to zero indicating no effect of Reynolds number beyond about 1000. At this and higher Reynolds numbers the flow is fully turbulent. There is a slight effect of impeller type as a function of Reynolds number with less streamlined impellers having lower transitional values. The values of the Power number in all regions depends on the impeller shape with again the least streamlined having the largest power number.

When there are no baffles, the laminar region for the baffled and unbaffled are identical to about the Reynolds number of 10. For unbaffled tanks the constant Power number region does not exist and the Power number continues to fall even above Reynolds numbers of 100,000. In the non-laminar region, a power law fit of the slope gives values between 0.12 and 0.20.

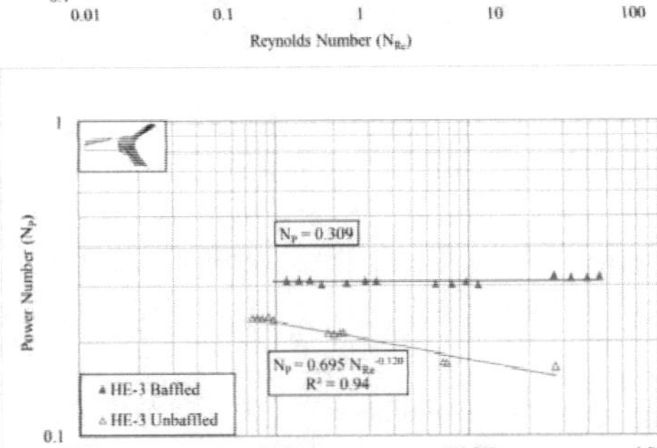

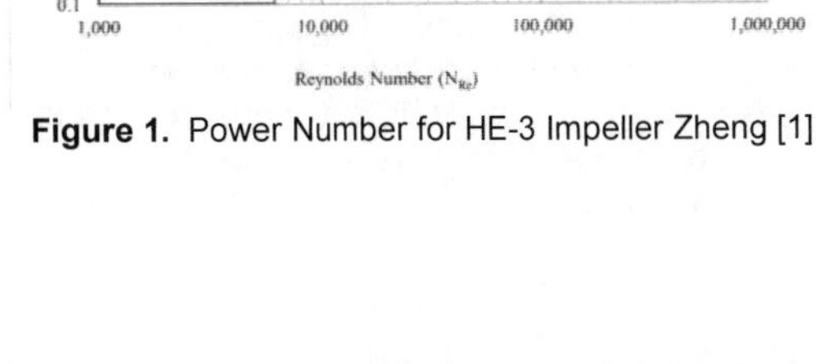

Figure 1. Power Number for HE-3 Impeller Zheng [1]

https://doi.org/10.52202/074122-0001

Vortex

There are a number of works characterizing the vortex depth as function of Reynolds number [2][3][4] Figure 2. The most common dependent variable is a Froude number based on the tip speed squared and with the dimension being the vortex depth. We have shown that the Inverse Froude number ($InvFr = \frac{Xg}{N^2D^2}$) reaches an asymptotic value at high Reynold's numbers. Since this variable is found to be constant above a Reynolds number of 40,000, then the absolute value of vortex depth is set by the impeller tip speed and does not depend on scale. This means that at equal tip speed a 10 cm vortex depth would exist in a one liter vessels and also in a 1000 liter vessel. Therefore, flow patterns will change with scale, unlike in a baffled tank. At a low Reynolds number of about 200 the vortex disappears. Another unusual characteristic of the vortex is that it is attached to the surface. Adding more fluid and raising the liquid level does no change to size of the vortex but only moves it up. If the conditions of liquid level and tip speed are right and the vortex touches the impeller there will be a large entrainment of gas. In many processes such as blending of viscous fluids this is undesirable.

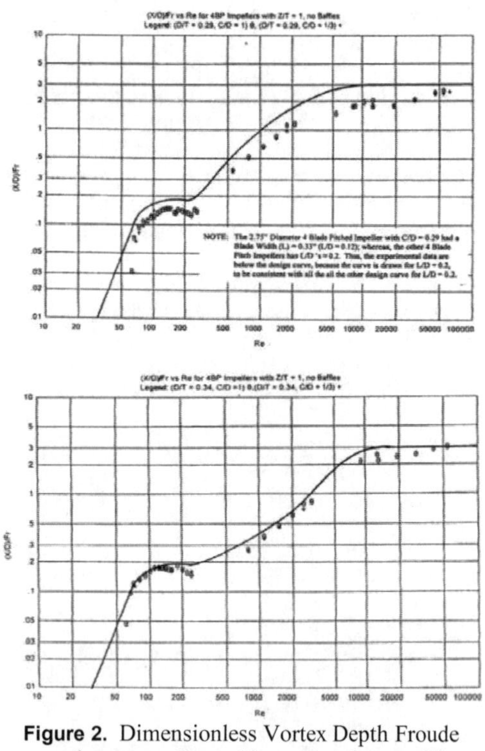

Figure 2. Dimensionless Vortex Depth Froude number versus Reynolds number Penney [3]

Blend Time

Unlike baffled systems there is very little data on blending in unbaffled tanks. No correlations exist similar to the work by Grenville [5,6] shown in Figure 3. and subsequent investigators which covers the range from highly turbulent down to Reynolds numbers close to or in the laminar range. In this work three zones can be observed similar to baffled Power number plots with turbulent, transitional and laminar sections. The transitional Reynolds number are not congruent as the Power number tracks the region of the impeller and the blend time tracks what goes on in the furthest regions of the tank and blend time correlations have dimensionless geometric ratios in the correlations such as impeller to tank diameter functions. In the turbulent flow region the dimensionless blend time is constant with Reynolds number and in the Transitional zone inversely proportional. Data suggest that at a Reynolds number between 100 and 200 the blending process becomes irreproducible with very high blend times if blending is achieved at all.

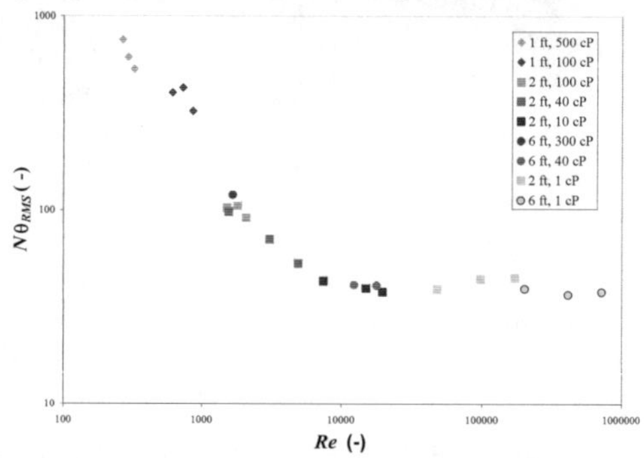

Figure 3. Dimensionless Blend Time versus Reynolds number Grenville [2]

Data by Brucato [7] suggest that at equal power the blend time in unbaffled tanks is about five times longer than in baffled tanks at high Reynolds numbers. Other data suggest at lower Reynolds numbers the baffles cause stagnation and the blend time increases by a similar amount. More data and correlations are definitely needed including information on the point at which baffles become an aid to blending rather than a hinderance.

Particulate Solids

Particulate solids heavier than the liquid are observed to move on the bottom of an unbaffled tank and can be lifted approximating off bottom suspension. However vertical lift is small and vertical distribution is very poor. Unbaffled tanks used for handling settling particulate solids are rare in industry but do exist is some specialized cases.

Conclusions

While there is substantial information on mixing in unbaffled tanks the most important missing information is a correlation for blending times in unbaffled tanks over a wide range of Reynolds numbers.

Coming Attractions

Experimental work at Rowan University is in progress in developing a blend time correlation for unbaffled tanks for several impellers over a wide range of Reynolds numbers with parallel blend time measurements on the same tank with using

https://doi.org/10.52202/074122-0001

a color change method. It is expected that much of this work will be available for the formal presentation in July 2023 in Porto.

Acknowledgments

Special thanks to the students working on this project: Nathaniel Nelson, Karly Amandeo, Austin Colon, Samual Goldberg

References

[1] Ma, Zheng, Impeller power draw across the full Reynodls Number spectrum (2014). MS Graduate Theses and Dissertations. 2014. advisor Kevin Meyers.

[2] W. Penney, R. and Spannell. Vortex Depth Pitched Blade Impellers. 2004 AIChE Meeting Cincinnati Ohio

[3] Markopoulos and E. Kontogeorgaki, Vortex depth in unbaffled single and multiple impeller agitated vessels Chem Eng Technol 18, 1995.

[4] S. S. Deshpande et al, *Chem Eng Sci*, 168, pp 495-506, 2017.

[5] S. Kresta et al. Handbook of Industrial Mixing , Chapter 9 Blending of Miscible Liquids, by R. Grenville and A. Nienow, John Wiley & Sons, 2004

[6] R. K. Grenville, Blending of Viscous Newtonian and Pseudo-Plastic Fluids, PhD Thesis, Cranfield University 1992.

[7] A, Busciglio, A. Brucato, F. Scargiali, Mixing dynamics in uncovered unbaffled stirred tanks, Chem Eng Journal 254, 2014

https://doi.org/10.52202/074122-0001

CFD Simulation of a Pulse Jet Mixed Vessel for Nuclear Waste Processing

Nienow
Session 2

R.V. Calabrese[1], *J. W. Kim*[1]. *(1) University of Maryland, College Park, MD 20742-2111, rvc@umd.edu , junkim@terpmail.umd.edu*

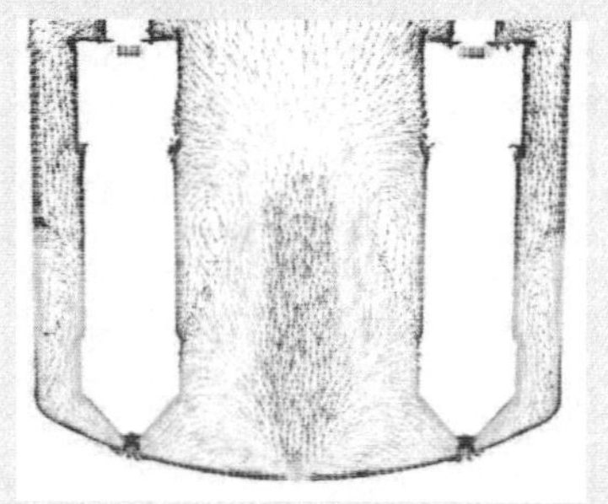

CFD prediction: Velocity field in a pulse jet mixed vessel during the PJM drive phase.

Pulse Jet Mixed (PJM) vessels are used to process nuclear waste due to their maintenance free operation. In this study we model the turbulent velocity field in water during normal PJM operation to gain insight into vessel operations and to evolve a strategy for further model development and validation. Well converged unsteady RANS (URANS) and LES predictions are compared to velocity measurements acquired for 3 test scenarios at 3 locations in a pilot scale vessel at the US DOE National Energy Technology Laboratory (NETL). The LES model well predicts the data provided that the operational pressure profile within the individual pulse tubes is considered. The URANS simulations failed to predict the data and exhibited significant differences from LES, particularly with respect to turbulence quantities. It is concluded that URANS is of limited use in this inherently transient flow. Exploitation of the LES model shows extremely complex flow patterns throughout the vessel with numerous local jets and mixing layer vortices.

Introduction

Pulse Jet Mixed (PJM) vessels have been used to process nuclear waste at Sellafield (UK) and are part of the flowsheet for the pretreatment and vitrification of high-level military waste stored in underground tanks at the Hanford Site in south-central Washington state (USA) [1, 2]. PJM vessels contain a series of downward directed pulse tubes that periodically discharge fluid into the vessel and are then refilled with vessel fluid, with the cycle being repeated until the process mixing/blending objective is met. During the drive phase, positive air pressure, supplied from outside the vessel's enclosure or "black cell", drives the fluid from the narrowed tips of the pulse tubes. These high intensity jets are discharged onto the vessel bottom, travel radially inward and merge to create a central up wash plume that mixes the vessel contents as it dissipates. During suction, negative air pressure pulls the vessel fluid back into the tubes. The air to operate the pulse tubes is supplied from the outside leaving the black cell free of moving parts and their maintenance.

Many Hanford vessels will process concentrated slurries with fast settling solids while others will remain relatively dilute. In this first study we model the turbulent velocity field for a pure liquid with the properties of water during normal PJM cycles to gain insight into vessel operations and to evolve a modeling strategy for process design and operator training. We compare unsteady RANS (URANS) and LES predictions to velocity measurements acquired at 3 locations and for 3 different test scenarios in a 1 m diameter laboratory scale vessel containing 4 pulse tubes (Figure 1) at the US Department of Energy's National Energy Technology Laboratory (NETL). The validated models are used to gain further insight into vessel dynamics. The suitability of URANS predictions to assess vessel operational requirements is tested by comparison to LES.

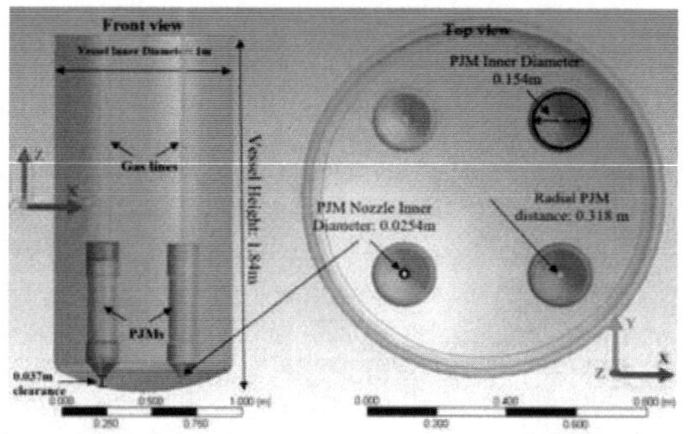

Figure 1. NETL laboratory scale vessel with 4 Pulse Jet Mixer (PJM) tubes.

Material and Methods

PJM Operation: Pulsed jet mixers are driven by a "jet pump pair" (JPP) which controls air flow to the pulse tubes (Figure 2). During the Drive Phase, compressed air enters the tubes and liquid is expelled from the jet nozzle into the vessel. The gas then decompresses during a short Vent Phase. During the Suction Phase, the liquid is drawn back into the pulse tube and the cycle is repeated. Figure 3 shows the measured JPP pressure profiles for the NETL vessel for test Scenario 1. With respect to the time axis, the drive phase (positive pressure) lasts about 1.6 s. The vent phase ends at about 2.3 s. While not fully shown, the suction phase (only slightly negative pressure since gravity aids in refill) is relatively long, lasting to 12 s. Similar data were acquired for all test scenarios.

Figure 3 shows that the pressure profiles are different in each pulse tube. This has significant consequences to model validation and process operation. The vessel geometry is symmetric so if the nozzle discharge velocity and drive duration are the same for each tube, the vessel can be divided into quadrants each containing one PJM. The velocity field would exhibit the same behavior in each quadrant, resulting in a "perfectly centered" upwash plume. Symmetry can be used to reduce

https://doi.org/10.52202/074122-0002

computational expense. During drive, the maximum in the upward vertical velocity is along the vessel centerline, facilitating the use of single point measurement probes. Since drive pressures are different in each tube, the PJM nozzle exit velocity and flow rate will be different for each emanating jet. The central upwash plume will not be symmetric about the vessel centerline and symmetry throughout the vessel will be broken, adding to computational expense, and complicating data interpretation. The Individual Pressure Profiles (IPP) of Figure 3 are required to validate the models by comparison to the NETL data. However, such pressure discrepancies can be eliminated. As a result, less computationally expensive scenarios based on the average pressure cycle (Figure 3, dotted line), referred to as the Unified Pressure Profile (UPP) were also exploited.

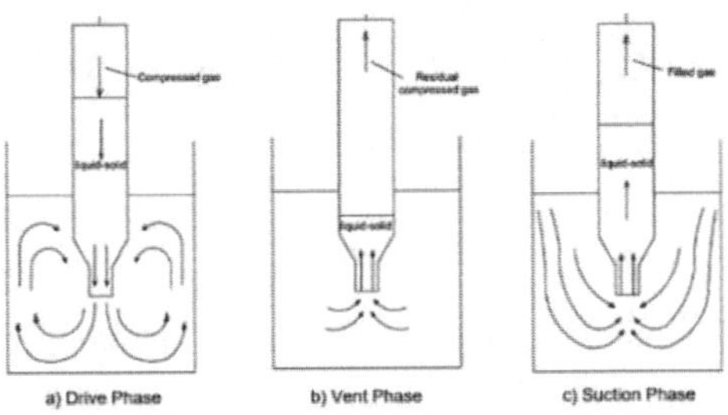

Figure 2. PJM Operating Phases.

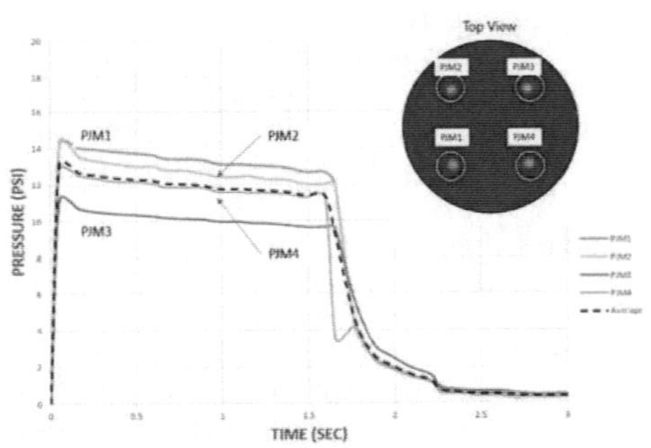

Figure 3. JPP operating pressure profiles for each PJM. Dotted line is Unified Pressure Profile (UPP) or average applied to all 4 PJMs. Test Scenario 1.

CFD Simulations: To account for fluid-air interactions within the PJM tubes and at the vessel's liquid-air interface, fully transient, 3-D Volume of Fluid (VOF) simulations were carried out with ANSYS Fluent. The URANS equations were solved using the realizable k-ε turbulence model with enhanced wall treatment. The LES simulations used the Wall-Adapting Local Eddy-Viscosity (WALE) sub-grid scale model with enhanced wall treatment. The effect of air compressibility was studied using the ideal gas law and it was found acceptable to treat the air as an incompressible fluid. A mesh independence study was conducted, and the final URANS and LES model settings are given in Table 1. To understand and analyze vessel dynamics we consider the cycle resolved velocity field produced by the pulsed jets. While the process is fully transient, one can capture a specific time in the drive/vent/suction cycle of Figures 2 and 3 for many realizations or PJM cycles, and ensemble average to construct an averaged vessel flow field at that time. URANS simulations do this directly. For LES, it was found that averaging over 100 cycles or realizations was sufficient to gain converged statistics. When all 4 PJMs operate with the same pressure profile (UPP), symmetry can be used to reduce the number of LES computed cycles to 25.

Table 1. URANS and LES simulation model settings

	RANS	LES
Mesh type	Polyhedral	Polyhedral
Cell count	1.8 million	12 million
Min. Cell Size	0.14 mm	0.34 mm
Max. Cell Size	12 mm	37 mm
# of cycles*	1 cycle	100 cycles**
Time Step Size	~1e-3 s	~4e-4 s

* cycles required to obtain converged statistics.
**25 cycles required if 4 PJMs operate synchronously.

NETL Data: Vertical velocity measurements at 3 locations were acquired using Valeport model 801 electromagnetic (EM) velocity probes, by synching the probe output to the PJM's JPP pressure profile. The probes are located off center between the PJM tubes at the vertical locations shown in Figure 4. Three test scenarios with different water depth, drive phase length and drive pressure were conducted, and more than 100 PJM cycles of data were acquired and used to construct the cycle resolved average velocity at each probe location for comparison to the RANS and LES predictions.

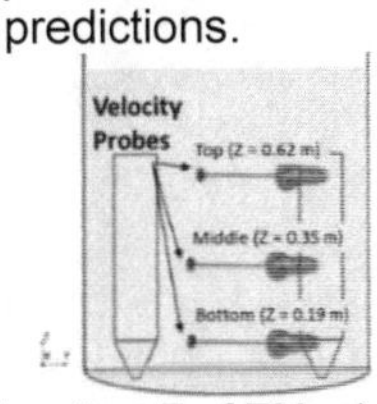

Figure 4. Vertical location, Z of EM velocity probes.

Results and Discussion

URANS and LES simulations were performed for all 3 test scenarios. Comparison to NETL data and model validation will be discussed and differences between converged URANS and LES predictions will be contrasted. Flow field details will be presented.

Model Validation: Figures 5 and 6 compare the cycle resolved EM probe data at the 3 vessel locations to the URANS and LES predictions, respectively with the independent PJM pressure profiles (IPP). URANS overpredicts top probe and underpredicts bottom probe velocities, while the LES simulation well-matches the data within statistical limits. Similar results were found with IPP for the other test scenarios. In all cases, simulations using a uniform pressure profile (UPP) failed to adequately predict the data. Figure 7 shows that IPP operation shifts the location of the upwash plume relative to the probes. However, the LES model with UPP remains valid for model exploitation.

https://doi.org/10.52202/074122-0002

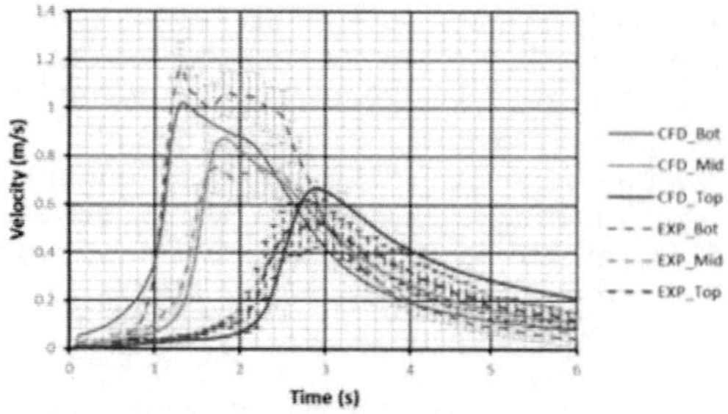

Figure 5. Cycle resolved average vertical velocity predicted by URANS IPP simulations (solid lines) and measured in NETL experiments (dashed lines). The dashed error bars are the 95% confidence intervals for the measurements. URANS only predicts averaged quantities. Test Scenario 1.

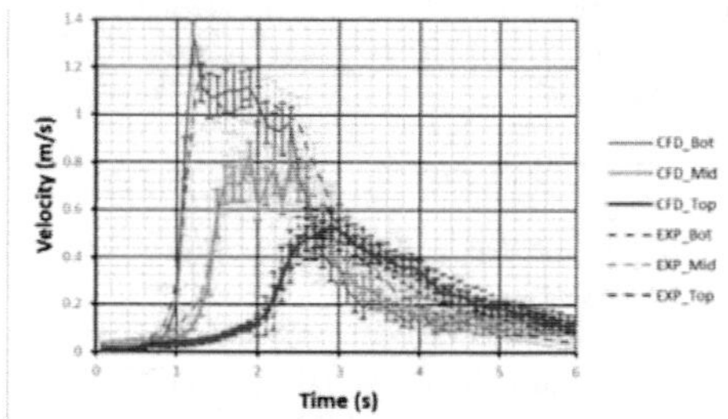

Figure 6. Cycle resolved average vertical velocity predicted by LES IPP simulations (solid lines) and measured in NETL experiments (dashed lines). The error bars are 95% confidence intervals. Dashed error bars apply to the data. Solid error bars apply to LES predictions. Test Scenario 1.

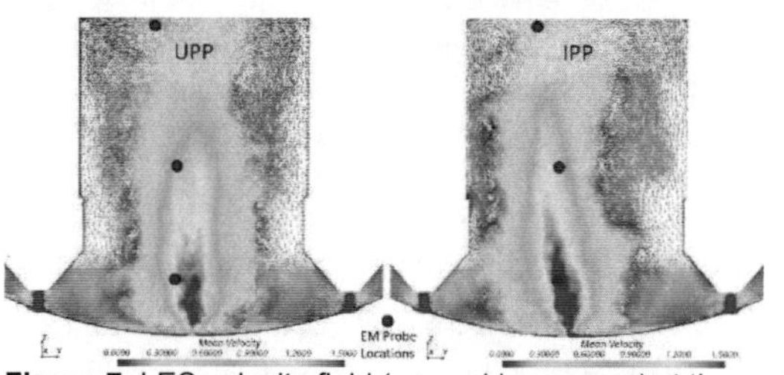

Figure 7. LES velocity field (ensemble average) at the end of the drive phase for UPP (left) and IPP (right) operation. Red dots show EM probe locations. Test Scenario 1.

LES-URANS Comparison: The simulations show complex flow patterns throughout the vessel. During drive, mixing layer vortices form between the central up wash plume and the pulse tube bodies and travel upward to the surface. There is weak downward flow close to the vessel walls. During suction, fluid is pulled downward from all vessel locations. There are numerous regions of flow reversal. The structural features produced by URANS are markedly different from LES. An example near the vessel bottom is given in Figure 8. For LES, distinct vortices appear on average but are "washed out" in URANS. Our study shows that differences in predicted velocity and deformation fields strongly impact model estimates of turbulent kinetic energy and energy production/dissipation rate throughout the vessel. In some regions, there are explainable order of magnitude discrepancies. URANS is of limited use for simulating these highly transient, multiscale

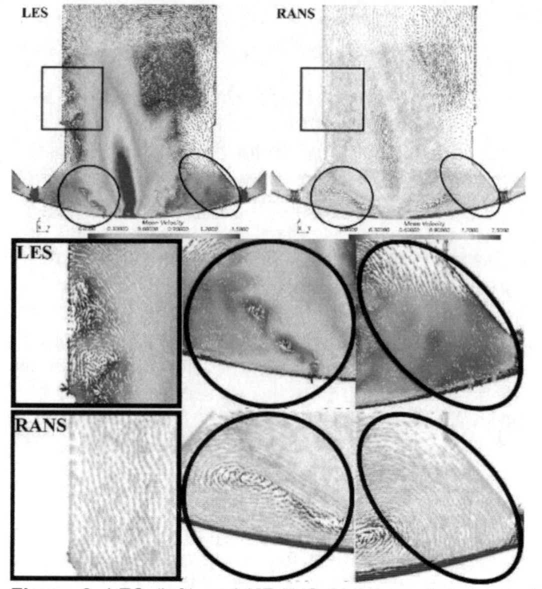

Figure 8. LES (left) and URANS (right) predicted velocity fields (ensemble average) at the end of the drive phase for IPP operation. Vortices are identified by black rectangles (left) and circles (right). Test Scenario 1.

flows. There are similar conclusions [3, 4] from other practical flow studies.

Flow Field Dynamics: A series of vector/contour plots and animations have been compiled to illustrate flow field dynamics, both transient (LES) and cycle resolved, for velocity and turbulence quantities. Quantitative estimates in various regions further illustrate the complexity of the flow and its resolution.

Conclusions

The simulations provide insight into flow field dynamics and support an evolving strategy for PJM vessel design, scale up, and operation. The LES simulation model was validated, at least for prediction of macroscopic quantities, by comparison to fixed point velocity data. Its adequacy to predict energy dissipation rate has not yet been assessed. For this flow, the URANS model is of limited utility.

Acknowledgments
US DOE National Energy Technology Laboratory provided partial funding. Univ. of Maryland provided HPCC resources.

References

[1] P. A. Meyer and A. W. Etchells, "Mixing with Intermittent Jets with Application in Handling Radioactive Waste Sludges," ChERD, **85**, 691-696, (2007).

[2] J. A. FORT, P. A. Meyer, J. A. Bamberger, C. W. Enderlin, P. A. Scott, M. J. Minette, P. A. Gauglitz, "Scaled testing to evaluate pulse jet mixer performance in waste treatment plant mixing vessels," Waste Management Assoc., WM2010 Conference, Pheonix, AZ (2010).

[3] D. Liu, "Comparison of laminar model, RANS, LES and VLES for simulation of liquid sloshing," Applied Ocean Research, **59**, 638-649, (2016).

[4] T. van Hooff, "On the accuracy of CFD simulations of cross-ventilation flows for a generic isolated building: Comparison of RANS, LES and experiments," Building & Environment, **114**, 148-165, (2017).

Measurement of pipe and stirred tank flows to understand the application of computational models in the transitional flow regime.

S1.1

***Georgina Wadsley*[1,2]**, *David F. Fletcher*[3], *Joelle Aubin*[2], *Andy Ingram*[1], *Waldo Rosales*[4], *Mark J.H. Simmons*[1].
(1) University of Birmingham, gkw059@student.bham.ac.uk. (2) Laboratoire de Génie Chimique (3) University of Sydney, (4) Unilever R&D, Port Sunlight Laboratory.

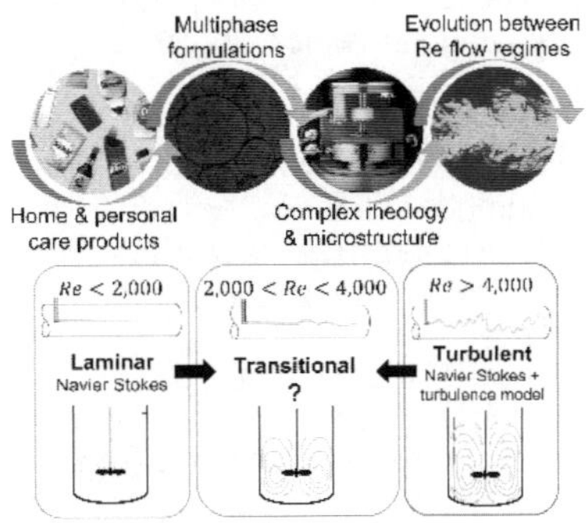

The potential for a dynamic turbulence closure model (SBES) to predict the hydrodynamics of transitional flow is assessed by comparing experimental and simulation data. While initial efforts were made for stirred tank flows, a simpler pipe geometry with an orifice plate was chosen for initial model validation. Unsteady flows of two Newtonian fluids, 50 wt% glycerol (μ = 5.55 mPa s) and 60 wt% glycerol (μ = 9.40 mPa s), were measured using Particle Image Velocimetry (PIV) over a range of Reynolds numbers (Re = 300 – 6,000). Results using the SBES model illustrate the dependence of flow behaviour on viscosity in the transitional flow regime despite scaling flow scenarios for the same Reynolds number, as identified using PIV data. Potential modifications of the SBES model are further explored to capture the correct balance between viscous damping and vortex generation, which is very sensitive in the transitional flow regime. The study will next apply knowledge gained to stirred tank flows using a similar investigative approach.

Introduction

Validated Computational Fluid Dynamics (CFD) simulations when applied to industrial flows are powerful predictive tools that facilitate rapid modification of existing batch mixing processes to increase efficiency, reduce energy consumption, and aid with scale-up and scale-down approaches. Over the last two decades, numerous experimental and numerical studies have explored the hydrodynamics of liquid flows in stirred tanks and pipes for a wide range of fully turbulent and laminar cases that are now applied in manufacturing industries [1, 2].

However, with the inclusion of high concentrations of thickeners and surfactants, many home, personal care, and food products develop a complex functional microstructure during mixing. This leads to an evolution of their rheology during manufacture. It is therefore possible for industrial fluids to vary from low to high apparent viscosity during processing as the microstructure builds, transitioning from turbulent, through transitional, to the laminar flow regime in the mixer. In the transitional flow regime, where current research is limited, hydrodynamic instabilities create unsteady flows that are difficult to predict using existing laminar or turbulence closure models, particularly in stirred tanks due to the complexity of blade synchronisation and periodic contributions [1, 3-5]. Existing turbulence closure models reported are not well adapted to transitional flows as the eddy viscosity hypotheses are designed for flows with high Reynolds number turbulence [1].

In this work, initial efforts were made on modelling transitional flows in a stirred tank geometry equipped with an A320 wide blade hydrofoil impeller, following the work of Mendoza et al. (2018) [1]. Mesh independent results were successfully achieved, however, analysis of data, in combination with preliminary experimental findings from Mendoza et al. proved to be highly complex due to periodic effects of the impeller and changes in flow regime spatially. Unwanted perturbations in the experimental set-up were also thought to have a significant impact on the transitional Reynolds number flows [1].

Pipe flows provide the benefit of a simpler geometry and more predictable local flow behaviour, which is beneficial during the initial stages of testing CFD closure models. Therefore, it was decided to reduce the complexity of the studied flow using a simple pipe mixer geometry and to minimise any unwanted vibrations to the system [2, 6]. Hence, the current study aims to advance the understanding of transitional flow and its characteristics in a pipe with a constant perturbation provided by a square edged orifice plate, encouraging transition to turbulence [2]. PIV measurements are used to identify flow phenomena and provide experimental data to validate the application of existing laminar and turbulence closure models in the transitional flow regime, determining their suitability and exploring potential modifications.

After, applying knowledge gained from the pipe flow study, the focus will return to exploring transitional flow in the stirred tank geometry with a similar investigative approach.

Material and Methods

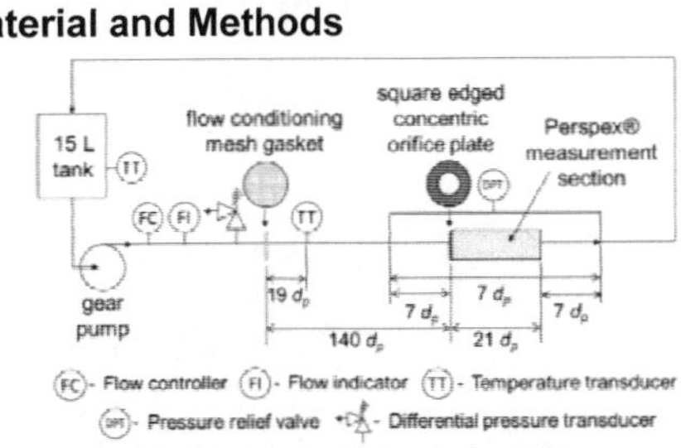

Figure 1. Schematic of pipe flow rig for PIV measurement.

A schematic of the pipe flow loop used for PIV measurement is presented in Figure 1, with lengths given in terms of pipe diameter (d_p = 15.75 mm).

https://doi.org/10.52202/074122-0003

Table 1. Summary of inlet velocities (V_{inlet}) for each Reynolds number flow scenarios using 50 and 60 wt% glycerol.

glycerol [wt%]	T [°C]	ρ [kg m^{-3}]	μ [mPa s]	Reynolds Number (Re)								
				300	400	500	750	1,000	1,500	2,000	2,500	6,000
50	18.7	1115.2	5.55	0.10	0.12	0.15	0.24	0.32	0.47	0.59	0.73	1.74
60	18.5	1132.0	9.40	0.16	0.21	0.27	0.42	0.56	0.65	0.90	1.17	3.10

A square edged orifice plate (diameter = 9.53 mm and thickness = 1 mm) was fitted before the measurement section, 140 d_p downstream from the inlet [2]. PIV measurements were carried out for a wide range of Reynolds number flows (Re = 200 – 6,000) with two Newtonian fluids of differing viscosity: 50 and 60 wt% glycerol. Flow measurements were taken for each fluid at known pipe inlet velocities (V_{inlet}), selected to scale for the same Reynolds number. A summary of flow scenarios is provided in Table 1.

To ensure full knowledge of the system and allow the equivalent flow scenarios to be set-up using CFD, the density (ρ) and viscosity (μ) of fluid samples taken before and after PIV measurement were measured offline at recorded process temperatures (T). Flow field measurements were also taken for the same flow scenarios without the orifice plate and at the inlet of the pipe.

ANSYS Fluent (version 2021R2) software was used to build transient CFD simulations for each experimental case, comparing a laminar and two existing turbulence closure models. A pipe geometry with an inlet length of 400 mm before the orifice plate and poly-hexcore mesh with ~300k elements was selected using a mesh independence study. The first turbulence model employed is a generalised k-ω shear stress transport model (GEKO). The second model is the stress blended eddy simulation (SBES), a dynamic model that switches between GEKO and Large Eddy Simulation (LES), employing the k-ω model at the wall eliminates the requirement for wall functions [7].

Results and Discussion

Velocity field data for all experimental cases show that a jet is formed on exit from the orifice plate with a resulting area of high velocity fluctuation at the location of jet breakdown. Distinct flow behaviours for each Reynolds flow regime are identified through plotting the time-averaged velocity magnitude normalised by the inlet velocity along the centreline of the pipe. In the laminar regime, the centreline velocity decays almost linearly with a constant jet spreading rate. As the Reynolds number of the flow increases the overall jet length from the orifice decreases, until the flow becomes fully turbulent. Then the length and characteristics of the normalised jet velocity become independent of Reynolds number.

For further investigation and comparison with CFD simulations, three cases were selected for each fluid to cover each flow regime: laminar (Re = 400), transitional (Re = 1,500) and turbulent (Re = 6,000). In the laminar and turbulent regime, plots of normalised time-averaged velocity magnitude scale exactly when normalised by V_{inlet} as the flow behaviour depends solely on Reynolds number. However, this phenomena is not apparent in the transitional regime, where the breakdown of the jet for the higher viscosity fluid occurs closer to the orifice plate than for the lower viscosity fluid.

As expected, in the laminar regime, a good prediction is achieved by the laminar model when comparing measured and modelled values of normalised velocity and normalised velocity fluctuations. The SBES model is also suitable for laminar flows yet is more computationally expensive. However, it is clear that the GEKO model is not suitable as it generates "artificial turbulence" leading to an earlier decay of the jet velocity compared with the PIV measurement. In the laminar regime, all CFD models investigated predict self-similar results between the two scaled flow scenarios with 50 and 60 wt% glycerol.

For the transitional flow scenarios (Re = 1,500) plots of normalised time-averaged velocity magnitude along the pipe centreline are presented in Figure 2a. While the SBES model predicts a longer jet length than measured with PIV, the sh-ape of the curves exhibiting decay of velocity are

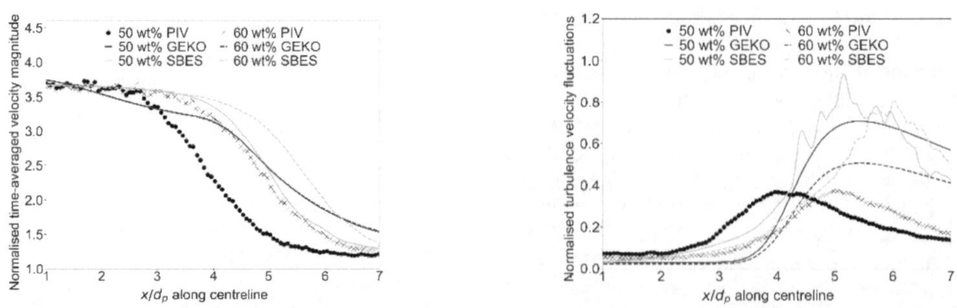

Figure 2. Comparison of (a) normalised time-averaged velocity magnitude and (b) normalised turbulence velocity fluctuations along pipe centreline at Re = 1,500 for glycerol solutions of 50 wt% (μ = 5.55 mPa s, V_{inlet} = 0.47 m s^{-1}) and 60 wt% (μ = 9.40 mPa s, V_{inlet} = 0.65 m s^{-1}) between measured PIV data and CFD predictions using GEKO and SBES models

https://doi.org/10.52202/074122-0003

very similar. Characteristic of transitional flows which are dependent on fluid viscosity, SBES predicts that the flow behaviour is not self-similar between the two flow scenarios, despite being scaled for the same Reynolds number. The jet length for the lower viscosity fluid is found to be shorter in both the PIV data and SBES prediction than for the high viscosity fluid. Not only does the GEKO simulation data illustrate a different shape and, thus, rate of velocity decay, but it predicts self-similar results for the two flow scenarios, which is not the case in the transitional flow regime. The laminar simulation results are not presented as a much longer jet length than observed in experiments due to lack of turbulence was predicted, making it unsuitable.

As displayed in Figure 2b, the area of high turbulence velocity fluctuations measured using PIV occurs much closer to the orifice plate than the SBES model predicts and at a lower level. However, there are some key similarities with the flow behaviour showing promise for further investigation of SBES. Again, the two flow scenarios are not self-similar despite scaling with Reynolds number and the area of high turbulence velocity fluctuations occurs closer to the orifice and at a slightly greater level for the lower viscosity fluid. It is also noted that experimentally, the flow enters with a slightly higher turbulence level than the simulated models. Therefore, by studying the flow without the orifice plate, the initial turbulence is evaluated and incorporated into the SBES simulation to test for a closer prediction.

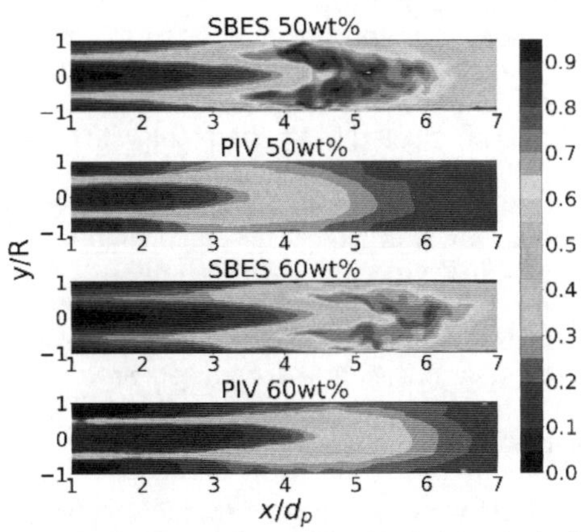

Figure 3. Contours of normalised turbulence velocity fluctuations plotted on pipe cross-section 1 d_p after the orifice plate, as region of interest within the flow domain. PIV and SBES data are compared at *Re* = 1,500 for 50 wt% (μ = 5.55 mPa s, V_{inlet} = 0.47 m s^{-1}) and 60 wt% (μ = 9.40 mPa s, V_{inlet} = 0.65 m s^{-1}) glycerol solutions.

The shape of the velocity jet and area of high normalised velocity fluctuations predicted using the SBES model (Figure 3) are also most similar to the PIV data, compared with other models, providing a good starting point for further investigation. Not representative of the PIV data in the transitional regime, the GEKO model predicts a very strong potential core with a triangular shape and no resolved turbulence. Indeed, for GEKO simulations the turbulence velocity fluctuation data is almost entirely modelled, whereas the simulations using the SBES model predict mostly resolved turbulence.

In the turbulent flow regime (*Re* = 6,000), both GEKO and SBES models provide a good prediction of the flow when compared with PIV data. Again, exact Reynolds number scaling is observed between the two scaled flow scenarios with 50 and 60 wt% glycerol for both models, as expected in the fully turbulent flow regime.

Conclusions

Experimental PIV data have been analysed and used for comparison with numerical simulations employing a laminar model and two existing turbulence closure models (GEKO and SBES) over a wide range of Reynolds number flows in order to determine the best approach to model transitional flow. Distinct flow behaviours have been identified for different flow regimes using PIV data. However, it is shown that even a simple orifice plate geometry reproduces complex flow behaviours. The SBES closure model is able to predict transitional flow features that are not observed using other models, where the normalised velocity and turbulence velocity fluctuations do not depend solely on Reynolds number. SBES also provides a greater simulation complexity but with reduced computational requirement compared with LES alone. However, the balance between viscous damping and vortex generation is very sensitive in the transitional flow regime. The simulation methodology is to be further improved, including exploring the effect of apparent differences in conditions at the orifice plate. More complex systems including stirred tanks and industrially appropriate flows are also to be explored.

Acknowledgments

GW is supported by the EPSRC Centre for Doctoral Training in Formulation Engineering (EP/S023070/1) at the University of Birmingham and Unilever.

References

[1] F. Mendoza, B. Lopes Bañales, E. Cid, C. Xuereb, M. Poux, D. F. Fletcher, J. Aubin, *Chemical Engineering Research and Design*, 132 (2018) 865-880.

[2] K.J. Winters, E.K. Longmire, *Experiments in Fluids,* 60 (2019) 1-17.

[3] M.B. Machado, K.J. Bittorf, V.T. Roussinova, A.M. Kresta, *Chemical Engineering Science,* 98 (2013) 218-230.

[4] F. Alberini, L. Liu, E. H. Stitt, M.J.H. Simmons, *Chemical Engineering Science,* 171 (2017) 189-203.

[5] Y. Zhang, Z. Gao, Z. Li, J.J. Derksen, *AIChE Journal,* 63 (2017) 3610-3623.

[6] B. Eckhardt, T.M. Schneider, B. Hof, J. Westerweel, *Annual Review of Fluid Mechanics,* 39 (2007) 447-468.

[7] F. Menter, A. Hüppe, A. Matyushenko, D. Kolmogorov, *Applied Sciences*, 11 (2021) 2459-2473.

https://doi.org/10.52202/074122-0003

Multi-objective impeller geometrical optimization in stirred tanks by using the mean age theory approach

S1.2

M Wu[1], *N Jurtz*[1], *L Hohl*[1], *M Kraume*[1] *(1) Chair of Chemical Process Engineering, Technical University Berlin, Ackerstr. 76, Berlin, Germany, wu.mei@campus.tu-berlin.de.*

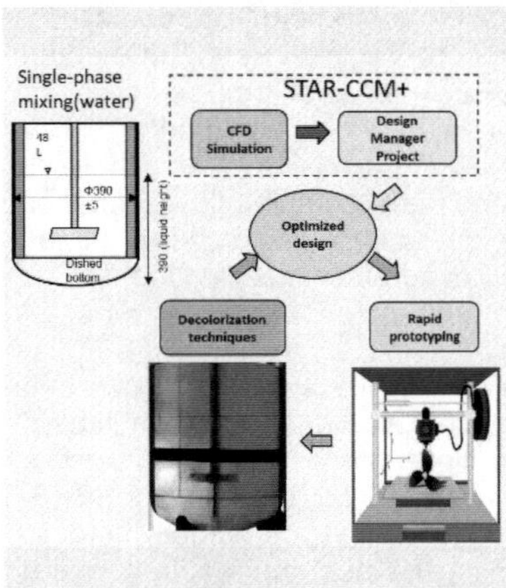

In this study, six impeller designs with varied geometrical parameters (blade diameter, angle, height, thickness, and number) are investigated numerically and experimentally. The Simcenter STAR-CCM+ by Siemens PLM Software is used and decolorization experiments are conducted. The accuracy of the Mean Age Theory (MAT) method in predicting mixing time for different axial-flow impellers is confirmed. Then, an impeller geometrical optimization study with six design parameters to find the best trade-off between mixing time and power consumption is conducted. The optimum design is printed by rapid prototyping and the mixing performance is studied. Good agreement is found between results from the MAT approach and experiment data. The MAT approach consumes only 3% of the computational time required by fully transient techniques, making multi-objective design optimization studies possible to develop new tailored stirrer designs.

Introduction

In the chemical industry, a failure to provide the necessary mixing may result in severe manufacturing problems. Mixing problems can be divided into three categories: blending, multi-phase, and reaction. Blending to homogeneity is the first mixing problem most people recognize. The time available to accomplish mixing (the time scale) and the required scale of homogeneity are two key variables to identify in the mixing problem [1]. In stirred tanks, the degree of mixing intensity mainly relies on the geometry of the selected impeller. To improve mixing efficiency while considering economic costs, impeller geometrical optimization has become one of the most preferred aspects in the mixing industry within the design process [2]. Today, the exponentially increasing computational power has enabled the combination of Computational Fluid Dynamics (CFD) and optimization algorithms for shape design optimizations, particularly for multi-objective problems [3].

The accuracy of mixing time prediction, and the corresponding numerical effort are two obstacles that increased the difficulty of conducting optimization studies efficiently. These issues have been investigated in previous work [4]: the accuracy of the mixing times of a standard Rushton turbine and Pitched-blade turbine predicted by the Mean age theory (MAT) approach was demonstrated and validated against experimental results. In addition, this steady-state method requires only small amount of the computational time compared to that from fully transient techniques. However, a broader range of different impeller designs needs to be examined to strengthen the accuracy of the MAT approach. Although a preliminary optimization study was conducted, more design parameters and larger design space are to be further investigated.

In this study, first, the mixing times of different impeller designs gathered from MAT are compared with experimental results. Then, based on a traditional impeller, a multi-objective geometrical optimization study is carried out using MAT. The aim is to find the optimal impeller with respect to the trade-off between different objectives. Finally, the optimum design gets manufactured by means of rapid prototyping, and the mixing performance is verified experimentally.

Material and Methods

For measurements of homogenization time (mixing times), a decolorization process by using a fast acid-base (H_2SO_4 and NaOH) indicator reaction is conducted in a stirred tank with distilled water. The tank has a diameter of T and a height of H in 390 mm. To objectify visual observation, one camera with a fixed white balancing is used to record the process. EKATO image analysis software is utilized to quantify the colours by HUE thresholds, and areas with values between 200°and 0° are defined as unmixed areas [5]. A 2N torque sensor is employed for the torque measurements.

For the numerical modelling, the Reynolds-averaged continuity and momentum equations for incompressible fluids in turbulent motion are solved, by applying the realizable k-epsilon turbulence model. The Multiple Reference Frame (MRF) model is used to describe the impeller motion. To apply the steady-state MAT approach, a source term gets implemented into a convection-diffusion equation to govern spatially averaged age distributions. The 95% dimensionless mixing time is within the range: (5~6)σ, where σ is the standard deviation of the mean age probability distribution function. The study of mesh independence on the power number calculated by torque and the mean age

https://doi.org/10.52202/074122-0004

mixing time are conducted. More detailed information can be found in [4].

In this study, the power number and the MAT predicted 95% dimensionless mixing time of six impellers are compared to experimental data. Then, a multi-objective optimization study is performed, using the SHERPA algorithm. Six design parameters are investigated, as shown in Table 1. The rotation rate is 400 rpm, and the Re number is between 91077.48 and 306121.53, indicating a fully turbulent flow condition.

Table 1. Design parameters.

Parameter	Range	Resolution
Blade diameter D [a]	0.3T~ 0.55T	10
Blade height H	0.1D ~ 0.25D	5
Blade angle A	30°~ 90°	13
Blade thickness t	0.01D ~ 0.04D	5
Off-bottom clearance C	0.15T ~ 0.5T	8
Number of blades N	3 ~ 6	4

[a] unit: mm.

Results and Discussion

As shown in Figure 1, labels at the x-axis are the name of designs, with a letter represents corresponding design parameter in Table 1, the number indicates the respective value, e. g. D156 means blade diameter of 156 mm.

Power numbers from steady-state simulations within 10% of experimental measurements.

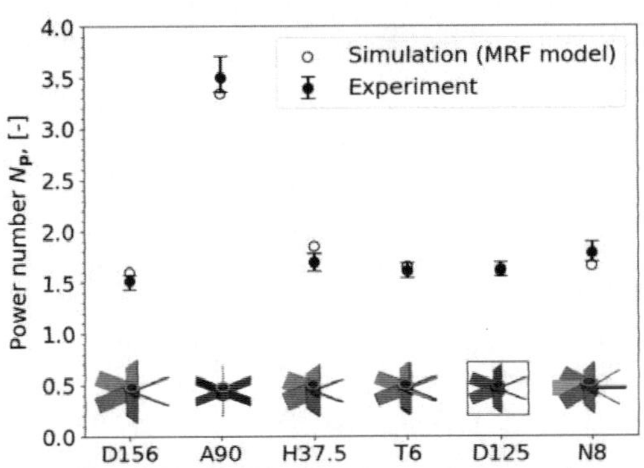

Figure 1. Comparison of power numbers from experiment and MAT approach for Pithed-blade turbines.

For mixing time comparison, the experimental data and the maximum (6σ) and minimum (5σ) dimensionless mixing time from the MAT approach are shown in Figure 2. The MAT mixing times show good agreement with the experimental results of different Pitched-blade turbines, while there is a higher deviation when the angle of blades was 90°.

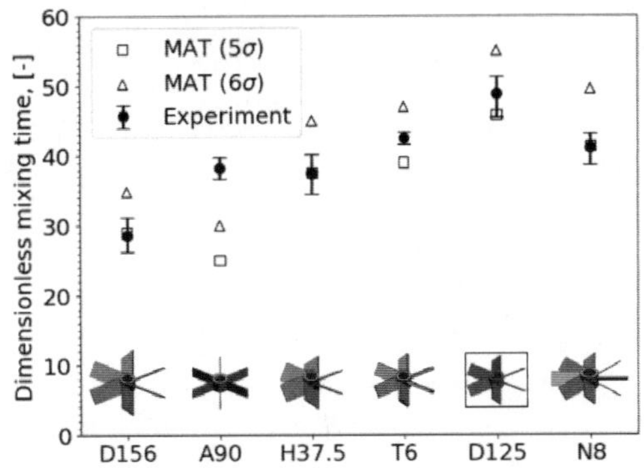

Figure 2. Comparison of D_{mix} from experiment and MAT approach for Pithed-blade turbine

This is in accordance with the conclusion of previous studies [4]: the deviation of MAT is higher when a radial flow impeller is used, since the radial flow forms two compartments within the vessel, whereas the axial flow impeller forms an overall bottom-to-top circulation loop.

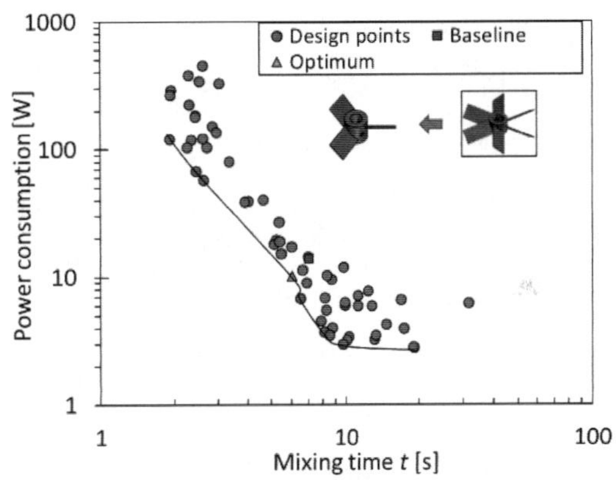

Figure 3. Pareto plot

In this study, sixty different designs based on pitched-blade turbine are carried out in an optimization study. In Figure 3, the relationship between power input and mixing time is presented in a Pareto plot. As can be observed a shorter mixing time implies a higher power input. The designs on the black solid line are better than other points for at least one objective and not worse for the others, also known as the Pareto-front designs.

One of these designs was chosen for experimental validation. From Figure 4 and 5, the good agreement between MAT method and experiment in case of mixing time are obtained.

Table 2 summarizes the dimensions of the optimum design, which was located at the position with a lower off-bottom clearance, and has a larger diameter, blade height and thickness, but a lower blade angle and a smaller number of blades.

https://doi.org/10.52202/074122-0004

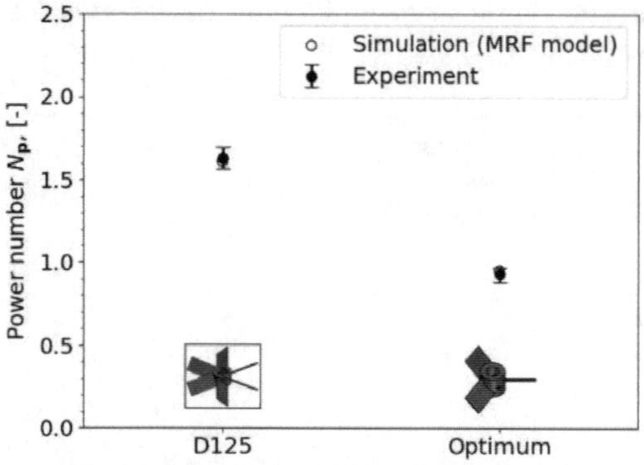

Figure 4. Power numbers of baseline and optimum design.

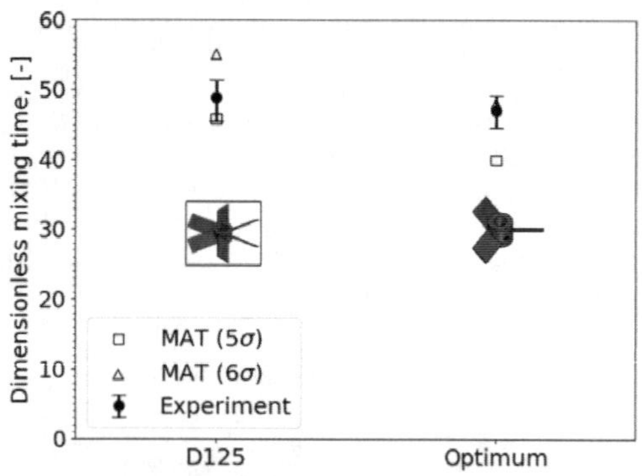

Figure 5. D_{mix} of baseline and optimum design.

The numerical optimization study using the MAT approach took 4 to 5 days on an 8-core computer, which is only 3% of the computation time for transient-based optimization..

Table 2. The dimensions of baseline and optimum design.

Parameter	Baseline	Optimized design
Blade diameter *D*	125	128.7
Blade height *H*	25	32.175
Blade angle *A*	45	40
Blade thickness *t*	2	2.25
Off-bottom clearance *C*	125	97.5
Number of blades *N*	6	3

Conclusions

1. The MAT approach was further validated by predicting blend time for different Pitched-blade turbines.
2. The steady state approach reduces the computational effort significantly.
3. The optimum design has lower power input, but better mixing performance compared to the baseline design.

References

[1] A. Kukukova, J. Aubin, S. M. Kresta, *Chemical engineering research and design*, 87(4) (2009) 633.
[2] S. Derakhshan, M. Pourmahdavi, E. Abdolahnejad, A. Reihani, A. Ojaghi, *Computers & Fluids* 81 (2013) 145.
[3] E. Zitzler, L. Thiele, *IEEE transactions on Evolutionary Computation* 3(4) (1999) 257.
[4] M. Wu, N. Jurtz, A. Walle, M. Kraume, *Chemical Engineering Science* 263 (2022) 118109.
[5] M. Kolano, J. Danke, M. Kraume, *Biomass and Bioenergy* 152 (2021) 106180.

Numerical Study on the Patterned Membrane Inducing Chaotic Mixing for Reducing Concentration Polarization in a Filtration Process

S1.3

Jo Eun Park[1], *Tae Gon Kang*[1], *Seon Yeop Jung*[2]. *(1) Korea Aerospace University, 76 Hanggongdaehak-ro, Goyang-si, Gyeonggi-do, Republic of Korea, p.joeun@gmail.com, tgkang@kau.ac.kr. (2) Dankook University, 152 Jukjeon-ro, Yongin-si, Gyeonggi-do, Republic of Korea, seon27@dankook.ac.kr.*

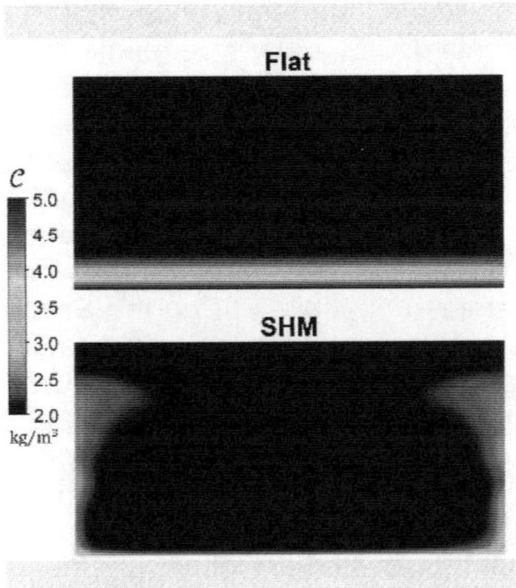

We introduced a staggered herringbone pattern (SHP) as a membrane surface topology to mitigate concentration polarization by inducing chaotic advection. To demonstrate the feasibility of the proposed patterned membrane, numerical simulations were conducted, enabling us to understand the flow and fouling characteristics of the membrane with SHP. A commercial CFD software based on the finite volume method is employed to solve the convection-diffusion equation to obtain the concentration distribution, which is coupled with the continuity and steady Navier-Stokes equations. The two counter-rotating flows repeating periodically in the feed flow direction resulted in chaotic advection, back-transporting the foulants away from the membrane surface. The proposed patterned membrane induces effective stirring of the fluid, subsequently leading to enhanced permeate flux due to reduced concentration polarization compared to that of a flat membrane.

Introduction

Chaotic mixing is an attractive way to efficiently reduce membrane fouling or concentration polarization in membrane filtration processes [1]. Membrane filtration process is employed as a unit process in various areas, e.g., environmental, chemical, biological, pharmaceutical, and food industries. Especially, crossflow filtration is widely used in such applications. Contaminants, such as organic materials and minerals filtered by the membrane, are accumulated at the membrane surface during the continuous operation of the process, which is called membrane fouling that leads to a decline in filtration performance. In this regard, many studies are being conducted to mitigate membrane fouling. One of the anti-fouling techniques is patterned membrane surfaces [2, 3]. In this study, a staggered herringbone pattern (SHP) [4], inducing chaotic advection, is employed as a membrane pattern. In this study, the flow and mass transfer characteristics due to the SHP will be investigated by numerical simulations. In addition, flux enhancements via chaotic advection is demonstrated as well.

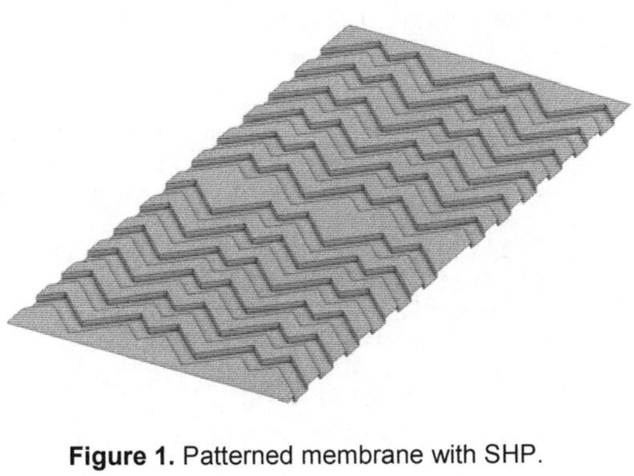

Figure 1. Patterned membrane with SHP.

Numerical Methods

To demonstrate the feasibility of the proposed patterned membrane, we conducted numerical simulations, enabling us to understand the in-depth flow in the filtration system and the concentration of foulants on the membrane with SHP. A commercial CFD software based on the finite volume method, ANSYS Fluent 18.1, is employed to solve the convection-diffusion equation to obtain the concentration distribution coupled with the continuity and steady Navier-Stokes equations. Using the velocity field, we also plotted Poincaré sections to understand the dynamical systems of the flows due to the SHP.

Results and Discussion

In a flat membrane, contaminants are easily accumulated on the membrane surface due to the permeate flux directed normally to the membrane surface. However, two repeating asymmetric rotational flows and chaotic advection caused by the patterned membrane with SHP can be a hydrodynamic method to mitigate fouling and concentration polarization [5]. In this filtration system, the rotational flows play a key role of sweeping the membrane surface so that contaminants are back-transported by upwelling flows. The flows due to the SHP can lead to chaotic advection because the two flow patterns intersect to each other [6], as shown in Figure 2.

Figure 3 depicts the Poincaré section for the flow system of the SHP with the Reynolds number $\mathrm{Re} = 100$. The Poincaré section not only represents a three-dimensional flow structure on a two-dimensional plane by the paths of tracer particles but can also qualitatively ascertain how chaotic a flow system is [7]. In the dynamical system theory, a flow system is chaotic if the trajectories of the particles along the flow become

https://doi.org/10.52202/074122-0005

disordered, resulting in a homogeneous distribution of the tracer particles.

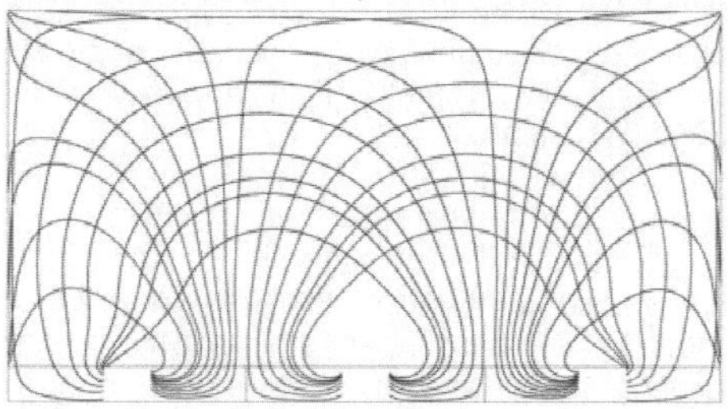

Figure 2. Intersection of streaklines in SHP when $\mathrm{Re} = 100$.

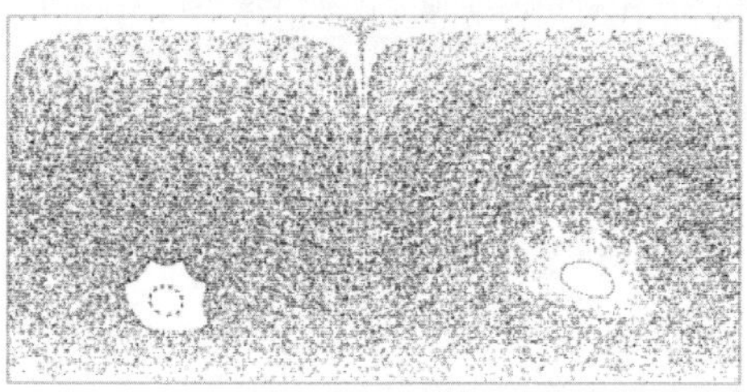

Figure 3. Poincaré section for the flow system of the SHP when $\mathrm{Re} = 100$.

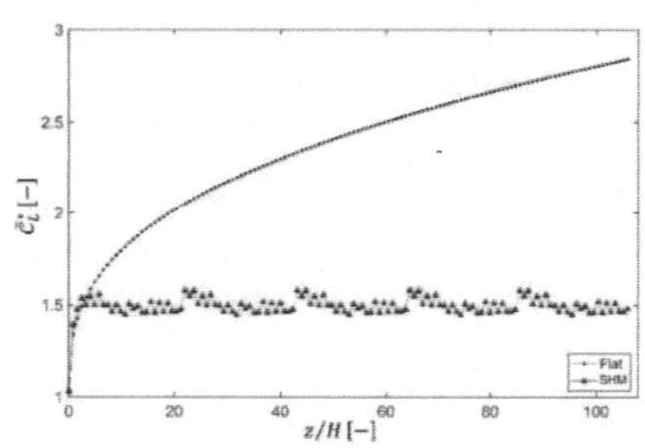

Figure 4. CP modulus in the z-direction.

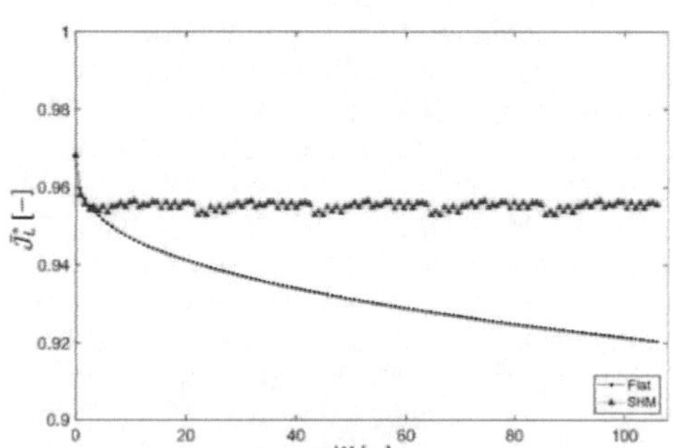

Figure 5. dimensionless permeate flux in the z-direction.

On the other hand, a flow system is regular (not chaotic) if the particles form stable trajectories. In the case of the SHP, two small islands representing unmixed regions are visible, but most regions are chaotic when $\mathrm{Re} = 100$ (see Figure 3).

Figure 4 shows the distribution of CP modulus, which is dimensionless wall concentration defined as $\mathcal{C}^* = (\mathcal{C}_w - \mathcal{C}_p)/(\mathcal{C}_0 - \mathcal{C}_p)$, where $\mathcal{C}_w$ is the wall concentration, $\mathcal{C}_p$ the concentration of the permeate region, and $\mathcal{C}_0$ the feed concentration. Figure 5 represents dimensionless permeate flux defined as $\mathcal{J}^* = \mathcal{J}/\mathcal{J}_0$ on the membrane surface when $\mathrm{Re} = 100$. In the flat membrane, the concentration of foulant continues to increase along the flow direction. In the membrane with SHP, the CP modulus does not increase beyond a certain level due to the efficient stirring, which clearly demonstrates anti-fouling using the SHP. We also plotted the permeate flux as an indicator of the filtration performance. In the flat membrane, the permeate flux decreases as the foulants accumulate, but in the patterned membrane with SHP, the permeate flux is maintained along the flow direction since the concentration is limited to a threshold value as depicted in Figure 4.

Conclusions

We introduced the staggered herringbone pattern (SHP) inducing chaotic advection as a surface pattern of a membrane in crossflow filtration. The flow and mass transfer characteristics were numerically investigated to understand the effect of the patterned membrane with the SHP on stirring and membrane fouling. The filtration system with the SHP can generate chaotic advection due to the two asymmetric rotational flow patterns with crossing streaklines. Compared to the flat membrane, it is confirmed that the patterned membrane with the SHP can lower the foulant concentration at the membrane and maintain the permeate flux below a specific level via chaotic advection and upwelling flows.

Acknowledgments

This research was supported by a National Research Foundation of Korea (NRF) grant funded by the Korea government (MSIT) (No: 2022R1A2C1007886)

References

[1] S.Y. Jung, H.I. Jung, T.G. Kang, K.H. Ahn, *AIChE Journal*, 66 (2020) e16792.

[2] M. Xie, W. Luo, S.R. Gray, *Water research*, 124 (2017) 238-243.

[3] Y.J. Won, J. Lee, D.C. Choi, H.R. Chae, I. Kim, C.H. Lee, I.C. Kim, *Environmental science & technology*, 46 (2012) 11021-11027.

[4] A. D. Stroock, S. K. Dertinger, A. Ajdari, I. Mezic, H. A. Stone, G. M. Whitesides, *Science*, 295 (2002) 647-651.

[5] S.Y. Jung, J.E. Park, T.G. Kang, K.H. Ahn, *Micromachines*, 10 (2019) 836.

[6] M. Speetjens, G. Metcalfe, M. Rudman, *Applied Mechanics Reviews*, 73 (2021) 3.

[7] K.T. Kim, J.E. Park, S.Y. Jung, T.G. Kang, *Membranes*, 11 (2021) 724.

https://doi.org/10.52202/074122-0005

Experimental analysis of the fluid flow in the vessel of the blade-free centrifugal mixer based on the heat balance equation

S1.4

Y. Komoda[1], H. Utsunomiya[1], N. Ohmura[1], Y. Nakamura[2], F. Takaoka[2]. (1) Kobe University, 1-1 Rokkodaicho, Nada-ku, Kobe, Hyogo, JAPAN, komoda@kobe-u.ac.jp. (2) SHASHIN KAGAKU Co., Ltd, 7-2-10 Nojihigashi, Kusatsu, Shiga, JAPAN.

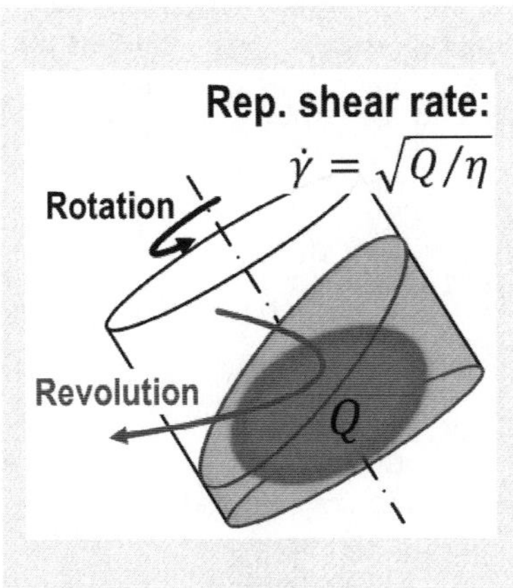

A planetary centrifugal mixer is frequently used to disperse particles into highly viscous materials. A shear rate applied to the materials is crucial in understanding the dispersion process but has yet been measured by experimental techniques. In the present study, focusing on the temperature rise due to viscous dissipation, we tried to measure the mean shear rate from the heat balance equation. The temperature rise of the test fluid and gas phase in an isolated vessel was measured during agitating. We discussed the effect of various operating conditions on the mean shear rate. Furthermore, we newly introduced the dimensionless number *Ut* to express the dominance of the centrifugal force against viscous force. As a result, the flow state could be classified into two types: centrifugal force dominant or viscous force dominant. Then, the strategy to increase the shear rate for each flow condition has been proposed.

Introduction

A planetary centrifugal mixer is a device that can efficiently mix or knee highly viscous or sometimes solid-like fluids without using impeller blade by revolving a self-rotating vessel. However, since the vessel position is not fixed and the rotating axis is tilted against the revolving axis, it is difficult to directly observe the fluid flow inside the vessel under practical operating conditions. Therefore, the relationship between the operating conditions and the flow state is studied mainly by numerical simulations [1][2][3] and rarely by experiment [4]. Furthermore, heat management often becomes a problem for this mixer because fluid is easily heated up due to sizeable viscous dissipation, but cannot install a jacket cooling system. We then focused on the fact that the temperature rise during agitation depends on the fluid viscosity and shear rate applied to the fluid. In other words, adequate measurement of temperature change should allow for calculating a shear rate based on a heat balance equation. In the present study, we established an experimental system to measure temperatures to estimate shear rate by heat balance equation and investigate the effect of operating conditions and fluid viscosity on the representative shear rate. Finally, we propose dimensionless numbers to classify the factors dominating the fluid flow.

Experiment setup and Analysis

50ml of the test fluids (silicone oil, viscosity 1-300 Pa·s) in a plastic vessel was agitated using planetary centrifugal mixers (Kakuhunter SK-300SII, SK-400TR, SHASHIN KAGAKU Co., Ltd.). The inner radius of the vessel, *r*, is 28.5 mm, whereas the revolution radii, R, 58 and 87 mm. Additionally, SK-400TR equips a function to adjust the angular speed ratio of the self-rotation to the revolution α from -0.2 to -2.0. The negative speed ratio means that the self-rotational direction is opposite to the revolution. The plastic vessel is inserted into the other larger plastic vessel using a spacer to minimize the heat flow at the bottom and side wall. Two temperature-measuring devices are placed in the inner and outer lids of the outer vessel. The temperature changes of the test fluid and the gas phase in the inner vessel were measured wirelessly using a thin-film type thermocouple and a non-contact infrared thermometer. The experimental setup is shown in **Fig.1**.

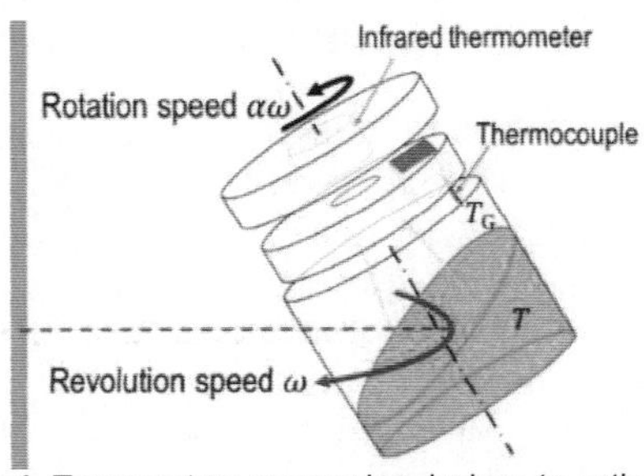

Figure1. Temperature measuring devices to estimate the representative shear rate in a vessel.

The heat balance equation for the test fluid can be expressed as Eq. (1).

$$\rho C_P \frac{dT}{dt} = \frac{UA}{V}(T_G - T) + \eta\dot{\gamma}^2 \quad (1)$$

where T and T_G are the temperatures of the test fluid and gas phase in the inner vessel, ρ, C_P, V, and η are the density, specific heat capacity, volume, and viscosity of the test fluid, and $\dot{\gamma}$ is the shear rate applied to the test fluid. A and U are the heat transfer area and the overall heat transfer coefficient at the gas-fluid interface. U is expressed by Eq (2) using the boundary heat transfer coefficients h_G and h_L for the gas and fluid sides, respectively.

https://doi.org/10.52202/074122-0006

$$\frac{1}{U} = \frac{1}{h_G} + \frac{1}{h_L} \quad (2)$$

The boundary heat transfer coefficient for the gas phase, h_G, is assumed to be constant because of the roughly constant physical properties of air and the small effect of airflow on the boundary. Therefore, h_G is estimated by Eq. (3) proposed by Özisik [5] and was in the range from 22 to 34 W/(m^2 K).

$$\frac{h_G d}{k} = 0.23\left(\frac{Vd}{\nu}\right)^{0.6} \quad (3)$$

On the contrary, the boundary heat transfer coefficient for the test fluid, h_L, is estimated from a famous relationship for agitated vessels among dimensionless numbers using the thermal conductivity k and other parameters that appeared in Eq. (1). The constant value of C has been determined so that Eq. (4) reasonably predicts the reference value of h_L for the oil having the same viscosity.

$$\left(\frac{h_L d}{k}\right) = C\left(\frac{\rho(\omega/2\pi)d^2}{\eta}\right)^{\frac{2}{3}}\left(\frac{\eta C_P}{k}\right)^{\frac{1}{3}} \quad (4)$$

Typical temperature changes of the gas and liquid and the related values are summarized in **Fig.2**. The viscous dissipation term (the 2nd term of RHS of Eq. (2)) can be estimated by subtracting the 1st term of RHS from LHS of Eq. (2). Fluid viscosity at each temperature was estimated using the equation provided by the supplier, and then the variation of shear rate applied to the fluid was obtained, as shown here. Finally, the mean shear rate was determined by averaging the shear rate for 20 min.

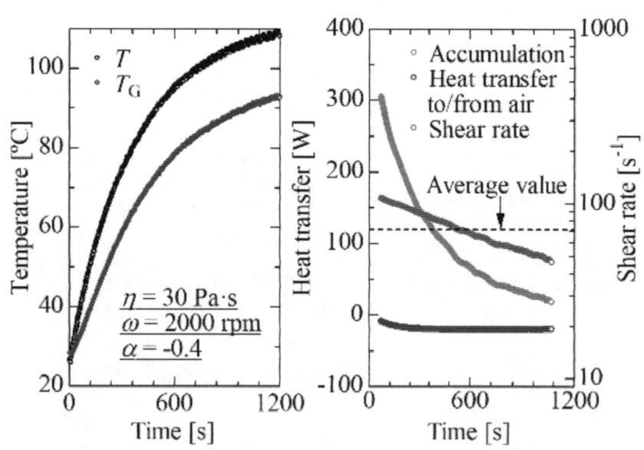

Figure 2. Raw temperature data and heat balance to calculate shear rates.

Results and Discussion

Fig.3 plots the effects of the revolution speed, revolution radius, and fluid viscosity on the representative shear rate at a constant speed ratio of -0.4. The shear rate always exhibited a linear relationship with the revolution speed. This relationship seems reasonable because the rotational motion of the vessel wall directly induces shear flow. No effect of revolution radius was observed for a low-viscosity fluid, and the linear relationship predicts zero shear rate in a vessel with no revolution and rotation. However, the shear rate increased for a larger revolution radius for fluids with higher viscosity. Furthermore, the shear rate was expected to be zero even though the vessel rotates at a fixed speed. We directly observed fluid flow in the vessel by a strobe scope technique and confirmed that the test fluid showed a solid-like rotation attached to the vessel wall. Therefore, a sufficiently large centrifugal force is required for highly viscous fluids to generate shear flow in a vessel.

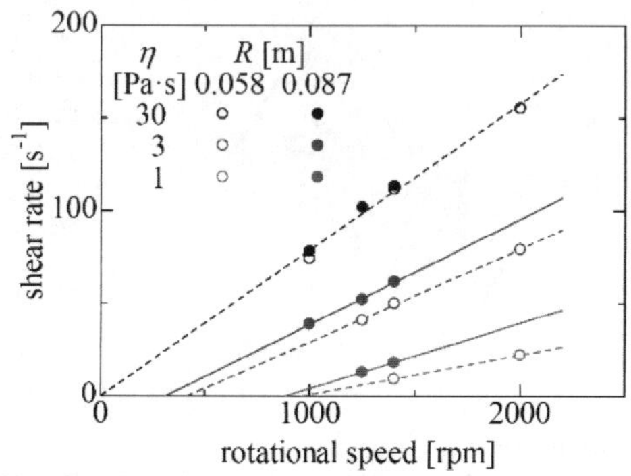

Figure 3. Representative shear rates as functions of revolution speed (angular speed ratio α = -0.4)

We consider that the factors dominating the fluid flow are the centrifugal force caused by the planetary motion and the shear force acting on the vessel wall. Similarly to the force balance in a planetary ball milling device [6], the acceleration of the centrifugal force can be expressed by Eq. (5). The first and second terms are the centrifugal forces caused by the revolution and self-rotation of the vessel, respectively. The third term is a Coriolis force. In contrast, a representative shear force is expressed by the product of the interfacial area of shear applied, the representative shear rate, and the fluid viscosity. Then, we defined the dimensionless number *Ut*, which is the ratio of the centrifugal forces to the viscous force expressed by Eq. (6) using the angular velocity of the vessel as a representative shear rate.

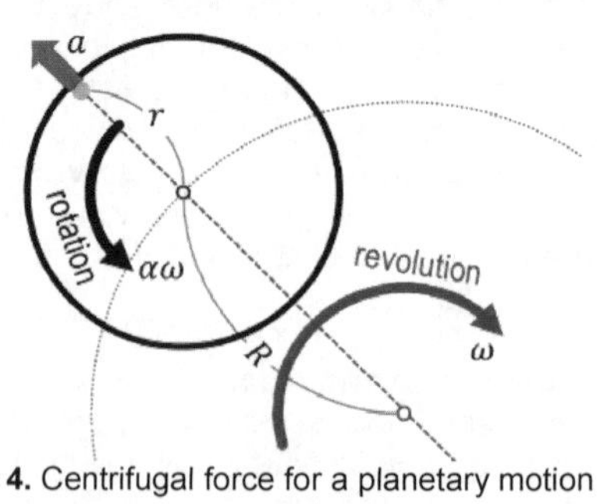

Figure 4. Centrifugal force for a planetary motion vessel

$$a = (R + r)\omega^2 + r(\alpha\omega)^2 - 2r\omega \cdot (\alpha\omega) \quad (5)$$

$$Ut = \frac{mR\omega^2\{r/R \cdot (1-\alpha)^2 + 1\}}{A\eta(\alpha\omega)} \quad (6)$$

$$\bar{\gamma} = \dot{\gamma}/\alpha\omega \quad (7)$$

where m and ρ are the mass and viscosity of test fluids, and A is the interfacial area at the fluid-solid during agitating. Additionally, the measured mean shear rate was reduced by the representative shear rate as denoted by Eq. (7).

The effects of agitating conditions on the relationship between the shear rate and *Ut* were investigated using test fluids having various viscosities. However, the definition of dimensionless shear rate in Eq. (7) was suitable only for larger *Ut*, and the plots were scattered in the smaller *Ut* regime. At large *Ut* conditions, a larger centrifugal force acts on test fluids. As a result, the test fluid is forced to push over the vessel wall, and the shear stress at the wall suppresses the fluid flow. Therefore, a large *Ut* regime is considered a "viscous dominant regime," and the shear rate should be normalized by the self-rotation speed of the vessel. In contrast, insufficient centrifugal force acts on test fluids at small *Ut* conditions. In other words, the intensification of fluid flow requires a larger centrifugal force in the "centrifugal force dominant regime." Since the revolution radius is sufficiently larger than the rotation one, the shear rate in this regime should be normalized by the angular velocity of the revolution. We then replot all experimental data in **Fig.5** using two representative shear rates depending on *Ut*. It is revealed that all data falls on two lines, corresponding to viscous and centrifugal force dominant regimes. The relationships are expressed by Eqs (8) and (9).

$$\frac{\dot{\gamma}}{\alpha\omega} \propto Ut^{\frac{1}{4}} \quad (Ut \geq 10^1) \tag{8}$$

$$\frac{\dot{\gamma}}{\omega} \propto Ut^{\frac{5}{4}} \quad (Ut < 10^1) \tag{9}$$

Although the mean shear rate can be reasonably fitted with the exponential functions of *Ut*, the exponents significantly differs with dominant factors. In the viscous dominant regime, the exponent is only 0.25, and a larger speed ratio is more effective in increasing the shear rate rather than a larger revolution radius. In contrast, the exponent of Ut in the centrifugal force dominant regime suggests that a large revolution radius helps applying high shear to the test fluid.

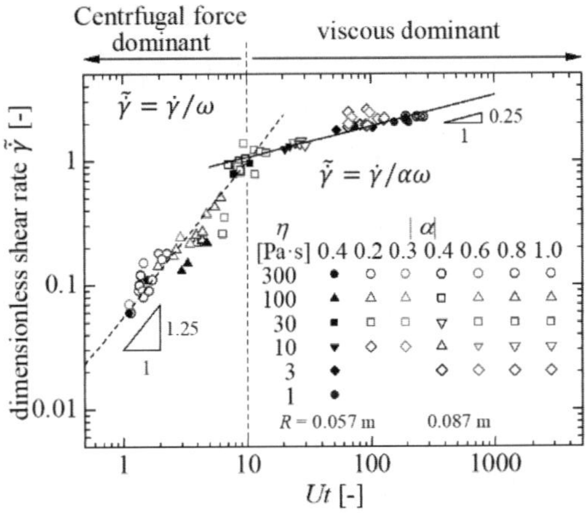

Figure 5. Relationship between dimensionless number Ut and dimensionless shear rate of the planetary centrifugal mixer

Conclusions

In the present study, the mean shear rate applied to the test fluid in the vessel of a planetary centrifugal mixer has been measured based on the heat balance equation. A less viscous fluid can flow at very slow rotational motion, whereas highly viscous fluid exhibits a solid-like rotation, and no shear deformation was applied. Referring to the centrifugal force applied in a ball milling device, a new dimensionless parameter *Ut*, the ratio of the centrifugal force to viscous force, has been introduced. The flow regime could be classified into two types based on *Ut*: viscous dominant and centrifugal dominant. At large *Ut* conditions, the mean shear rate is governed by viscous force and can be significantly increased by the self-rotation speed. On the other hand, a larger mixer with a large revolution radius effectively applies a higher shear rate to test fluid at Ut less than 10, which is the centrifugal force dominant regime.

References

[1] Chergui, N., Lateb, M., Lacroix, E., Dufresne, L., CFD study of flow dynamics in a blade free planetary mixer (BFPM) – A qualitative flow study, *Chem. Eng. Res. Des.*, 102, 2015, 100-115.

[2] Son, K. J., A numerical study of the influence of rheology of cohesive particles on blade free planetary mixing, Korea Aust. Rheol. J., 30(3), 2018, 199-209

[3] Shen, B., Zhan, X., Sun, Z., He, Y., Long, J., Li, X., PIV experiments and CFD simulations of liquid–liquid mixing in a planetary centrifugal mixer (PCM), Chem. Eng. Sci., 259, 2022, 117764.

[4] Yamagata, T., Sugisawa, J., Fujisawa, N., Experimental study on laminar mixing in planetary mixer, Exp. Fluids, 62:28, 2021

[5] Özisik, M., Basic heat transfer. New York: McGraw Hill, 1977

[6] Schliz, J., "Internal Kinematics of Tumbler and Planetary Ball Mills: A Mathematical Model for the Parameter Setting" Mater. Trans., 39(11), 1988, 1152-1157

https://doi.org/10.52202/074122-0006

On the Just Suspended Speed in Widely Graded Slurries S1.5

J. J. Giacomelli[1,2], R. K. Grenville[1], B. Jarrett[3], H. E. A. Van den Akker[2]. (1) SPXFLOW, 1221 East Main St., Palmyra, 17078 USA, Jason.Giacomelli@spxflow.com (2) University of Limerick, Bernal Institute, Limerick, V94 T9PX, Ireland. (3) Rowan University, Glassboro, 08028, USA.

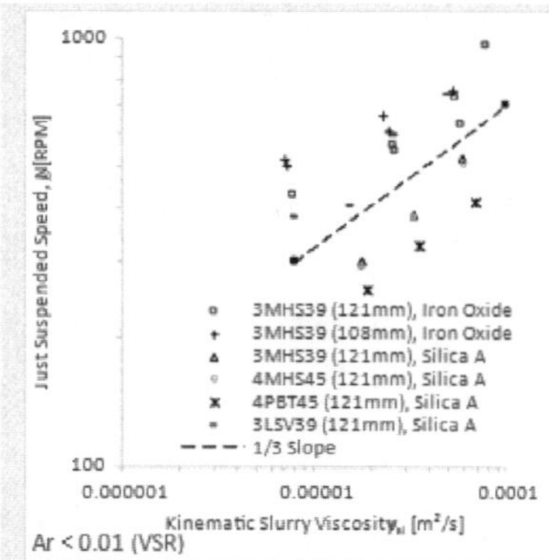

Figure 4. N_{JS} vs. ν_{sl} in the viscous regime with 1/3 reference slope per Equation 6.

Typical correlations to calculate an impeller's just suspended speed are based upon experiments with slurries with narrow particle size distributions and concentrations smaller than 20-30%v/v in which suspension is due to turbulent eddies in the inertial subrange. In the minerals processing industry, however, particles are often widely graded with a significant portion of them (≈>50%) being smaller than 75 μm. While at elevated concentrations (>20-30% v/v), these types of slurries can become viscous and even non-Newtonian, we do not have an applicable correlation. New suspension experiments are presented for three model widely graded slurries over concentrations of 1 < x_v < 50%v/v. As the concentration increases, the suspension mechanism changes from inertial subrange turbulence to the viscous regime. We present a preliminary correlation for the viscous suspension regime, for which one is not currently available. This correlation exploits a suspension viscosity due to the presence of the fines.

Introduction

The suspension of solid particles in stirred vessels is an important mixing duty. The minimum speed to operate an impeller with the view to producing a suspension is known as the just suspended speed, N_{JS}. The energy balance family of correlations developed by Baldi et al. [1], Davies [2], and Buurman et al. [3] has recently been improved by Grenville et al. [4] by reformulating it into two dimensionless groups, the Eddy Reynolds, Re_E, and Archimedes, Ar, numbers. Using data developed by the FMP consortium of the British Hydrodynamic Research Group (BHRG), Grenville et al. [4] produced a useful design correlation and demonstrated that two suspension regimes exist.

The first suspension regime corresponds to suspension by turbulent eddies in the inertial sub-range (ISR). Suspensions within this regime are characterized by freely settling, relatively large and dense particles. The second regime occurs when particles mainly interact with eddies in the viscous subrange (VSR). The boundary between these regimes was approximately identified at $Ar \approx 10^{-2}$ [4].

As demonstrated by Grenville et al. [4] and Giacomelli et al. [5] [6], the Re_E- Ar design equation approach has been shown to be successful and effective for the high-Ar /ISR suspension cases. Now, we desire to develop a matching correlation for the low-Ar/VSR range to handle cases common to the minerals processing industry which fall outside the ISR systems.

The FMP/BHRG dataset comprises tests with viscosity modifiers to increase (liquid) viscosity. Slurries, however, can also reach viscous states through a combination of particle size (distribution) and concentration. As particle sizes become smaller and more widely graded, spanning several decades with multimodal characteristics, and concentrations increase, particle-particle interactions begin to produce a suspension viscosity. This is particularly common in minerals processing in which various comminution processes produce widely graded slurries with a large fraction of fines. While the definition of 'fines' is not straightforward; the slurry transport industry has been classifying particles with diameters smaller than 75μm as fines [7].

Cases with high concentrations of narrowly graded large particles have been studied before. Buurman [8] tested concentrations up to 60%v/v, Davies [2], Ayranci and Kresta [9], and Myers et al. [10] all tested concentrations up to 20-30%/v/v. In all these cases, however, the particles were relatively large, at least larger than 75μm, and narrowly graded, to the effect that the slurries did not develop viscous behaviour, although Buurman predicted and observed a 'crowding' behaviour for high concentrations.

Combining widely grading particle sizes (including fines) and high(er) concentrations brings us into unknown territory as to (1) how to predict when a slurry will produce viscous or non-Newtonian behaviour, and (2) how to modify the available low-Ar/VSR correlation [4] such that it also describes viscous and non-Newtonian slurries. To address these issues, we carried out an experimental program in which we tested three widely graded model solids samples with varying fractions of material sub-75μm and across a wide range of concentrations (1-50% v/v).

Theory

The energy balance between turbulent eddies and particles when cast in the form of the aforementioned dimensionless numbers results in two forms depending on which portion of the turbulent spectrum the eddies originate. The details of these derivations can be found in [4]. For the ISR case, $u' \propto (\varepsilon l_e)^{1/3}$, which results in:

https://doi.org/10.52202/074122-0007

$$Re_E \propto Ar^{1/2} \quad 1.$$

while in the VSR, $u' \propto \eta_k(\varepsilon/\nu)^{1/2}$, resulting in:

$$Re_E \propto Ar^{1/3} \quad 2.$$

where Re_E and Ar are defined as follows:

$$Re_E = N_P^{1/3}\frac{ND^2}{\nu}\left(\frac{d_{43}}{D}\right)^{4/3} \quad 3.$$

$$Ar = \frac{gd_{43}(\rho_s - \rho_l)}{\rho_l \nu^2} \quad 4.$$

and the specific power input per mass, ε, is defined as $N_P N^3 D^5/V$. When the Re_E - Ar is reduced to its dimensionless form to compute N_{JS}, we find two forms depending on the suspension regime [4], [6]:

$$N_{JS}\Big|_{\substack{ISR \\ Ar>10^{-2}}} \propto \frac{d_{43}^{1/6}}{N_P^{1/3}D^{2/3}}\left(\frac{g\Delta\rho}{\rho}\right)^{1/2}\left(\frac{V^{1/3}}{D}\right) \quad 5.$$

$$N_{JS}\Big|_{\substack{VSR \\ Ar<10^{-2}}} \propto \frac{1}{N_P^{1/3}D^{2/3}}\left(\frac{\nu}{d_{43}}\frac{g\Delta\rho}{\rho}\right)^{1/3}\left(\frac{V^{1/3}}{D}\right) \quad 6.$$

Here it should be clarified that the kinematic viscosity term is made up of the slurry viscosity and slurry density:

$$\nu \equiv \nu_{sl} = \frac{\mu_{sl}}{\rho_{sl}} = \frac{\mu(x_v, \dot{\gamma})}{x_v\rho_s + (1 - x_v)\rho_l} \quad 7.$$

where $\dot{\gamma}$ denotes the shear rate. In the ISR regime, viscosity cancels out which is consistent with expectations in the fully turbulent regime. However, this is not the case in the VSR regime. The impeller power number N_P is known to be a function of viscosity and Reynolds number in the transitional and laminar regimes.

Material and Methods

For the just suspended experiments, a 305mm diameter (T) flat bottomed vessel, common to the minerals processing industry, was selected. The vessel is equipped with four T/12 wide baffles offset from the wall by T/60. Four impellers manufactured by Philadelphia Mixing Solutions, LLC, were selected for this research, a narrow blade hydrofoil (3LSV39), a 3 and 4 blade version of a wide blade hydrofoil (3MHS39/4MHS45), and a common pitched blade turbine (4PBT45).

The just suspended impeller speed, N_{JS}, was measured per the Zwietering definition, i.e., when *no particle was observed stationary on the vessel base for longer than 1-2s.* The impeller power input at N_{JS}, P_{JS} was determined by measuring the impeller reaction torque, Γ, according to $P = 2\pi N\Gamma$ (see Giacomelli et al. [5] for details).

Three test solid materials were selected to simulate the widely graded nature of those found in the minerals industry, *viz.* an iron oxide, DENSFIL (from Prince International Corporation), and two silica materials, SILCOSIL and MINUSIL, silica A and B (from US SILICA), respectively. The particle size distributions were measured using a Beckman Coulter LS13320 Universal Liquid Module (ULM). The physical properties and particle size distribution of the three materials are given in Table 1 and Figure 1, respectively. Tap water at ambient conditions was used as the process fluid.

Table 1. Physical properties of the three test materials.

Parameter	Iron Oxide	Silica A	Silica B
ρ_s [kg/m³]	4535	2828	2829
d_{43} [μm]	40	82	14
$\Delta\rho/\rho$ [-]	3.5	1.8	1.8
$x_v\|_{d<75}$ [v/v]	87%	57%	99%
φ_{max} [v/v]	94%	92%	84%

A Brookfield RST-SST rheometer with Couette geometry was used to characterize the rheology of the various concentrations of the solids. To accommodate the narrow Couette gap, particles larger than 125μm were removed. The shear stress was correlated by using the most appropriate rheological model. The parameters of the model were correlated with x_v to allow precise calculation of the apparent viscosity at a given N_{JS} - x_v combination.

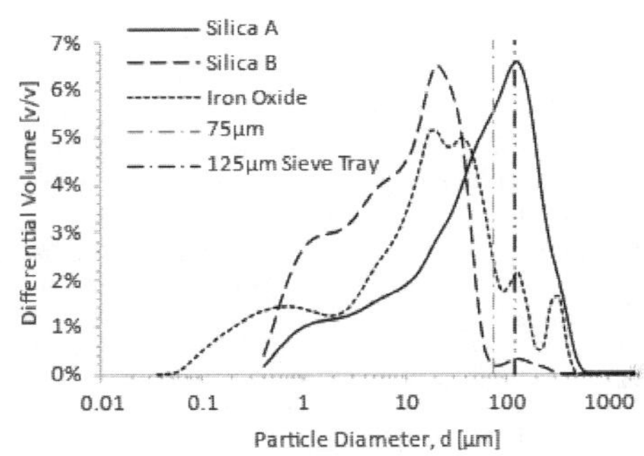

Figure 1. Particle size distributions of the three materials chosen for this investigation. 75μm and below is considered 'fines'. 125μm corresponds to the sieve tray used to prepare the material to characterize in the rheometer.

Results and Discussion

[At this juncture, experimental data have been collected for the iron oxide and silica A materials, however work continues on the silica B material.]

First, between 1 and 5%v/v, the slurries are clearly free settling despite the widely graded PSD. The specific N_{JS} power increases and a power law response is observed with respect to x_v (Figure 2). After 5%, the specific N_{JS} power decreases until a minimum is reached at 20%v/v, yet the slurries still appear freely settling. This type of response is not seen with larger, narrowly graded materials in stirred vessels [4] [8] [9] [2] but is often observed in pipeline slurry transport [7] [11]. While the two test materials have different characteristics (Table 1), the response to concentration is similar.

Both test materials then show a steep increase in power demand beyond 20%v/v as viscosity became evident. Testing was considered complete after 45%-50% v/v as not all impeller configurations were capable of mixing at this concentration. The smaller D/T cases were less effective at the higher

https://doi.org/10.52202/074122-0007

concentrations compared to the larger D/T cases. In several cases, a second impeller was used spaced one impeller diameter above the bottom impeller which improved mixing the most viscous concentrations.

By using the effective impeller shear rate according to Metzner and Otto [12], $\dot{\gamma} = k_s N$, for defining ν_{sl}, the iron oxide and silica slurries were observed to collapse on the Re_E vs. Ar (Figure 3). The boundary or slope change between ISR and VSR is clearly identified at $Ar \approx 10^{-2}$ which agrees with [4].

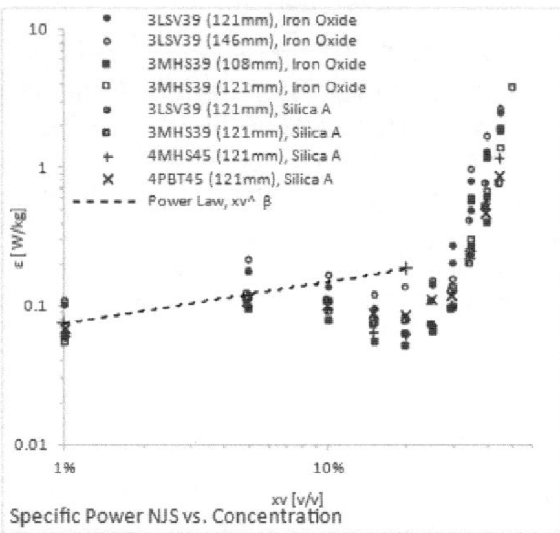

Figure 2. Power input per unit mass at NJS vs. concentration for (4) impeller types and (2) of the test solids. Typical power law response overlain for reference. In all cases, the power input exhibits a minimum at ≈20%v/v before a steep rise in power. The power law is obviously inadequate at describing this behaviour.

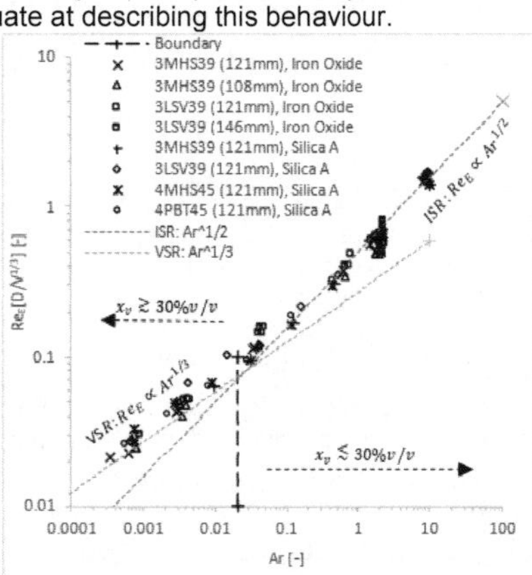

Figure 3. Re_E vs. Ar for two test solids and four impeller types. This plot shows the apparent regime change at Ar = 10^{-2}.

A regression analysis was used to verify that the slope on Ar really was 1/3 for $Ar < 10^{-2}$ and to find preliminary values for the proportionality constant in the VSR correlation, Equation 6. Figure 4 plots N_{JS} versus ν_{sl} for the viscous regime further confirming the 1/3 slope as Equation 6 predicts. As soon as viscosity becomes the dominant characteristic of the slurry, a variable for concentration does not seem to be required to correlate the data. However, the ISR data for widely graded slurries will require a more sophisticated concentration model to describe the features of Figure 2 for x_v < 30%v/v.

Conclusions

The effect of solids concentration on N_{JS} is a complex function of concentration and particle size. A solids concentration effect is captured by the concept of a slurry viscosity when the $Ar < 10^{-2}$. The VSR/ISR boundary has been observed in two solid samples across four impeller types and the boundary occurs approximately at Ar = 10^{-2}. The ISR correlation for widely graded slurries ($Ar > 10^{-2}$) requires a more sophisticated model to predict the effect of concentration on N_{JS}. Preliminary correlation coefficients for four impeller types are presented.

References

[1] G. Baldi, R. Conti and E. Alaria, “Complete Suspension of Particles in Mechanically Agitated Vessels,” *Chem. Eng. Sci.,* pp. 21-25, 1978.

[2] J. T. Davies, “Particle Suspension and Mass Transfer Rates in Agitated Vessels,” *Chemical Engineering and Processing,* pp. 175-181, 1986.

[3] C. Buurman, G. Resoort and A. Plaschkes, “Scaling-Up Rules for Solids Suspension in Stirred Vessels,” *Chem. Eng. Sci.,* vol. 41, no. 11, pp. 2865-2871, 1986.

[4] R. K. Grenville, A. T. C. Mak and D. A. R. Brown, “Suspension of Solid Particles in Vessels Agitated by Axial Flow Impellers,” *Chem. Eng. Res. Des.,* pp. 282-291, 2015.

[5] J. J. Giacomelli, R. K. Grenville and H. E. Van den Akker, “On the suspension of large and dense solid particles in turbulent stirred vessels,” *Chem. Eng. Res. Des.,* vol. 180, pp. 318-332, 2022.

[6] J. J. Giacomelli, R. K. Grenville and H. E. Van den Akker, “Bridging the gap between solids suspension theory and equipment design,” *Chem. Eng. Res. Des.,* 2022.

[7] W. Parzonka, J. M. Kenchington and M. E. Charles, “Hydrotransport of solids in horizontal pipes: Effects of solids concentration and particle size on the deposity velocity,” *Can. J. Chem. Eng.,* vol. 59, pp. 291-296, 1981.

[8] C. Buurman, “Stirring of concentrated slurries: A semi-empirical model for complete suspension at high solids concentrations and 5m³ verification experiments,” *IChemE Symposium Series No. 121,* pp. 343-350, 1990.

[9] I. Ayranci and S. Kresta, “Design rules for suspending concentrated mixtures of solids in stirred tanks,” *Chem. Eng. Res. Des.,* vol. 89, pp. 1961-1971, 2011.

[10] K. J. Myers, E. E. Janz and J. B. Fasano, “Effect of solids loading on agitator just-suspended speed,” *Can. J. Chem. Eng.,* vol. 91, pp. 1508-1512, 2013.

[11] R. G. Gilles, J. Schaan, R. Sumner, M. J. McKibben and C. A. Shook, “Deposition velocities for newtonian slurries in turbulent flow,” *Can. J. Chem. Eng.,* vol. 78, pp. 704-708, 2000.

[12] A. B. Metzner and R. E. Otto, “Agitation of Non-Newtonian Fluids,” *AIChE Journal,* vol. 3, no. 1, 1957.

[13] E. L. Paul, V. A. Atieno-Obeng and S. M. Kresta, Handbook of Industrial Mixing, Hoboken: Wiley & Sons, INC, 2004.

https://doi.org/10.52202/074122-0007

The impact of impeller flow types on the performance of batch mixing miscible liquids in turbulently stirred vessels, an experimental study

S1.6

S S Mirfasihi[1], *W Basu*[1], *P Martin*[1], *T L Rodgers*[1], *A Kowalski*[2], *C P Fonte*[1], *A Keshmiri*[1] *(1) The University of Manchester, Oxford Road, Manchester M13 9PL, United Kingdom, sorosh.mirfasihi@manchester.ac.uk (2) Unilever R&D Port Sunlight Laboratory, Bebington CH63 3JW, United Kingdom*

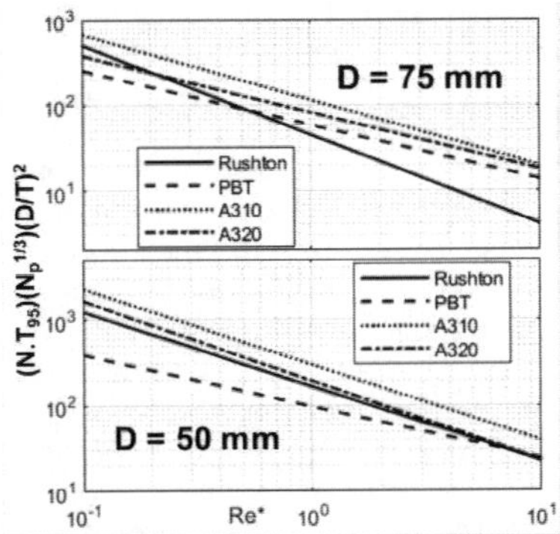

Graphical abstract - Comparison of power-law trends across impeller types.

The mixing dynamics of highly viscous additives in a less viscous bulk is of immense practical interest, yet common models for mixing time prediction, such as Grenville's correlation [1], do not offer sufficiently accurate estimates in such cases. The current study seeks to assess the effect of additives' viscosity ratio in combination with various impeller types on dimensionless mixing time using a set of Electrical Resistance Tomography (ERT) experiments. We consider 2 sizes each of 4 types of impeller geometry – Rushton Turbine, Pitched Blade Turbine, A310 and A320 – and compare their mixing time trends across a wide range of additive viscosity values.

Introduction

The commonality of blending miscible liquids with different physical properties in many chemical, pharmaceutical and food industries as well as in the production of fast-moving consumer goods is indisputable. However, a lack of adequate knowledge of such mixing processes often results in significant waste of money and energy in these sectors. The current upsurge in digital manufacturing promises to address such suboptimal process configurations using computational/predictive frameworks, and this in turn has led to growing demand for reliable validation data.

As part of the CAFE4DM [2] research initiative, we are currently performing a multifactorial study to assess the effect of various parameters such as additive viscosity, impeller speed, volume of additive etc. on mixing time to develop a model capable of mixing time prediction with improved accuracy. Having presented the results from other phases of this study corresponding to the impact of additive viscosity, impeller speed, and volume fraction variation before [3], the authors would like to take this opportunity to focus on the impact of different impeller types on mixing time. While a small number of past studies [4, 5] have evidenced a variation in mixing time with impeller geometry i.e., shape and dimensions, a rigorous experimental characterization of the effect across a wide range of relevant mixing conditions such as additive viscosity, speed etc. is not known from the literature and no analytical models are known to satisfactorily describe this dependence either. In the following sections, we present the methodology of the study including details of the experimental setup and the algorithms used for arriving at mixing time estimates, empirical results from this phase of the study and a consideration of the next steps to be taken for an improved understanding of the effect of impeller flow types on mixing time.

Material and Methods

The experimental setup used in this study consists of a standard baffled cylindrical vessel (diameter, T = 0.15 m) with a centrally mounted impeller. The data acquisition setup for real-time detection of fluid properties comprises of an ITS v5r ERT (Electrical Resistance Tomography) instrument [6] and a host computer. Figure 1 below contains a diagram of the key elements of this setup.

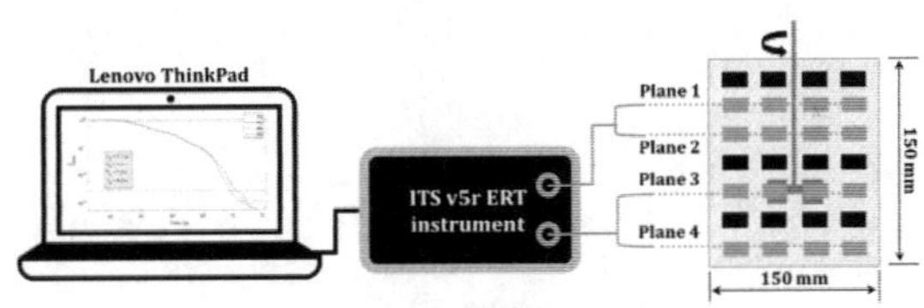

Figure 1. Experimental setup

For each experiment, the tank was filled with tap water up to a volume, V = 2.65 litre as the bulk liquid, the stirrer motor activated at speed = N rpm, and 3 ml of an additive was introduced from above the liquid surface using a truncated plastic syringe. The 4 additives (listed below in table 1) consisted of highly concentrated aqueous mixtures of glucose syrup with added salt for increased conductivity to enable ERT.

Table 1. Additive liquid viscosities

Additive index	1	2	3	4
Viscosity (Pa.s)	138.83	38.40	15.60	0.001

The speed, N, was selected to be fast enough to achieve an impeller Reynolds number in the turbulent regime. A total of 8 impellers were used in this study, comprising of 2 sizes each of 4 different impeller types (Table 2). Each such experiment, with a unique combination of impeller and additive, was repeated 5 times for determining the corresponding mixing time values with statistical confidence.

In order to determine the mixing time in each experiment, the ERT instrument was connected to 4

https://doi.org/10.52202/074122-0008

sets of 8 electrodes embedded in the walls of the mixing vessel was used. The instrument is controlled from the host computer over a USB connection and acquires adjacent pairwise trans-impedance measurements at a rate of 312 frames per second across the 4 active planes (shown in green in Figure 1) by applying ~0.5 mA current at 9.6 kHz. The trans-impedance values obtained using the ERT setup for each experiment were then processed to obtain dynamic estimates of the concentration distribution throughout the vessel. Next, the convergence rate of the concentration distribution to homogeneity was characterized using the mixing index, σ_{RMS}, as per the following equation [7]:

$$\sigma_{RMS}(t) = \sqrt{\frac{1}{N_e}\sum_{k=1}^{N_e}\left(\frac{c_{t,k}-c_{0,k}}{c_{\infty,k}-c_{0,k}} - 1\right)^2} \quad (1)$$

where,

N_e = number of elements,

$c_{t,k}$ = k^{th} measurement/concentration at time 't',

$c_{0,k}$ = k^{th} measurement/concentration prior to the additive being introduced, and

$c_{\infty,k}$ = k^{th} measurement/concentration after mixing has progressed to completion.

Finally, the 95% mixing time, T_{95}, was determined as the time taken for $\log(\sigma_{RMS}^2)$ to decrease from the instant of additive introduction and settle below $\log(1-0.95)^2 = \log(0.05)^2 = -2.6$, as described in [7].

Results and Discussion

Figure 2 plots the distribution of dimensionless mixing time against modified impeller Reynolds number (Re*) as used previously by [8]:

$$Re^* = \rho N D^2/\mu_a \quad (2)$$

where μ_a is the added liquid viscosity in Pa.s (see Table 1), D is impeller diameter (m), and ρ is the bulk liquid density (in kg/m^3). It has been shown by multiple studies [4, 8, 9] that at high enough Re* numbers, i.e., in the stirrer-controlled regime, the dimensionless mixing time is independent of the viscosity ratio and can be well predicted by the Grenville's correlation (within some tolerance) given H =T, the correlation is stated below,

$$NT_{95} = 5.2\, N_p^{-1/3}\, (D/T)^{-2} \quad (3)$$

where N_p is the impeller power number (see Table 2). Having said that, there also exists such Re* threshold below which in the additive controlled regime where the Grenville's correlation is no longer valid to be used for mixing time estimation of miscible liquids with high enough viscosity ratios as demonstrated in the past [8]. Given this, multiple power law correlations have been proposed in Table 2 to account for predicting the mixing time in such regimes. Also, by incorporating the power number in the analysis of dimensionless mixing time as done in the graphical abstract, cross impeller comparison is made possible, from the same graph, it can be seen that the correlated mixing time fits get closer to each other as the power number compensates for the variance in the power draw for each impeller design.

Results gathered from all impellers confirm an agreement between the correlated mixing time in Eq. 1, and experimentally obtained values in the stirrer-controlled regime (within the +/-10% bounds indicated by dotted lines: blue for D = 75 mm, gold for D = 50 mm). From this correlation, it can be confirmed that indeed the mixing time is inversely proportional to the impeller diameter throughout both additive-controlled and stirrer-controlled regimes. In addition, the NT_{95} vs Re* trends were seen to conform fairly to power-law models (indicated by solid lines in Figure 2). For each type of impeller, the dimensionless mixing times were seen to be lower on average for D = 75 mm (*blue*) than for D = 50 mm (*gold*).

The results can also be interpreted based on the pumping efficiency of each impeller. As shown in Figures 2a-d, the most efficient impeller does not necessarily enhance the mixing the most. However, there seems to be a direct correlation between a higher flow number and faster absolute mixing times, which is seen to hold true for all impellers across the data set.

Table 2. Comparison of impellers, Flow number and Power number values have been taken from [10].

Impeller Specification					Mixing time trends	
Impeller	**D/T**	**Power number (Np)**	**Flow number (Nq)**	**Pumping Efficiency (Nq/Np)**	**Power law fit**	**Goodness of fit, R^2**
A310	1/2	0.30	0.52	1.73	$NT_{95} = 114.6 \cdot (Re^*)^{-0.77}$	0.84
	1/3	0.30	0.52	1.73	$NT_{95} = 298.4 \cdot (Re^*)^{-0.89}$	0.71
A320	1/2	0.70	0.66	0.94	$NT_{95} = 81.7 \cdot (Re^*)^{-0.66}$	0.78
	1/3	0.70	0.66	0.94	$NT_{95} = 193.7 \cdot (Re^*)^{-0.93}$	0.97
PBT	1/2	1.50	0.80	0.53	$NT_{95} = 58.7 \cdot (Re^*)^{-0.63}$	0.74
	1/3	1.50	0.80	0.53	$NT_{95} = 97.6 \cdot (Re^*)^{-0.60}$	0.84
Rushton	1/2	5.00	0.65	0.13	$NT_{95} = 45.2 \cdot (Re^*)^{-1.05}$	0.95
	1/3	5.00	0.65	0.13	$NT_{95} = 166.6 \cdot (Re^*)^{-0.87}$	0.88

https://doi.org/10.52202/074122-0008

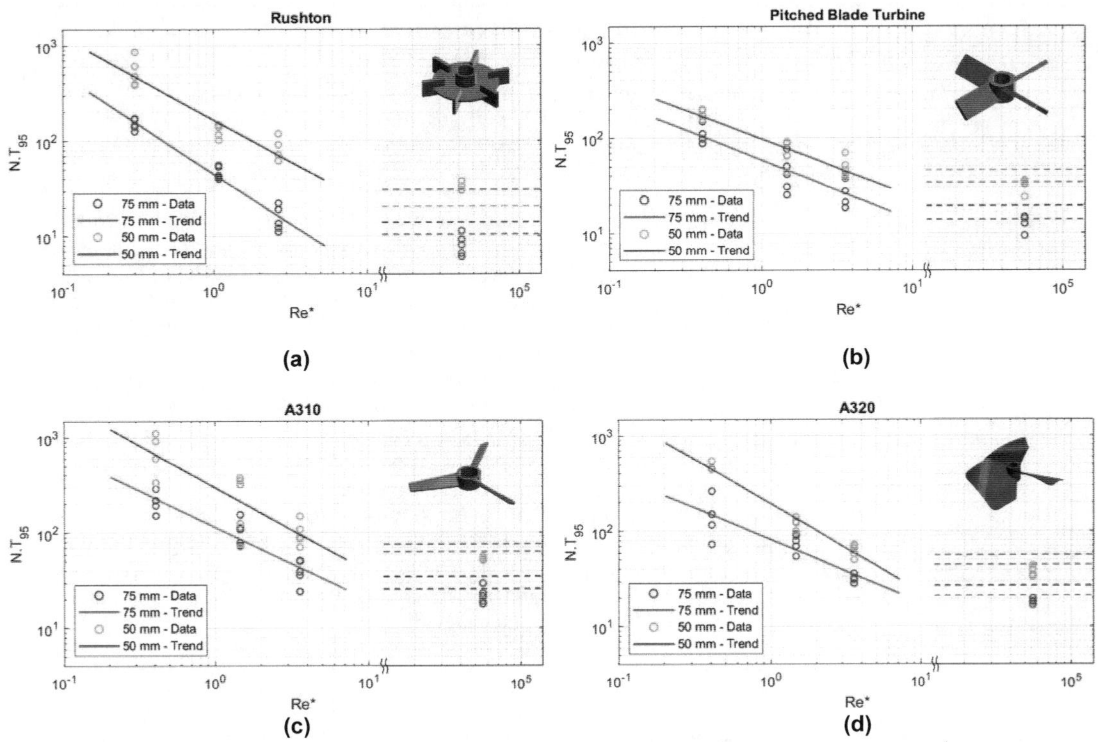

Figure 2 a-d. For each impeller (*shown inset in the plot windows*), dimensionless mixing time was observed to decrease with a reduction in additive viscosity in the additive-controlled regime (*indicated with a white background*). Despite the scatter in the observed mixing times (*indicated by bubbles*).

Conclusions

Cross comparison of impellers in the mixing of miscible liquids with high viscosity contrasts has been presented. The results conform well to the theoretical understanding of mixing mechanics. The flow number is a good predictor of well-designed impeller for achieving lower mixing times, in combination with power number, the pumping efficiency could indicate the most energy efficient yet performant impeller design for this application. To aid with the understanding of mixing time estimates in the additive controlled regime, regression analysis obtained power law fits for each impeller design.

Acknowledgments

The authors would like to acknowledge Unilever and the CAFE4DM consortium for funding and technical input.

References

[1]. Grenville RK (1992) Blending of viscous Newtonian and pseudo-plastic fluids. Cranfield Institute of Technology.

[2]. "CAFE4DM - The University of Manchester." https://www.cafe4dm.manchester.ac.uk/ (accessed Oct. 18, 2022).

[3]. S. S. Mirfasihi, W. Basu, T. L. Rodgers, C. P. Fonte, A. Keshmiri, P. Martin and A. J. Kowalski, "An experimental study using Electrical Resistance Tomography on the influence of additive viscosity on mixing time in turbulently stirred vessels," 26th International Congress of Chemical and Process Engineering 2022. https://2022.chisa.cz/.

[4]. Jones PN, Özcan-Taşkin GN (2005) Effects of physical property differences on blending. Chem Eng Technol 28:908–914.

[5].Rodgers TL, Cooke M, Siperstein FR, Kowalski A (2009) Mixing and dissolution times for a cowles disk agitator in large-scale emulsion preparation. Ind Eng Chem Res 48:6859–6868.

[6]. "v5r (Electrical Resistance Tomography) | ITS Ltd." https://www.itoms.com/products/v5r-electrical-resistance-tomography/ (accessed Oct. 11, 2022).

[7] Paul EL, Atiemo-obeng VA, Kresta SM (2003) HANDBOOK OF INDUSTRIAL MIXING Edited by.

[8] Burmester SSH, Rielly CD, Edwards MF (1991) The mixing of miscible liquids with large differences in density and viscosity. 7th Eur. Congr. Mix.

[9].Rodgers TL, Cooke M, Siperstein FR, Kowalski A (2009) Mixing and Dissolution Times for a Cowles Disk Agitator in Large-Scale Emulsion Preparation. Ind Eng Chem Res 48:6859–6868.

[10].Grenville, Richard & Giacomelli, Jason & Brown, David & Padron, Gustavo. (2017). Mixing: Impeller performance in stirred tanks. Chemical Engineering. 124. 42.

https://doi.org/10.52202/074122-0008

Impact of Mixing on Cellulase Production by the Filamentous Fungi *Trichoderma Reesei: Scale-down Experiments and Modelling* B1.1

Augier Frédéric[1], *Roque G. Tamiris*[1,2], *Ben Chaabane Fadhel*[1], *Nienow Alvin*[3], *Béal Catherine*[2] *(1) IFP Energies Nouvelles, 1 et 4 avenue de Bois-Préau, 92852 Rueil-Malmaison, France, frederic.augier@ifpen.fr (2) AgroParisTech, UMR SayFood, 12 place de l'Agronomie, 91120 Palaiseau, France (3) School of Chemical Engineering, University of Birmingham, Edgbaston, Birmingham B15 2TT, United Kingdom.*

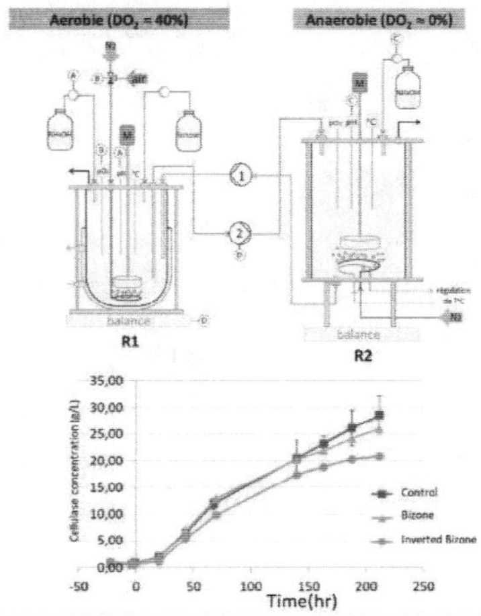

The scale-down method was applied on the enzyme production by the filamentous fungus *Trichoderma Reesei* to study the impact of heterogeneities of dissolved oxygen on the culture and its yield. Indeed, substrate gradients may appear when the mixing time in industrial bioreactors becomes higher than the characteristic time of consumption of the considered substrate.

A single bioreactor under dynamic (cyclic) aeration conditions and two bi-zone configurations were used to mimic the frequent exposition of the microorganisms to poorly oxygenated environments. A simple metabolic model was also developed to help the understanding of experimental results.

Results showed that enzyme production is affected by oxygen starvation, but that the presence of small bubbles trapped into the fermentation broth considerably mitigates the lack of oxygen and helps the microorganisms to maintain their activity in poorly aerated areas of large-scale bioreactors.

Introduction

Bioethanol production from lignocellulosic biomass (as wheat straw or wastes from agriculture) is an attractive source of biofuels as it is does not compete with food. But it requires a large amount of enzymes (cellulases) to hydrolyze this biomass into glucose, which will then be fermented into alcohol. Cellulase are secreted by the filamentous fungi *Trichoderma reesei* [1], during a strict aerobic fermentation divided into steps. First, the fungi growths in batch mode for 50hr approximately, using glucose as a carbon source, until reaching the desired biomass concentration (between 15 to 40 g/L). Then, a fed-batch culture is conducted under lactose limitation, to induce the enzyme production by the fungi. This step can last hundreds of hours to achieve important concentrations of enzymes in the fermentation broth.

Whereas fungi growth and enzyme production are rather well controlled and efficient at lab- or pilot-scale, several potential issues must be tackled to insure a safe scale-up of the process to bioreactors of hundreds of cubic meters.

First, due to the complex morphology of the fungi, the broths exhibit a strong non-Newtonian rheology which negatively affects the oxygen mass transfer [1]. Consequently, an intensive stirring is required to maintain the dissolved oxygen in the safe range of concentrations (> 2-3 mg/L).

Secondly, as the hydrodynamic stress impacts the morphology, the rheology, the growth and production rates, during the culture [2, 3], it is primordial to pay particular attention to the agitation technology and the operating conditions.

Finally, and this is the main object of the proposed study, the homogeneity of substrates and dissolved oxygen concentrations becomes more and more difficult to insure while increasing the size of bioreactors. As the mixing time increases inexorably with the scale of the bioreactor [4], it can exceed some characteristic times of substrate consumption. This induces the apparition of heterogeneities of concentration, beginning with the dissolved oxygen that can be hard to keep homogeneous in some large bioreactors [4].

In this context, the principal objective of the presented study was to evaluate the sensitivity of *Trichoderma reesei* heterogeneities of dissolved oxygen. It is proposed to reproduce at laboratory scale some mixing defaults representative, or even bigger than those expected at very large scale, following the so-called "scale-down" approach [5].

Material and Methods

Experiments

Two types of experiments were used to mimic heterogeneities of dissolved oxygen: a single bioreactor operating under periodic aeration conditions and a bi-zone bioreactor. The 2.5L bioreactor that was operated under cyclic aeration is schematized in Figure 1. The bi-zone bioreactor was constituted by one aerated (oxygenated) and one deoxygenated bioreactor, both being connected by pumps to force the circulation of the broth between them (see the graphical abstract). This system was composed of two vessels of 2 L and 0.5 L (liquid volume). A first set of experiments was conducted with an aeration of the bigger vessel and a continuous deoxygenation of the smaller vessel, using N_2 injection. Operating and stirring conditions allowed to operate the reactors with 40% and almost 0% of oxygen

https://doi.org/10.52202/074122-0009

saturation in the two vessels, respectively. The average residence time of the liquid was 11.5 min in the aerated reactor and 2.5 min in the unaerated one.

The opposite configuration, called inverted bi-zone bioreactor, was set up to attribute the bigger volume (and residence time) to the unaerated vessel thus reinforcing the oxygen starvation, as compared to the previous configuration.

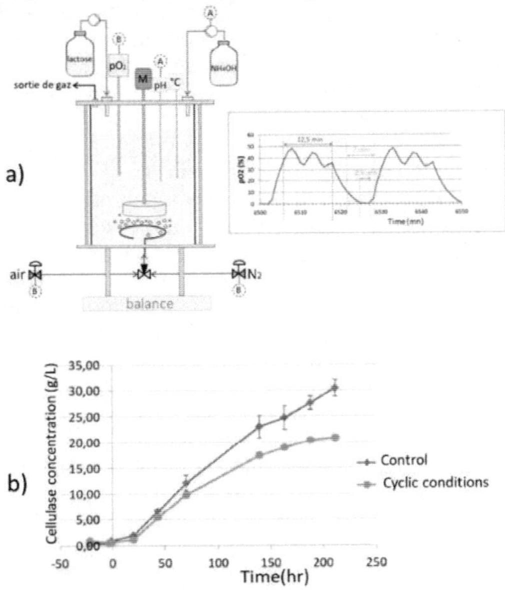

Figure 1. Bioreactor operated under cyclic aeration. a) Scheme, b) Cellulase production.

The third configuration used a single bioreactor operated under periodic aeration conditions composed of cycles of 27 min: 12 min of high dissolved oxygen concentrations, then 15 min of unaerated conditions, effectively including 7 min at very low oxygen concentration (< 10%), and 2.5 min with no dissolved oxygen.

All experiments were conducted in the fed-batch production step of the fermentation (i.e. during the cellulase production step), after a first growth step conducted under homogeneous oxygenation. A reference "control" culture was also performed in a single bioreactor at constant dissolved oxygen of 40% of the saturation.

Modelling

A simple model was used to help the analysis of experimental results. The model was developed on a pseudo-kinetic approach with no consideration of eventual adaptation nor relaxation time required by microorganisms to adapt their metabolism to any change of environment. Similarly, the effect of residence time distribution in the reactors was not taken into account. The growth and the production metabolism models consider the effect of limitations of substrate and dissolved oxygen, and the inhibition of the growth in case of high substrate concentration. The following equations detail the growth rate μ (hr^{-1}), the enzyme production rate q_p ($g_P \cdot g_X^{-1} \cdot hr^{-1}$) as functions of substrate [S] and dissolved oxygen concentrations [O_2], and kinetics constants.

$$\mu = \mu_{max} \cdot \frac{[S]}{K_{S\mu}+[S]} \cdot \frac{[O_2]}{K_{O_2\mu}+[O_2]} \quad (1)$$

$$q_p = \left((q_{pmax} - q_{pmin}) \cdot \frac{K_{iXP}}{K_{iXP}+[X]} + q_{pmin}\right) \cdot \frac{[S]}{K_{SP}+[S]} \cdot \frac{[O_2]}{K_{O_2P}+[O_2]} \cdot \frac{K_{iSP}}{K_{iSP}+[S]} \quad (2)$$

Both μ and q_p are potentially limited by substrate and the dissolved oxygen concentrations following Michaelis-Menten kinetics. The production rate can also be inhibited by high concentrations of substrate, as considered in the term $\frac{K_{iSP}}{K_{iSP}+[S]}$. Related rates of oxygen and substrate consumption are deduced from μ, q_p, and stoichiometric constants, and completed with the maintenance mechanism contribution m_s).

$$q_{O_2} = \frac{\mu}{Y_{X/O_2}} + \frac{q_P}{Y_{P/O_2}} + \left(Y^m_{O_2/S} \cdot m_S\right) \quad (3)$$

$$q_S = \frac{\mu}{Y_{X/S}} + \frac{q_P}{Y_{P/S}} + m_S \quad (4)$$

The model also includes mass balances in and between the bioreactors, as well as oxygen mass transfer between gas and liquid, not detailed here (see [4] for details). Figure 2 reports the previous validation of the simple metabolic model on a culture conducted under stable aeration conditions. During the culture, the fed-batch of substrate started at 45 hr when the substrate concentration fell below 1 g/L. The switch of the metabolism from growth to protein (enzymes) production was well predicted by the model.

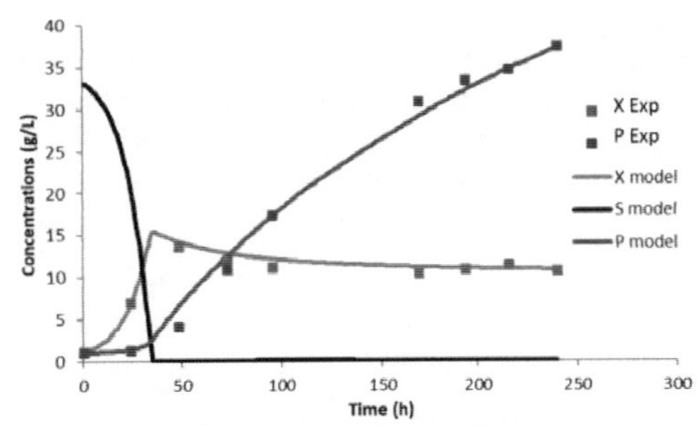

Figure 2. Validation of the metabolic model on a pseudo steady-state fermentation.

Results and Discussion

All results were compared in terms of average protein (cellulases) production rate q_P. On Figure 3, experimental results are reported in orange and results obtained with the initial model in green. The periodic aeration configuration exhibited a strong loss of enzyme production (-60%). The model slightly underestimates this loss (-40%) but the agreement was considered as acceptable. The first bi-zone set-up was not impacted by the poor aeration in the small reactor. This result was well predicted by the model since the dissolved

https://doi.org/10.52202/074122-0009

oxygen that circulates between the two bioreactors was sufficient to maintain the enzyme production in the unaerated vessel.

Conversely, the inverted bi-zone configuration exhibited a stronger loss of q_P (-34%), even still less pronounced than with the periodic aeration configuration. This result was very surprising as the majority of the broth was exposed to almost zero dissolved oxygen, which is considered as a very unproductive oxygenation in standard lab conditions. Then, from the first modelling of the inverted bi-zone experiments (in green on Figure 3), a stronger underestimation of the enzyme production was expected due to the lack of oxygen.

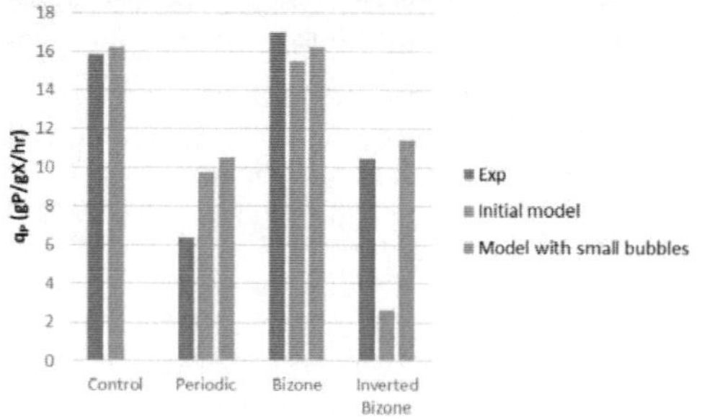

Figure 3. Enzyme productivity: comparison between experimental and numerical results.

However, small bubbles of approx. 150 µm were observed in the connecting lines between the two bioreactors. This indicates that a non-negligible transport of oxygen from the aerated to the unaerated reactor can be suspected. This gas hold-up in the lines has been estimated in the range of 25%. This phenomenon has been added to the model, while considering an instantaneous equilibrium between the liquid and the small bubbles trapped in the broth. The results obtained with the new model including the small bubbles predicted well the enzyme production, as reported on Figure 3 in blue. Numerical results showed that the small bubbles act as "oxygen tanks" and release oxygen to the liquid when the dissolved oxygen decreases.

Conclusions

A scale-down approach has been performed experimentally to evaluate the sensitivity of the enzyme production by the filamentous fungi *Trichoderma reesei* to heterogeneities of dissolved oxygen. Experimental results obtained under periodic aeration showed that the culture of the fungi is sensitive to oxygen starvation, and that this result can be explained qualitatively with a simple modelling approach based on dissolved oxygen limitation.

Conversely, experiments done in a bi-zone bioreactor showed a small sensitivity to a long residence time spent in the unaerated zone. Experimental observations and appropriate modeling have shown that small bubbles trapped in the fermentation broth play a role of oxygen carrier and induce a micro-aeration of the unaerated bioreactor. This phenomenon strongly mitigates the impact of heterogeneities of dissolved oxygen that may appear in large scale bioreactors. This result paves the way for the development of very large bioreactors as the fungi *T. reesei* appear to be quite tolerant to local lacks of oxygen.

References

[1] J.C. Gabelle, E. Jourdier, R. Licht, F. Ben Chaabane, I. Hénaut, J. Morchain, *Chemical Engineering Science,* 75 (2012), 408.

[2] N. Hardy, F. Augier, A. W. Nienow, C. Béal, F. Ben Chaabane, *Chemical Engineering Science* 172 (2017), 158.

[3] T. Roque, J. Delettre, N. Hardy, A. W. Nienow, F. Augier, F. Ben Chaabane, C. Béal, *Chemical Engineering Science*, 232 (2021) 116353

[4] V. Cappello, C. Plais, C. Vial, F. Augier F., *Chemical Engineering Science*, 229 (2021) 116033.

[5] A. Lara, E. Galindo, O. Ramírez, L. Palomares, *Molecular Biotechnology*, 34 (2006), 355

https://doi.org/10.52202/074122-0009

Fluid Dynamics and Power Consumption in a Single Use Stirred Tank Adopted in the Pharmaceutical Industry

B1.2

Federico Alberini[1*], *Giuseppina Montante*[1], *Francesco Maluta*[1], *Alessandro Paglianti*[1], *(1) Department of Industrial Chemistry "Toso Montanari", University of Bologna, viale Risorgimento 4, 40136 Bologna, Italy, (*) federico.alberini@unibo.it.*

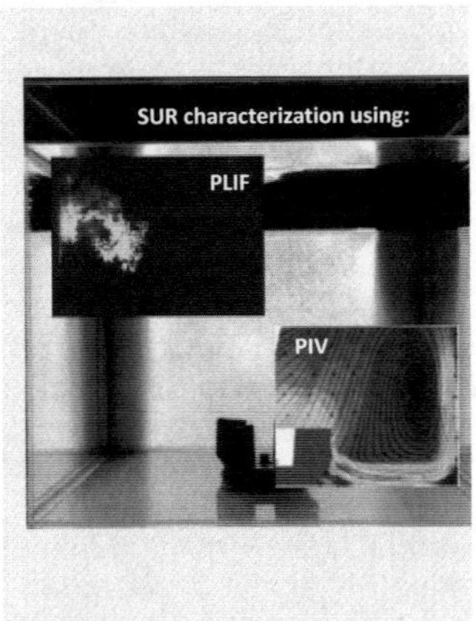

Single use bioreactors are recurrently adopted within the pharmaceutical industry, since they offer a number of advantages with respect to traditional bioreactors. Generally, they consist of a stainless steel container with a bag and an integrated impeller for handling (mixing stage), storage (post mixing stage I) and transfer (post mixing stage II) of formulated products. In this work, one of the single use bioreactor types, the Flexel Levmixer by Sartorius, is investigated. It is a vessel with a square section stirred with a bottom impeller, for which little information are available in the literature, thus making troubleshooting and optimization difficult. In order to address this issue, a combination of experimental techniques, namely Particle Image Velocimetry, Planar Laser Induced Fluorescence and torque sensors based on strain gauges, are selected for the investigation aimed at the understanding of the mixing performances with different filling levels and impeller speeds.

Introduction

Within the pharmaceutical industry, formulated liquid products are often manufactured in batch vessels and ensuring good mixing is a critical factor. In such manufacturing processes, different mixing operations can be required [1], as, for instance, miscible liquids blending, immiscible liquids dispersion [2], solid dissolution [3]. In a pharmaceutical context, a critical aspect for the manufacturing of formulated liquid products is to ensure containment and sterilization. These aspects are often quite challenging and entail very high costs and risk associated with cleaning, maintenance, and subsequent validation of the mixing systems. In the last decade, the use of single use bioreactors (SUB) has been increasing [4]. This is mainly driven by the great advantages that SUBs provide to the users, including flexibility, scalability, productivity, and ease of cleaning [4]. Generally, the vessel consists of bag of varying volume, depending on the production scale. As new technologies, SUB have been designed with several configurations [5] differing in the vessel geometry (parallelepiped, cylindrical or bag shape) in the impeller type and its installation in the vessel (top entering or bottom entering types). These features, together with the uncommon mixing set up (parallelepiped geometry, unbaffled and bottom agitated), make such devices attractive compared to standard vessels. However, due to the very scarce information available, their optimization is challenging. In this work, we have selected one of the available commercial SUBs, the Sartorius Flexel Levmixer, which consists of an unbaffled vessel of squared cross section stirred with a bottom entering radial impeller. The purpose of this work is to investigate the single-phase fluid dynamics characteristics of the Flexel Levmixer under different operating conditions, focusing on the effect of the filling level (corresponding to a volume range of 35-50 L) and speed of the impeller (50-200 rpm). Particle Image Velocimetry (PIV) for the fluid velocity measurement, Planar Laser Induced Fluorescence (PLIF) for the mixing time characterisation and a torque sensor for the power consumption determination have been used for the data collection at the different conditions.

Material and Methods

A cubic vessel of volume equal to 50 L and with the same geometrical characteristics of the Flexel Levmixer (square base, 395mm x 395mm) has been used for the investigation. The differences with respect to the original vessel are: the walls material, which is Perspex instead of plastic, to ensure optical transparency, and the coupling between the impeller and the motor, for which a standard shaft connected to the motor is used to accurately measure the impeller power consumption, instead of a magnetic levitating system. For agitation, a 3D printed version of the original impeller is mounted (D=126 mm). The adopted 3D printer is a Guider Ⅱ s (Flashforge, China). The schematic of the rig is depicted in Figure 1. The motor (i) is connected to a variable frequency driver, which allows a precise control of the impeller speed. The motor is connected to the impeller with a dismountable shaft, equipped with couplings to allow the integration of the Kistler torque sensor type 4502A0.5HA with a nominal range of measurement of 0.5 Nm.

The 3D PIV measurements have been collected in the bottom quarter of the tank (v), close to the impeller (iv), while the PLIF measurements has been collected in the top quarter of the vessel at 1 cm distance from the liquid level (vi). Both measurement zones lay in a vertical plane located mid-way between two lateral vessel walls. The images acquisition was performed by two HiSenseMKII digital cameras (1344×1024 pixels CCD) mounted on two Scheimpflug adapters. One camera was perpendicular to the laser sheet and the other with

https://doi.org/10.52202/074122-0010

20° angle and both provided with an optical filter to cut the ambient light. The laser sheet was obtained by a pulsed Nd:YAG Litron laser (light wavelength equal to 532 nm, frequency equal to 15 Hz, energy of 65 mJ/pulse) controlled and synchronized with the camera by the Dantec Dynamics "Timer Box" system. The data acquisition and processing were handled by the "Dynamics Studio" software.

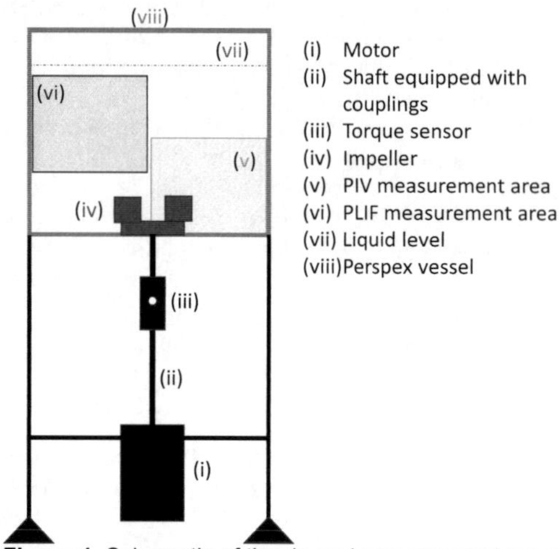

Figure 1. Schematic of the rig and measurement areas.

Results and Discussion

Selected snapshots of the vessel working with different liquid levels and impeller speeds are summarized in Table 1, in order to provide an overview of the different qualitative fluid dynamics features in the single-phase stirred tank. Considering that the recommended operational speed, N, ranges from 50 rpm up to 200 rpm, from the snapshots shown in the first row with a volume of 50 L it is clear that a vortex forms on the free liquid surface at 100rpm, at 150rpm it reaches the impeller, and above 150rpm, air is from the vortex is entrained in the liquid. As it can be observed from the snapshots in the second row of Table 1, the vortex shape changes depending on the filling level, at constant impeller speed. At the lowest filling level (volume of liquid 35L) bubbles are dispersed in the liquid. Increasing the liquid volume, both the amount of bubbles and the vortex size decrease. Indeed, as in conventional unbaffled tanks, the speed at which the vortex reaches the impeller decreases when the liquid level is reduced [6].

Depending on the type of application, the formation of the vortex and the subsequent air entrainment should be avoided. For this reason, a detailed characterization of the relationship between the liquid level and the impeller speed in the range of conditions typically adopted in industrial applications will be discussed in the work and a comparison with the results collected in standard geometry unbaffled tank will be performed.

To obtain quantitative information on the behaviour of the mixing system, velocity data have been collected to characterize the flow field induced by the impeller. In Figure 2a, the ensemble averaged velocity magnitude on the measurement plane obtained in a selected run at the impeller speed N=50rpm and the volume of liquid V_L=50L is presented. Observing the streamlines, it is apparent that the impeller discharges the fluid towards the bottom of the tank and once the discharged fluid hits the side wall, the flow is forced towards the top of the vessel generating a recirculation loop, which pushes the liquid from the top of the vessel towards the impeller. Comparing the different investigated configurations, the overall flow pattern is consistent (formation of the recirculation loop as showed in Figure 2a), but the presence of the vortex (increasing N) tends to squeeze the recirculation loop.

The mixing time is also analysed to identify the mixing performances of the investigated system, and the effect of the operative conditions. The results obtained following the homogenization dynamics of a pulse of a concentrate solution of water and Rhodamine 6G injected just below the free liquid surface are analysed. A total number of 1000 frames have been recorded for each test at a frequency of 1 Hz. The location of the measurement area has been selected to monitor the zone further away from the impeller, hence the zone where the mixing is less intense. Each test has been repeated three times for statistical consistency. Once collected, the images have been post processed for obtaining the grey scale values of the images to evaluate the Coefficient of Variation, CoV, for each single image as:

$$CoV = \sqrt{\frac{\sum_{i=1}^{N}\left(\frac{c_i}{\bar{c}}-1\right)^2}{N-1}} \qquad (1)$$

Where N is the number of pixels in each image, $\bar{c}$ is the average pixel grey value on the plane and c_i is the local grey value. Then, using the CoV_0, that is the CoV value obtained just before the tracer addition, and CoV_{final}, that is the value after the achievement of the complete tracer homogenization, the non-dimensional CoV_{Norm} is evaluated as:

$$CoV_{Norm} = \frac{CoV_i - CoV_{final}}{CoV_0 - CoV_{final}} \qquad (2)$$

To compare the different configurations a threshold value of CoV_{Norm}=0.05 is then considered to estimate the mixing time [7]. A typical time trace of the CoV_{Norm} obtained from the raw data post-processing and subsequent noise removal based on the FFT analysis of the data is shown in Figure 2b.

The results ,shown in Figure 2b, represent the trend of the CoV_{Norm} vs time for the same selected run at N=50rpm and V_L=50L. In this case the CoV_{Norm}, is constantly below the threshold value after ~55 seconds and then slowly, the system approaches the perfect mixing conditions.

Table 1. Snapshots of the investigated SUB at different operative conditions. First row: constant liquid volume and variable impeller speed; second row: constant impeller speed and variable liquid volume.

N	**50 rpm**	**100 rpm**	**150 rpm**	**200 rpm**
Constant filling level (V=50 L)				
V	**35 L**	**40 L**	**45 L**	**50 L**
Constant impeller speed (N=200 rpm)				

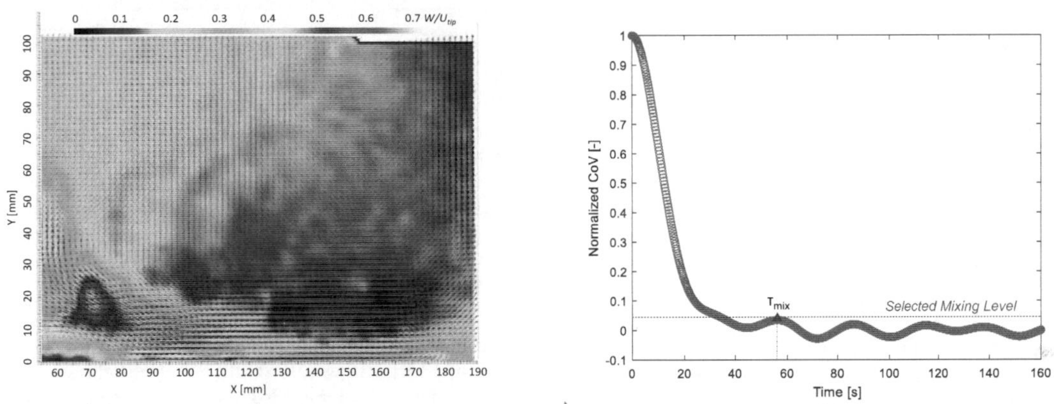

(a) (b)

Figure 2. Example of experimental data for the characterisation of the mixing system: a) 3D PIV, b) PLIF results.

Conclusions

The characterisation of an industrially relevant mixing system which mimics a Flexel Levmixer used in the pharmaceutical industry shows that the fluid dynamics and mixing characteristics are strongly influenced by the formation of a central vortex, which limits the maximum velocity at which the system can operate. The overall fluid flow structure induced by the bottom impeller is analysed and a quantitative way to compare the mixing time in such systems is proposed.

The detailed characterization of the relationship between fluid filling volume and impeller speed with velocity field, mixing time and power consumption reported in this investigation allows to obtain important information for guiding the selection of operating conditions and for devising scale-up rules.

Acknowledgments

The contribution of Mr Luca Cavazzini in carrying out part of the experimental program is gratefully acknowledged.

References

[1] Hickey AJ, Ganderton D. Pharmaceutical Process Engineering, CRC Press, Second Edition, 2016.
[2] Souto EB, Cano A, Martins-Gomes C, Coutinho TE, Zielińska A, Silva AM.. Bioengineering. 2022 Apr 1;9(4).
[3] Bai, G., Armenante, P.M. (2008) Pharmaceutical Research, 25, 320-336.
[4] de Boulard, A., Kienle, K. (2022) Chemie-Ingenieur-Technik, 94, 1962-1967.
[5] Samaras, J.J., Micheletti, M., Ding, W. (2022) Annual Review of Chemical and Biomolecular Engineering, 13, 73-97.
[6] Scargiali, F., Busciglio, A., Grisafi, F., Tamburini, A., Micale, G., Brucato, A. (2013) Industrial and Engineering Chemistry Research, 52, 14998-15005.
[7] Busciglio, A., Montante, G., Paglianti, A. (2015) Chemical Engineering Research and Design, 102, 42-56.

https://doi.org/10.52202/074122-0010

Fluid Flow and Mixing in a Novel Intermittently Rotating Bioreactor for CAR-T Cell Therapy Manufacturing B1.3

G. Atanasova[1], *M. Micheletti*[1], *A. Ducci*[2]. *(1) Dept. of Biochemical Engineering, University College London, Gower Street, London WC1E 6BT, UK, gergana.atanasova.16@ucl.ac.uk. (2) Dept. of Mechanical Engineering, University College London, Gower Street, London WC1E 6BT, UK, a.ducci@ucl.ac.uk.*

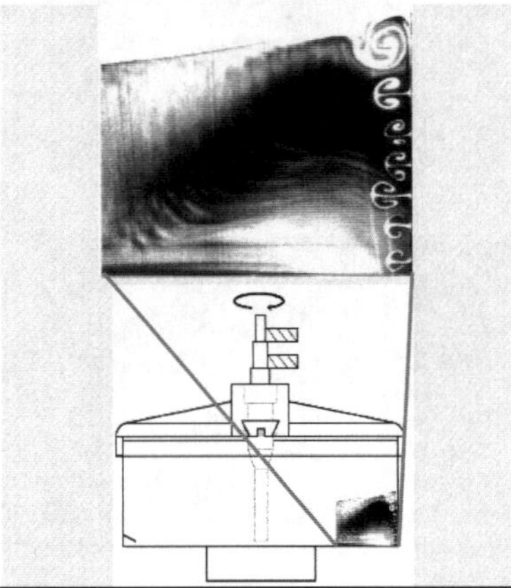

This work explores the mixing and fluid dynamics in a novel bioreactor currently used for the automated manufacturing of CAR-T cell therapy, a personalised cell-based gene therapy that offers a single-dose cure for several forms of advanced blood cancer. The vertical-cylinder bioreactor has no impeller, nor a sparger, and rotates intermittently around its central axis to mix the cell broth. To the best of our knowledge, no engineering characterisation has been conducted to date for such a system in a bioprocessing context. Moreover, there are no studies in literature addressing the transient unstable flows produced by intermittent rotation in a partially filled cylinder. PIV, PLIF and reflective flakes experiments were used to study the flow and mixing in this bioreactor under different operating conditions. An intermittent rotation pattern was designed that utilises specific features of the transient flow to optimise the mixing performance.

Introduction

The first two decades of the twenty-first century mark the beginning of a new era in Medicine with the licensing of the world's first advanced therapy medicinal products (ATMPs). ATMPs represent a class of innovative gene- or cell-based therapeutics which offer a curative single-dose treatment to serious diseases and injuries with no previously known cure. However, the complex nature of these medicines often results in their unaffordable prices and high risk of process failure. To address the manufacturing challenges and decrease costs, efforts have been made to design novel fully automated platforms that can accommodate the entire manufacturing process in one closed system.

One such platform that is gaining interest in clinical and industrial settings is the CliniMACS Prodigy® (Miltenyi Biotec) [1]. It is an all-in-one agitated system, tailored specifically to the manufacturing of Chimeric Antigen Receptor (CAR) T-cell therapy, a personalised cell-based gene therapy used for the treatment of several types of late-stage leukaemia. At the core of its functionality is the CentriCult™ chamber, a single vertically rotating cylinder serving both as a bioreactor for culturing cells and as a centrifuge for primary recovery. It has no impeller, nor a sparger. The cylinder rotates intermittently around its central axis and, if required, can be programmed to switch the direction of rotation to further enhance the mixing performance [2].

While mixing and flow dynamic characterisation has yet to be published for a bioreactor with a similar mixing operation, the flow in a rotating cylinder has been examined by other disciplines, such as ballistics, aerospace engineering and oceanography. Three distinct transient flow phases can be identified during the operation of the bioreactor: spin-up, spin-down or spin-over, depending on whether the rotation rate of the chamber is increased, decreased or the direction of rotation is reversed, respectively. Each of the three transient processes is characterised by different secondary flow. When the chamber is initially set in rotation from rest, within the first few revolutions, a quasi-steady boundary layer, namely the Ekman layer, develops at the bottom of the cylinder and plays an essential role in the spin-up process. At this point, viscous effects are confined to this layer and the bulk of liquid remains inviscid. Rings of fluid forming in the Ekman layer are pushed radially outward due to centrifugal forces, and then upward into the vertical boundary layer of the sidewall. Meanwhile, non-rotating fluid immediately above the Ekman layer is drawn into it to complete the circulation. In that sense, the Ekman layer acts as a pump, drawing low angular momentum fluid from the interior and putting it in rotation, a mechanism known as Ekman pumping. When the entire body of fluid is flushed through, the Ekman layer decays, and the fluid approaches the new state of solid body rotation. This process occurs on the Ekman timescale [3].

Spin-down, on the other hand, is vastly different from spin-up. At sufficiently high Reynolds, the curvature of the sidewall creates an environment for centrifugal instabilities in the vertical (sidewall) boundary layer, known as Taylor-Görtler vortices, which take the form of counter-rotating pairs and occur along the full length of the liquid height (as seen in the Graphical Illustration above). The vortex wavelength and onset time depend on the Reynolds number and can be robustly predicted [4]. The vortices exist in this structured form very briefly, after which the boundary layer becomes turbulent. This accelerates the dissipation of kinetic energy [5]. Meanwhile, the Ekman layer draws fluid axially downward along the sidewall and radially inward, acting in an opposite manner to spin-up. The layer thickness increases until it reaches the free surface, which again occurs on

https://doi.org/10.52202/074122-0011

the Ekman timescale [6]. While spin-up and spin-down have been characterised to an extent, the problem of spin-over in the context of a rotating cylinder is very complex and a description of the flow is yet to be published [7].

The type of flow generated in this intermittently rotating bioreactor has no precedent in bioprocessing. Spin-up is a convective process, however it is expected to contribute very little to small mixing scales in the reactor. Most of the mixing is expected to occur during spin-down, which enhances both micromixing (in the vertical boundary layer after it becomes turbulent) and macromixing (due to Ekman convection). The amount of energy available for dissipation during spin-down depends on the duration of the spin-up step before. However, the starting state of the fluid may not always be rest or solid body rotation. Such motion, combined with the presence of the free surface, results in a complex fluid dynamics problem which has not been studied extensively so far.

The aim of this work is to examine the transient flows during spin-up from rest, followed by spin-down to rest in the bioreactor, to clarify the impact of angular velocity, angular acceleration/ deceleration and time spent in spin-up on the flow and generation of instabilities, and to design an optimal rotation pattern for the chamber, which will improve the mixing performance, leading to productivity improvements and lower costs.

Material and Methods

Experiments were carried out in an optically accessible mimic of the CentriCult™ chamber, with radius R=49mm and liquid height H_L=25mm, thus matching the aspect ratio of the bioreactor at maximum working volume. Time-resolved Particle Image Velocimetry (PIV) experiments, with neutrally buoyant Polyamide Seeding Particles having mean diameter of 5 µm, were carried out in both the vertical and horizontal cross sections of the cylinder. The temporal resolution varied for different *Re* and different planes between 200µs (5000fps) and 3ms (333fps). The final 8 x 8 pixels spatial resolution after processing varied between 0.28mm (pixel size = 3.5×10^{-5} m) and 0.73mm (pixel size = 9.1×10^{-5} m).

Planar Laser-Induced Fluorescence (PLIF) was utilised as a qualitative technique to visualise the Taylor-Görtler vortices in the vertical plane, as shown in the Graphical Illustration at the top of page 1, as well as to evaluate the mixing kinetics. For visualisation purposes, 1.5mL of 0.01mM solution of rhodamine 6G in Milli-Q® water was injected around the entire circumference of the wall during solid body rotation. The temporal resolution varied between 1ms (1000fps) and 5ms (200fps), and the spatial resolution was 9.1×10^{-5} m/pixel. For mixing experiments, 50µL of 0.5mM solution was injected at the centre and close to the bottom at rest. The temporal resolution was 16.7ms (60fps), and the field of view included the entire vertical cross-section from the sidewall to the axis of rotation (1.3×10^{-4} m/pixel).

Reflective flakes (RF) Silver Pearl Lustre Pigments (from Cornelissen) were used to visualise the Taylor-Görtler vortex tubes on the wall of the cylinder, and to measure their wavelength and onset time for different operating conditions. Table 1 shows the range of experimental conditions studied with each technique.

Table 1. Range of experimental conditions studied

Parameter	Range	Technique
Re [-]	2500 – 25000	PIV, PLIF, RF
Fr [-]	0.01 – 1.07	PIV, PLIF, RF
Ω' [rad/s^2]	2π – 100π	RF
T_{SU} [-] [a]	0.17 – 1.04	PIV

[a] Indicates spin-up Ekman time before the rotation was stopped and the resulting spin-down was studied.

Results and Discussion

Firstly, spin-up from rest to solid body rotation and spin-down from solid body rotation to rest were studied. Horizontal plane PIV measurements of the tangential velocity were used to find the time at which spin-up was complete for different Reynolds and Froude numbers. The characteristic Ekman time required to complete the process increases with the increasing impact of the deforming free surface on the flow. A similar observation was not made for spin-down.

Vertical plane PIV measurements of the axial and radial component of the velocity were obtained for spin-down from solid body rotation to rest. The data was used to calculate the normalised instantaneous vorticity in the plane which is shown in Fig. 1 for different Reynolds numbers at different instants in the dimensionless Ekman time, *T*.

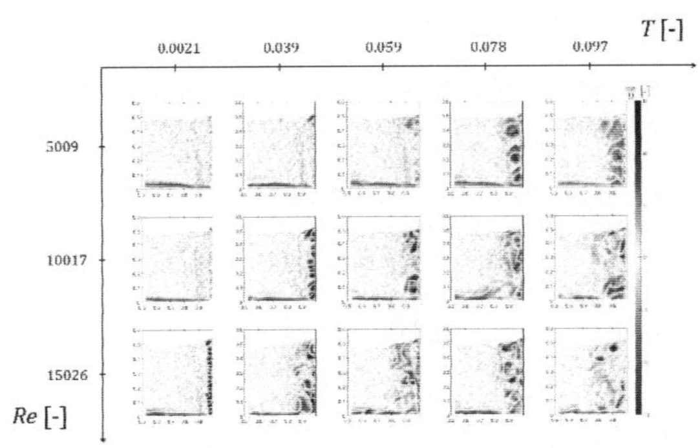

The different-sized structured pairs of counter-rotating Taylor-Görtler vortices can be seen at *T*=0.078, *T*=0.039 and *T*=0.021 from low to high *Re* (Figure 1). The instabilities first appear small and are seen to briefly grow before they lose shape and the flow transitions to turbulence. The dissipation of the turbulence facilitates micromixing in the vertical boundary layer region, to which the unstable flow is confined. The

https://doi.org/10.52202/074122-0011

turbulent boundary layer grows radially inward in time, as can be seen from figure 1, while the expansion of the Ekman layer occurs over longer timescales and cannot be seen in this figure.

In order to determine the intensity of instabilities and show how they decay in time, several vortex identification methods were tested, and the λ_2 technique was found most effective. It relies on the fact that a pressure minimum is present at the core of a vortex and this can be identified by two negative eigenvalues of the matrix $S^2 + A^2$, where S and A are the symmetric and antisymmetric components of the velocity gradient tensor [8]. The space-average instantaneous vorticity, ω_θ, of all points in the plane for which $\lambda_2 < -6.5\left(\frac{2\Omega}{\pi}\right)^2$ was found and multiplied by the area of the cross-section pertinent to those points, A_{ω_θ}, to give the total instantaneous vorticity of the instabilities in the cross-section. Fig. 2 shows how the vorticity varies with Ekman time for the range of Re studied.

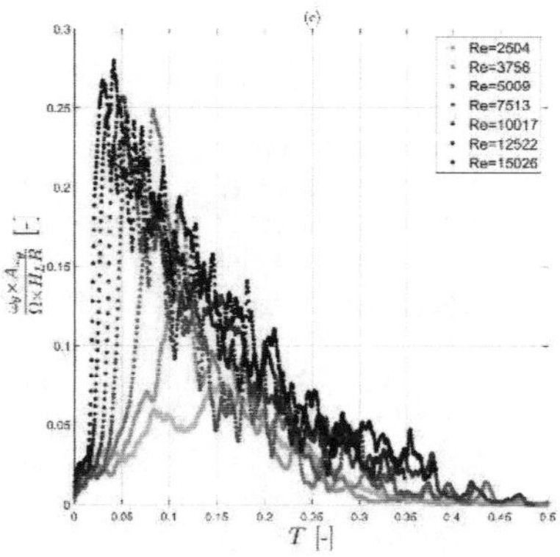

Figure 2. Total vorticity of the instabilities in the cross section for spin-down from solid body rotation to rest for different *Re*, as a function of Ekman time; all curves are obtained independently for two experimental repeats, and only the average is plotted.

Information about the timescale of vorticity decay relative to the spin-down timescale was extracted for each *Re*. For example, at *Re*=15026 and *T*=0.34, 90% of the vorticity has decayed, while the fluid core still has 45% of its original angular velocity. This information is essential when selecting the optimal duration of a spin-down step in the rotation programme, since the instabilities contribute significantly to the micromixing and not much mixing occurs in the bulk.

An analogous analysis was carried out on spin-down to rest from different levels of spin-up. Shorter spin-up times result in less kinetic energy being transferred to the fluid. This, in turn, resulted in weaker instabilities during spin-down. The drop in the intensity of instabilities during spin-down to rest due to incomplete spin-up was quantified and will be presented in this paper. For the shortest spin-up time tested, about 10% of the kinetic energy of solid body rotation was transferred to the fluid, while the vorticity of the instabilities during spin-down was less than 1% of that for spin-down from solid-body rotation.

Finally, mixing experiments were carried out for a rotation pattern comprising of spin-up followed by spin-down. The two steps were looped until mixing was deemed complete. The contribution of each step to the overall mixing was analysed. As expected, spin-up was found to have insignificant contribution to mixing, while most of the mixing occurred during spin-down and was a result of the thickening of the Ekman layer but also of the dissipation of the instabilities. The impact of injection timing and position were also studied, as well as different rotation patterns such as spin-over and multiple spin-down steps (not to rest) after a single spin-up.

Conclusions

Several features of the transient flow during spin-up and spin-down contribute differently to mixing in the bioreactor. While spin-up contributes very little to the overall mixing, it is necessary to input kinetic energy into the fluid, part of which will be converted to turbulent kinetic energy due the onset of centrifugal instabilities, a feature which contributes to micromixing in the vertical boundary layer. The dissipation of that turbulence in the vertical boundary layer is the primary mechanism of mixing in the early stages of spin-down, while in late stages, when the vorticity has decayed below detectability, the primary mechanism is through the convective thickening of the bottom boundary layer. It is clear that both stages are necessary to achieve complete mixing in the bioreactor. Ultimately, different mixing strategies can be designed based on the objective of the cell culture step.

Acknowledgements

The author would like to express her sincerest gratitude to the Engineering and Physical Sciences Research Council (EPSRC) and the Future Targeted Healthcare Manufacturing Hub for the financial support making this project possible.

References

[1] U. Mock, L. Nickolay, B. Phillip, G.W.K. Cheung, H. Zhan, I.C.D. Johnston, A.D. Kaiser, K. Peggs, M. Pule, A. J. Thrasher, W. Qasim, *Cytotherapy*, 18 (2016) 8.

[2] C. Priesner, K. Aleksandrova, R. Esser, N. Mockel-Tenbrinck, J. Leise, K. Drechsel, M. Marburger, A. Quaiser, L. Goudeva, L. Arseniev, A.D Kaiser, W. Glienke, U. Koehl, *Human Gene Therapy*, 27 (2016) 10.

[3] *The Theory of Rotating Fluids,* H.P. Greenspan, Cambridge University Press, 1968.

[4] M. Kim and C.K. Choi, *Chemical Engineering Science*, 61 (2006).

[5] F. Kaiser, B. Frohnapfel, R. Ostilla-Monico, J. Kriegseis, D. E. Rival, D. Gatti, *Journal of Fluid Mechanics*, 885 (2019).

[6] O. Savas, *Journal of Fluid Mechanics*, 234 (1992) 3.

[7] P. Duck, M. Foster, *Annual Reviews Fluid Mechanics*, 33 (2001).

[8] J. Jeong, F. Hussain, *Journal of Fluid Mechanics*, 285 (1995).

Prediction of Blend Time in a Large-Scale Viral Inactivation Vessel Used in a Commercial Biopharmaceutical Plant

B1.4

Piero M. Armenante,[1] *Chadakarn Sirasitthichoke,*[1,2] *Duc Hoang,*[2] *Poonam Phalak,* [2] *Barak Barnoon,*[2] *Ishaan Shandil*[2] *(1) New Jersey Institute of Technology, Otto H. York Department of Chemical and Materials Engineering, Newark NJ 07102, USA, piero.armenante@njit.edu; (2) Bristol Myers Squibb Company, Manufacturing Science and Technology Department, Devens, MA, 01434, USA, chadakarn.sirasitthichoke@bms.com; ishaan.shandil@bms.com.*

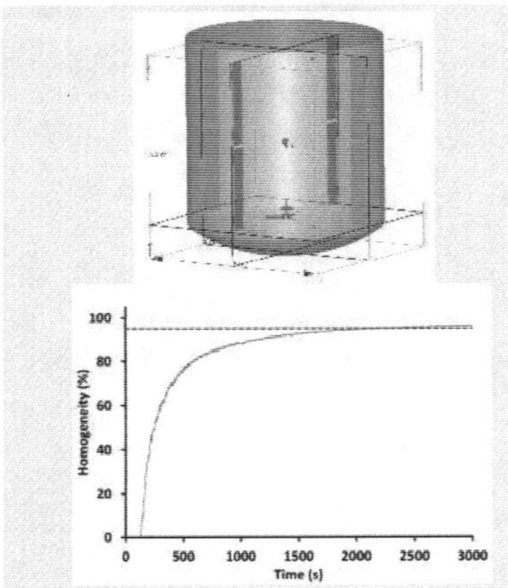

Across the biopharmaceutical industry, viral inactivation (VI) of protein solutions is commonly carried out to eliminate the cytotoxicity resulting from possible residual viral contamination. This step requires adding a specific agent to the solution and allowing sufficient time to inactivate the viruses. In this work a CFD model based on the use of the Lattice Boltzmann Method (LBM) was used to predict the blend time needed to homogenize the viral inactivation agent in an actual full-scale commercial VI vessel equipped with an HE-3-impeller. Different addition modalities (batch vs. continuous) were considered. Although direct experimental progress of the blending process was not possible because of manufacturing restrictions, the degree of homogeneity measured at the end of the process confirmed that the viral inactivation agent was appropriately dispersed, as computationally predicted here. The results of the work were used to support an actual production process at the company biomanufacturing site.

Introduction

Production of therapeutic proteins is a resource-intensive process. Despite the major advantages of using Chinese Hamster Ovary (CHO) cells, viral contamination from different foreign sources such as a trace level of viruses from adventitious agents from the supplemented medium is inevitable [1]. Consequentially, biopharmaceuticals derived from mammalian tissues, such as CHO cells, are exposed to the risk of viral contamination from endogenous retroviruses or minute viruses from mice. As a result, viral inactivation (VI) followed by additional removal steps is crucial to prevent the risk of pathogen contamination and ensure viral safety.

Triton X-100 is commonly used as a non-ionic detergent-treated solvent for viral inactivation since it chemically interrupts the interactions between the molecules in the lipid coating of the viruses [2]. Insufficient blending of Triton X-100 with the protein pool solution, potentially resulting in inhomogeneity in the system, may result in transferring viral activity to the downstream process, which can introduce pathogenic cytotoxicity in the final product viruses [3]. Therefore, the homogenous dispersion of the buffer solution is a critical step to ensure the efficacy of the VI process.

In this work we used a CFD model to determine the required blend time of a binary fluid at a specific agitation speed during a typical industrial viral inactivation process in an industrial stirred commercial vessel. We additionally used this approach to gain in-depth insight into the hydrodynamics of the system. The results obtained here were not only critical to predict the blend time in an actual full-scale VI production process at a Bristol Myers Squibb facility, but they can also provide general guidance to industrial practitioners in the biopharmaceutical manufacturing process facing similar issues.

Computational Approach

The Lattice Boltzmann Model (LBM) (M-Star CFD (Version 3.0.19, M-Star Simulations, LLC) was used to conduct simulations to predict the system hydrodynamics, impeller power dissipation, and blend time to achieve a predefined level of homogeneity of the fluid mixture [4,5].

The geometrical configuration used in the CFD simulations was identical to that of an actual full-scale system used for the virus inactivation step in a production downstream process at Bristol Myers Squibb (vessel diameter: 3.05 m; liquid height: 3.2 m; impeller type: HE3; impeller diameter: 0.64 m).

In the simulations, the initial fluid contained in the mixing vessel was water with a volume of 20.7 m^3. Then, 0.8 m^3 of 20% Triton X-100 solution was numerically added, batchwise, to the top of the vessel. Two modes of additions were used, the first consisting of numerically spreading all the Triton X-100 on top of the water (Addition *A*), starting the agitation, and computationally monitoring the evolution of the blending process, and the second (Addition *B*) consisting of adding the same volume of Triton X-100 numerically placed in a computational cylinder 1.30 m diameter and 0.6 m length immersed near the surface of the water. All simulations were conducted at 30, 70, 140, and 200 rpm.

Additionally, in order to simulate the continuous Triton X-100 addition used in the actual viral inactivation process, the same amount of Triton X-100 was added to the vessel over a 50-minute period while maintaining the agitation speed at 140 rpm.

https://doi.org/10.52202/074122-0012

Figure 1. Time evolution of Triton X-100 concentration across the fluid for a single rapid addition of Triton X-100 at the surface (*Addition A*) for N=140 rpm.

To predict the blend time in this system, the concentration of the added Triton X-100 solution was numerically calculated at each location in the vessel. Then, for each time, the space-averaged concentration in the entire vessel and its standard deviation were calculated, and the coefficient of variation, *CoV*, was predicted as a function of time.

Here, the blend time to achieve 95% homogeneity was considered appropriate for this process. Therefore, the time at which the *CoV* reached 5% (corresponding to a 95% uniformity level, U_{95}) was defined as the 95% blend time, θ_{95}.

Results and Discussion

Figure 1 show the evolution of the Triton X-100 concentration across the vessel at different times, as predicted by the CFD simulation when Triton X-100 was rapidly added in a single shot at the top of the fluid (*Addition A*) and the impeller was rotating at 140 rpm. Under these conditions, the predicted 95% blend time θ_{95} (i.e., for 5% *CoV*), was achieved at about 56 s (Figure 2).

A simple correlation to predict the blend time was obtained by assuming an inverse proportionality between θ_{95} and N [6,7], i.e.:

$$N\theta_{95} = constant \qquad (1)$$

Regressing the predicted θ_{95} (in s) values vs. N in (in rps) resulted in the constant being equal to 135.75. The corresponding plot of Equation 1 using this constant is shown in Figure 3. Clearly, there is significant agreement with the predicted θ_{95} values, especially for higher agitation speeds. The experimentally derived blend time correlation (Equation 2) of Grenville [8] was additionally used to calculate the blend times:

$$Po^{1/3} N\theta_{95} \frac{D^2}{T^{3/2} H^{1/2}} = 5.2 \qquad (2)$$

where H/T~1. Figure 3(b) presents a comparison between the CFD-predicted blend times and those obtained with Equations 1 and 2.

In the simulation of the actual process, Triton X-100 was computationally continuously added to the vessel over a 50-minute period while agitating at 140 rpm.

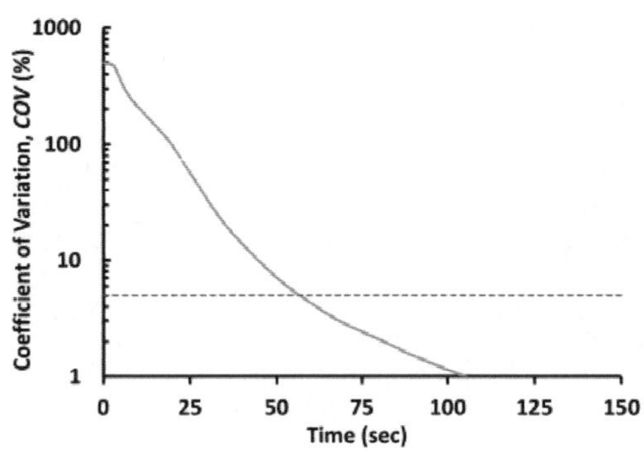

Figure 2. *CoV* evolution as a function of time for a single rapid addition of Triton X-100 at the surface (*Addition A*) for N=140 rpm.

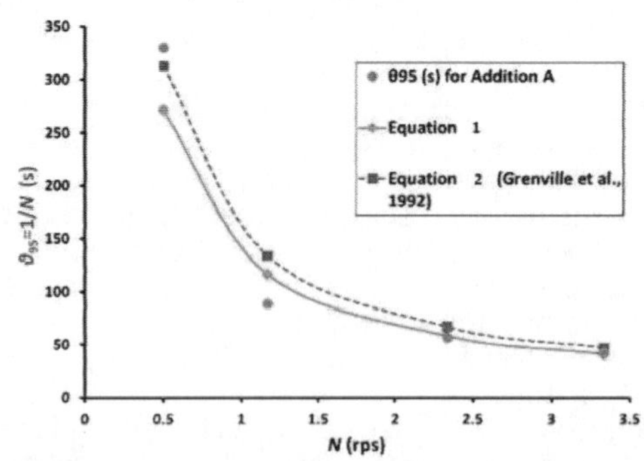

Figure 3. (a) Predicted time evolution of Triton X-100 *CoV* at different agitation speeds in the full-scale system for *Addition A*; and (b) Comparison between the corresponding CFD-predicted blend times and the blend times obtained with correlations from Equation 1 and Equation 2.

Under these conditions, the time required for the system to achieve 95% homogeneity was calculated to be about 36 min, as shown in figure next to the abstract (*Graphical Abstract*), indicating that even before the completion of the addition process the Triton X-100 was blended in the liquid bulk to the desired homogeneity level (5% *CoV*). It should be noted that this blend time is much longer than the corresponding blend times for a single-shot additions of Triton X-100 (Figure 4 and Table 1).

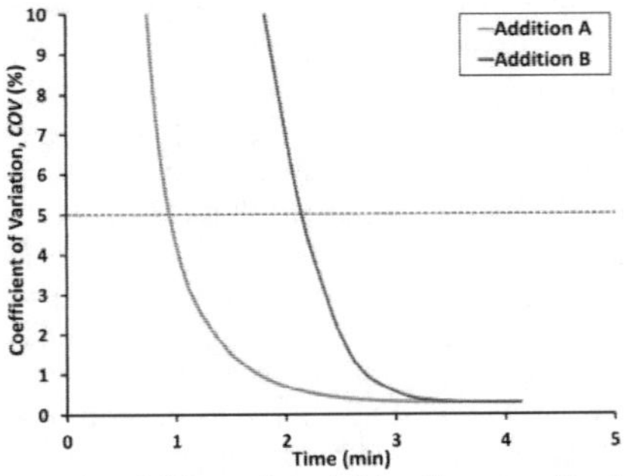

Figure 4. CFD-predicted blending results times as a function of addition location of Triton X-100 in the full-scale-system operated at 140 rpm for *Addition A* and *Addition B*.

The results of this study can be used to explain the rationale for the way in which the viral inactivation step in the full-scale system was conducted, especially considering that the blend time in the full-scale vessel could not be monitored directly during the Triton X-100 addition since this is a drug substance manufacturing process subject to GMP compliance restrictions. In the actual process the vessel was mixed for an additional 20-minute period after the completion of the Triton X-100 addition (50 minutes), bringing the total time of agitation to 70 minutes, i.e., twice the predicted blend time (36 minutes), as a safety factor. During this extra 20 minutes, samples could be withdrawn, and in fact this was done at 2-minute intervals at a location near the bottom of the vessel. Each sample was analyzed to determine the Triton X-100 concentration from triplicate absorbance measurements at 276 nm (the wavelength at which the maximum absorbance of Triton X-100 solution can be observed) using a spectrophotometric method based on the linear calibration that had been previously established between the Triton X-100 concentration and absorbance at this wavelength. The experimental Triton X-100 concentrations in all samples did not change over the 20-minute time period of observation (average deviation from average<0.6%), indicating full dispersion of the Triton X-100 throughout the liquid in the vessel, thus indirectly validating the computational predictions.

Table 1. CFD-predicted blend time for the mixing of water and Triton X-100 initially located at Addition *A* and Addition *B* for different operating conditions.

Agitation Speed, *N* (rpm)	θ_{95} (s) for Addition *A*	θ_{95} (s) for Addition *B*
30	329.6	n/a
70	88.7	969.1
140	56.6	128.7
200	42.8	74.1

Conclusions

In this work, we showed that a computational approach to determine blend time can be especially significant in all those cases in which experimental measurements cannot be easily conducted, as in the case of the full-scale process studied in this work. The computational blend time predictions reported here were used to confirm the correctness of the approach used in the actual viral inactivation step used in a production facility subject to GMP compliance requirements. We expect that similar computation approaches will find significant applications in pharmaceutical the industry and will become widely accepted as a result.

Acknowledgments

The authors would like to thank Bristol Myers Squibb colleagues Zizhuo Xing, Weixian Shi, Wei Chen, Gang Huang, Steven Traylor and Donald Gibbs for their input and discussion. A special thanks to John Thomas from M-Star Simulation, LLC for technical support related to M-Star.

References

[1] Garnick, R. L. (1998). Raw materials as a source of contamination in largescale cell culture. *Developments in Biological Standardization*, 93, 21–29.
[9]

[2] Chen, D., Luo, W., Hoffman, J., Huang, L., Sandefur, S., Hall, T., Murphy, M., & O'Donnell, S. (2020). Insights into virus inactivation by polysorbate 80 in the absence of solvent. *Biotechnology Progress*, 36(3), e2953. https://doi.org/10.1002/btpr.2953

[3] Brown, F., Lubiniecki, A. (1996). *Viral safety and evaluation of viral clearance from biopharmaceutical products*. Karger.

[4] Sirasitthichoke, C., Teoman, B; Thomas, J., Armenante, P. M. (2022), Computational Prediction of the Just-Suspended Speed, Njs, in Stirred Vessels Using the Lattice Boltzmann Method (LBM) Coupled with a Novel Mathematical Approach, *Chem. Eng. Sci.*, 2022, 251, 117411.

[5] Thomas, J., Sinha, K., Shivkumar, G., Cao, L., Funck, M., Shang, S., Nere, N. K. (2021). A CFD digital twin to understand miscible fluid blending. *AAPS PharmSciTech*, 22(3), 91. https://doi.org/10.1208/s12249-021-01972-5

[6] Bai, G., Armenante, P. M., & Plank, R. V. (2007). Experimental and computational determination of blend time in USP dissolution testing apparatus II. *J. Pharm. Sci.*, 96(11), 3072-3086. doi:10.1002/jps.20994

[7] Fasano, J. B., Bakker, A., & Penney, W. R. (1994). Advanced impeller geometry boosts liquid agitation. *Chemical Engineering*, 101, 8.

[8] Grenville, R. K. (1992). Blending of viscous Newtonian and pseudo-plastic fluids; http://hdl.handle.net/1826/4362

https://doi.org/10.52202/074122-0012

Multiscale DPD – CFD Approach for Modelling Rheological Behaviour of Complex Fluids Applied to a Mixing Process CF1.1

F. De Roma[1], *D. Marchisio*[1], *G. Boccardo*[1], *M. Bouaifi*[2], *A. Buffo*[1]. *(1) DISAT, Institute of Chemical Engineering, Politecnico di Torino, C.so Duca degli Abruzzi 24, Torino 10129, Italy, francesco.deroma@polito.com. (2) Centre de Recherche et d'Innovation de Lyon, Solvay, 85 Avenue des Frères Perret, BP 62, Saint-Fons Cedex 69192, France.*

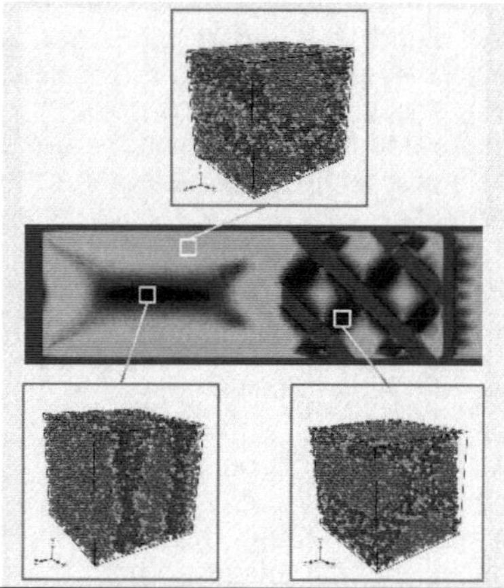

Unit operations that involve the usage of complex fluids are widespread, but the difficulties in modelling the rheology of these fluids are far from being overcome. Coarse-grain (CG) computational tools such as Dissipative Particle Dynamics (DPD) may help to obtain useful information of these systems at the micro and mesoscale level. In this work we propose to combine such information with a macroscale Computational Fluid Dynamics (CFD) description, setting up a multiscale approach in which DPD works as a rheometer, providing rheological behaviour used in CFD. The investigated system is a mixture of water and SLES under laminar fluid dynamics regime. The results show that the mixture correctly exhibits a shear-thinning behaviour, and the pressure drops across the mixer agree with the ones of similar systems. While the feasibility of the coupling approach is shown, further studies on DPD are needed to improve the accuracy and the predictability of the methodology.

Introduction

Complex fluids are widely used in industrial applications, but their modelling turns out to be challenging from a rheological point of view. They consist of a mixture of at least two phases, like colloidal dispersions or solutions of macromolecules. One of their main features is the capability to undergo self-assembly, that leads to the formation of different microstructures, mostly influenced by the concentration of the various phases and by the fluid dynamics conditions. The resulting fluid often exhibits non-Newtonian behaviour, and its apparent viscosity depends on both the composition and the shear rate applied to the fluid.

Among the applications of complex fluids, the self-care industry uses different mixtures of water and surfactants to obtain products with the features requested by the consumer market. The complex non-Newtonian behaviour constitutes a great challenge in the design of mixing processes, due to the absence of accurate models to predict the viscosity of the mixture during the different phases of the process.

An aid to the modelling of such processes may come from micro and mesoscale computational methods, as more accurate and less demanding tools are being developed over time. Dissipative Particle Dynamics (DPD) is a coarse-grain (CG) computational technique that has been proven to be able to qualitatively reproduce a variety of microstructures for complex fluids [1]. Its main advantage is the possibility to reduce the computational resources needed to simulate a system at the mesoscale, while retaining a certain degree of chemical specificity, thanks to its CG nature. For this reason, DPD simulations had a role in the study of polymer melts, colloids, and water – surfactants systems. Moreover, preliminary attempts to retrieve transport properties from DPD, such as viscosity and diffusivity, have been reported in the literature, even with difficulties related to the loss of degrees of freedom associated with the CG procedure [2, 3].

Therefore, having a good CG model of the complex fluid is required so to be able to extract viscosity information from DPD mesoscale simulations, and to build a rheology model that can be used for macroscale Computational Fluid Dynamics (CFD) simulations [2].

In this work we apply a multiscale approach based on Gaussian Process Regression [2] to model the whole mixing process of a mixture of water and Sodium Lauryl Ethoxy Sulphate (SLES), typically employed in self-care items production, such as soaps and shampoos. In particular we focused on the dependence of the viscosity of the complex fluid on the shear rate, thus highlighting the local differences in terms of apparent viscosity inside the mixing device.

Material and Methods

The multiscale framework used in this work is based on the combination of three different methods: the mesoscale simulations, the regression process, and the macroscale simulations.

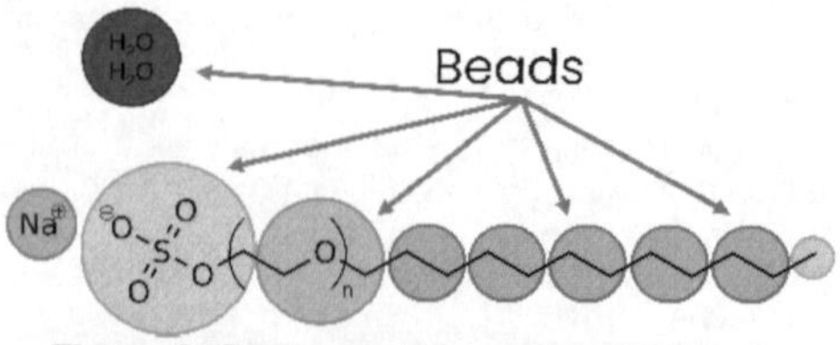

Figure 3. Coarse-grained representation of a molecule of Sodium Lauryl Ethoxy Sulphate (SLES) and a bead of water

https://doi.org/10.52202/074122-0013

The mesoscale simulations are conducted with the software LAMMPS, using the DPD technique. First, a CG model is developed (Figure 1): the molecules are depicted with beads, entities that represent groups of atoms, and interacts through three different pairwise forces, reported below:

$$\vec{F}_C = a_{ij}\left(1 - \frac{|\vec{r}_{ij}|}{r_c}\right)\hat{r}_{ij} \quad (1)$$

$$\vec{F}_D = -\gamma\left(1 - \frac{|\vec{r}_{ij}|}{r_c}\right)^2 (\hat{r}_{ij} \cdot \vec{v}_{ij})\hat{r}_{ij} \quad (2)$$

$$\vec{F}_R = \sigma\left(1 - \frac{|\vec{r}_{ij}|}{r_c}\right)\frac{\xi_{ij}}{\sqrt{\Delta t}}\hat{r}_{ij} \quad (3)$$

These are the mathematical expressions of the conservative force ($\vec{F}_C$), the dissipative force ($\vec{F}_D$) and the random force ($\vec{F}_R$), representing the (soft) repulsive interaction between beads, the dissipation caused by their relative motion and a random component capable of restoring the energy in the system according to the fluctuation-dissipation theorem. In this set of equations, a_{ij} is the repulsive parameter, different for various types of beads, $\vec{r}_{ij}$ is the distance between the centres of mass, $\vec{v}_{ij}$ is the relative velocity, r_c is the cut-off radius, distance beyond which the interactions are negligible, γ is the friction coefficient for the dissipative force, σ is the amplitude parameter of the random force and ξ_{ij} is a Gaussian random scalar variable, with zero mean and unitary variance.

The repulsive parameters a_{ij} are obtained through the water/octanol partition function method [4,5]. Bonds between beads are modeled using spring potentials.

The repulsive parameters are validated by partial reproduction of the water/SLES phase diagram [6], through equilibrium DPD simulations.

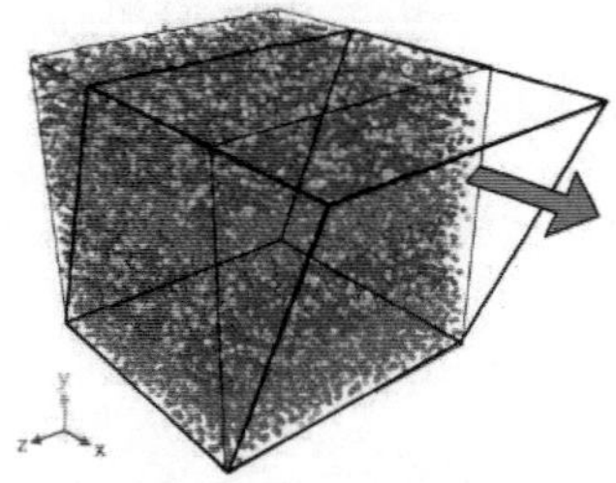

Figure 4. Example of shear deformation applied to the simulation box. Bead colours correspond to the ones reported in Figure 1

The apparent viscosity of the system, μ, is calculated using non-equilibrium simulations with Lees – Edwards boundary conditions, that apply a deformation to the box in the xy plane as in Figure 2, leading to a linear velocity profile for $v_x(y)$, hence to a constant shear rate, $\dot{\gamma}$.

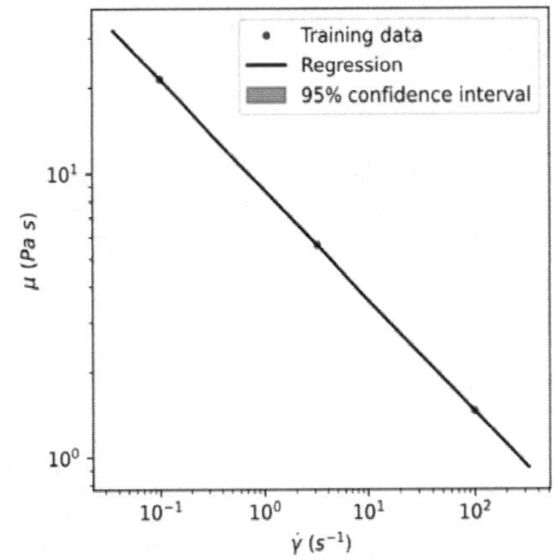

Figure 5: Rheogram obtained from the GPR model for the viscosity of the mixture. The confidence interval is shown, but it is narrow enough to be hidden by the representation of the regression curve.

The apparent viscosity is then extracted from simulations at different shear rates and the results are fed to a script that implements Gaussian Process Regression (GPR), to build a rheogram, as the one in Figure 3. The GPR employs a Bayesian approach for the regression process, in which the addition of a pair of values ($\dot{\gamma}$, μ) updates the viscosity prediction for untested shear rate values. Additional DPD simulations are performed until a criterion on the quality of the regression is satisfied, based on the a-priori selected interval of confidence associated with the viscosity predictions. It is worth remarking that the outcome of the GPR are not the fitted parameters of a certain prescribed rheological law, but the parameters of a sum of selected kernel functions. In such a way, the particular of nature of the rheological behaviour can come automatically from the CG simulation, with no knowledge necessary of the form of the underlying rheological law.

Figure 6: Example of standard Sulzer SMX static mixer

Once the GPR viscosity model is satisfactorily built, a call function has been implemented in OpenFOAM, the CFD software used to perform the macroscale simulations. The investigated geometry is a standard Sulzer SMX static mixer, shown in Figure 4. The current model is centred around the influence of the shear rate, leaving the role of composition for future works. Hence, under laminar regime, the viscosity field inside the domain is monitored together with the pressure

https://doi.org/10.52202/074122-0013

drop across the mixer, that constitutes the macroscopic information of interest.

Results and Discussion

The correct reproduction of micellar and lamellar microstructures at different compositions, as reported in the water/SLES phase diagram, confirmed the quality of both the coarse-grained model and the interaction parameters (see Figure 5).

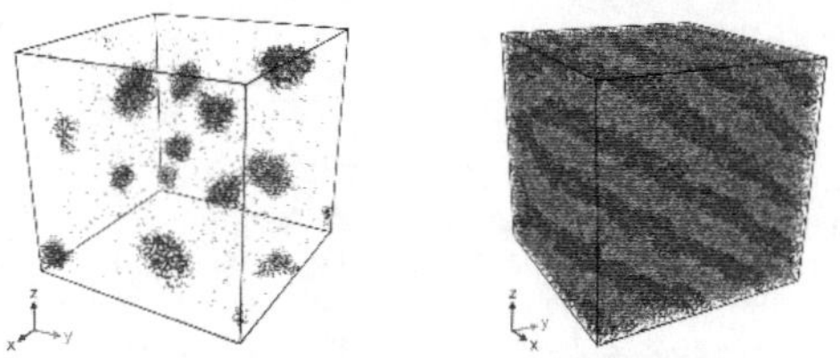

Figure 7: DPD reproduction of water/SLES mixture. (a) Micellar microstructure for low surfactant concentration and (b) lamellar microstructure for high surfactant concentration.

Hence, the second step was the extraction of the apparent viscosity values from the non-equilibrium DPD simulations. The GPR model obtained from these results shows the same trend of a power law, the model usually adopted to approximate the rheology of this kind of mixtures at constant composition. Moreover, the method requested only three DPD simulations to satisfy the constraint on the interval of confidence, which is very narrow according to the predictions over a span of two order of magnitude for the shear rate, as presented in Figure 3.

From Figure 6 it is possible to see how the dependence of viscosity is correctly implemented from the model, showing the shear-thinning behaviour typical of these fluids. In fact, the apparent viscosity is smaller in high shear rate regions, close to the mixing sections, while it is higher in low shear regions. The evaluation of the pressure drops leads to results of the same order of magnitude when compared to similar fluids in similar equipment and analogous operating conditions. An exact comparison with the simulated fluids was not possible due to lack of experimental data at the present date.

(a)

(b)

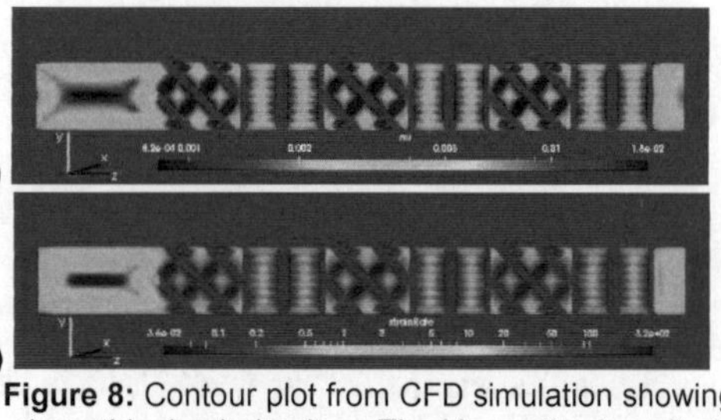

Figure 8: Contour plot from CFD simulation showing the shear thinning behaviour. The kinematic viscosity (a) and the local strain rate (b) are shown.

Conclusions

In this work we showed that it is possible to apply a coupled DPD – CFD multiscale method to a realistic industrial application. The DPD was confirmed to be an excellent tool for the qualitative reproduction of microstructures, but a deeper investigation on some of the model parameters will result beneficial to obtain more accurate quantitative results.

The key role of the GPR is to provide a good quality regression with a limited amount of input data, in order to reduce the number of DPD simulations, that are the most expensive computational step.

Hence, future studies should revolve on the research for the optimal parameters and conditions to perform DPD simulations, in order to reduce the sources of uncertainty for the calculation of the apparent viscosity. To conclude, it is possible to obtain accurate results for modelling unit operations in laminar regime involving complex fluids once a robust CG description of the fluid is developed and available.

Acknowledgments

We acknowledge the CINECA award under the ISCRA initiative, for the availability of high performance computing resources and support.

References

[1] H. Droghetti, I. Pagonabarraga, P. Carbone, P. Asinari, D. Marchisio, *The Journal of Chemical Physics*, 149 (2018) 184903.

[2] L. Zhao, Z. Li, B. Caswell, J. Ouyang, G. E. Karniadakis, *Journal of Computational Physics*, 363 (2018) 116-127.

[3] N. Lauriello, J. Kondracki, A. Buffo, G. Boccardo, M. Bouaifi, M. Lisal, D. Marchisio, *Physics of Fluids*, 33 (2021) 073106.

[4] R. L. Anderson, D. J. Bray, A. S. Ferrante, M. G. Noro, I. P. Stott, P. B. Warren, *The Journal of chemical physics*, 147 (2017) 094503.

[5] R. L. Anderson, D. J. Bray, A. Del Regno, M. A. Seaton, A. S. Ferrante, and P. B. Warren *Journal of Chemical Theory and Computation*, 14 (2018) 2633-2643.

[6] H. Li, L. Dang, S. Yang, J. Li, H. Wei, *Colloids and Surfaces A: Physicochemical and Engineering Aspects*, 495 (2016) 221-228.

Numerical Simulations of Immiscible Liquids Using Mesoscopic Techniques

CF1.2

Marco MF Ferrari[1], *Alexandra AK Komrakova*[2], *Gianluca GB Boccardo*[1], *Daniele DM Marchisio*[1], *Antonio AB Buffo*[1]. *(1) Politecnico di Torino, Corso Duca degli Abruzzi 24, Torino, Italy, marco.ferrari@polito.it. (2) University of Alberta, 9211 116 Street NW, Edmonton, Canada.*

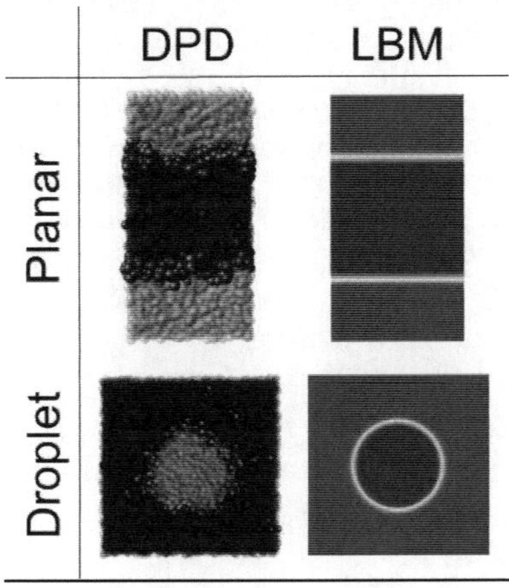

This work focuses on the application of mesoscopic techniques, e.g., Dissipative Particle Dynamics (DPD) and Lattice Boltzmann Method (LBM), for the simulation of interfacial oil-water systems in presence of surfactants. Equilibrium properties were analysed for the planar interface and droplet configuration. A scaling scheme was employed to show that typical domain formations of multi-component systems can be performed in DPD at different levels of coarse-graining. A simplified DPD model is capable of properly reproducing the drop of interfacial tension observed with more detailed mesoscale simulations. Moreover, the oil-in-water droplet system was studied under simple shear flow by means of DPD and LBM. Finally, a comparison of the results obtained with both mesoscopic models was performed, highlighting the advantages and limits of each technique. Hence, this work will pave the way for a better understanding of breakage and coalescence of oil droplets to develop large-scale models of emulsion production process.

Introduction

Computer modelling techniques are widely used to enhance the comprehension of the way the molecules organize themselves in a liquid, especially when experimental evidence is hardly available due to the difficulty in isolating individual chemical species [1]. Among molecular techniques, Dissipative Particle Dynamics (DPD) is a well-established method for simulating soft matter systems at the mesoscale level of description [2]. DPD is a coarse-graining technique designed for modelling various fluid systems. In particular, this method is well-suited for studying multi-component systems such as interfacial systems, which have been largely investigated due to its remarkable applications to industrial cases, such as for food engineering research [1,3]. Indeed, DPD has been successfully employed to analyse both static and dynamic properties even with amphiphilic and protein molecules acting as surfactants [1,3,4].

On the other hand, Lattice Boltzmann Method (LBM) has been widely used to model multiphase flows [5]. Although both DPD and LBM are successfully employed in modelling mesoscopic systems, they are conceptually different. Indeed, the former is a particle-based model solving Newton's second law of motion for a system consisting of *N* particles, while the latter considers the fluid consisting of fictive particles (particle populations). Such particles perform consecutive propagation and collision processes over a discrete lattice, thus solving the discrete-velocity Boltzmann equation, that approximate the Navier-Stokes equation in the nearly-incompressible limit.

In this work, applications of DPD to oil-water interfacial systems are investigated, including a coarse-graining procedure for a surfactant molecule referring to our previous work [1]. A combined coarse-graining and scaling procedure were tested for planar interfaces with and without surfactants and the main findings are, eventually, compared with the previous work. An example of simulating a droplet configuration is also illustrated and discussed.

In addition, based on the DPD study, the LBM is tested to compare the two mesoscopic models of immiscible liquids in presence of surfactants. This comparison is based on both equilibrium and dynamic properties. In particular, the reduction of the interfacial tension at increasing surfactant concentration, the droplet deformation and the inclination angle under simple shear rate at increasing Weber number (We) were studied.

Therefore, the main novelty of this work is to study the effects of scaling the DPD model to different coarse-graining levels by conserving the equilibrium properties of interfacial systems and note the similarities and differences between the DPD and LBM when dealing with immiscible fluids.

Material and Methods

Each DPD particle represents a molecular cluster (a molecule fragment or a group of solvent molecules) rather than an individual atom. The force acting on each bead contains three parts, which are all pairwise additive: (1) conservative, (2) dissipative, and (3) stochastic (random) forces [2]. When changing the level of coarse-graining for the DPD particles, their number is accordingly scaled and their size (radius of interaction) is adjusted in order to keep the relative overlap of the interacting particles constant. Hence when a system with many DPD beads is mapped onto one with fewer but larger and heavier particles, the interaction parameters have to be changed to maintain the overall system properties [6]. In order to compare DPD simulations with different coarse-

https://doi.org/10.52202/074122-0014

graining levels, functions of the scaling ratio ϕ are identified and these scaling expressions refer to relations between the respective parameters of two systems with different coarse-graining levels [6]. This scaling approach was tested by reproducing a ternary system made by water, oil, and protein surfactant, which was investigated in our previous work [1], thus labelled here as the reference model for ϕ = 1. The general idea is therefore to scale the DPD model up to the protein molecule level by maintaining the appropriate differences between the three phases. Here the protein molecular level of detail will be lost but favouring instead the mutual repulsion with the remaining two phases (oil and water). The new coarse-graining level was decided in order to represent the protein molecule as a single DPD particle or as two bonded beads (Figure 1). DPD simulations of the ternary interfacial system were performed in an orthorhombic box, where the starting configuration consists of two symmetrical interfaces due to the periodic boundary conditions applied in the three directions. The 50/50 oil-to-water bead ratio was again kept constant, and both the number of water and oil beads filling the simulation box was adjusted to keep the same overall number density when the protein beads were also added in the DPD box. Indeed, simulations were performed to study equilibrium proprieties of the interfacial system, such as the interfacial tension, at increasing protein surfactant concentration. An illustrative test was also conducted by simulating an oil droplet in water bulk in presence of protein surfactants at equilibrium in order to investigate the capability of the parameterization employed and the scaling procedure for an additional interfacial system setup.

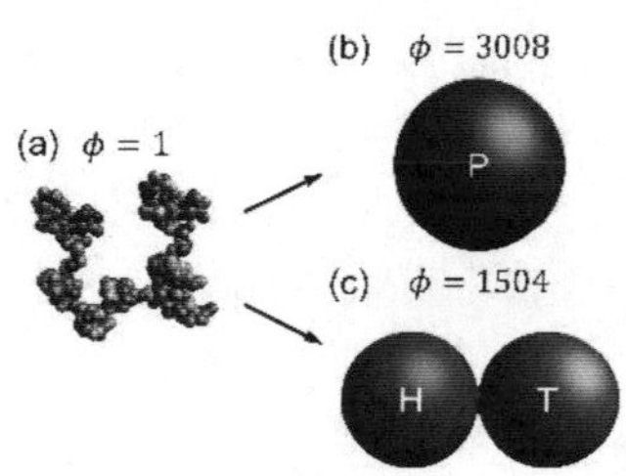

Figure 1. Schematic representation of the coarse-grained protein molecule in the reference DPD model [1] (a) and in the upscaled DPD model ((b) and (c)).

As regards LBM, both the thermodynamic and hydrodynamic description of a realistic ternary system consisting of oil, water, and a surfactant solute is represented by the phase-field model [7]. Here the same model is employed, therefore further details can be found in the literature. In this case, the surfactant is modelled as a typical non-ionic single-chained amphiphilic molecule with a hydrophilic head and a hydrophobic tail for both techniques. First, the equilibrium properties were analysed to check if the LBM approach can capture the same results of DPD simulations [8] and, eventually, to determine the parameter set then employed for the simple shear flow LBM study. Equilibrium LBM simulations are performed in a cubic 3D domain with two planar interfaces equidistant in x direction. The periodic boundary conditions are imposed at all the boundaries. The parameter set found will be then tested for simulating one droplet in shear flow using LBM, eventually comparing results with those of DPD simulations [8]. Being the droplet radius R = 20, the simulation domain is shown in Figure 2. Due to the symmetry of the problem only half of the drop has been simulated, while the boundary conditions are the same as those used in the work of Komrakova et al. [9].

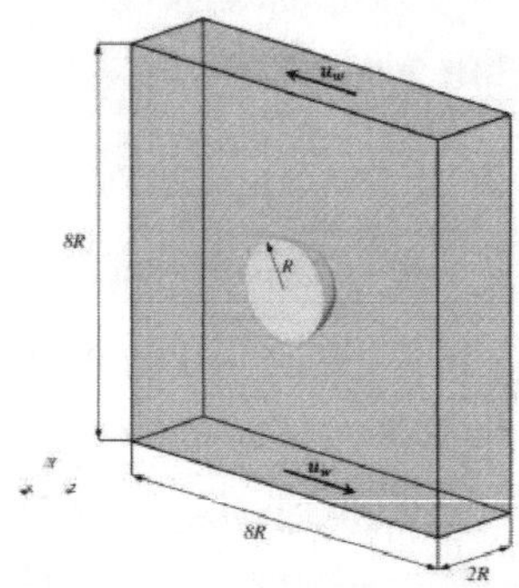

Figure 2. Simple shear flow LBM simulation domain with boundary conditions: x = 0 and x = 8R are periodic boundaries; y = 0 and y = 8R are no-slip walls moving with constant velocity u_w; z = 0 and z = 2R are the symmetry planes. At t = 0 the drop has a spherical shape.

Results and Discussion

Figure 3 represents one of the most interesting results of this work. It reports the interfacial tension as a function of the surfactant (protein) surface number density by comparing the reference results for ϕ = 1 with those obtained in this work for different coarse-graining ratio ϕ. As it can be seen, a very good agreement is achieved for both ϕ values investigated here at increasing protein surface number density. According to a simple DPD parameters fine-tuning, it is possible to preserve the interfacial tension trend in an upscaled DPD model with surfactant (protein) molecules. In both cases, the interfacial tension decreases as the protein surface density increases until it reaches a minimum value at the saturation of the interface. As it is shown, a further increase in protein surface concentration has almost no effect on the interfacial tension, which is a typical behaviour of an interfacial system stabilized by surfactant proteins [10]. Figure 4 shows an illustrative example of using the scaling procedure to simulate an oil droplet in water bulk in presence of surfactants. Similarities and differences between the two coarse-graining ratio cases are investigated. In both of them, it is important to highlight that the equilibrium configuration as a single droplet is observed due to the parameterization and the scaling procedure employed. This can demonstrate once again that the scaling relations in the DPD framework are able to describe different structural conformations.

Regarding LBM simulations, as already pointed out in previous works [7], the model can correctly reproduce both the Langmuir and Frumkin isotherms since simulations results follow the analytical predictions. Figure 5 shows the comparison of the interfacial tension trend at increasing surfactant concentration ψ_0 between the DPD [8] and LBM simulations. Both Langmuir and Frumkin isotherms were tested, and a very good agreement was achieved between the two mesoscopic techniques. The goodness of curve fitting was evaluated by computing the Root Mean Squared Error (RMSE). Similar accordance was also obtained for dynamic properties, namely the deformation and inclination angle of a single oil droplet under simple shear flow.

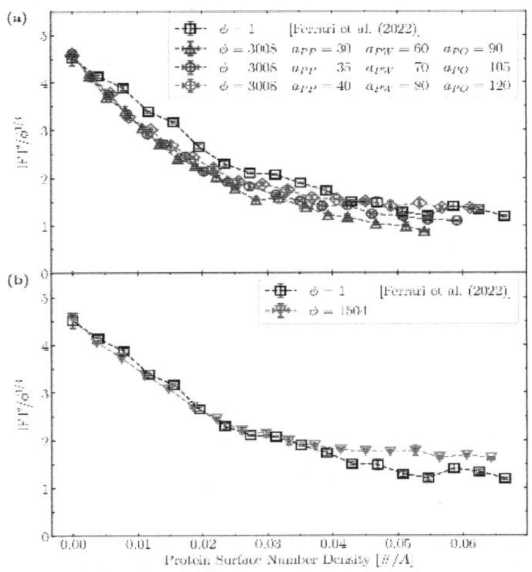

Figure 3. Interfacial tension as a function of the protein surface number density, comparing between reference results with $\phi = 1$ and for different coarse-graining ratio ϕ ((a) and (b)) and different DPD parameterization (coloured symbols).

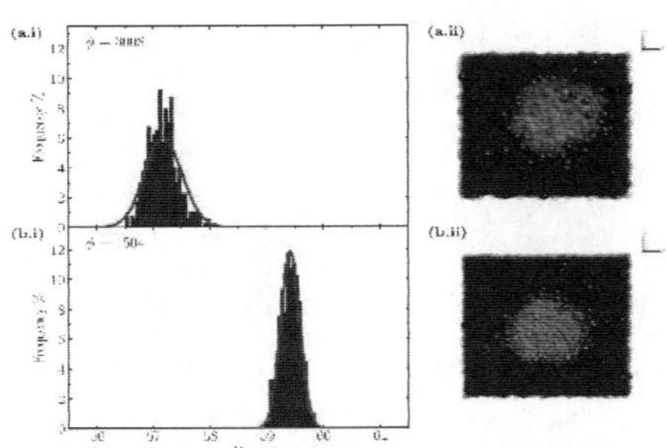

Figure 4. Time-frequency distributions (blue histograms) of the radius of gyration value of an oil droplet covered by surfactant molecules in water bulk for different coarse-graining ratio and the relative Gaussian fitting curves (red lines) ((a.i) and (b.i)). Corresponding snapshots of clipped simulation boxes are shown in (a.ii) and (b.ii).

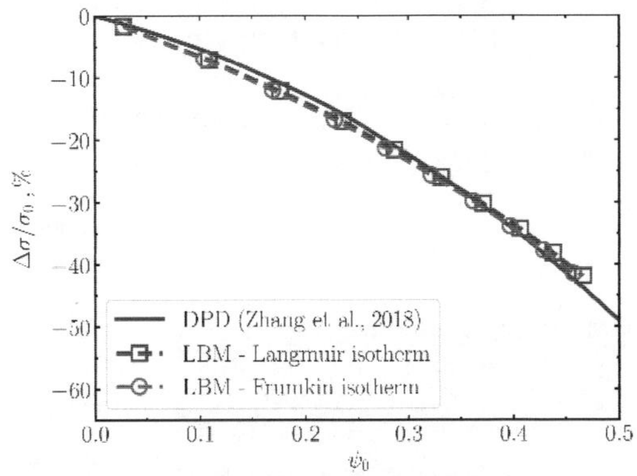

Figure 5. Comparison of interfacial tension trend at increasing surfactant concentration between DPD and LBM simulations.

Conclusions

When dealing with interfacial systems which are one of the most successful applications of both DPD and LBM, they tend to exhibit a typical length scale due to the domain formation. In this work, we showed that, if an interfacial system can be simulated on a small scale, the scaling of interactions does not prevent a simulation on a larger scale unless specific issues are dealt with. Moreover, a comparison of the two methods was carried out, highlighting the possibility to transfer information between these simulation techniques. Therefore, this work will pave the way for developing large-scale models of emulsion production processes.

Acknowledgments

The VIMMP project has received funding from the European Union's Horizon 2020 Research Innovation Programme under Grant Agreement n. 760907.

References

[1] M. Ferrari, J.-W. Handgraaf, G. Boccardo, A. Buffo, M. Vanni, D. L. Marchisio, *Physics of Fluids*, 34 (2022) 021903.

[2] R. D. Groot, P. B. Warren, *The Journal of Chemical Physics*, 107 (1997) 4423–4435.

[3] M. Ndao, F. Goujon, A. Ghoufi, P. Malfreyt, *Theoretical Chemistry Accounts*, 136 (2017) 21.

[4] A. Khedr, A. Striolo, *Journal of Chemical Theory and Computation*, 14 (2018) 6460–6471.

[5] H. Liu, Y. Zhang. *Journal of Computational Physics*, 229 (2010) 9166–9187.

[6] R. M. Füchslin, H. Fellermann, A. Eriksson, H.-J. Ziock, *The Journal of Chemical Physics*, 130 (2009) 214102.

[7] Y. Shi, G. H. Tang, L. H. Cheng, H. Q. Shuang. *Computers & Fluids*, 179 (2019) 508–520.

[8] Y. Zhang, J. Xu, X. He. *Molecular Physics*, 116 (2018) 1851–1861.

[9] A. E. Komrakova, O. Shardt, D. Eskin, J. J. Derksen. *International Journal of Multiphase Flow*, 59 (2014) 24–43.

[10] V. B. Fainerman, E. H. Lucassen-Reynders, R. Miller, *Advances in Colloid and Interface Science*, 106 (2003) 237–259.

https://doi.org/10.52202/074122-0014

Lattice Boltzmann Study on Surfactant-laden Drop Behaviour in Shear Flow

CF1.3

Z. A. Chen[1], *P.A Tsai*[1], *A.E Komrakova*[1]. *(1) University of Alberta, 116 St & 85 Ave, Edmonton, Canada, zhe8@ualberta.ca*

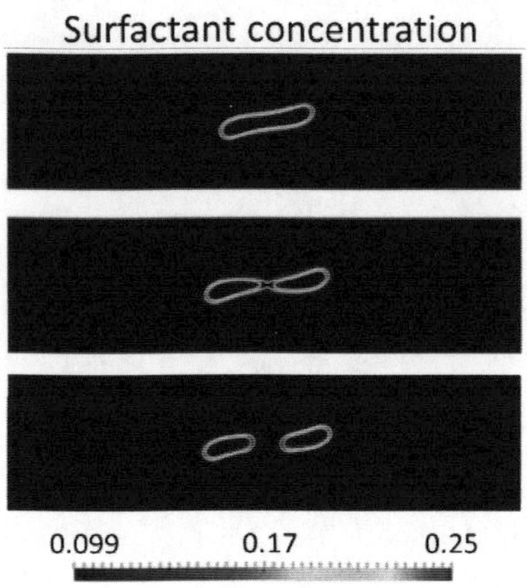

We study the behaviour of a surfactant-laden drop in a simple shear flow using a phase-field lattice Boltzmann method. The proposed phase-field model is based on modified Ginzburg-Landau free energy functional to account for non-ionic soluble surfactants. First, we modelled the planar interface and static drop equilibration to validate the method. Results show that the proposed method can accurately capture surfactant distribution and is in favour of lower spurious velocities. We then studied the influence of capillary number, bulk surfactant load, and the combined effect of surfactant and viscosity ratio between the drop and the continuous phase on drop deformation and breakup. The results indicate that drop deformation increases with the capillary number and bulk surfactant concentration increase. Also, the combined effect of the viscosity ratio between the liquids and the presence of surfactant determines the topological changes of the drop (achievement of equilibrium deformation or breakage). This work will pave the way for a better understanding of surfactant-laden drop behaviour in emulsion production.

Introduction

Modelling the behaviour of two immiscible liquids stabilized by a soluble surfactant (i.e., emulsion) is a challenging numerical task. A physically realistic representation of such a system requires resolving the highly nonlinear interplay between hydrodynamics, the interfacial topology change, and the surfactant transport [1]. In this study, we developed and applied a diffuse interface phase field method to perform time-resolved three-dimensional simulations of emulsions. The governing system of equations consists of the Navier-Stokes equations coupled with two transport equations. An Allan-Cahn equation governs the transport of the phase field parameter that accounts for the two immiscible liquids. The Cahn-Hilliard-type equation describes the transport of the second-order parameter that represents the concentration of surfactant. The terms of these transport equations were determined based on a free energy functional. A widely used Ginzburg-Landau functional for the pure system was modified to account for the surfactant [1][2][3].

Several phase-filed models have been proposed to simulate emulsions. The governing system of equations was solved by traditional numerical techniques [4][5] and more recent mesoscale strategies such as the lattice Boltzmann method (LBM) [1][2][3][6]. Here, we extended the LBM for pure system [7][8] to accommodate the presence of surfactant. This variation of the method has a complex lattice structure (D3Q27) and the weighed multi-relaxation-time collision operator that enhances the model's accuracy and stability.

First, we validate the model using the simulations of the planar interface for which analytical surfactant distributions exist. We subsequently validate the numerical results concerning the effect of surfactant on a circular drop during its equilibration. Finally, we perform simulations of a surfactant-laden drop in a simple shear flow. We study the effect of capillary number, bulk surfactant load, and the joint effect of surfactant and viscosity ratio between the drop and the continuous phase.

Methodology

The governing equations of a surfactant-laden emulsion include the continuity equation, the momentum equation, the conservative phase field equation, and a transport equation for surfactant [1][7]:

$$\nabla \cdot \boldsymbol{u} = 0,$$

$$\rho\left(\frac{\partial \boldsymbol{u}}{\partial t} + \boldsymbol{u} \cdot \nabla \boldsymbol{u}\right) = -\nabla p + \nabla \cdot (\mu[\nabla \boldsymbol{u} + (\nabla \boldsymbol{u})^{\mathrm{T}}]) + \boldsymbol{F}_s,$$

$$\frac{\partial \phi}{\partial t} + \nabla \cdot (\phi \boldsymbol{u}) = \nabla \cdot M_\phi \left(\nabla \phi - \frac{1 - 4(\phi - \phi_0)^2}{\xi} \frac{\nabla \phi}{|\nabla \phi|}\right),$$

$$\frac{\partial \psi}{\partial t} + \nabla \cdot (\psi \boldsymbol{u}) = \nabla \cdot (M_\psi \nabla \mu_\psi).$$

where ϕ is the phase parameter, and ψ is the surfactant concentration. The effect of two immiscible phases and surfactant is incorporated into the momentum equation via the interfacial tension force, which is calculated as follows [1]:

$$\boldsymbol{F}_s = \mu_\phi \nabla \phi + \mu_\psi \nabla \psi$$

The chemical potentials μ_ϕ and μ_ψ are obtained using the variational derivation of the free energy functional with respect to ϕ and ψ as:

$$\mu_\phi = 4\beta\phi(\phi - 1)(\phi - 0.5) - (\kappa - d\psi)\nabla^2\phi + d\nabla\phi \cdot \nabla\psi + w(\phi - \phi_0)\psi^2 - e\psi,$$

$$\mu_\psi = k_B T[\ln \psi - \ln(1 - \psi)] - c\psi + \frac{w}{2}(\phi - \phi_0)^2 - \frac{d}{2}(\nabla\phi)^2 - e(\phi - \phi_0).$$

The lowering of interfacial tension depends on the excess amount of surfactant on the interface.

https://doi.org/10.52202/074122-0015

An equation of state (EOS) is used to calculate this decrease [1]:

$$\sigma(\psi_0) - \sigma_0 = \alpha\left[k_B T \ln(1-\psi_0) + \frac{c}{2}\psi_0^2\right],$$

where σ_o is the surface tension of a surfactant-free system, and coefficient α is a model-dependent parameter. The detailed LBM description to solve the macroscopic governing equations can be found in our previous studies [7][8].

Numerical results and discussion

Model validation: planar interface

We analyzed the capability of the model to resolve realistic Langmuir and Frumkin adsorption at the planar interface. The change of the surfactant concentration along the interface for three different bulk surfactant concentrations $\psi_b = \{5\times10^{-2}, 1\times10^{-1}, 2\times10^{-1}\}$ is shown in Fig.1. Two different adsorption isotherms overlap at small bulk surfactant concentrations. With the bulk surfactant concentration increase, the isotherms no longer coincide. This observation aligns with the results of Liu and Zhang [1] and proves that Frumkin adsorption needs to be considered for the high bulk concentration of surfactant.

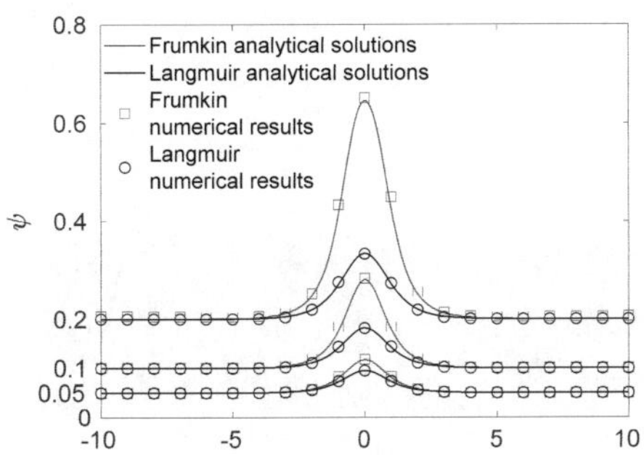

Figure 1. Surfactant concentration along the interface in the case of Langmuir and Frumkin adsorption (the comparison between numerical and analytical results).

Model validation: drop equilibration

The interfacial tension decreases due to the surfactant adsorption at the interface. To check the capability of our model to capture this effect, we simulated an equilibration of a circular surfactant-laden drop. The drop with a radius of $R_0 = 20$ lattice units (lu) is located at the centre of the fully-periodic domain of $80 \times 80 \times 80$ lu. The physical and numerical parameters were set the same as in the planar interface for Frumkin adsorption. Fig.2 illustrates the relationship between surface tension decrease $\Delta\sigma/\sigma_o$ and interface surfactant concentration $ln(1-\psi_o)$. The numerical results (symbols) agree well with the analytical solution (solid curves) by fitting $\alpha = 0.9806$ into the EOS: a higher surfactant load leads to a higher decrease in the interfacial tension. Numerically originated, spurious velocities are measured to be around 10^{-6}, which is lower than the commonly reported range 10^{-4} to 10^{-5} [1], as shown in Fig. 3.

Drop in a simple shear flow

We study the effect of capillary numbers on drop deformation and breakup. The parameters are kept the same as in the previous cases. Bulk surfactant concentration $\psi_b = 0.1$ was set, and the capillary number Ca varied in a range from 0.05 to 0.32, where Ca is defined based on drop radius, the shear rate between walls, continuous fluid viscosity, and initial surface tension. The comparison between the shape of the surfactant-laden drop in equilibrium at different Ca is shown in Fig.4. We find that surfactant-laden drops undergo a more significant deformation at the equilibrium when Ca is higher. The corresponding re-circulations inside the drop (white curves) and the surfactant concentration distributions for $Ca = 0.25$ are shown in Fig.5. The concentration of surfactant at the drop tips (highest curvature) is higher than those at other locations. This observation aligns with the previous study [3].

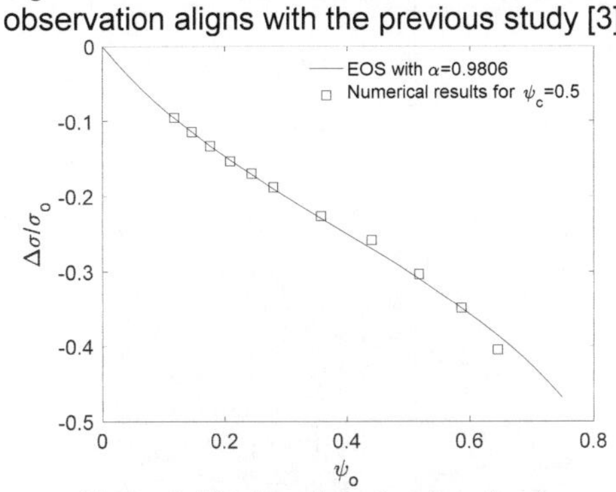

Figure 2. The lowering of surface tension as a function of ψ_o.

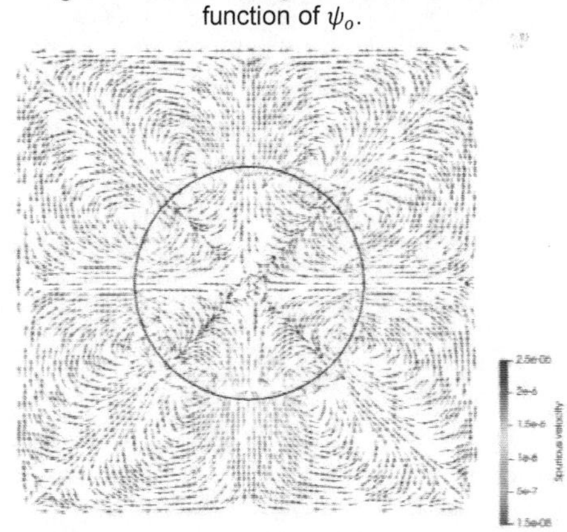

Figure 3. Spurious velocity around the drop.

The effect of surfactant concentration on drop deformation is also studied. We perform simulations with different bulk loads of surfactant $\psi_b = \{0.5\times10^{-2}, 1\times10^{-1}, 2\times10^{-1}\}$ at $\mathrm{R}e = 0.0625$ and $\mathrm{Ca} = 0.2$. The results are shown in Fig. 6. Higher bulk surfactant concentration leads to a higher deformation.

Finally, we study the combined effect of surfactant and viscosity ratio between the drop and the continuous phase ($\lambda = \mu_d/\mu_c$). The increase of λ hinders drop deformation and increases the stability of the drop [9]. Therefore, the presence of surfactant and the increase in viscosity ratio have opposing effects on drop

 https://doi.org/10.52202/074122-0015

deformation and breakup. For $Re = 0.0625$, $\mathrm{Ca} = 0.29$, $\psi_b = 1 \times 10^{-1}$, the effective capillary number based on actual interfacial tension achieved by the system with surfactant at equilibrium becomes $Ca_e = 0.4$. This value is lower than the critical capillary number of ~0.41 for the pure system at $\lambda = 1$ [10], which means at the capillary number below critical, the drop should not break. However, when the surfactant is present, the drop breaks at $Ca_e = 0.4$ when $\lambda = 1.05$ as shown in Fig.7.

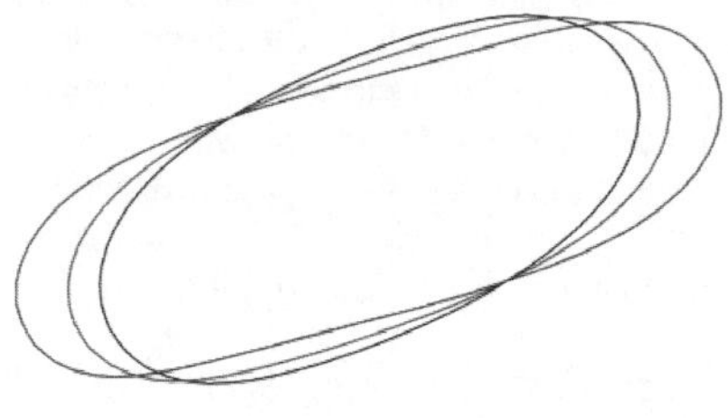

Figure 4. The shape of the drop at equilibrium: the black curve represents $Ca = 0.18$, the red curve represents $Ca = 0.22$, and the blue curve represents $Ca = 0.25$. The bulk surfactant concentration was set at 0.1 for each case.

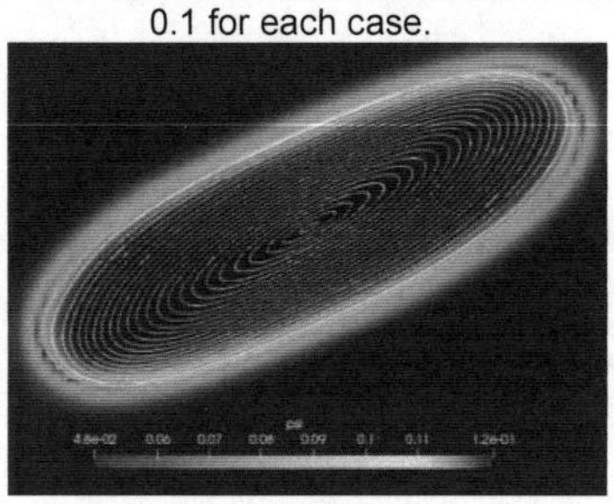

Figure 5. The internal re-circulations (white curves) and surfactant concentration distribution at drop equilibrium for $Ca = 0.25$.

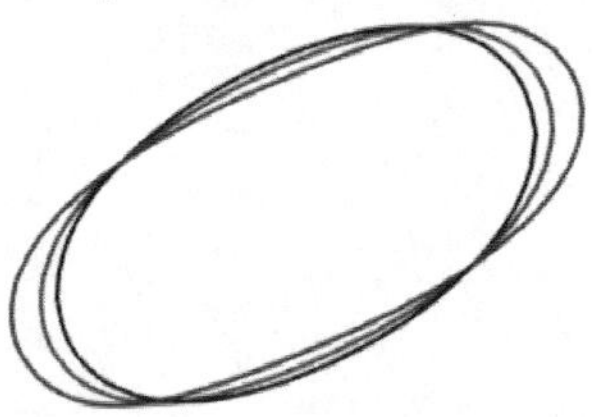

Figure 6. The drop shape at equilibrium: the black curve represents $\psi_b = 0.05$; the red curve represents $\psi_b = 0.1$; and the blue curve represents $\psi_b = 0.2$.

Conclusions

We developed and applied a diffuse interface lattice Boltzmann method to simulate a system of immiscible liquids with soluble surfactants. The method was validated using the interfacial profile of the planar interface and static drop with analytical solutions. Surfactant-laden drop behaviour in shear flow was studied. We varied the capillary number and bulk surfactant load and looked at the combined effect of surfactant and viscosity ratio. The presence of surfactant increases drop deformation and lowers the critical capillary number promoting drop breakup when the viscosity ratio between the liquids increases.

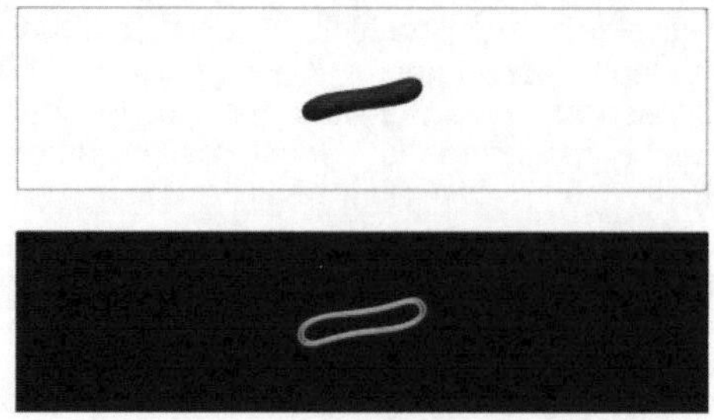

(a)

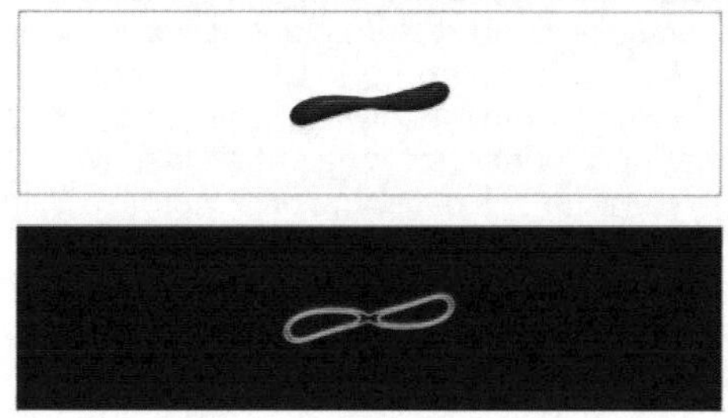

(b)

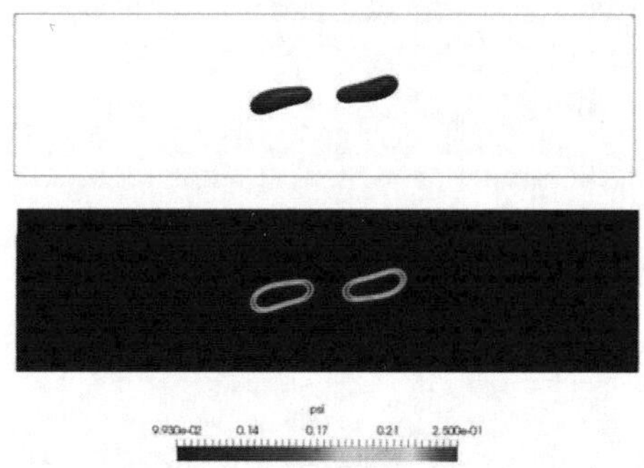

(c)

Figure 7. (a) Time step=30; (b) Time step=46; (c) Time step=49. Upper: three-dimensional drop shape; Bottom: ψ profile.

References

[1] Liu, Haihu, and Yonghao Zhang. Journal of Computational Physics 229.24 (2010): 9166-9187.
[2] Zong, Yajing, et al. Physics of Fluids32.12 (2020): 122105.
[3] Shi, Y., et al. Computers & Fluids 179 (2019): 508-520.
[4] Soligo, Giovanni, Alessio Roccon, and Alfredo Soldati. Journal of Fluid Mechanics881 (2019): 244-282.
[5] Engblom, Stefan, et al. Communications in Computational Physics 14.4 (2013): 879-915.
[6] Timm, Krüger, et al. Cham, Switzerland: Springer International Publishing AG (2016).
[7] Chen, Zhe Ashley, et al. International Journal of Multiphase Flow147 (2022): 103863.
[8] Huang, Fenglei, et al. "Numerical study of drop behavior in a pore space."Chemical Engineering Science 233 (2021): 116351.
[9] Komrakova, A. E., et al. International Journal of Multiphase Flow 59 (2014): 24-43.
[10] Li, Jie, Yuriko Y. Renardy, and Michael Renardy. Physics of fluids 12.2 (2000): 269-282.

https://doi.org/10.52202/074122-0015

Experimental study of the mixing and flow properties of fractal oscillating grids in water.

B2.1

Valentin Musy[1,2], *Anne-Lise Hantson*[1], *Diane Thomas*[1], *Jean-Christophe Baudez*[2], *Tom Lacassagne*[2].
(1) Chemical and Biochemical Process Engineering Unit, Faculty of Engineering, University of Mons, 7000 Mons, Belgium. (2) IMT Nord Europe, Institut Mines Télécom, Univ. Lille, Center for Energy and Environment, F-59000 Lille, France.

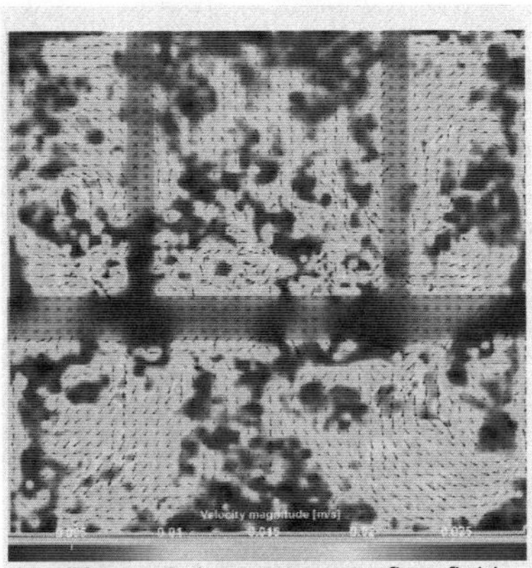

Figure 0. Instantaneous flow field generated by a simple Cartesian oscillating grid.

If a biochemical process deals with micro-organism cultures, the mixing device used in the bioreactor should promote all the transfer phenomena (nutrients, gas-liquid, light, ...) while limiting the mechanical stress induced on the micro-organisms. With this comment in mind, we propose to design a photobioreactor (PBR) stirred by fractal oscillating grid systems for the culture of shear-sensitive microalgae. Oscillating cross-barred grids generate almost isotropic and homogeneous turbulence with a weak (almost null) mean flow. Grids with fractal geometries were used as fixed grids to generate downstream, isotropic and homogeneous turbulence in wind and water channels. Therefore, our work is the first implementation of a fractal oscillating grid flow. Flow characterisation experiments namely Planar Laser Induced Fluorescence (PLIF) and PIV (Particle Imaging Velocimetry) are performed, with the aim to determine the best conditions (grid pattern, oscillation frequency and amplitude) *a priori* for microalgae cultures.

Introduction

The promotion of transfers within a reactor is often the keystone of the enhancement of a chemical or biochemical process. In the case of microorganism cultures within a bioreactor, this transfer intensification should not be accompanied by a shear stress increase inducing mechanical stress on the microorganisms leading to their lysis. This is especially true for the culture of some shear-sensitive microalgae (red microalgae, diatoms, dinoflagellates, ...). The culture of this kind of photosynthetic micro-organisms can be performed in several designs of photobioreactors (PBR) with different and geometries to capture CO_2 and produce interest products (lipids for biodiesel production, proteins, pigments or carbohydrates for health, food or cosmetic purposes). Each PBR configuration or geometry offers a compromise between intense mixing and high shear stress. Among the classical configurations, we can mention the raceway ponds, the stirred tank, the tubular, the flat plate, the bubble column and the airlift PBR **[1 – 5]**.

The aim of this PhD work is to design an innovative PBR stirred by a fractal oscillating grid system for the culture of shear-sensitive microalgae. This type of system is classically composed by a cross-barred grid (Cartesian mesh) vertically oscillating with a specific frequency ν and amplitude (peak to peak) A around an equilibrium position z_{eq} in a fluid medium contained in a prismatic tank (Figure 1). The grid oscillatory motion induces turbulence in the fluid medium whose properties can be monitored thanks to the oscillation and grid geometry parameters. The generated turbulence is considered nearly isotropic and homogeneous and with almost no mean flow; the flow is therefore virtually shear-free according to the generated turbulence. Oscillating grid systems were used to study interfacial and stratified flows and particles sedimentation and flocculation. More recently, they were used to study transfers in micro-organism cultures and biofilm degradations **[6 – 7]**.

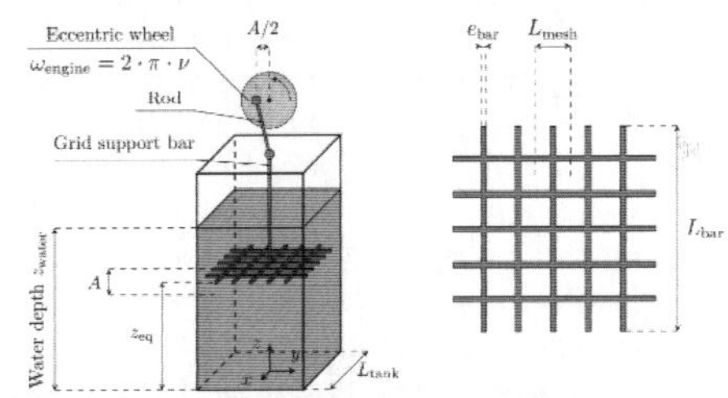

Figure 1. Simplest configuration of an oscillating grid system in water.

Grids with fractal geometries were used as fixed grid in wind or water tunnels to generate a downstream, isotropic and homogeneous turbulence **[8]**. Such grids are designed in an iterative way by the repetition of a base pattern - composed of N bars of length L_0 and width e_0 - over several size scales. The bar length and width for a pattern at the iteration i are linked to those for the same pattern at the iteration $i-1$ respectively by ratios ratio_L and ratio_e. These ratios are kept constant iteration after iteration. In our work, a classical Cartesian grid is used as a reference case and 4 fractal grids (with 3 fractal patterns: Cartesian, square and "I") constitute the core of our study.

Before any microalgae culture, the first part of our work aims to perform a parametric study of the properties of the flow generated by the investigated fractal oscillating grid systems. Based on this study, the best *a priori* configuration of fractal oscillating grid system will be selected to perform microalgae cultures.

https://doi.org/10.52202/074122-0016

Material and Methods

To study the properties of the fractal oscillating grid flow, 2 optical flow visualisation methods – PLIF (Planar Laser Induced Fluorescence) and PIV (Particle Imaging Velocimetry) – have been implemented. In our case, both methods involve a vertical laser sheet generated by a continuous laser Dantec Dynamics RayPower and a scientific camera LaVision Imager M-lite 5M. Our parametric study is summarised in the Table 1 and focuses on the variation of the grid type and of the oscillation frequency and amplitude; each combination is investigated. The grid geometries and features are summarised in the Table 2.

Table 1. Parameters levels for the parametric study of the fractal oscillating grid flow.

ν levels [Hz]	A levels [m]	Grid type
0.5	0.02	Simple Cartesian
1.0	0.035	Fractal Cartesian
2.0	0.05	Fractal square pattern
		Fractal "I" pattern square fitting
		Fractal "I" pattern rectangle fitting

With the PLIF method, our aim is to evaluate the mixing time in the PBR stirred by oscillating grid as a function of the above-mentioned parameters (ν, A and grid type) by recording the dispersion of a fluorescent dye (in our case, rhodamine 6G). Classically, calibration between the intensity of the local fluorescence radiation and the local dye concentration is performed. However, in our case and due to the wide range of dye concentrations (between injection concentration 10^{-3} [mol/L] and the final concentration when homogeneity is reached 10^{-8} [mol/L]), such a calibration was not possible, and fluorescence absorption phenomena arise, limiting the method to qualitative considerations.

With the PIV method, 50 [μm] polyamide particles (reflecting the vertical laser sheet radiation) are dispersed in the fluid medium. Two-dimensions vector fields of fluid velocity are obtained after image and PIV processing by several Matlab codes (image normalisation, grid masking algorithm) and by PIVlab (a Matlab app dedicated to PIV treatments). With the PIV method, several vertical laser planes, some across the grid bars and some across the grid bars and holes, are investigated to evaluate the potential inhomogeneity of the flow. The obtained vector fields from PIV measurements are analysed each velocity field is broken down into a mean flow contribution, an oscillatory contribution and purely turbulent fluctuations according to a triple decomposition method **[9]**.

Results and Discussion

For the evaluation of the mixing time according to the grid types and the oscillations parameters, the results for each grid can be gathered by defining an oscillation-based Reynolds number Re_{osc} (see below). Figure 2 depicts the expected decrease of the mixing time as Re_{osc} increase (meaning that ν and/or A increase).

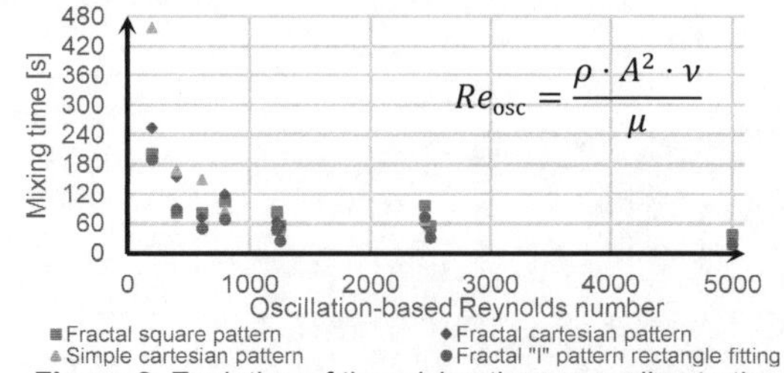

Figure 2. Evolution of the mixing time according to the oscillation-based Reynolds number for each grid.

The fractal nature seems to enhance the mixing performances at low Reynolds number although it is risky to conjecture on which grid seems to perform better at high Reynolds number. To answer this question, the repeatability of each point should be evaluated. Thus, the Figure 3 shows the variation of the mixing time according to Re_{osc} for the simple Cartesian grid with each measurement point repeated 5 times (the error bars depict the standard deviation).

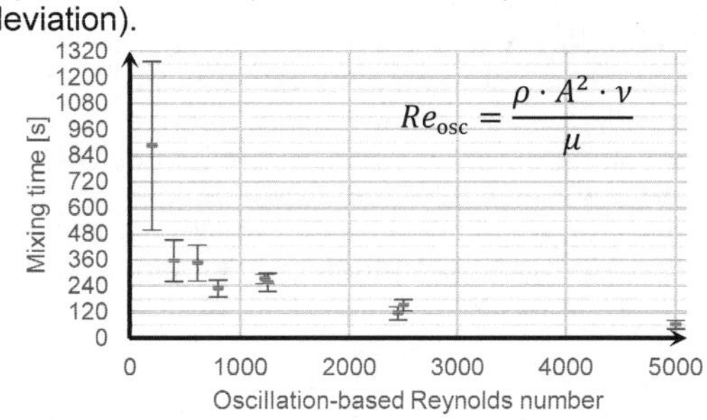

Figure 3. Evolution of the mixing time according to the oscillation-based Reynolds number for the simple Cartesian grid with 5 replications for each measurement point.

The poor repeatability, illustrated by these important error bars, can be attributed to an important non repeatable mean flow and a relative precision about the position of the injection point. It thus appears that the mixing time could not be the most relevant indicator to compare the grids with each other. Indeed, for our final application (culture of shear-sensitive microalgae), flows properties (turbulence variables, shear rate, ...) could be more interesting to compare the grids performances. Therefore, we decided to focus our work on PIV measurements.

Figure 4 depicts the flow decomposition into a mean contribution, an oscillatory contribution (RMS values) and purely turbulent fluctuations (RMS values) for a statistically permanent flow generated by 1000 complete oscillations of the simple Cartesian grid with $\nu = 0.5$ [Hz] and $A = 0.05$ [m]. For each z coordinate, the associated mean value of the RMS value of the turbulent fluctuations can be compared to the classical Hopfinger and Toly relations (Figure 4) **[6 – 7]**. Despite favourable set-up conditions for a strong mean flow (large oscillation amplitude compared to the grid mesh size, water volume as high as it is wide) and significant side effects near the free surface and the tank bottom, our results remain in a good agreement with the Hopfinger and Toly's decay laws.

Table 2. Grid geometries and features used for the parametric study of the fractal oscillating grid flow.

Features	Simple Cartesian	Fractal Cartesian	Fractal square pattern	Fractal "I" pattern square fitting	Fractal "I" pattern rectangle fitting
Pattern design	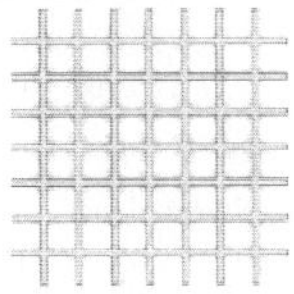	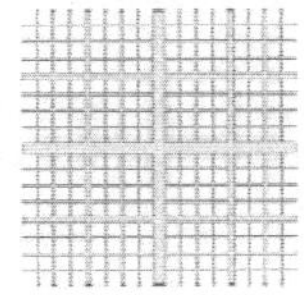	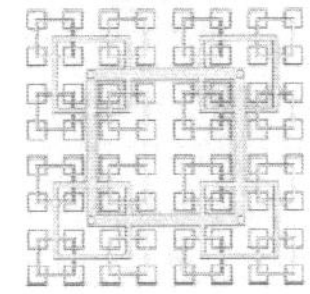	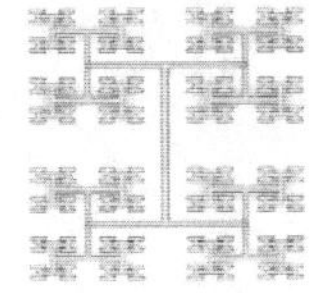	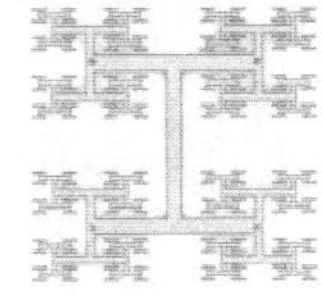
N	2	2	4	3	3
L_0 [mm]	234.0	234.0	132.9	140.8	145.1
e_0 [mm]	5.5	9.5	8.7	4.9	12.1
ratio_L	0.488	0.480	0.500	0.452	0.426
ratio_e	1.000	0.500	0.500	0.901	0.526

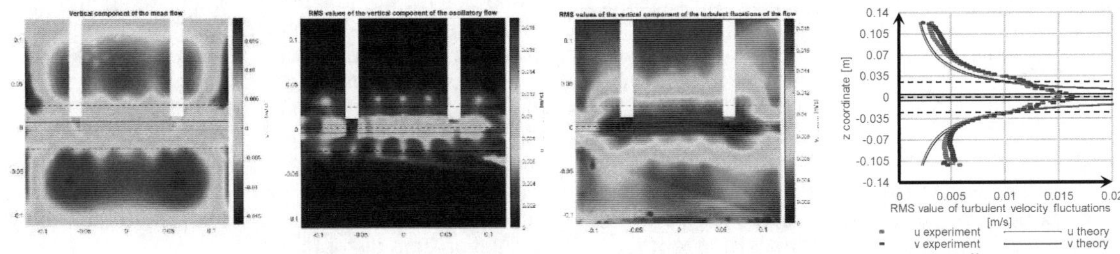

a) b) c) d)

Figure 4. Decomposition of the vertical velocity component of the flow generated by the simple Cartesian grid with $\nu = 0.5$ [Hz] and $A = 0.05$ [m]. (a) Mean flow. (b) Oscillatory flow (RMS values). (c) Turbulent fluctuations (RMS values). (d) Turbulence decay along the z coordinate (comparison with the Hopfinger and Toly relations).

Conclusions

The aim of this project is the design of a photobioreactor stirred by a fractal oscillating grid system for shear-sensitive microalgae culture. To define the best *a priori* oscillating grid configuration (grid pattern, oscillation parameters) for microalgae culture, flow visualisation and characterisation are firstly performed by PLIF and PIV. Due to the important variability of our PLIF results for the evaluation of the mixing time, we have decided to focus on the PIV measurements. Due to the important treatment time to analyse the PIV images and vector fields, the complete parametric study is not already achieved at the time of writing these lines. However, this study is going to be completed before March 2023 (the beginning of the first microalgae cultures).

Acknowledgments

Financial support from the University of Mons, IMT Nord Europe and French National Research agency (grant n° ANR-22-CE51-0015-01) are gratefully acknowledged.

References

[1] W. Levasseur, P. Perré, V. Pozzobon, *A review of high value-added molecules production by microalgae in light of the classification*, *Biotechnol. Adv.* vol.41 p. 107545, July 2020.

[2] C. Wang, C. Q. Lan, *Effects of shear stress on microalgae – A review*, *Biotechnol. Adv.* vol. 36 n°4 p. 986 - 1002, July 2018.

[3] N. Seyed Hosseini, H. Shang, J. A. Scott, *Biosequestration of industrial off-gas CO_2 for enhanced lipid productivity in open microalgae cultivation systems*, *Renew. Sustain. Energy Rev.* vol. 92 p. 458 - 469, September 2018.

[4] J. Fu, Y. Huang, Q. Liao, A. Xia, Q. Fu, X. Zhu, *Photobioreactor design for microalgae: A review from the aspect of CO_2 transfer and conversion*, *Bioresour. Technol.* vol. 292 p. 121947, November 2019.

[5] R. N. Singh, S. Sharma, *Development of suitable photobioreactor for algae production – A review*, *Renew. Sustain. Energy Rev.* vol. 16 n° 4 p. 2347 - 2353, May 2012.

[6] J. Yan, N.-S. Cheng, H.-W. Tang, S. K. Tan, *Oscillating-grid turbulence and its applications: a review*, *J. Hydraul. Res.* vol. 45 n° 1 p. 26 - 32, January 2007.

[7] E. J. Hopfinger, J.-A. Toly, *Spatially decaying turbulence and its relation to mixing across density interfaces*, *J. Fluid Mech.* vol. 78 n° 1 p. 155 - 175, November 1976.

[8] D. Hurst, J. C. Vassilicos, *Scalings and decay of fractal-generated turbulence*, *Phys. Fluids* vol. 19 n° 3 p. 035103, March 2007.

[9] T. Lacassagne, A. Lyon, S. Simoëns, M. El Hajem, J.-Y. Champagne, *Flow around an oscillating grid in water and shear-thinning polymer solution at low Reynolds number*, *Exp. Fluids* vol. 61 n° 1 p. 15, January 2020.

https://doi.org/10.52202/074122-0016

Experimental Investigation of Circulation Times with Lagrangian Sensor Particles in a 15,000 L Acrylic Glass Bioreactor

B2.2

***Sebastian Hofmann*[1]*, Lukas Buntkiel*[2]**, *Paramveer GopalSingh*[1], *Jürgen Fitschen*[1], *Marko Hoffmann*[1], *Sebastian Reinecke*[2], *Uwe Hampel*[2,3], *Michael Schlüter*[1]. *(1) Institute of Multiphase Flows, Hamburg University of Technology, Hamburg, Germany, sebastian.hofmann@tuhh.de. (2) Helmholtz-Zentrum Dresden-Rossendorf, Dresden, Germany, l.buntkiel@hzdr.de. (3) Institute of Power Engineering, Dresden University of Technology, Dresden, Germany.*

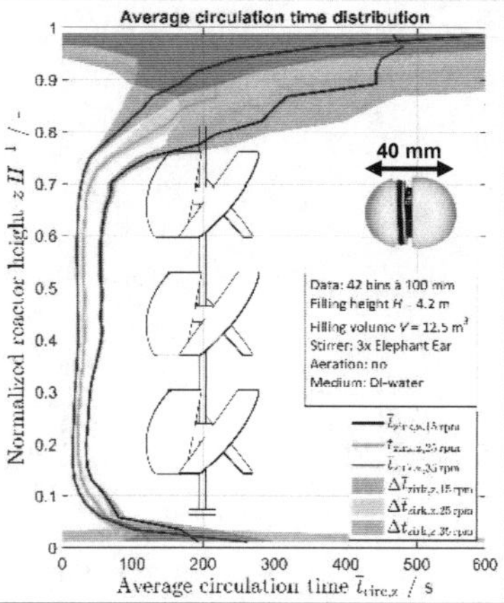

In this study, Lagrangian Sensor Particles (LSP) are applied to a 15,000 L acrylic glass bioreactor in order to measure circulation times to show the formation of compartments in an industrial-scale bioreactor. The LSPs are equipped with a pressure sensor from which their axial position, axial velocity distribution and circulation time distribution can be derived based on the measurement of the hydrostatic pressure. In a first step, the LSP's axial distribution in the reactor is determined. Results show a homogeneous distribution in the center region for three different impellers frequencies, however, a lower probability density at the reactor bottom and top. Moreover, velocity profiles demonstrate a distinctive upward and downward flow profile of an axial-pumping impeller type. Eventually, the circulation time distribution depicts an upper, a center and a lower compartment. Single data sets are further analyzed to discuss their correlation with global mixing time measurements with phenolphthalein for the same setup.

Introduction

Industrial-scale bioreactors with working volumes up to 22,000 L are often subjected to transient cultivation conditions caused by spatial and temporal gradients in mixing [1]. Those heterogeneities lead to a deterioration of the microorganism's cellular performance, and thus to a reduction in product quality and quantity [2]. However, biopharmaceutical fabrication of vaccines or monoclonal antibodies depend on precise and reliable process control and output, which bears a high societal relevance. Moreover, newest studies on quorum sensing and cell-cell interactions show the impact during dynamically changing environments [3]. Therefore, cell-bioreactor interactions must be as well investigated thoroughly to understand the population behaviour in the turbulent flow field during a fermentation in order to reveal major process improvements.

One approach to obtain a deeper insight into complex flow patterns and the formation of compartments is the decolorization method by means of phenolphthalein in a 15,000 L acrylic glass bioreactor (Fig. 1(a)). This method was successfully demonstrated by Rosseburg et al. at the Institute of Multiphase Flows at the Hamburg University of Technology [4]. However, the lack of optical access into generic bioreactors requires an alternative analysis of the global and local mixing performance. Industrially used fixed probes at the reactor wall can neither resolve gradients spatiotemporally nor the cell's residence time in different zones of the reactor.

That is the reason why "Traveling along the Lifelines of Single Cells" is another approach introduced by Lapin et al. [5], yet with a numerical method. This Lagrangian analysis is also experimentally scrutinized in a recent study, showing promising results on laboratory-scale by means of Lagrangian Particle Tracking. [6] On production-scale, mobile Lagrangian Sensor Particles (LSPs) [7,8] offer crucial data alongside their trajectories throughout the reactor as axial velocity profiles and circulation time distributions.

In this study, an experimental method to determine flow patterns and circulation times in a 15,000 L bioreactor by means of LSPs is presented. The results are discussed regarding their axial distribution and data density and further compared to mixing time trials for the same setup.

Material and Methods

Lagrangian Sensor Particles (LSPs) with the diameter of d_p = 40 mm (Shell material: casted Polyoxymethylene, Fig. 1(b) and Graphical Abstract) are applied to an identical replica of an industrial-scale bioreactor made of acrylic glass (Filling volume V = 12,500 L, filling height H = 4.2 m, reactor diameter d_1 = 2.0 m, Fig. 1(a)). The LSPs are developed by the Helmholtz-Zentrum Dresden-Rossendorf (HZDR, Germany) and comprise a microcontroller, a data logging unit, a pressure sensor unit and batteries.

Neutral buoyancy of the LSPs is gained by twisting the two shelfs, and thus altering their volume so that a particle density ρ_p is fitted to $\Delta\rho = (4 \pm 2)$ kg m^{-3} in deionized water (T = 22 °C, ρ_L = 997.8 kg m^{-3}). Based on the measurement of the hydrostatic pressure p with the pressure sensor unit (Recording frequency f = 50 Hz), the data deliver the LSP's vertical residence probabilities, axial velocities and vertical circulation times in predefined reactor slices (h_{bin} = 100 mm, Tab. 1).

The bioreactor is equipped with a three-stage Elephant Ear stirrer (downward pumping) and four baffles (α = 90 °). The impeller frequencies used are n ϵ {15, 25, 35} rpm, which corresponds to power

https://doi.org/10.52202/074122-0017

Table 1. Evaluation of the pressure sensor data.

Axial position z in the reactor	$z = f(p, \rho_L)$
Average vertical velocity $\bar{v}_z$ for reactor height z	$\bar{v}_z = \frac{\lvert dz(t) \rvert}{dt}$
Average circulation time $\bar{t}_{circ,z}$ for reactor height z at time i	$\bar{t}_{circ,z} = t_{z,i} - t_{z,i-1}$
Uncertainty Δ for $\bar{x}$ with sample size N	$\Delta_{\bar{x}} = \sqrt{\frac{1}{N(N-1)} \sum_{i=1}^{N} (x_i - \bar{x})^2}$

inputs as P $V^{-1} \epsilon$ {4, 20, 54} W m^{-3}. At each setting, three LSPs are applied to the reactor.

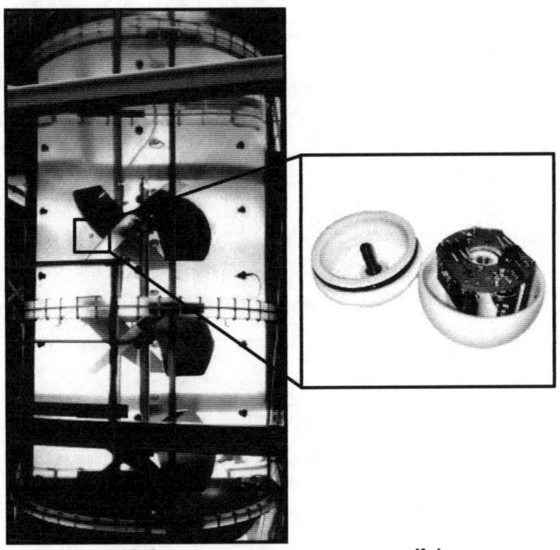

(a) **(b)**

Figure 1. (a) 15,000 L acrylic glass bioreactor with an upper and lower compartment and (b) Lagrangian Sensor Particles (diameter d_p = 40 mm) equipped with a pressure sensor.

Results and Discussion

Although neutral buoyancy is established, the LSPs need to be characterized regarding their axial distribution throughout the reactor. For that reason, Figure 2 depicts the axial probability density distribution for the three used impeller frequencies of 15 (blue), 25 (green) and 35 rpm (red) and respective uncertainties (Tab. 1).

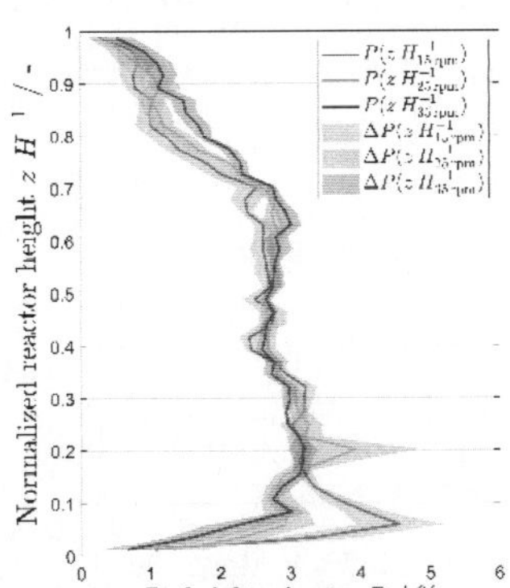

Figure 2. Vertical probability densities (solid lines) and respective uncertainties Δ (shaded areas).

Even though their density is close to deionized water, in the case of 15 rpm, the LSPs tend to accumulate with up to 4 % probability density at the reactor bottom up to a normalized reactor height of approx. 0.15. From that height on, they distribute more equally, similar to 25 and 35 rpm until a normalized height of 0.7. In this region, the three impeller stages are located (see Graphical Abstract). For all three impeller frequencies, above that height, the probability density decreases until lower than 1 % at the fluid surface at the very top.

Utilizing the derivative of the vertical position with respect to the time (Recording frequency f = 50 Hz, Δt = 0.02 s), positive (solid lines) and negative (dashed lines) axial velocities are calculated depending on whether the LSPs flow upwards near the reactor wall or downwards in the vicinity of the impellers (Fig. 3). All velocity profiles show a similar behaviour as for $|\bar{v}|_{z,neg}$ two peaks at reactor heights of approx. 0.2 and also at 0.7 and for $\bar{v}_{z,pos}$ one peak at the reactor height of approx. 0.2. The reason is the downward pumping direction of the axial-pumping impeller. Both described locations are either the lowest impeller edge of the lower stage or the highest impeller edge of the upper stage, respectively. At $z\,H^{-1}$ = 0.2 the LSPs have accumulated their highest downward flow speed (solid line) before they are redirected at the reactor bottom back to the top with a decreasing velocity alongside the baffles. Afterwards, at $z\,H^{-1}$ = 0.7 they are promptly incorporated back into the downward fluid flow.

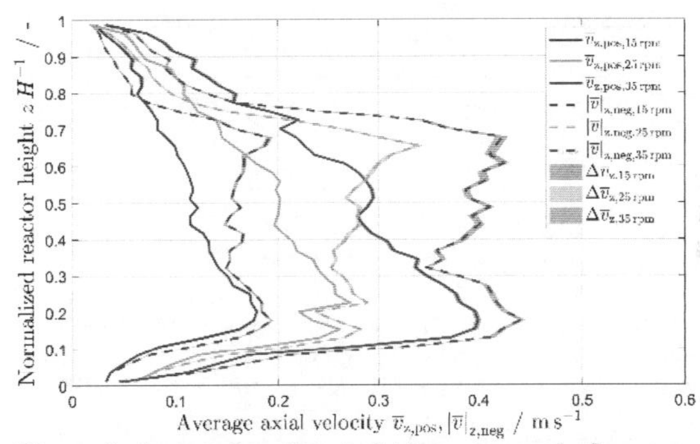

Figure 3. Average positive (solid lines, upwards flow near reactor wall) and negative (dashed lines, downwards flow in the vicinity of the impellers) axial velocity distributions with respective uncertainties Δ (shaded areas).

Eventually, the circulation time distribution over the reactor height (see Graphical Abstract) is depicted similarly as in Reinecke (2013) [9]. For all three impeller speeds, three major flow regions can be obtained: Firstly, a center regime, at which the times are the lowest of approx. 45 s (15 rpm), 30 s (25 rpm) and 20 s (35 rpm). Secondly, a bottom and a top regime are present with circulation times approx. 10-fold higher than the corresponding ones in the center regime. The respective total average axial circulation times are shown in Table 2. They are by a factor of 1.3 to 1.8 lower than the measured global mixing times with phenolphthalein.

A more detailed analysis of circulation time profiles at selected reactor heights (Fig. 4) reveals log-normal distributions. Due to the partly sparse data point availability (Fig. 2), large values strongly

https://doi.org/10.52202/074122-0017

influence the average calculation, albeit not being

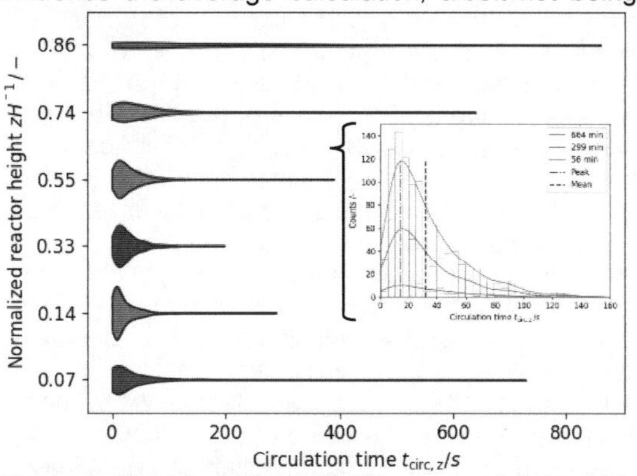

Figure 4. Circulation time distribution at six reactor heights for 25 rpm. Inside graph: Circulation time distribution for increasing recording times at $z\,H^{-1}$ = 0.55 (25 rpm). The peak of the green distribution (red line) and mean value (black line) are indicated. Data larger than 160 s are not shown.

necessarily representative.

As an alternative approach, utilizing the value of the corresponding peak at each distribution, the influence of large values can be diminished. The results are shown in Table 2 as peak averaged circulation times. For that reason, the inside graph in Figure 4 depicts the histogram of the circulation time distribution at $z\,H^{-1}$ = 0.55 with increasing recording time. The smoothed curves are the probability density function estimated by a kernel density estimation. Considering 10 h overall recording time as significant for this reactor height, $z\,H^{-1}$ = 0.86 would require over 22 h to obtaining a similar distinctive peak. Peak averaged circulation times result in a factor of 3.2 to 4.1 lower than the measured global mixing times (Tab. 2).

Conclusions

Lagrangian Sensor Particles, which are equipped with a pressure sensor, are able to detect compartments in industrial-scale bioreactors, without necessarily having optical access. Recorded data deliver beside an axial probability density also axial velocity profiles and most importantly axial circulation time distributions at three used impeller frequencies. The average value of the latter depends strongly on data point availability and the evaluation method. Gained results imply a dependency between the total average circulation time and the global mixing time.

Table 2. Number of data points, total recording times and comparison of total average circulation times, peak averaged circulation times and measured global mixing times with phenolphthalein.

Impeller frequency / rpm	Total number data points / -	Total recording time / min	Total average circulation time $\bar{t}_{circ,z,tot}$ / s	Peak averaged circulation time $\bar{t}_{circ,z,tot}$ / s	Global mixing time $\bar{t}_{mix,global,90}$ / s
15	$1.2\cdot10^6$	426	61 ± 2	24	81
25	$1.2\cdot10^6$	427	38 ± 1	16	65
35	$0.7\cdot10^6$	247	27 ± 1	14	45

Acknowledgments

The authors gratefully acknowledge the German Research Foundation (DFG, project number 427899833) for funding within the priority program SPP 2170 *InterZell*. Parts of the work of the HZDR are funded by the BMEL in the projects SensOMix (FKZ 2219NR390) and Neobio (FKZ 22032618) as well as from the Impulse and Networking Fund of the Hemholtz Association for the Clean Water Technology Lab (HIL-A02).

References

[1] C. Haringa, A.T. Deshmukh, R.F. Mudde, H.J. Noorman, Euler-Lagrange analysis towards representative down-scaling of a 22 m^3 aerobic S. cerevisiae fermentation, Chemical Engineering Science. 170 (2017) 653–669. https://doi.org/10.1016/j.ces.2017.01.014.

[2] A.R. Lara, E. Galindo, O.T. Ramírez, L.A. Palomares, Living With Heterogeneities in Bioreactors: Understanding the Effects of Environmental Gradients on Cells, MB. 34 (2006) 355–382. https://doi.org/10.1385/MB:34:3:355.

[3] S. Mukherjee, B.L. Bassler, Bacterial quorum sensing in complex and dynamically changing environments, Nat Rev Microbiol. 17 (2019) 371–382. https://doi.org/10.1038/s41579-019-0186-5.

[4] A. Rosseburg, J. Fitschen, J. Wutz, T. Wucherpfennig, M. Schlüter, Hydrodynamic inhomogeneities in large scale stirred tanks – Influence on mixing time, Chemical Engineering Science. 188 (2018) 208–220. https://doi.org/10.1016/j.ces.2018.05.008.

[5] A. Lapin, D. Müller, M. Reuss, Dynamic Behavior of Microbial Populations in Stirred Bioreactors Simulated with Euler−Lagrange Methods: Traveling along the Lifelines of Single Cells, Ind. Eng. Chem. Res. 43 (2004) 4647–4656. https://doi.org/10.1021/ie030786k.

[6] S. Hofmann, C. Weiland, J. Fitschen, A. von Kameke, M. Hoffmann, M. Schlüter, Lagrangian sensors in a stirred tank reactor: Comparing trajectories from 4D-Particle Tracking Velocimetry and Lattice-Boltzmann simulations, Chemical Engineering Journal. 449 (2022) 22. https://doi.org/10.1016/j.cej.2022.137549.

[7] L. Buntkiel, S. Reinecke, Richtungsaufgelöste Messung von Beschleunigungen mit Sensorpartikeln in industriellen Prozessbehältern, in: 6. Sensoren für Verfahrenstechnik, 16. Dresdner Sensor-Symposium 2022, Dresden, Germany, 2022: pp. 51–56. https://doi.org/10.5162/16dss2022/6.3.

[8] S.F. Reinecke, L. Buntkiel, R. Kipping, U. Hampel, Process characterization in industrial vessels by flow-following sensor particles, Meas. Sci. Technol. 33 (2022) 095106. https://doi.org/10.1088/1361-6501/ac75af.

[9] S.F. Reinecke, Instrumentierte Strömungsfolger zur Prozessdiagnose in gerührten Fermentern, PhD Thesis, Dresden University of Technology, 2013.

A Closer Look at Nonuniform Dispersion Loop Reactors

B2.3

Lars Uby[1]. *(1) Xylem, Gesällvägen 33, SE-174 87 Sundbyberg, Sweden, lars.uby@xylem.com.*

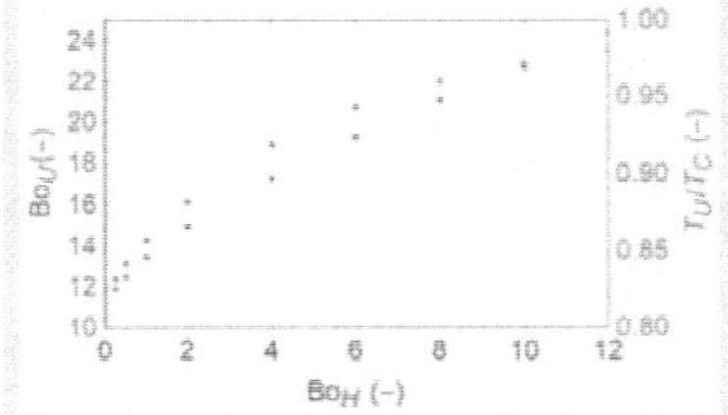

Bodenstein number and circulation time corrections for nonuniformity.

Dispersion nonuniformity caused by gassing in loop reactors is measured for cases where a single tracer pulse self-interacts (bites its tail). Two discrete levels of dispersion and their distribution, related to the size of the gassing section, are assumed. The impact of these on the correction to the uniform-dispersion interpretation of tracer data in terms of Bodenstein number and circulation time is demonstrated by enhanced accuracy numerical simulation and a novel approach to parameter fitting. Full-scale test and CFD data support the use of the nonuniform axial dispersion model. Applications in mass transfer measurement and reaction selectivity enhancement are anticipated.

Introduction

An accurate account of mass transfer and reactions in loop reactors requires characterization of mixing, in terms of the loop circulation time and dispersion coefficient. Nonuniform dispersion has previously been addressed for systems where the variance of a recurring tracer pulse could be estimated and variance additivity could be used to deduce local dispersion rates, e.g. [1].

However, loop reactors come in a variety of executions, such as slender tubes, airlift reactors and oxidation ditches, all differing in their mixing characteristics.

A general approach to simple but detailed mixing characterisation of loop reactors is given here. It extends the work in [2] (based on uniform dispersion) motivated by deficiencies in oxygen transfer rate measurement practice, chiefly the ignoring of mixing time scales in data analysis [3]. The present work is part of a project targeting reduced oversizing of loop reactor equipment in wastewater treatment by increasing the accuracy of mass transfer rate estimates. Improved reaction selectivity could possibly be achieved in advanced loop reactors using these findings.

Material and Methods

A full-scale wastewater loop reactor, Figure 1, was tested for circulation time and dispersion coefficients using online fluorometer (YSI) readings of the concentration of advected-dispersed Rhodamine WT slugs at air flow rates from 0 to 0.83 (reported here) Normal m^3/s.

Analytical fitting of tracer data to a 1D periodic advection-dispersion fundamental pulse provided dispersion coefficients based on a uniform approach [4]. The aerated cases were numerically fitted based on the unaerated dispersion coefficient for the unaerated zone and the coverage rate of the air diffuser/sparger grid to yield corrected circulation times and dispersion coefficients in the aerated zone. The fitting function was obtained by repeatedly solving the nonuniform 1D advection-dispersion equation with a moderately advanced pulse initial condition using the QUICK scheme for the advection term and 2nd order central differencing for the dispersion term. The first tracer peak was excluded from fitting since the test tracer distribution with non-ideal initial pulse conditions requires some time to reach universality.

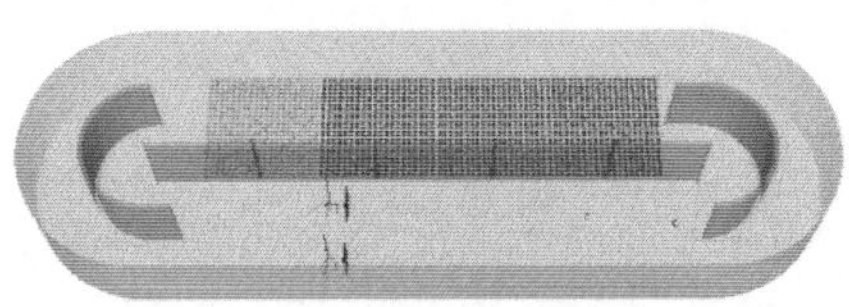

Figure 1. Reactor loop of 98.5 m length and 5000 m^3. Greyed-out air diffusers were not active. Probe locations are shown as spots.

Inflation of the tracer test data c by subtraction of the terminal concentration (massless data $\tilde{c}$) and multiplication by primarily a time exponential resolved even late peaks and permitted complete use of at least four recurring tracer peaks, which substantially improved the fit. The massless initial condition was used for the numerical solution and the same inflation applied after solving.

In reverse, generation of nonuniform dispersion tracer data by numerical simulation of an array of system characteristics and of six combinations of injection and probe locations and fitting to the analytical uniform advection-dispersion function produced a mapping used for further analysis.

The resulting advection-(nonuniform-)dispersion operator can be used for 1D modelling of mass transfer and reactions.

For easy numerical handling and comparison between systems, a dimensionless formulation is used. The loop coordinate is normalized with loop length L (thus $0 \le x \le 1$) and on obtaining the circulation time T_C it is used for time normalisation for any further study, making velocity equal to 1 (L/T_C). The high and low dispersion coefficients $D_{H,L}$ are represented by the inverse of the respective Bodenstein numbers, $Bo_{H,L} = L^2/(T_C D_{H,L})$.

https://doi.org/10.52202/074122-0018

Local Peclet numbers $Pe_{L,H} = Bo_{L,H}\ (D_H/L)$ can be obtained knowing that the hydraulic diameter is $D_H = 9.6$ m (depth 6.0 m) and the loop length is $L = 98.5\ m$.

Fully 3D Computational Fluid Dynamics (CFD) simulations were used to generate similar tracer data for reactor I, and to further corroborate the use of the nonuniform axial dispersion model. CFD++ 19.5 with a realizable k-ε model and (two-way coupled) Euler Dispersed Phase air model was used. Solid boundaries of the closed reactor were no-slip with y+ < 100, and the water surface was a flat slip surface with air escape.

Results and Discussion

Figure 2 shows a reasonable agreement between inflated test and CFD tracer data on one hand and fitted models on the other.

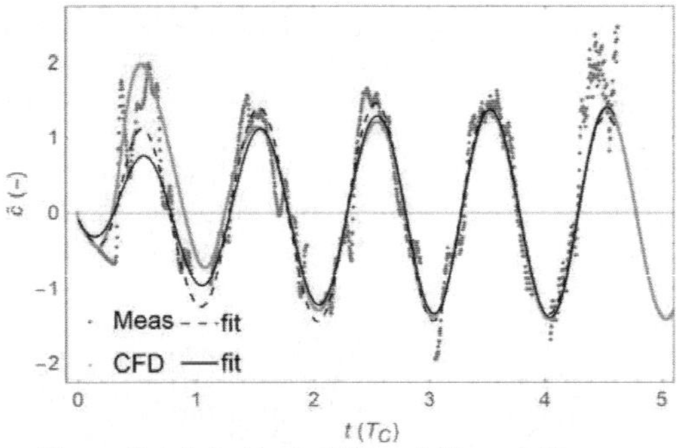

Figure 2. Inflated test, CFD and 1D model tracer response curves for the uniform case.

The numerically obtained array demonstrated a very accurate mapping between "uniform" and "non-uniform" loop times (T_U, T_C) and dispersion coefficients, based on the high dispersion zone coverage but regardless of tracer injection point and probe positions. Such mathematical "homogenisation" is previously known for long wavelength (compared to the zone lengths) signals, but these results for tracer pulses are similarly intuitive, just like a black-and-white raster can appear as uniformly gray when looked at from a distance.

This data demonstrates that applying addition of variances (i.e. dispersion coefficients) to deduce the high-level value from the "unaerated" and "uniform aerated" values of the dispersion coefficients can cause errors of tens of %s. The "uniform" circulation time value of the aerated case can fall more than 40% short of the true value. Large ratios of the two zonal dispersion coefficients increase these errors, but primarily the error in dispersion coefficient arises because of self-interaction of the test pulse across one or several periods, e.g. when Bo << 200.

The loop time estimate deficit is maximised for medium coverage rates of the high dispersion zone, when none of the two zones completely dominates the uniform appearance. Typical corrections are shown in the Graphical Illustration, valid for the reactor under study, with $Bo_L = 39$.

A more complex distribution of high-level dispersion zones shows little difference to the above unless coarse-graining is reduced, e.g. by introducing numerous alternating high/low-dispersion zones of a relatively small size. For repeat patterns, the analysis can be applied to a single repeat element.

The reduction of the full 3D momentum and tracer equations to the 1D advection-dispersion equation requires the dispersion coefficient to capture not only molecular and turbulent diffusion (as given by e.g. RANS or LES methods of CFD), but also the reduction of a complex 3D flow field into a constant unidirectional velocity. The latter turns out to be crucial in nontrivial geometries and may override even the turbulent dispersion.

The localisation of high-level dispersion is indicated in a section of a CFD velocity solution in Figure 3. Medium-high turbulent dispersion is seen down-stream of the impellers, but the buoyancy driven upwelling above the diffusers, in the diffusion peak area, adds a different level of dispersion by 3-dimensionality of the flow.

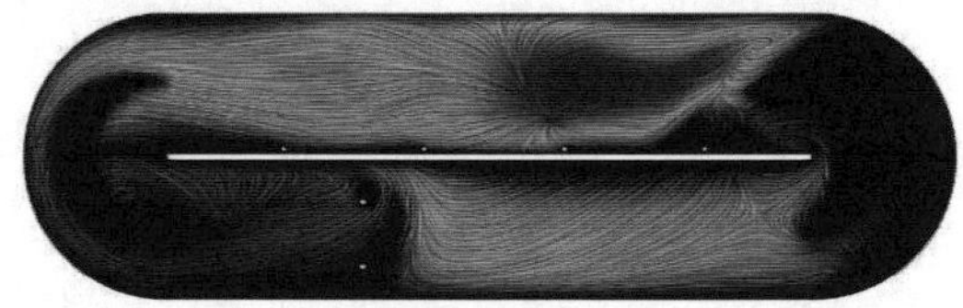

Figure 3. Surface velocity field colored by turbulent viscosity, red corresponding to the maximum 160 Pa s.

Conclusions

A method for quantifying nonuniform loop dispersion was developed to cover cases with strong dispersion causing tracer pulse self-interaction. The method of inflation assisted in resolving data for strong dispersion systems, and numerical solution of the nonuniform advection-dispersion differential equation was sufficiently fast to permit parameter fitting. Corrected dispersion coefficients and circulation times are thus obtained and will be helpful in improving the accuracy of mass transfer testing.

Acknowledgments

Tracer test preparation and data capture were carried out together with Mr. M. Fahlgren. CFD simulations were carried out by Mr. A. Monokrousos.

References

[1] H.-J. Warnecke, *Acta Biotechnol.*, 9(2) (1989) 111.
[2] K. Nakamachi et Al., *J. Water Environ. Technol. 10(3) (2012), 229.*
[3] L Uby, *Water Research 157 (2019), 415.*
[4] R.M. Voncken, D. B. Holmes, H.W. den Hartog, *Chem. Eng. Sci. 19 (1964), 209.*

Refractive index matching for PIV measurements of liquid-solid flow in a stirred tank bioreactor used for mesenchymal stem cell culture

B2.4

***M. Madani**[1], A. Delafosse[1], D. Toye[1]. (1) Chemical Engineering Department, University of Liege, Liege, Belgium, m.madani@uliege.be, angelique.delafosse@uliege.be, dominique.toye@uliege.be.*

(a). Raw PIV image, Cytodex-1/water

(b). Raw PIV image, PMMA/NH_4SCN

Raw PIV images for α_S=0.5 v%

A liquid-solid two-phase flow is characterized via Particle Image Velocimetry PIV inside a stirred tank type bioreactor STR used for stem cells expansion. The reactor of a working volume of V_L=0.7 L is equipped with HTPGD impeller type. The liquid-solid system investigated inside the reactor is a refractive index matched one (RIM) composed of NH_4SCN as a solution and PMMA microparticles for the solid phase. Using the PIV-RIM technique, mean and turbulent velocities of the liquid phase were characterized in presence of solid for solid volume fractions α_S up to 2-3 v% range, at N=150 rpm which corresponds to a homogeneous suspension state.
This influence of varying solid fractions on the liquid phase has been studied and compared to single flow (without microparticles). The first assessment showed no significant effect of solids on the liquid part in terms of mean flow and turbulence.

Introduction

Mesenchymal stem cells (MSCs) have been the focus in regenerative medicine and cell therapy due to their differentiation abilities and immunosuppressive properties. To produce sufficient quantities of MSCs for clinical applications, an expansion step is necessary, and many studies are underway to develop scalable systems for culturing these cells.

One such system involves using small-size spherical microbeads, e.g., commercial Cytodex-1, called microcarriers, suspended in stirred tank reactors (STRs). Since cells are anchorage dependent, the microcarriers provide a surface for the MSCs to adhere to, which is necessary for cell growth. However, it is important to maintain optimal suspension conditions (just suspended state) in the reactor to limit the hydromechanical stresses that the cells are exposed to (collisions and shear stress), while ensuring a homogeneous medium and avoiding aggregations of microcarriers.

To be able to further understand the stresses encountered in the reactor, developing and validating a detailed hydrodynamic two-phase model is the core in such study. Particle image velocimetry PIV – a commonly used technique for single phase flow – is set to characterize the liquid-solid flow for mean and turbulent quantities. In a preliminary test, Cytodex-1 ($d_s \approx 160$ µm; $\rho_s \approx 1020$ kg/m3) which is often used in cell culture, were suspended in a small STR (V = 1 L) at a rotation speed N = 200 rpm. The solid volume fraction used was up to maximum α_s=0.5 v% which is low compared to typical microcarrier volume fraction used in MSCs cell culture (2-3 v%). The PIV images taken were blurred with no vision of the seeding/tracing particles (graphical illustration – (a)), as microcarriers tend to scatter or diffuse the laser light, even with low solid fraction, resulting in problems to extract the necessary data [1].

To account for the visibility issues in PIV and to minimize light scattering of the laser sheet, it was necessary to shift to a "model system" of liquid and solid, with adequate optical properties: refractive index RI matching technique. The effect of this shift could be visualized in the raw images of PIV in the graphical illustration – (b), as the tracing particles of the flow could be spotted and tracked.

Material and Methods

The challenge in the new system lies in respecting the original liquid-solid one properties in terms of density difference (≈ 20 kg/m^3) and viscosity range, so that the hydromechanical stresses encountered in the system are reproduced and can be measured within the desired levels of solids volume fraction α_S. Based on literature [2], an alternative liquid-solid system candidate is studied. Poly(methyl methacrylate) PMMA (Cospheric), optically clear microparticles, were considered a good substitute [3], alongside Ammonium Thiocyanate NH_4SCN salt solution.

The refractive index of NH_4SCN solution (RI_L) was measured with an Abbe refractometer, whereas the solution density (ρ_L) and viscosity (μ_L) were assessed via Lovis 2000 M/ME viscometer. As for PMMA particles, a volumetric pycnometer and an optical microscope were used to determine the solid density (ρ_S) and diameter (d_S) respectively.

The stirred tank bioreactor STR used in this work a hemispherical bottom shape tank made of transparent glass and equipped with 4 probes as shown in figure 1. The working volume of the reactor is V_L=0.7 L in which the fluid height corresponds to H=0.082 m and the tank diameter T=0.12 m. An HTPGD (Pierre Guerin ®) axial 3-blade type impeller is used with a down pumping mode, with an off-bottom clearance C/T =0.5 and

https://doi.org/10.52202/074122-0019

an impeller diameter D/T=0.5.

Regarding the PIV setup, it consists of a laser source (YAG 2x65mJ/532nm) illuminating the stirred tank reactor that's placed inside an aquarium filled with paraffin oil, whereas a CMOS camera (FlowSense EO 4M, 16-25 Hz) captures the images of the laser sheet passing through the reactor. The images are then processed via Dantec Dynamics and Matlab to account for mean and fluctuating velocities.

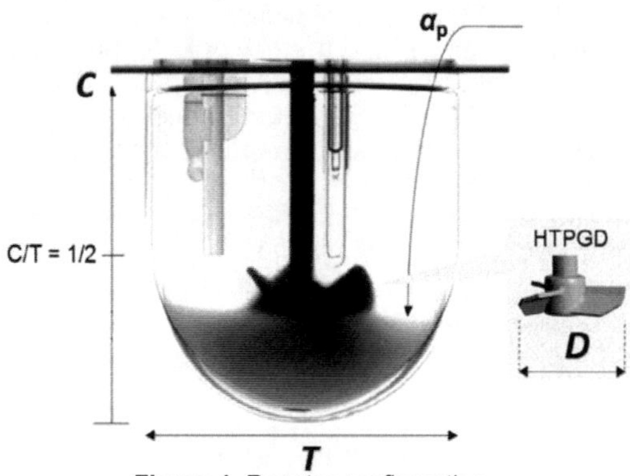

Figure 1. Reactor configuration

Results and Discussion

PMMA particles (RI_S≈1.49) were matched with NH_4SCN solution of approximately 55.1 wt% (RI_L≈1.47) around a temperature T≈22 °C. The properties of the PMMA/NH_4SCN system are close to those of the Cytodex-1/Water system (Table 1), in terms of density difference, viscosity and particle diameter, which also means that the same ranges of Reynolds number (Re = 5000-6000) and Stokes number are retained. The optical properties of the matching refractive index system were first assessed by microscopy as shown in figure 1. While PMMA particles in water appear nearly opaque, they are being matched with the light background when in NH_4SCN solution.

Table 1. Properties of the two liquid-solid systems.

Properties	PMMA/NH₄SCN	Cytodex-1/Water
$\rho_S - \rho_L$ (Kg/m³)	≈60	≈20
μ_L (mPa.s)	≈1.5	≈1
d_S (µm)	≈168	≈160

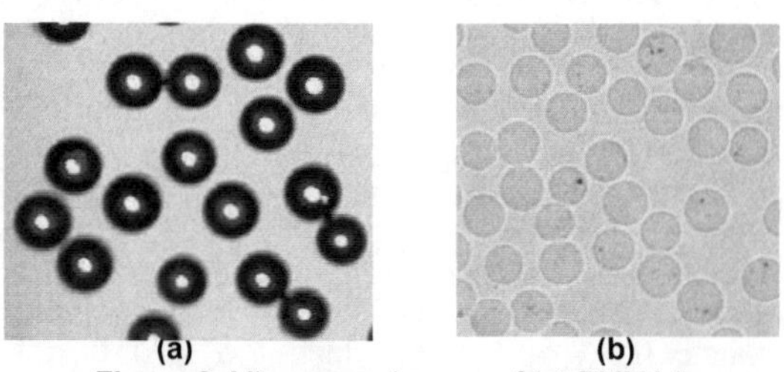

Figure 2. Microscope images of (a) PMMA in presence of water and (b) PMMA/NH_4SCN RI matched system.

The first test of PMMA/NH_4SCN solid-liquid system via PIV technique, operated at N=150 rpm that insures homogeneous mixing with no entrainment of air at the top surface of the liquid, showed good promises. The volume fraction reached was α_S≈ 3%v which is 5 times greater than that resolved from cytodex-1/water one. The good optical properties (RI matched) allowed to process the raw data, for mean velocities – normalized by (ND) – as can be shown in the contour plot in figure 3. For further analysis, vertical profile cut just in front of the blade tip at x≈31.5 mm, could be done to follow the variation of the mean velocity along height Z. This plot presented in figure 4 shows almost no variation with the increasing of solid volume fractions. Besides, the investigation on the fluctuation velocity term also revealed the same for turbulent kinetic energy TKE, as the solids doesn't affect the turbulence phenomenon inside.

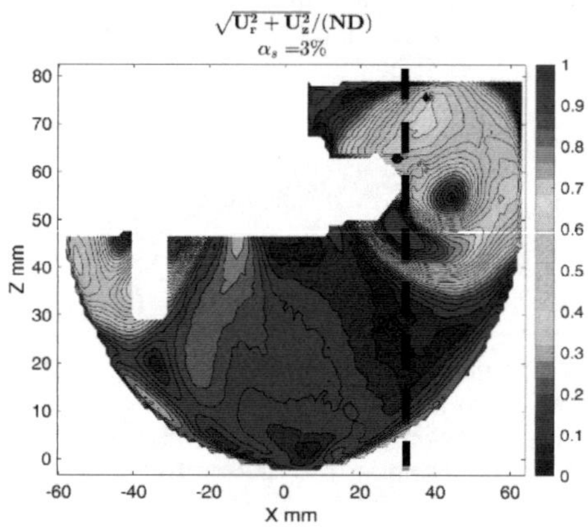

Figure 3. Normalized mean flow at α_S=3 v%

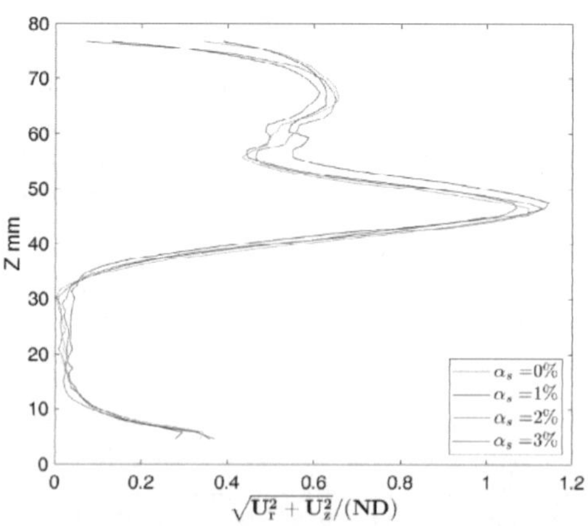

Figure 4. Normalized mean flow as function of height Z of the reactor at X≈31.5 mm

Conclusion

Preliminary assessment of the PMMA-NH_4SCN system offers potentials for the PIV experiments. Measurements were recorded at the desired working conditions (i.e., solid volume fractions close to 2-3%, as used in MSC cultures), which allowed the evaluation of mean and turbulent velocities of the liquid phase in presence of solids.

Next step is to validate the results by

experimental repetitions on one hand and estimate the mean and fluctuating properties of the solid phase on the other hand.

Acknowledgments
The authors are extremely grateful to the FRS-FNRS for their financial support of the research project PDR-OptiReac4MSC.

References

[1] M. Madani, A. Delafosse, D. Toye, *Refractive index matching for PIV measurements of liquid-solid flow in a stirred tank bioreactor used for mesenchymal stem cell, Toulouse,* France, 2022, 18
[2] D. Borrero-Echeverry, B. C. Morrison, *Aqueous ammonium thiocyanate solutions as refractive index-matching fluids with low density and viscosity*, 57 (7), 2016, 1-5.
[3] S. F. Wright, I. Zadrazil, C. N. Markides, *A review of solid–fluid selection options for optical-based measurements in single-phase liquid, two-phase liquid–liquid and multiphase solid–liquid flows*, 58 (9), 2017, 1-39.

 https://doi.org/10.52202/074122-0019

Hydrodynamic Characteristics and Comparison of Scale-up Approaches in Anaerobic Stirred Digesters

B2.5

F. Maluta[1], F. Alberini[1], A. Paglianti[1], ***G. Montante[1(*)]****.(1) Dipartimento di Chimica Industriale "Toso Montanari", Alma Mater Studiorum – Università di Bologna, viale Risorgimento 4, 40136 Bologna, Italy. (*) giuseppina.montante@unibo.it*

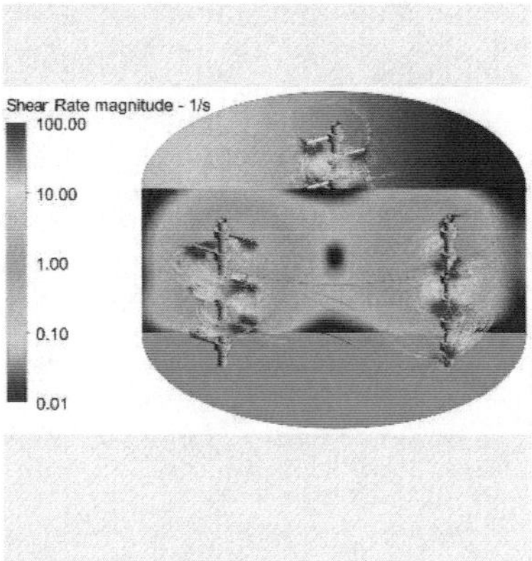

A stirred tank of geometry coincident to an industrial anaerobic digester adopted at large scale for the production of biogas is investigated by Computational Fluid Dynamics considering the physical properties of a sample of the fluid taken from the biogas production plant. The computational model is validated by comparing the predicted power consumption of the three top-entering agitators with that measured by a strain gauge technique in a laboratory stirred tank of very similar geometry. The comparison of typical variables affecting the biogas production, such as dead zones and shear rates, predicted in two vessel sizes at different agitation speeds allows a detailed evaluation of change of scale rules. The fluid dynamics analysis can contribute to improve the equipment design, to optimize the energy requirement and to avoid failure of the biogas production due to poor or unproper fluid mixing.

Introduction

Biogas production has become widespread in several countries as a mean to produce energy from waste materials with relatively low added value. Some of the advantages of the biogas production through anaerobic digestion are related to its flexibility in converting very diverse carbon sources, leading to a straightforward integration with different food and agricultural productions [1].

On the other hand, several aspects related to anaerobic digestions still need to be optimized in order to improve the efficiency of this energy production route [2]. For instance, it is known that mixing is one of the major cost items in biogas plants and one of the factors which affects the process efficiency. To characterize the effect of mixing on the biogas production several indicators and criteria have been proposed, such as the identification of dead zones [3], threshold values on the shear rate [2] and the analysis of the velocity gradients [4], to name a few. The fluid dynamics in the digester is tightly affected from the fluid non-Newtonian rheology. The evolution of the fluid rheology and the variation of the "degree of mixing" needed in the different stage of the digestion [5] are aspects which are not usually taken into account in the design of a biogas plant, and they are difficult to derive a priori. In this regard, one of the challenges for future biogas plants is indeed related to becoming more and more flexible and this goal can be achieved by improving the modelling of the digestion process. A further challenge is related to the scale-up criteria needed to extend the lab-scale results to the production scale, which is often performed by means of global parameters.

In this work, starting from the operating conditions of a production plant, the effects of the change of scale rules are addressed by Computational Fluid Dynamics (CFD) simulations of two identical stirred vessels of different size.

Material and Methods

The simulated stirred tanks, which geometry is depicted in Figure 1, are two identical unbaffled vessels of volume equal to 0.04 m^3 and 0.32 m^3, obtained from the scale down of an industrial digester for biogas production, with an operating volume of 1471 m^3.

The smaller vessel, of diameter, T, equal to 0.49m, is very similar to the experimental stirred tank investigated by Alberini et al. [6]. The bigger vessel is obtained doubling all the dimensions of the first. The two vessels are named T49 and T98 in the following. Agitation is provided with three vertical shafts, each of them supporting six blades inclined at 45° and - 45° with respect to the horizontal plane. As for the agitator size, D, the blade swept diameter is considered. Full geometrical details can be found elsewhere [6].

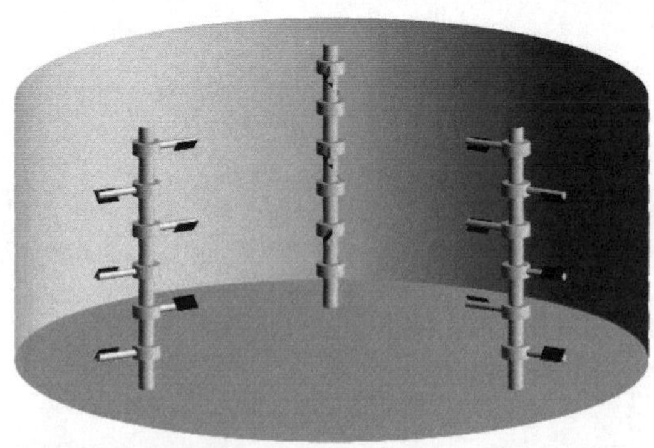

Figure 1. Geometry of the stirred tanks. The blades are coloured in black.

The simulations are based on the numerical solution of the unsteady, incompressible, isothermal Navier-Stokes equations, which read as:

$$\nabla \cdot (\rho \boldsymbol{u}) = 0 \quad (1)$$

$$\frac{\partial \boldsymbol{u}}{\partial t} + \nabla \cdot (\rho \boldsymbol{u}\boldsymbol{u}) = \nabla \cdot \left(\mu(\nabla \boldsymbol{u} + \nabla \boldsymbol{u}^T)\right) - \nabla P \quad (2)$$

Where the symbols have the usual meaning. As for

https://doi.org/10.52202/074122-0020

the fluid physical properties, the density, ρ, of 980 kg/m^3 and the apparent viscosity, μ, are obtained from the direct measurement on a sample taken from the production digester. In particular, the sample apparent viscosity follows the correlation of Ostwald-de-Waele:

$$\mu = k_s \times \dot{\gamma}^{n-1} \quad (3)$$

With $\dot{\gamma}$ being the shear rate magnitude, k_s the consistency index and n the flow behaviour index. The values of k_s and n are equal to 63.56 Pa s^n and 0.282, respectively, as measured at the digestion temperature of 40°C.

The equations are numerically solved running the finite volume CFD code ANSYS FLUENT 2020 R2. The computational grids consist of about 2.4 and 4.4 million polyhedral cells for the T49 and the T98, respectively. The grid is refined close to the blades and the final size is based on a preliminary grid dependency analysis on both the vessel sizes. For the coupling between the rotating and the steady part of the domain, the Sliding Mesh method is adopted, since, in case of unbaffled vessels stirred with eccentric agitators, steady approaches lead to unphysical results, due to the strong dependency of the solution on the location of the moving blades with respect to the vessel wall [7]. The pseudo steady state solution is achieved after a few seconds of physical blade rotation, with time steps corresponding to a blade rotation of 5° and 25 iterations per time step. The results discussed in the following are either angle resolved or ensemble averaged on 72 time steps corresponding to one complete blade revolution.

Results and Discussion

To estimate the simulation accuracy, the T49 filled with Newtonian and pseudo-plastic fluids stirred at different agitation speeds, N, was simulated and the power consumption, P, was obtained from the torque on the blades, the torque on the static walls and the integral of the mean flow energy dissipation rate, obtaining very similar values. The resulting power numbers, Np, are compared with the experimental values [6] in Figure 2.

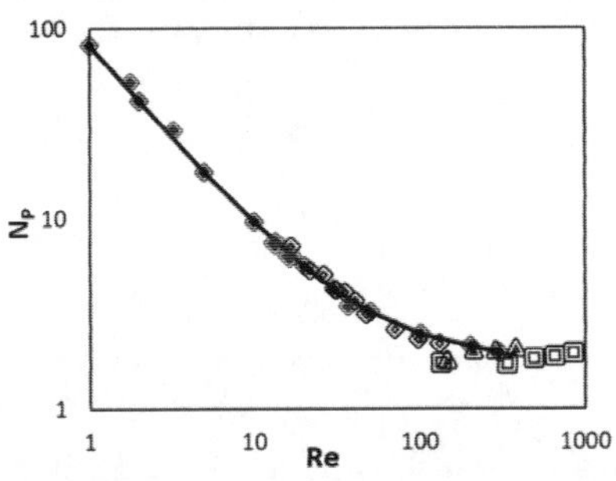

Figure 2. Experimental (black symbols) and numerical (red diamonds and black line) power curve for a single shaft.

The modified rotational Reynolds number, Re, in Figure 2 is calculated adopting the apparent viscosity based on the Metzner and Otto method, with the constant K=11, as:

$$Re = \frac{\varrho N^{2-n} D^2}{k_s K^{n-1}} \quad (4)$$

As can be observed, the predicted power numbers very closely follow the experimental power curve in the laminar regime with deviations lower than 10%.

Having obtained reliable predictions of the power consumption, the computational model is adopted for comparing the fluid dynamics characteristics of the T49 and the T98 vessels, with the fluid in the digesters stirred at blade speeds corresponding to different change of scale rules. In particular, the following variables have been maintained constant with respect to that of the production digester, which power consumption is in the range 6.6-7.2 KW and the shaft rotational speed is in the range 10.5-12.5 rpm: the blade tip speed, v_{tip} , the power consumption per unit volume, P/V, the average shear rate, $\dot{\gamma}_{av}$, as obtained from a literature correlation [8] and the apparent Reynolds number. The evaluation of the different change of scale rules is based on the comparison of variables affecting the biogas production and the power consumption.

The maps of the shear rate magnitude on one horizontal section obtained with the four criteria close to one of the impellers at fixed position are shown in Figures 3 and 4 for the T49 and the T98, respectively. As can be observed, the same shear rate is obtained only at equal P/V, that is at equal N in the laminar regime, thus confirming the Metzen-Otto relationship. The correlation of Wichterle et al. for maintaining constant the average shear rate [8] leads to similar, but not completely overlapped shear rate distributions. At equal Re and v_{tip}, lower shear rates are obtained in the higher scale, since N is lower. For both the scales examined here, the shear rate is much lower than the threshold of $5s^{-1}$, which was identified as the limit at which the abrasion of anaerobic sludge granules determines a decrease of methane production [2]. Worsening or improvement of the production due to the passing of the shear rate threshold at increasing production size can be expected depending on the adopted change of scale rule. The distribution of the shear rate magnitude and of the shear stress may help in determining whether critical values identified in laboratory biogas production are exceeded at large scales, in a significant volume percentage or not. Apparently, keeping N constant in the laminar regime ensure the same specific power consumption and the same shear rate. Being the local velocity magnitude higher at the higher scales as confirmed by the results in Figure 5, the P/V criterion should reduce the feedstock sedimentation or floatation at large scale.

The presence of poor mixed zones in the system and their weight with respect to the total volume is also considered adopting the criterion proposed by Bridgeman [3].

https://doi.org/10.52202/074122-0020

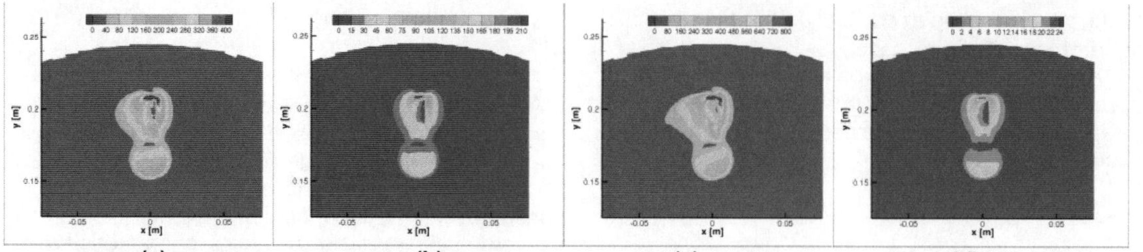

(a) (b) (c) (d)

Figure 3. Map of the shear rate close to one of the blades as obtained from the simulations keeping constant (a) v_{tip} (b) $\dot{\gamma}_{av}$; (c) Re; (d) P/V. T49, horizontal vessel section at the lower blades elevation.

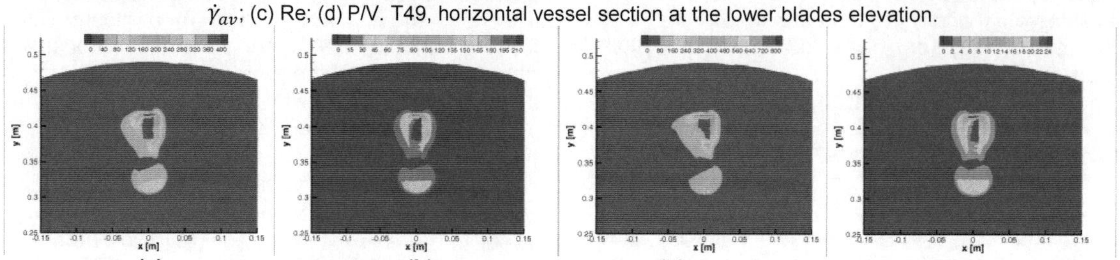

(a) (b) (c) (d)

Figure 4. Map of the shear rate close to one of the blades as obtained from the simulations keeping constant (a) v_{tip} (b) $\dot{\gamma}_{av}$; (c) Re; (d) P/V. T98, horizontal vessel section at the lower blades elevation.

With the Bridgeman criterion, in the so-called dead zones the velocity is lower than 5% of the reactor maximum velocity, which in the stirred reactor coincides with v_{tip}. The results reported by means of example in Figure 6 show that with the constant v_{tip} criterion almost 98% of the volume falls in the selected dead zone definition.

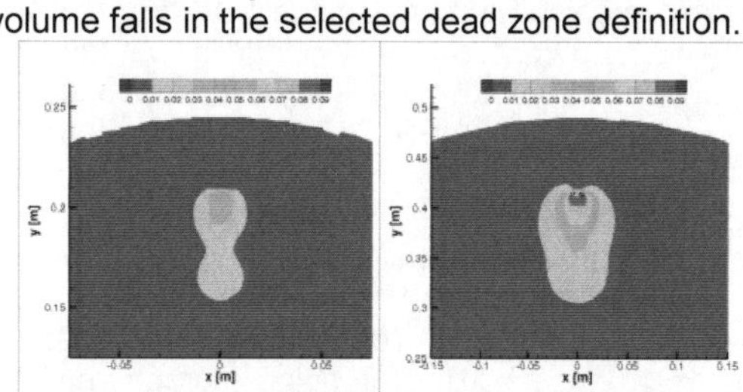

(a) (b)

Figure 5. Map of the in plane velocity close to one of the blades, at constant P/V. (a) T49 (b)T98.

Conclusive results into the most effective change of scale rules will be based on additional analysis, including the requirements for achieving equal feedstock distribution.

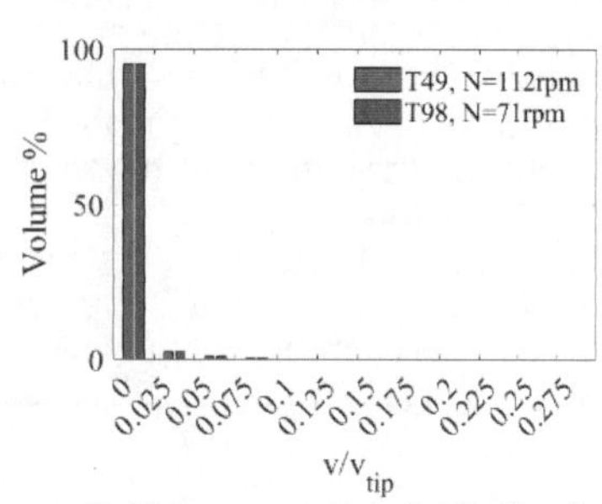

Figure 6. Volume percent distribution based on the local dimensionless velocity, at constant v_{tip}.

Conclusions

The mixing of an anaerobic slurry in a typical anaerobic digester geometry is investigated by CFD simulations. Due to the relatively high apparent viscosity of the pseudo-plastic fluid, the digester operates in the laminar regime. Different change of scale rules are assessed and the simultaneous analysis of the mixing state of the system by means of widely adopted local and global variables, helps in characterizing the digester hydrodynamics. Guidelines for optimization of the industrial operation considering the energy demand and the hydrodynamics characteristics affecting the biogas production are obtained.

Acknowledgments

This study was carried out within the Agritech National Research Center and received funding from the European Union Next-GenerationEU.

References

[1] S. Theuerl, C. Herrmann, M. Heiermann, P. Grundmann, N. Landwehr, U. Kreidenweis, A. Prochnow, *Energies.* 12 (2019) 396.

[2] A. Lebranchu, S. Delaunay, P. Marchal, F. Blanchard, S. Pacaud, M. Fick, E. Olmos, *Bioresour. Technol.* 245 (2017) 1139–1147.

[3] J. Bridgeman, *Adv. Eng. Softw.* 44 (2012) 54–62.

[4] R. Sindall, J. Bridgeman, C. Carliell-Marquet, *Water Sci. Technol.* 67 (2013) 2800–6.

[5] J. Lindmark, E. Thorin, R. Bel Fdhila, E. Dahlquist, *Renew. Sustain. Energy Rev.* 40 (2014) 1030–1047.

[6] F. Alberini, F. Maluta, A. Paglianti, G. Montante, *ACS Eng. Au* 3, 2 (2023) 102–113.

[7] G. Montante, A. Bakker, A. Paglianti, F. Magelli, *Chem. Eng.Sci.*, 61 (2006) 2807-2814.

[8] Grenville, R.K., Nienow. A.W. 2004. Blending of Miscible Liquids, in "Handbook of Industrial Mixing: Science and Practice" p.518.

A rapid method for flow patterns estimation of stirred vessels using 2-D network of zones (NoZ) model

MT1.1

Y. Bai[1]*, C.P. Fonte*[1]*, A. Kowalski*[2]*, T.L. Rodgers*[1]*. (1) University of Manchester, Oxford Rd, Manchester, United Kingdom, yuqi,bai@manchester.ac.uk. (2) Unilever R&D Port Sunlight Laboratory, Quarry Road East, Bebington, United Kingdom*

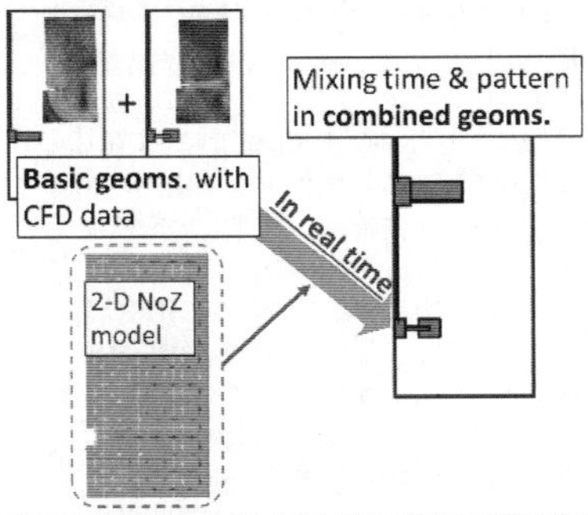

A general methodology for quick flow pattern estimations was created for designing turbulent stirred tank mixers, basing on 2-D network-of-zones model. Given flow field data (e.g. CFD velocity fields) in several more commonly-studied basic stirred tank configuration with various type of impellers, the radial-axial flow patterns were estimated and the mixing time calculated in almost real time (approx. 5 mins) for any arbitrary combined stirred tank geometry with varying impeller size and clearances. Mixing times predicted via this method were consistent with CFD passive scalar transport predictions with various impeller type, counts, and clearances within a deviation range of ±35%.

Introduction

International corporations in the field of daily care and food manufacture tend to have a wide range of mixing devices over different cites or product lines. As the product line-ups change frequently with the market demands, a frequently question is that whether existing mixing devices from other production lines, says impellers of different types and sizes, could be re-assembled into geometries capable for mixing requirements in new product lines. The number of candidate vessel geometries would be overwhelming especially when more than one impeller of varying sizes and types were installed in the vessel, which would make it almost impossible to carry out case-by-case time-consuming studies such as CFD or pilot-scale laboratory works on each possible combination of impeller geometries before the designing decision needs to be finalised.

In this work, a rapid method for mixing time and core flow pattern estimation was set up using two dimensional network-of-zones [1] (2-D NoZ) model, so that a "Yes/No" decision could be made in almost real time to cross out the majority of impossible candidate geometries, before further evaluations such as CFD or pilot scale laboratory studies are carried out. This enables considering designs from a broader geometrical range without excessive increase in time consumptions for refined case studies.

Material and Methods

As shown in the Graphical Illustration, this work aimed to estimate mixing time and core flow patterns of stirred tank geometries with translated, resized or multiple impellers (referred as "target geometries" below) in real time starting from known flow profile data of only a few more widely studied geometries (referred as "known geometries" below). The rapid method raised in this work consisted of three core methodologies: the 2-D NoZ model, mixing time prediction algorithm based on the 2-D NoZ model, and the non-iterative estimation of flow patterns.

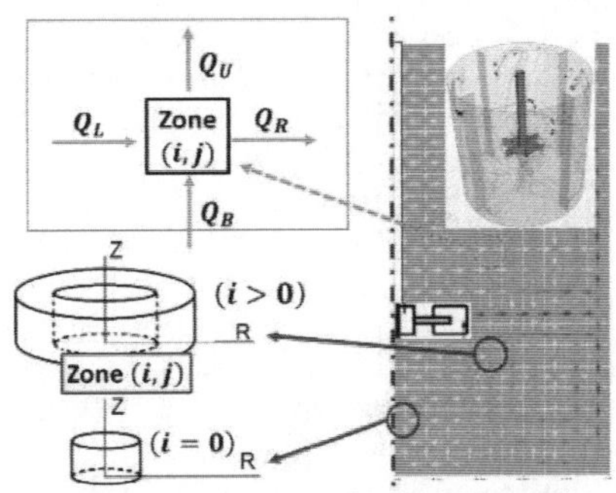

Figure 1. Schematics of 2-D NoZ model.

The 2-D NoZ model coarse-grained the fluid domain in a stirred vessel geometry into a structured grid of cylindrical and annular cylindrical zones. The number of the zones were limited (typically around 200) so that a trade-off between resolution for mixing time prediction and demand for computational resource was considered. The topology among the zones were simplified into two directional (radial-axial) only by azimuthal averaging within each zone. Flow patterns in both radial and axial directions between two connected zones were conserved as radial or axial flowrates by integrating velocities on radial or axial direction on orthogonal inter-zonal boundaries respectively, so that mass was always conserved in each of the zones under steady-state operating condition. The 2-D NoZ model served as both a compressed storage format for stirred vessels (typically a .csv file with several hundred rows for each geometry), and also the flow pattern variables to drive the mixing time prediction algorithm. For the known geometries, a 2-D NoZ model was generated by integrating the known flow pattern data from a more refined solution, e.g. a CFD flow pattern (referred as CFD-NoZ model below). A schematic figure of the 2-D NoZ model was shown in Figure 1.

Mixing time prediction under fully turbulent regime were based on convection-driven passive scalar transport among all zones. Each zone was considered homogeneous at each time step. An additional global constant Q_d with a dimension of volumetric flowrate was specified in transport

https://doi.org/10.52202/074122-0021

equation to approximate inter-zonal transport based on turbulent diffusion. Mixing time to achieve 95% homogeneity was defined by monitoring the root mean square dimensionless variation of dimensionless concentrations [2] while solving Equation 1.

$$V_i \frac{dC_i}{dt} = \sum_{i,x}(\delta_+ Q_{i,x} + Q_d) * C_x - (\delta_- Q_{i,x} + Q_d) * C_i$$

(Equation 1)

In Equation 1 δ_+, δ_- were indicator functions of flowing into and out from zone i respectively, whilst $Q_{i,x}$ denoted flowrate between zone i and each of its connected neighbourhood zone x. As Equation 1 depended on the volume and flowrates but not on the shape of the zones, the mixing time predicted in 2-D NoZ model using cylindrical or annular cylindrical was identical to NoZ model using a real 2-D flow with uniformed thickness as long as all the inter-zonal flowrates were identical with the 2-D NoZ model while setting all zonal volumes equivalent to cylindrical or annular cylindrical zones accordingly. Based on this equivalence, a 2-D NoZ model could be reversibly projected to a real 2-D flow domain, where a continuous and second order differentiable scalar stream function $\psi(r,z)$ enabled the equation of continuity to be satisfied spontaneously [3]. Conservation of momentum was converted into another scalar function vorticity $\omega = -\nabla^2\psi$, where the conversion could be solved using matrix method non-iteratively.

Vorticity value blocks at key positions (e.g. surrounding impellers and baffles) were cut and re-assembled in a void 2-D domain (according to vessel sizes of target geometries) by translating to corresponding positions in target geometries after $\{\psi, \omega\}$ fields were converted from CFD-NoZ models of the known geometries. Vorticity values at rest of the void domain were then solved by $\nabla^2\omega \approx 0$ non-iteratively using matrix method, assuming an isotropic propagation of vorticity. The stream functions and hence velocity profiles in target geometries could then be solved in real time, before converted to 2-D NoZ model and ran the mixing time predictions. A schematic diagram of non-iterative estimation of flow patterns was shown in Figure 3.

Results and Discussion

The method raised in this work was tested using 3 exemplar D=T/3, C=T/3 standard baffled tanks with single impeller as known geometries, the impeller type of which were 6-bladed Rushton disc, 4-bladed pitched-blade turbine downwards (PBTD) and Lightnin A320 respectively. A group of 21 standard baffled tanks in total (9 single impeller tanks, 9 co-axial dual impellers tanks with large clearance C2=T/2 and 3 co-axial dual impellers tanks with medium clearance C2=T/3) were selected as target geometries, the impeller sizes of which ranged from D=T/4 to D=T/2. For comparison, CFD simulations of the target geometries were conducted on the target geometries using standard k-epsilon model (SKE) and multiple reference frame (MRF) method with grid sensitivity tested. The CFD mixing times were obtained from passive scalar transport modelling of a hypothetical dye fed on top of the tank. All tanks were filled with water and the impeller Reynolds numbers (for the co-axial dual impeller tanks, the impeller Reynolds numbers of the larger impeller) were specified at a constant of 40000 so that the flows were fully turbulent.

Figure 2 showed that despite 4 outliers, the dimensionless turbulent mixing times predicted by CFD were consistent with prediction from the current rapid method within an error range of ±35% in most target geometries. The average deviation of the mixing time prediction by current rapid method from the CFD prediction is around +10%. The four outliers were all geometries in which interactions between solid boundaries were strong, for instance a large (D=T/2) PBTD interacting with tank wall or C2=T/3 dual PBTDs interacting each other.

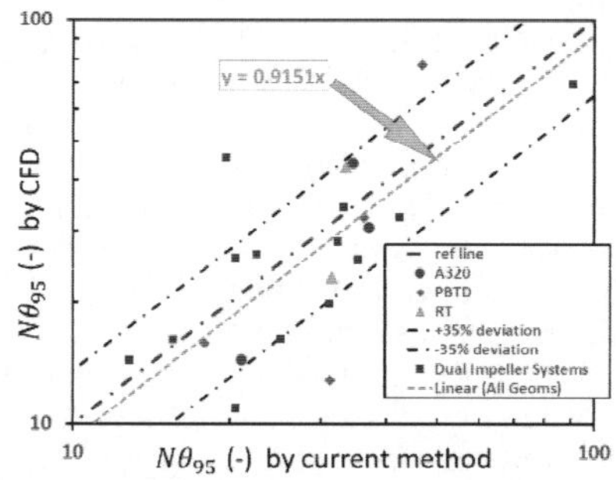

Figure 2. Dimensionless mixing time comparison among all target geometries.

Table 1 showed the comparison between computational performances of the CFD method and the current rapid method on mixing time prediction. The SKE+MRF were one of the faster CFD methods compared to other more complicated turbulent models (e.g. Reynold Stress Transport and Large Eddy Simulation) or sliding mesh solution, hence the passive scalar transport solved on a SKE+MRF CFD was a representative of preliminary evaluation methods on stirred vessels before further studies on specific geometries. By compromising with a larger standard deviation for same level of confidence, the current rapid method managed to predict mixing time with equivalent level of accuracy as the SKE+MRF CFD in almost real time with considerably less number of elements.

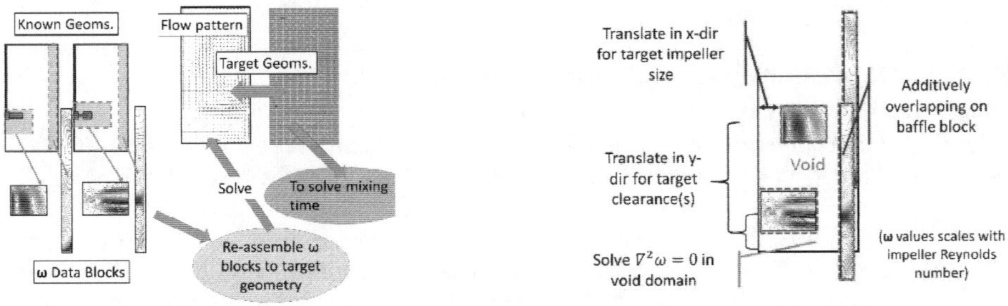

(a) (b)

Figure 3. Schematics of non-iterative estimation of flow patterns: (a) overall flow diagram (b) re-assembling of vorticity blocks to target geometry.

Table 1. Computational requirement metrics of CFD and current rapid method.

Method	CFD (SKE, MRF + passive scalar)	Current rapid method
Average cells / elements /zones per target geometry	Approx. 2-4 million	Approx. 200 zones (2-D NoZ) and approx. 20 thousand ($\{\psi, \omega\}$ domain)
Total time consumption from setup per target geometry	Approx. 1 day on 4 parallel processors, Star-CCM+	Approx. 5 mins on serial processor, Python 3
$N\theta_{95}$ percentage deviation from laboratory correlation[a] , on 9 single impeller tanks	Mean $\mu = -22\%$ Standard Deviation $\sigma = -12\%$	$\mu = -10\%$ $\sigma = -30\%$

[a] Grenville's correlation [4] in fully turbulent regime.

Conclusions

The rapid method developed in this work was defined with a clear routine starting from only a few simple stirred tank geometries with CFD data. The method was able to generate mixing time estimations and flow patterns in almost real time for theoretically unlimited number of stirred tank geometries without doing preliminary CFD studies on each of the geometries, as long as the geometries were consisted with same type of impellers as the simple stirred tank geometries with CFD data. This enabled the corporations with wide ranges of mixing devices to explore more combinations and variations of existing impellers and vessels for better mixing performances without spending a long time on preliminary CFD studies case by case.

The rapid method was validated with a total number of 21 geometries as objectives starting from 3 different types of both radial and axial flow impellers as inputs. Mixing time estimated by the rapid method was consistent with mixing time by CFD prediction with SKE+MRF passive scalar transport in most tested geometries within a percentage error of ±35%. Though larger error ranges were required to achieve same level of confidence, the method was accurate enough for taking over the duty of preliminary CFD studies in crossing-out impossible geometrical variances during mixer designs prior to further computational or laboratory studies.

These developed flow patterns can now be used in NoZ models with more complex calculations, e.g. crystallisation, reactions, and droplet break-up [5]. This overcomes the previous difficulties of the use of NoZs in the fact that the flow patterns had to be generated by hand and flows had to be estimated.

Acknowledgments

The authors acknowledge the help, support and financial aids from the EPSRC and Unilever for the industrial case award and the EPSRC Project: Centre in Advanced Fluid Engineering for Digital Manufacturing (EP/R00482X/1).

References

[1] T. L. Rodgers, F. Siperstein, R. Mann, T. York and A. Kowalski, *Meas. Sci. Technol.*, 22 (2011) 104014.

[2] J. James, M. Cooke, A. Kowalski, T.L. Rodgers, *Chem. Eng. Res. Des.*, 124 (2017) 321.

[3] G. K. Batchelor, *J. Fluid Mech.*, 1(1956) 177.

[4] *Handbook of Industrial Mixing.* E.L. Paul, V.A. Atiemo-Obeng and S.M. Kresta, New Jersey, Wiley, 2003.

[5] E. K. Nauha, Z. Kálal, J. M. Ali and V. Alopaeus, *J. Chem. Eng.*, 334 (2018) 2319.

 https://doi.org/10.52202/074122-0021

Use of Positron Emission Particle Tracking to Assess Mixing of a Lithium-ion Anode Slurry in an Eirich Mixer

MT1.2

Sam D. Hare[1], *Dominik Werner*[1], *C. R. Kit Windows-Yule*[1], *Tzany Wheldon*[1], *Emma Kendrick*[1], *Mark J. H Simmons*[1] *(1) University of Birmingham, Edgbaston, Birmingham, UK, s.d.hare@bham.ac.uk*

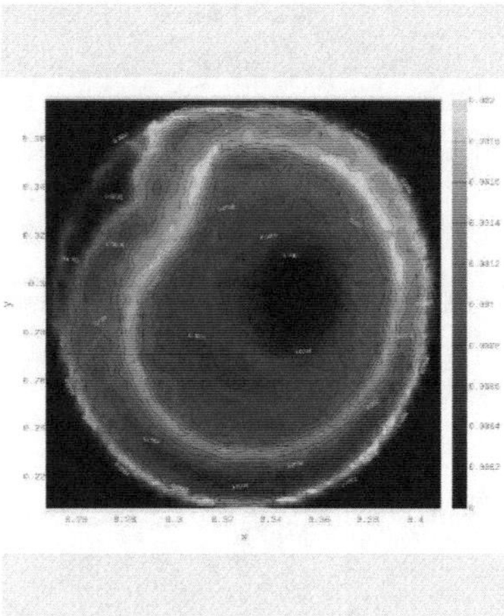

The mixing of lithium-ion electrode slurries is a key processing step during manufacture of batteries, and poor mixing can have a dramatic effect on final cell electrochemical performance. The slurries are multi-component, high solids fraction with a complex, evolving non-Newtonian rheology. In this study, Positron Emission Particle Tracking (PEPT) is used to determine the fluid dynamics and dispersion of a model graphite-based anode slurry in an Eirich EL1 mixer. This mixer features a scalable design (1-12,000 L) consisting of a rotating outer pan, scraper and internal rotor. For experiments performed at different internal rotor tip speeds, with fixed outer pan speed, the dispersion dynamics change with the emergence of a high-dispersion region due to interactions between the flows caused by the rotating pan/scraper and the rotor (shown in graphical illustration). Increasing rotor tip speed increased both mixer effectiveness (average mixer dispersion) as well as the active mixing volume, which will be used to inform scale-up criteria.

Introduction

Lithium-ion battery manufacturing is a complex, multistage process. Traditionally, the impact of each processing step is often only considered in terms of its sufficiency for the subsequent step, with the final performance of the cell only given scrutiny once manufactured[1]. Thus, issues with upstream processing are difficult to diagnose and require significant amounts of time for troubleshooting. There is thus significant scope to develop physical and chemical models to better understand the impact of each process step, in particular the formulation and mixing of the electrode materials to form a slurry in the early part of the process, before they are coated onto a metal substrate to form a battery component[2].

Significant challenges exist due to the variety and scale of the mixers used, and the sensitivity and complexity of the materials being blended into battery electrode slurries. Common lab-scale mixers used include dual asymmetric (Eirich EL1 with a star rotor and rotating mixing pan) and planetary (Thinky ARE-250 rotating 200-2000 rpm) designs, which are believed to provide good levels of mixing for small volumes of material (< 1-5 L). The Eirich mixer is commonly used at scale-up facilities such as UKBIC due to models between 1 – 12,000 L. However, the mixing of a new formulation is nevertheless done via legacy knowledge or trial and error, rather than by consideration of mixing fundamentals and mechanistic understanding is limited.

The mixing of anode slurries is complicated by the initial wetting of the hydrophobic graphite powder (95 wt% of solids within the slurry)[3]. This tends to be more of an issue with non-contact planetary mixers whereas the dual-asymmetric mixer has a wall-scraper to assist in the breakup of the dry powder. The next step involves homogeneous dispersion of carbon black conductive additive throughout the slurry, which requires good levels of both dispersive and distributive mixing as the nanoparticulate carbon agglomerates easily. The main way of overcoming this issue is to ensure a sufficient mixing time as well as minimising any dry powder premixing. Otherwise, the carbon black coats the graphite, leaving less free to form conductive chains throughout the slurry once coated onto the electrode substrate.

The rheology is a critical property for the final slurry as this parameter affects its coatability onto the electrode substrate. The slurry needs to possess the correct consistency to be able to flow through the coater and then hold its shape once coated. The rheology is a strong function of the slurry microstructure, which is governed by both the formulation and the mixing. Poorly mixed inks tend to have a higher consistency due to agglomerates not being fully broken down, and particles being poorly covered with polymer binder.

Characterisation of the slurry mixing is therefore a key step. The complex non-Newtonian rheology of battery slurries prohibits the use of transparent mimic fluids suitable for measurements of mixing using optical methods. However, Positron Emission Particle Tracking (PEPT) provides the ability to track a radioactive particle in opaque systems and mixers using the actual slurries. In this work, PEPT is used to determine the fluid dynamics and dispersion for a model graphite-based anode slurry in a nominal 1 L capacity Eirich EL1 mixer.

Material and Methods

Slurries of 50% w/w graphite, and 1.5% carboxymethyl cellulose (CMC) in water were made by pre-dispersing the CMC (Ashland BVH8) using a Turbula mixer, and then mixing together the graphite and CMC solution (Re ~0.8 – 3.8). An Eirich EL1 mixer (nominal capacity 0.3 – 1L) equipped with a pin-type rotor was used, at rotor tip speeds of 2, 6

https://doi.org/10.52202/074122-0022

and 10 m/s (500 – 2400 rpm), with the mixing pan rotation set constant at 0.5 m/s, Figure 1. The vessel was filled with 0.35 L of fluid to a height of 15 mm, to which 350 g of graphite was added, resulting in approx. 0.5 L of slurry. Due to the complexity of the enclosed mixer design, it was not possible to measure the power input into the fluid using a torque meter.

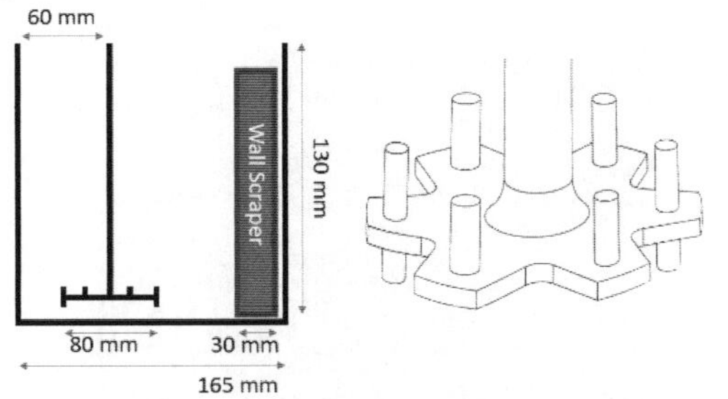

Figure 1. Schematic of Eirich EL1 mixer. Gap between scraper and wall 3.5mm, impeller clearance from base 15mm.

For the PEPT measurements, the Eirich mixer was placed between the 2 detectors of a modified ADAC Forte camera, so that the mixing pan was in the centre of the detectors, with the detectors then moved as close as possible to give maximum data rate and minimise scatter.

An ion-exchange resin of approximately 300 µm and a similar density to the fluid and solids was used to measure the flow during mixing.

The dispersive mixing rate was calculated as described by Windows-Yule et al.[4] in which tracer locations with a similar “start point” are grouped together and their spread measured after a given timestep. The mixer effectiveness (ME) is then quantified by averaging all the dispersion values across the mixer for that timestep.

$$ME = \frac{1}{N_P}\sum_{k=1}^{N_C} \sigma_k^2 \, n_{p,k} \qquad (1)$$

where N_p is the number of passes through all N_c cells, σ_k is the cell dispersion value, and n_p the number of passes in each cell *k*.

Results and Discussion

Values of ME were calculated from approximately 45 minutes of data at each value of rotor tip speed, to build up a sufficient body of tracer locations to ensure ergodicity of the tracer particle and to minimise the impact of any short periods when the tracer became stuck to a surface in the mixer.

Figure 2 shows the mixer effectiveness (ME) for 3 different tip speeds; the plateau value for ME depends on the volume occupied by the tracer during mixing and is limited by the physical dimensions of the vessel. The variation in values between rotor speeds is due to the material being sprayed further up the sides of the vessel as the rotor speed increases, as well as the void around the rotor head increasing with rotor speed.

At very short timesteps all rotor speeds have a similar ME but between 1 s and 2 s, the 10 m/s speed has a higher ME value. Both 2 m/s and 6 /s speeds have a similar ME value initially. However, they diverge at around 1.5 s where the 2 m/s dataset has a gradient change, and the 6 m/s continues increasing at the same rate until it has a similar gradient change at around 3 s.

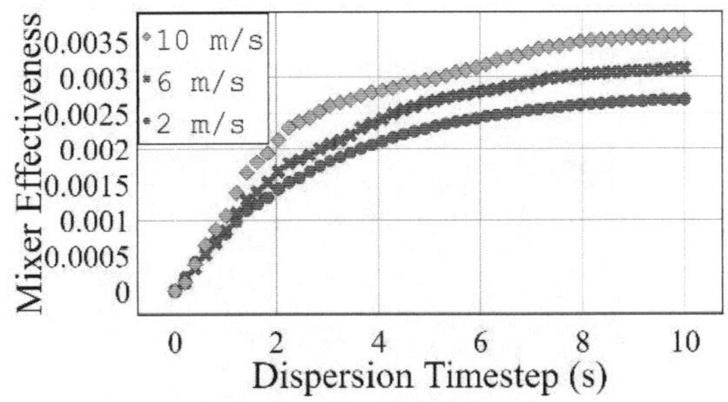

Figure 2. Mixer effectiveness values for the Eirich mixer at different rotor speeds.

The end point of the linear regime is indicative of distributive mixing occurring within the vessel as it shows that higher levels of dispersion are still occurring as the tracer is allowed to travel further. Declines in ME at shorter timesteps may indicate the presence of complex flow patterns that inhibit the tracer moving from one region of the vessel to another.

Whilst the overall value of ME is a useful metric, it is a complex function of the fluid dynamics/mixing intensity and the active mixing volume. The degree of dispersion and fluid dynamics were thus assessed locally to devise criteria for optimal operation.

Figure 3 shows heatmaps of dispersion values for the 2 m/s rotor speed at 1 s and 5 s timesteps. At shorter timesteps, there are two discrete regions of activity within the mixer. The first, around the rotor head, is relatively uniform, but with a region of slightly higher dispersion in the narrower section where the rotor is closer to the pan wall, due to a higher tracer velocity in this region (thus allowing it to travel further in a fixed timestep). The second region of mixing is at the pan wall, with a particular “hotspot” around the fixer wall scraper, this is due to the significant disruption of fluid flow caused by the scraper.

As the dispersion time step is lengthened these two regions grow and eventually meet one another, giving rise to the inner ring of high dispersion shown in Figure 3b. For some periods during mixing, the tracer can become trapped within the outer ring zone for a short period of time, which is much more prevalent at lower mixing speeds. This is due to the outer shear regime becoming progressively smaller as the power output from the rotor increases.

The complex flows arising from these two regions as a function of both mixer speed and dispersion time lead to areas of lower dispersion in the mixer that aren’t necessarily captured by the overall value of ME. However, the distribution of dispersion values for each dispersion timestep and mixer speeds can be plotted to show how the different regions of the

 https://doi.org/10.52202/074122-0022

mixer behave, which when combined give the ME value. Figure 4 shows how the distribution of dispersion values for a 10 s timestep vary between a rotor speed of 2 and 10 m/s, the number of zones in the dispersion calculation are kept the same for each mixing speed.

Whilst there appears to be a similar number of very low dispersion areas at both rotor speeds, the majority of these can be attributed to the mixer scraper and the void around the mixer head. At 10 m/s there is a more gaussian distribution of dispersion values, whilst for the 2 m/s speed, there is an extended plateau.

This difference in distributions shows that at higher rotor speed, shows that more regions of the mixer are "active" as well as more regions having more intensive mixing. Which together give rise to an overall increase in mixer effectiveness for the higher mixing speed.

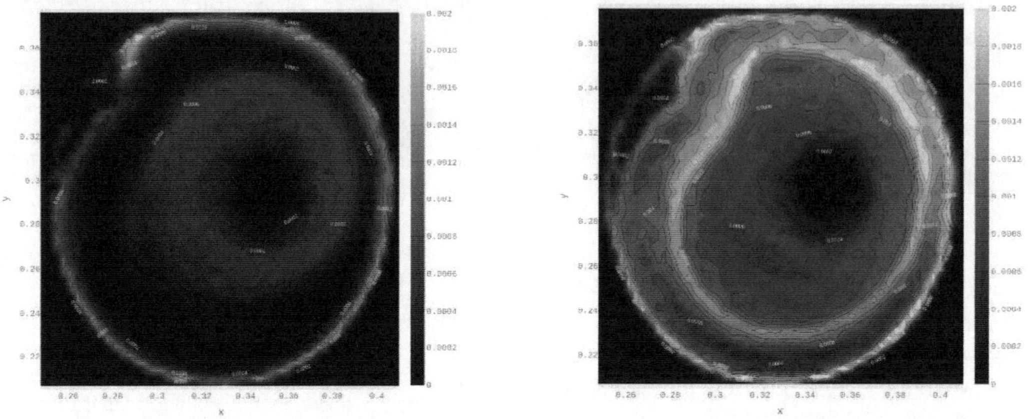

(a) (b)

Figure 3. Heatmap of dispersion values for different dispersion timesteps, 1 second (a) and 5 seconds (b) for the Eirich mixer at 2 m/s rotor tip speed. Dead zone in both images is the location of the rotor head.

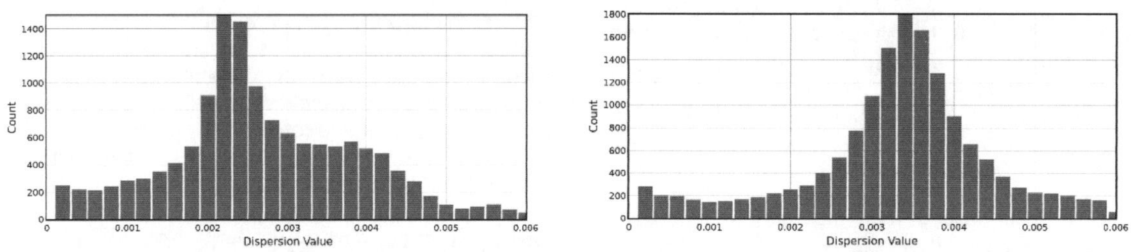

(a) (b)

Figure 4. Distribution of dispersion values for a 10s timestep in the Eirich mixer for two different rotor speeds, 2 m/s (a) and 10 m/s. (b)

Conclusions

PEPT has been used to characterise the mixing and dispersion of a model Li-ion anode slurry in an Eirich EL1 mixer. Two main regions of dispersive mixing were identified due to the motion of the internal rotor head and the wall scraper next to the edge of the rotating mixing pan. As the rotor speed is increased the overall mixer effectiveness also increases. Mixer effectiveness is a measure of active mixing area and degree of mixing intensity. The active mixing area increases with rotor speed, as well as the intensity of mixing.

The two zones of mixing, the one around the rotor and the one generated by the wall scraper/rotating pan come into contact and cause an inner ring of increased dispersion where they meet, with the tracer sometimes becoming trapped in the outer regime at lower mixing speeds. By considering the transitions between both these shear regimes as well as mixer effectiveness, the optimal mixing speed that balances both dispersive and distributive mixing can be ascertained.

Acknowledgments

Special thanks to Leonard Nicuşan for his assistance with the PEPT library codebase. This work was supported by the Faraday Institution NEXTRODE project (faraday.ac.UK; EP/S003053/1, FIRG015).

References

[1] H. Bockholt, M. Indrikova, A. Netz, F. Golks, A. Kwade, *Journal of Power Sources.* 325 (2016) 140–151.

[2] C.D. Reynolds, P.R. Slater, S.D. Hare, M.J.H. Simmons, E. Kendrick, *Materials & Design.* 209 (2021) 109971.

[3] J.-H. Lee, U. Paik, V.A. Hackley, Y.-M. Choi, *J. Electrochem. Soc.* 152 (2005) A1763

[4] *The fundamentals of PEPT.* K. Windows-Yule, L. Nicuşan, M.T. Herald, S. Manger, D. Parker, IOP Publishing, 2022.

On the Validation of Laminar Flow Computational Mixing Models with Positron Emission Particle Tracking

MT1.3

Roberto Hart-Villamil[1], Andy Ingram[1], Kit Windows-Yule[1], Santoshkumar Gupta[2], Waldo Rosales[3], Adam Kowalski[3]. (1) University of Birmingham, rxh659@student.bham.ac.uk. (2) Hindustan Unilever R&D, Bangalore. (3) Unilever R&D, Port Sunlight Laboratory.

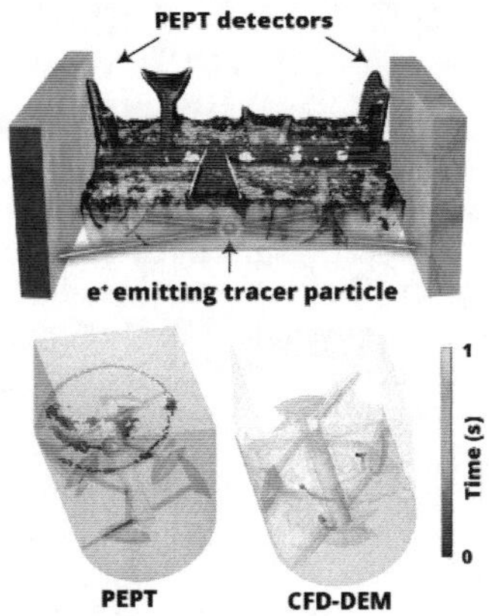

A methodology for the validation of a CFD mixing model using Positron Emission Particle Tracking (PEPT) was applied to a horizontal laminar stirred tank containing Newtonian glycerol. Particle trajectories recorded using PEPT were compared to two one-way particle-coupled CFD models (CFD-DEM, and CFD-Lagrangian) through time-averaging and binning into a grid of 3D voxels. A cell-by-cell comparison of particle velocities between the computational models and PEPT highlighted the comparative strength of the CFD-DEM model (R-squared = 0.78) over its Lagrangian (R-squared = 0.23) counterpart. Furthermore, using PEPT as a validation data set, the R-squared value can be used as a cost function to guide an autonomous calibration of an N-dimensional CFD parameter space. Further investigations will apply this methodology to calibrate non-Newtonian laminar CFD models.

Introduction

The prediction of complex mixing in industrial processes is a universal challenge that requires a strong understanding of the underlying hydrodynamics involved. Validated Computational Fluid Dynamics (CFD) models offer a flexible and robust alternative to lab-scale experimentation. Validating a CFD model requires a comparison of statistics between the computational and the lab environment. Attempts at validation in literature often rely on torque agreement or 1 dimensional velocity profiles. More thorough optical techniques such as Particle Image Velocimetry (PIV), and Laser Doppler Velocimetry (LDV) can reconstruct 2D or 3D fluid flow-fields [1,2] and are best suited for transparent single or dilute binary phase systems [3].

The fluids encountered in consumer goods industrial processing environment often contain a wide range of additives that when blended yield formulations with complex rheologies. In the case of dense opaque multi-phase flows, tomographic techniques such as Positron Emission Particle Tracking (PEPT) [4], Electrical Resistance Tomography (ERT) [5], and Computer Automated Radio Particle Tracking (CARPT) [6] are extensively employed. These methods allow for un-invasive reconstruction of three-dimensional fluid flow-fields.

In the PEPT technique, a radionuclide containing isotopes with a proton-rich nucleus (e.g. Na-22, F-18, Ga-68) is placed within the system under study and continuously undergoes β^+ decay. The emitted positron annihilates with an electron to produce two co-linear and anti-parallel 511 keV gamma rays. The penetrating gamma rays can be detected through industrial casing by the surrounding ADAC Forte Gamma Camera detectors (Figure graphical illustration). A spatial distribution of the radionuclide position is reconstructed by projecting each detected event uniformly along the gamma-ray path (termed 'lines of response' or LoRs). The LoRs can then be applied to triangulate the Lagrangian trajectory of the tracer through space (Figure graphical illustration).

Two central questions arise: 1) Whether PEPT can be applied to validate CFD mixing models with an evolving (dynamic) rheology and 2) Whether a validated CFD model can predict how these fluids mix in on an industrial scale. In the first stage of answering these questions, this work aims to show how PEPT can be applied as a powerful validation tool for viscous Newtonian CFD models operating in the laminar flow regime.

Material and Methods

A ploughshare mixer (Figure 1) was filled to different heights with glycerol of varying concentrations and mixed at different rotation rates (Table 1) within the laminar flow regime. PEPT was used to record the Lagrangian trajectory of a radiolabelled particle neutrally buoyant in the glycerol (Figure 2). Encased in a spherical 0.8 mm diameter ion-exchange resin, the F-18 tracer had a half-life of 110 minutes which provided for 4 hours of high-quality data. Beyond this timeframe the low tracer activity meant that fewer points would be available to reconstruct the trajectory and the measurement error in position would increase [7].

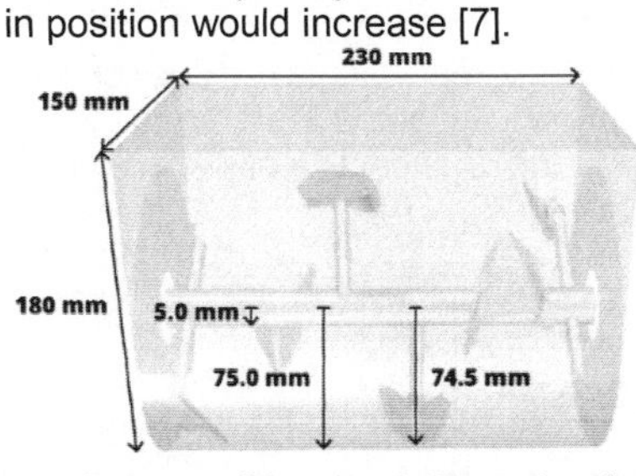

Figure 1. Diagram of the ploughshare mixer for PEPT experiment and CFD simulation.

https://doi.org/10.52202/074122-0023

Table 1. Summary of the conditions investigated in the ploughshare mixer using PEPT and simulated in CFD.

Glycerol	Rotation rate	Re	Fill[a]
[wt.%]	[rpm]	[-]	[% vol.]
100	33	10	100
100	33	10	40
100	40	15	100
96	21	15	100

[a] The liquid height at full-fill was 16.8 cm.

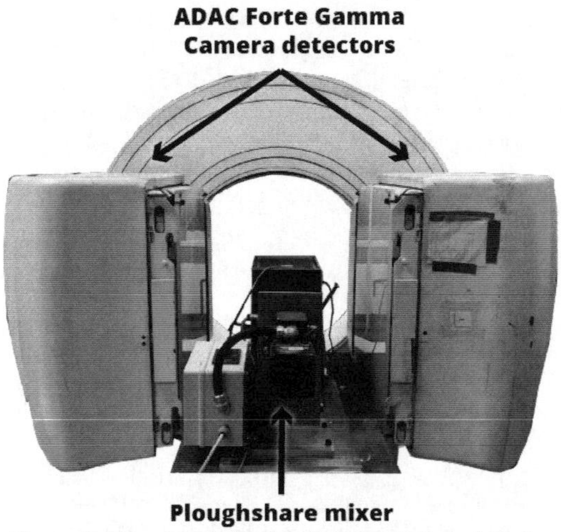

Figure 2. Experimental setup: the ploughshare impeller is positioned parallel to each PEPT detector face.

Each PEPT experiment ran for approximately 4 hours to gather sufficient Lagrangian data. Using STAR-CCM+, the same conditions were replicated in a Finite Volume (FV) CFD simulation using the Volume of Fluid (VOF) and laminar flow model. The flow-field independent FV mesh contained 1.2 M polyhexcore cells and was divided into a static and rotating region per the sliding mesh approach. No-slip conditions were applied to all solid boundaries except from the roof of the mixer. The system comprised of glycerol (constant ρ and μ) and air (ideal gas), separated by an immiscible interface refined by the High-Resolution Interface Capturing (HRIC) scheme. The free surface was further refined using an Adaptive Mesh Refinement criterion.

To replicate the PEPT Lagrangian trajectories, the CFD model was one-way coupled with either 1650 spherical Lagrangian material particles (Lag) or Discrete Element Method (DEM) [8] particles seeded on a horizontal plane half-way up the tank. The DEM particle parameters were chosen to minimise the impact of particle-particle interactions.

Each simulation ran for 30 seconds with a 1° time-step resolution. Much like the PEPT experiment, the trajectory of each particle was recorded.

The Lagrangian trajectories from PEPT were computed using the PEPT-ML algorithm [9] and then post-processed identically to the CFD trajectories.

The instantaneous velocities were calculated for PEPT and CFD using the Savitzky-Golay filter [10] and each Lagrangian trajectory was then discretized into an Eulerian grid of voxels (Figure 3) and time-averaged. Each model could then be compared to the PEPT ground truth through a cell-by-cell parity plot (Figure 4).

Results and Discussion

Figure 3 illustrates the results from the Re = 10, Fill = 40 vol% run. There is a uniform distribution of tracer velocities between the shaft and the wall along the axial direction of both the PEPT and CFD-DEM description. The flow near the shaft is lower than in the central region of fluid due to a decreasing blade speed. In contrast, the red ring in the CFD-Lag model displays an overprediction in tracer velocities. This difference could be a result of the different contact models between DEM and Lagrangian particles. DEM uses a soft-contact approach in which particles are allowed to overlap, and the contact force is proportional to the overlap which itself is a function of the particle material properties and shape (e.g. Young's Modulus, Poisson's ratio), while the Lagrangian model computes a hard-contact approach using the coefficient of restitution rather than contact forces. The absence of a spin lift term may have also played a minor role in this over-prediction.

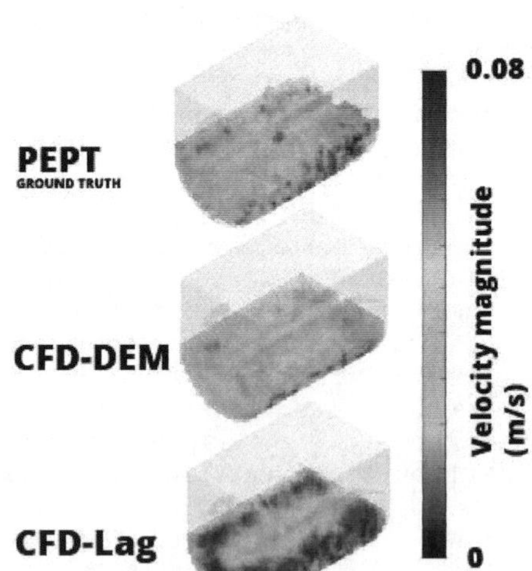

Figure 3. 3-D 40^3 voxel grid illustrating the particle velocity magnitude between the PEPT experiment and CFD models for the Re = 10, Fill = 40 vol % run.

This discrepancy is quantified in figure 4, where model strength is evaluated by plotting PEPT particle velocities against their respective CFD models. The Pearson R-squared statistic is used as a validation statistic since it applies a greater penalty to poorer models. In this run, the CFD-DEM model has an R-squared of 0.78, which suggests it is the more appropriate particle coupled CFD model for simulating this Newtonian laminar flow mixing environment.

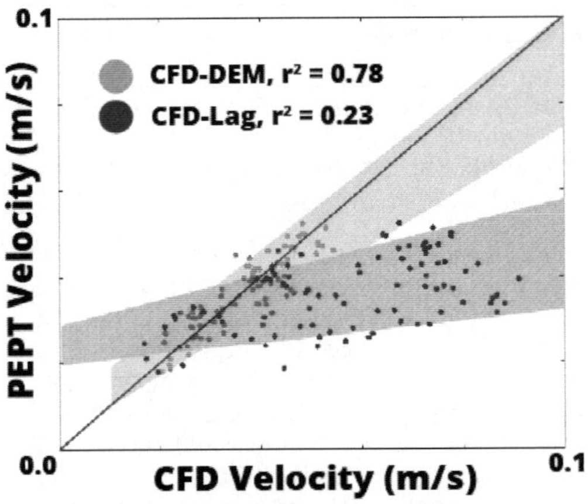

Figure 4. Parity plot representing the velocity comparison between PEPT and CFD-DEM (green) and CFD-Lagrangian (blue) models. The highlighted regions illustrate the 95% confidence interval. The 45° red line is the parity line.

Since this workflow only requires trajectory data from comparative systems, it is flexible to fluids, geometries, and parameter combinations. Furthermore, with sufficient PEPT datasets the Pearson R-squared can be used as a cost function to autonomously calibrate CFD models through iterative simulation. One such technique named Autonomous Calibration and Characterisation via Evolutionary Software (ACCES) [11], applies the CMA-ES evolutionary algorithm to efficiently explore a high dimensional parameter space and converge upon a global solution, i.e. the combination with the highest R-squared value.

Conclusions

The validation strategy described herein can automatically compare simulation data with a local description of experimental flow and provide a quantitative validation statistic. This case-study showcased the strength of the CFD-DEM model compared to its Lagrangian counterpart. Furthermore, the strengths of PEPT as a CFD validation tool are highlighted through its ability to uninvasively capture a velocity flow-field through industrial casing in three dimensions. The next steps for this workflow will involve autonomous calibration of CFD models using an evolutionary strategy. Further investigations will apply this workflow to the validation of non-Newtonian CFD models.

Acknowledgments

RHV is supported by the EPSRC Centre for Doctoral Training in Formulation Engineering (EP/S023070/1) at the University of Birmingham and Unilever.

Reference

[1] H.S. Yoon, K.V. Sharp, D.F. Hill, R.J. Adrian, S. Balachandar, M.Y. Ha, and K. Kar. *Chemical Engineering Science*, 56 (2001) 6635–6649.

[2] H. Singh, D. F. Fletcher, and J. J. Nijdam. *Chemical Engineering Science*, 66 (2011) 5976–5988.

[3] K. Cole, P. R. Brito-Parada, K. Hadler, D. Mesa, S. J. Neethling, A M. Norori-McCormac, and J. J. Cilliers. *Chemical Engineering Journal*, 433 (2022) 133819.

[4] J. Parker, C.J. Broadbent, P. Fowles, M.R. Hawkesworth, and P. McNeil. *Nuclear Instruments and Methods in Physics Research Section A: Accelerators, Spectrometers, Detectors and Associated Equipment*, 326 (1993) 592–607.

[5] C. Carletti, G. Montante, C. De Blasio, A. Paglianti. *Chemical Engineering Science*, 152 (2016) 478.

[6] R. Rammohan, A. Kemoun, M.H. Al-Dahhan, M.P. Dudukovic. *Chemical Engineering Science*, 56 (2001) 2629–2639.

[7] M. Herald, J. Sykes, D. Parker, J. Seville, T. Wheldon, C.R.K. Windows-Yule. *Nuclear Instruments and Methods in Physics Research Section A: Accelerators, Spectrometers, Detectors and Associated Equipment*, 1047 (2023) 167831.

[8] P. A. Cundall, O.D.L. Strack. *Geotechnique*, 29 (1979) 47-65.

[9] A.L. Nicuşan, C.R.K. Windows-Yule, *Review of Scientific Instruments*, 91 (2020) 13329.

[10] A. Savitzky, M.J.E. Golay, *Analytical chemistry* 36.8 (1964) 1627-1639.

[11] M. Herald, A. Nicuşan, T.K. Wheldon, J. Seville, C.R.K. Windows-Yule, *Scientific Reports*, 12 (2022) 19515.

https://doi.org/10.52202/074122-0023

Machine-Learning-Based Fully Automatic Foam Detection in a Stirred-tank Reactor

MT1.4

S.S. Hoseini[1], *E. Rundquist*[2], *M. Poux*[1], *J. Aubin*[1]. *(1) Laboratoire de Génie Chimique, Université de Toulouse, CNRS, INPT, France; seyedsalar.hoseini@toulouse-inp.fr; joelle.aubin@toulouse-inp.fr ; martine.poux@toulouse-inp.fr; (2) Process Studies Group, Technology and Engineering, Syngenta, Jealott's Hill International Research Centre, RG42 6EY Bracknell, Berkshire, UK, elin.rundquist@syngenta.com*

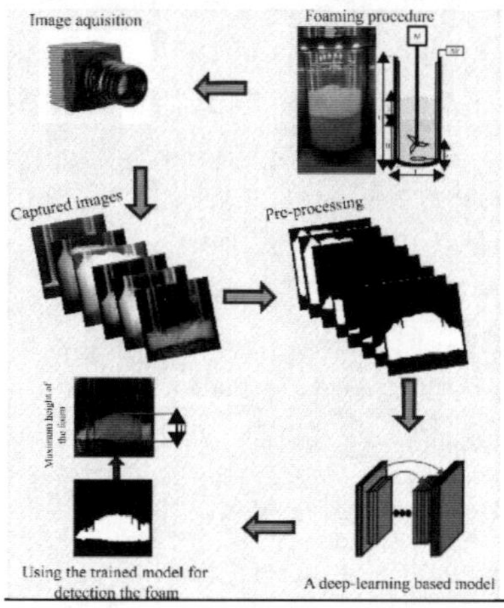

Foam formation in many chemical processes creates problematic operational issues in chemical reactors/bioreactors. As a result, the control and reduction of foam in chemical reactors in an important topic in industry. Detection and quantification of generated foam under different operating conditions in chemical reactors can provide important insight into the development of an appropriate foam control technology. In this study, a deep learning-based approach is developed to automatically detect and quantify foam in a stirred tank reactor under different operating conditions. After training the developed model, the results show that the Intersection on Union (IoU) is 92% and the pixel value prediction accuracy is 97%. Based on the superior ability of the trained model, the produced foam is automatically detected with high segmentation accuracy for different working conditions of the stirred tank reactor, then the geometric characteristics of the detected foam (such as maximum height and volume value) are quantified.

Introduction

Foam production can be observed in many natural, technical or scientific processes [1]. In this regard, foaming occurs frequently in chemical reactors. Foaming is more common in gas-liquid contacting operations using a reactor with gas sparging. The produced foam can create a variety of issues in such systems, including, but not limited to, lowering the working liquid volume, losing the accumulated reactants and products in the foam during foam escape, fouling of sensors, increased possibilities of contamination, and suppression of oxygen transfer [2]. Due to the adverse effects of foam, foam control is essential in many chemical processes. In the past years, different approaches have been employed for foam control, including physical methods (such as ultrasonic deformers, thermal foam breakers, electrical foam breakers), mechanical methods (such as nozzle foam breaker, syneresis-based foam breaker, impact spray foam breaker, cyclone type deformers, and foam breaker with rotating parts), chemical methods (using chemical antifoam agents), and combined action foam controller approaches [3]. It is generally understood that the first and most important step in using these foam control approaches is to closely monitor and accurately detect the presence of foam in chemical reactors. Accurate detection of the amount of foam can help to identify the effectiveness of the foam control method employed, to reduce the working time in mechanical and physical foam breakers and thus reduce the required power consumption in these devices, or reduce the amount of chemical antifoam required in a chemical foam reduction approach. During recent years, researchers as well as suppliers have offered a wide variety of detection methods (such as conductive liquid level switch [4], ultrasonic level switch [5], and Impedance probes [6]) to monitor and detect foam in chemical reactors. A review of the research conducted in the field of foam detection in chemical reactors shows that many of these detection systems are contact-based sensing strategies. As a result, the performance of the sensors is affected by the fluid (after the foaming procedure, material left on sensor surfaces can cause reduce the signal of the sensor), which usually leads to false positive signals. Furthermore, non-contact detection technologies, such as ultrasonic sensors, can be influenced by temperature, humidity, and false positive signals induced by agitation in chemical reactors. Unlike the above-mentioned methods, machine vision-based approaches are promising candidates for foam detection in chemical reactors and do not rely on contact-based sensing. Over the past decade, machine learning and deep learning technologies such as deep convolutional neural networks (CNNs) have made significant contributions to advances in machine vision. By training a deep learning-based model, using a large image dataset consisting of different operating conditions in chemical reactors, the foam geometrical characteristics can be well predicted. In this research project, we propose an intelligent detection method to automatically detect the amount of foam in a stirred tank reactor. In detail, we develop a deep machine learning-based approach to automatically detect the foam that is generated.

Material and Methods

Experimental section: The experimental setup consists of a stirred dished-bottom tank (T = 0.19 m) with a ring gas sparger (Ds = 0.33T) as shown in Figure 1. Three different impellers are

https://doi.org/10.52202/074122-0024

used: an axial flow hydrofoil (D = 0.5T), a Rushton turbine (D = 0.33T), and a Cowles disk (D = 0.29T). The experimental setup is used to produce foam under various operating conditions. To achieve this goal, the aqueous solution of Tween® 20 (ARCOS Organics) at concentrations of 0.1, 0.2, and 0.3 g/L is used. For each aqueous solution, combinations of three different rotational speeds of impellers (200, 375, 550 rpm) and three different gas flow rates (0, 30, 60 SLPM) are used to produce different types of foam in the stirred tank.

Imaging system and foam image acquisition: 2D gray-scale images of the foam are captured using a high-speed camera with a full camera-link interface (Mikrotron Eosens Camera Link series) at up to 500 frames per second at 1280 × 1024 pixels.

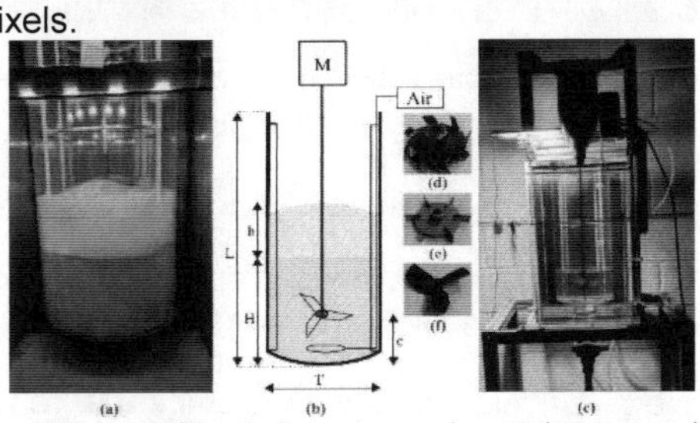

Figure 1. Fig. 1. Actual experimental setup and its schematic structure of the stirred-tank

Images labelling and data augmentation: In the current study, all images in the dataset are manually annotated. A total of 170 images, including backdrop and foam in one class, have been fully annotated using the online software "apeer". Each class is associated with a binarized image, where white represents the foam zone and black represents the background. Figure 2 displays some images of various types of foam obtained. In the next step, data augmentation techniques are then applied on the image dataset and annotated (mask) dataset. Finally, following the augmentation process, the training dataset contains about 1562 images.

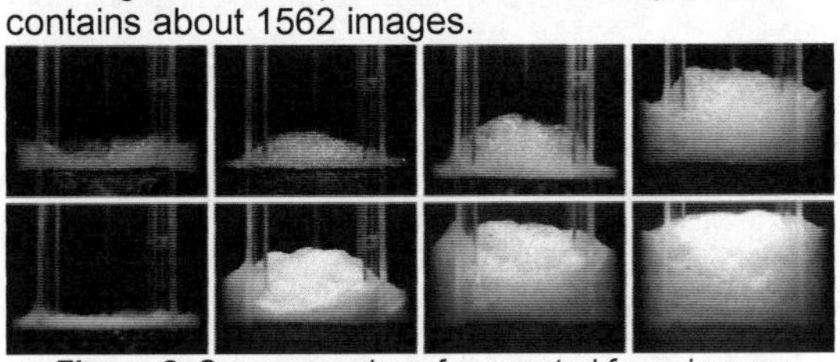

Figure 2. Some samples of generated foam images related to different experiment conditions

Segmentation network architectures: The deep learning network used in this study is based on CNN. CNNs, which are useful tools for semantic segmentation and image classification, are the most useful methods for segmenting images captured from the foam production process in chemical reactors. Figure 3 shows the comprehensive information of the U-net architecture that is used in the current study for the segmentation of the images. The U-net architecture is divided into two major parts: the encoder (on the left side) and the decoder (on the right side). The encoder part consists of five convolution modules.

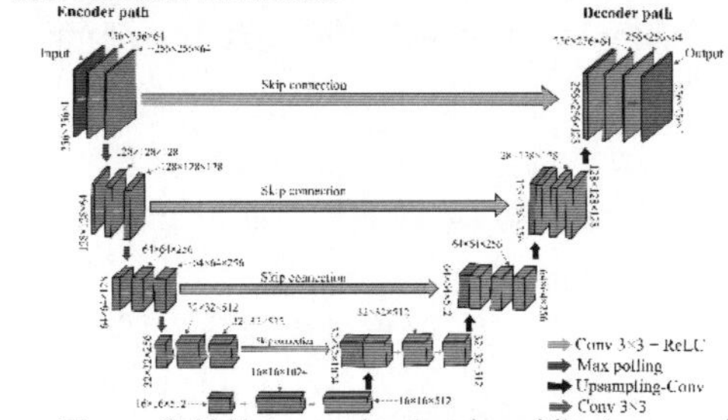

Figure 3. U-Net neural network architecture used to detect foam in collected images.

Network training procedure and evaluation: The following conditions are considered for training the models: (1) The optimizer of "Adam" is used to optimize networks; (2) The training rate of 1×10^{-3} is considered for training the networks; (3) The batch size and epoch are set to 8 and 100, respectively; (4) For the training process, the binary focal loss function (with $\gamma = 2$) is utilized, which evaluates the detection outcomes of the networks and quantify the errors. Figure 4 shows the network training and verification procedure. To measure and compare segmentation performance, overall pixel accuracy (Acc) and intersection over union (IoU) are utilized. The implementation of the image segmentation model is performed with Python 3.9 using Keras/Tensorflow version 2.3. The network is developed on a workstation equipped with Intel(R) Xeon(R) silver 4216 CPU @ 2.10GHz 2.10GHz (2 processors) and NVIDIA Quadro P1000 graphics processor (GPU).

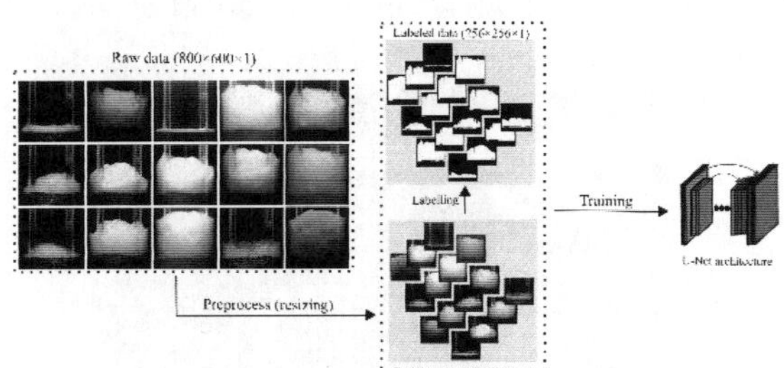

Figure 4. The trends of the suggested deep-learning-based techniques for detecting foam.

Results and Discussion

Overall model performance: The overall performance of the trained model based on the U-Net architecture is evaluated in the first stage. During the training process, three metrics, including pixel accuracy, Jaccard coefficient or intersection over union (IoU), and loss function, are evaluated. The performance of the trained U-net model for a dataset containing 1562 images of different classes of foams is shown in Figure 5. The results show that the trained model has good overall performance. The optimal value of IoU, pixel accuracy, and local loss, for the trained

https://doi.org/10.52202/074122-0024

model based on the standard U-net, are 92%, 97%, and 0.1%, respectively. By considering these results of overall performance, it can be concluded that the performance of the trained model is acceptable and that the results are reliable.

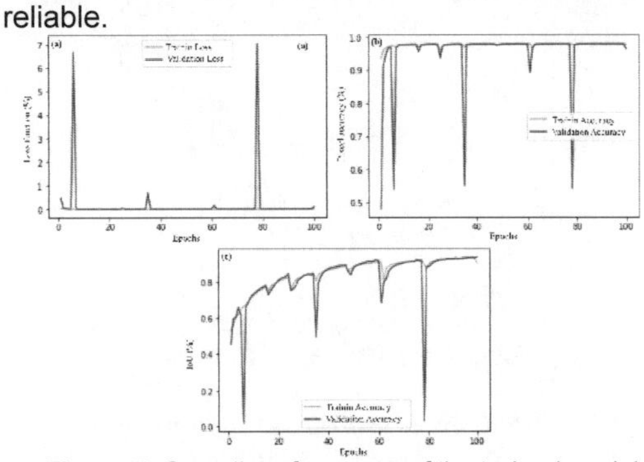

Figure 5. Overall performance of the trained model: (a) Loss function (%); (c) pixel accuracy (%); (b) IoU (%).

Deep-Learning-Based Automatic foam Detection: The performance of the trained model is thoroughly evaluated both qualitatively and quantitatively. The foam detection results, for different types of foam is shown in Figure 6. The results show that the segmentation performance of the trained model is acceptable and the model can recognize the target pixels (foam area) with high accuracy. To verify the advantage of the proposed deep machine learning method used in the current study, the conventional threshold-based image segmentation algorithms were applied to detect the foam pixels. As shown in Figure 6, the conventional approach cannot clearly separate the foam and background regions, while the trained model shows significantly improved performance in terms of pixel segmentation accuracy.

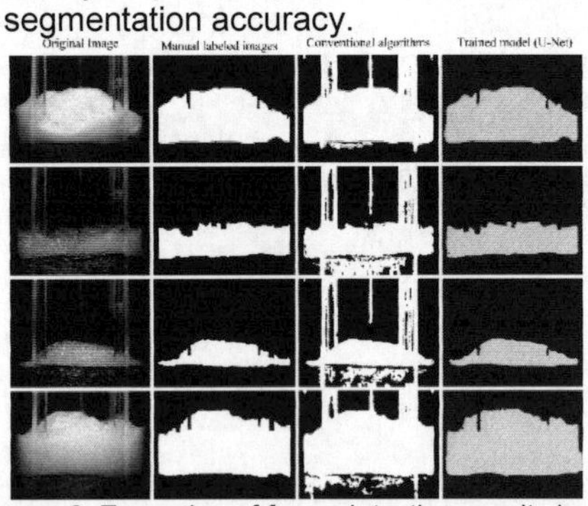

Figure 6. Examples of foam detection results based on the trained model

Quantification of geometrical parameters of detected foams: Based on the trained deep learning model, the maximum height of different types of foam is determined. The predicted maximum foam height for different types of foam is shown in Figure 7. The results show that using the trained model, based on the deep machine learning approach, it is possible to automatically calculate the maximum foam height in different conditions.

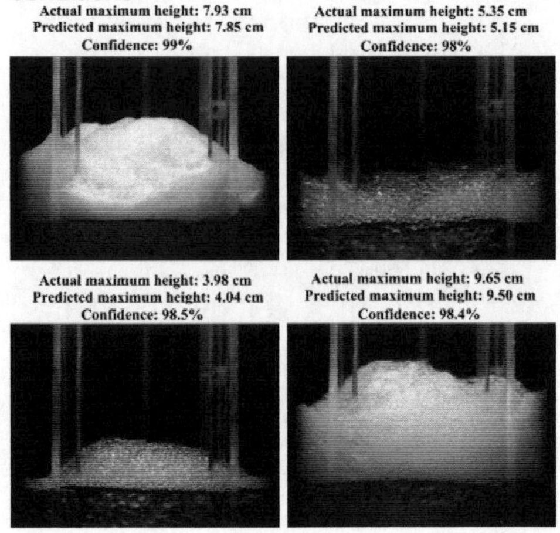

Figure 7. Quantification of geometrical characteristics of detected foams

Conclusions

In conclusion, this work highlights the effectiveness of advanced machine learning approaches for automatic detection and measurement of foam in chemical reactors. Furthermore, the current model performs effectively under various experimental and operational situations of the foaming process in chemical reactors, even when the corresponding data are not provided in the training dataset. Although the present model was developed focusing on the foaming process based on gas-liquid flows in a small stirred tank reactor, we believe that it can be extended to the foaming process in other multiphase flows and other sizes of stirred tank reactors.

References

[1] W. Drenckhan, A. Saint-Jalmes, The science of foaming, *Advances in Colloid and Interface Science* 222 (2015) 228-259.

[2] J. Aubin, D. Daniel, A. Deu, C. Raffel, W. Shaikh, E. Rundquist, P. Piccione, Effects of process conditions on foaming in stirred tanks, *16th European Conference on Mixing*, 2018.

[3] N. Ghildyal, B. Lonsane, N. Karanth, Foam control in submerged fermentation: state of the art, *Advances in applied microbiology* 33 (1988) 173-222.

[4] A. Brown, C. Isbell, S. Gallagher, P. Dodd, J. Varley, an improved method for controlling foams produced within bioreactors, *Food and bioproducts processing* 79(2) (2001) 114-121.

[5] A.K. Sahoo, S.K. Udgata, A novel ANN-based adaptive ultrasonic measurement system for accurate water level monitoring, *IEEE Transactions on Instrumentation and Measurement* 69(6) (2019) 3359-3369.

[6] M. Hofmann, M. Funke, J. Büchs, W. Mokwa, U. Schnakenberg, Development of a four-electrode sensor array for impedance spectroscopy in high content screenings of fermentation processes, *Sensors and Actuators B: Chemical* 147(1) (2010) 93-99.

[7] H.-J. Yoo, Deep convolution neural networks in computer vision: a review, *IEIE Transactions on Smart Processing and Computing* 4(1) (2015) 35-43.

Experimental Study and Turbulence Dissipative Scale Modelling of the Rapid Micromixing of Impinging, Paper-Thin Sheets of Liquids

MT1.5

Robert J. Demyanovich[1]*. (1) RJD Technologies, Inc., Seattle, WA, USA, bob@rjdtech.com.*

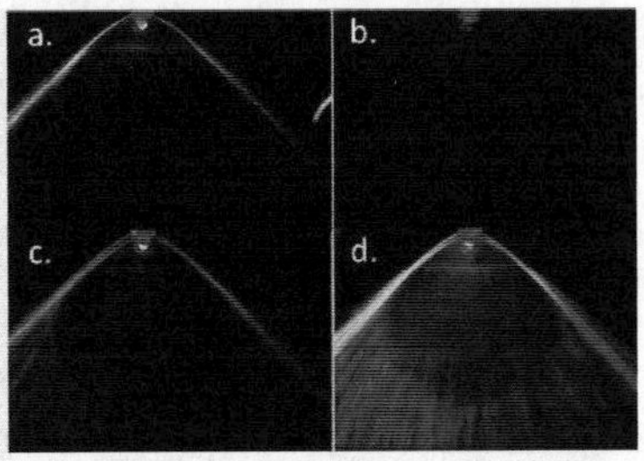

Impinging sheets of: a) base with excess acid leading to disappearance of fluorescence, b) acid with itself, c) pH7 with itself, d) base with itself. The sheet pH decreases as the reaction proceeds in Figure a. For Figures b through d, the pH remains constant throughout the mixed sheet.

The energy released by the impingement of thin liquid sheets is dissipated within approximately one turnover time of the largest eddies [3]. As a result, micromixing was studied using a simple model that assumes diffusion of components in slabs of fixed thickness, which are generated within the impingement zone. The experimental study followed the progress of an acid-base neutralization containing a fluorophore. A simulation, incorporating the micromixing model and a module for determining the fluorescence, calculated the diffusion slab thickness ($2L$) for experimental energy dissipation rates (ϵ) ranging from $5x10^3$ to $7.7x10^6$ W/kg. $2L$ was found to lie in the range of turbulence dissipative scales less than the Taylor microscale but greater than the Kolmogorov microscale. $2L$ is a function of $(\nu/\epsilon)^{1/2}$ but also depends on the large-scale eddy Reynolds number. The results indicate that the scalar diffusion thickness, $2L$, is approximately equal to the turbulence length scale, λ_d, at which most of the dissipation of turbulent kinetic energy occurs.

Introduction

The impingement of thin sheets of liquids has been shown to be an effective method for producing high energy dissipation rates (ϵ) yielding rapid micromixing [1]. Recent studies have shown that the amount of energy released is a function of the coefficient of restitution (COR) of the collision, and that the energy released is dissipated within the impingement zone [2,3]. However, the azo-coupling study [1] did not explicitly formulate a micromixing model for impinging sheets; instead, a version of the EDD (engulfment, deformation, and diffusion) model [4] was used that assumed a fixed striation thickness equal to the Kolmogorov microscale (η).

Since the impingement zone is the only area in which energy is injected and dissipated, turbulence subsides rapidly after the impingement zone (liquid residence time of 0.005 to 0.3 ms). Therefore, a simple micromixing model is presented that assumes the energy dissipated produces slabs of fixed thickness (within the impingement zone) in which diffusion is the only transport mechanism. Since turbulence is damped within one large-scale eddy turnover time, stretching of lamella or vortices is assumed to be negligible. The model assumes perfect macromixing in the impingement zone, which implies that a slab of A is next to a slab of B at all times. A slab thickness equal to η is not assumed.

For an acid-base neutralization, a simplified form of the governing equation is Fick's second law:

$$\frac{\partial c}{\partial t} = D\frac{\partial^2 c}{\partial x^2} \quad (1)$$

where concentration, c, is defined as:

$$c(x,t) = c_A(x,t) - c_B(x,t) \quad (2)$$

The boundary conditions are,

$$t > 0,\ x = \pm L \qquad \partial c/\partial x = 0 \quad (3)$$

where L is the half slab thickness. The initial conditions at $t = 0$ are:

$$c(x,0) = c_{A0} \quad for - L \le x < 0$$
$$c(x,0) = c_{Bo} \quad for\ \ 0 < x \le +L \quad (4)$$

Since the reaction is instantaneous and irreversible, it is diffusion controlled and A and B cannot coexist. With $c(x,t)$ defined by Eq. 2 and requiring that the diffusion coefficient (D) is identical for both A and B, the analytical solution is:

$$c(x,t) = \frac{c_{Ao} - c_{Bo}}{2} + \sum_{n=1}^{\infty} exp\left(-\frac{Dn^2\pi^2 t}{4L^2}\right)\frac{2}{n\pi} (c_{Ao} + c_{Bo}) sin\left(\frac{n\pi}{2}\right) cos\left[\frac{n\pi}{2L}(x+L)\right] \quad (5)$$

Figure 1 illustrates how the concentration profiles vary with time. Since flow within a mixed sheet (not the impingement zone) is along radial lines with no crossover of liquid, Figure 1 should be representative of the progress of the neutralization reaction along a radial line. This radial line flow offers the opportunity to visually observe the progress of the neutralization reaction within the mixed sheet.

In this study the neutralization of NaOH with HCl is followed via the change in fluorescence intensity of a fluorophore added to the reactants. The fluorescence intensity along a radial line is measured from photographs using ImageJ software. From the fluorescence intensity vs time profiles, the slab thickness is determined and related to dissipation microscales of turbulence.

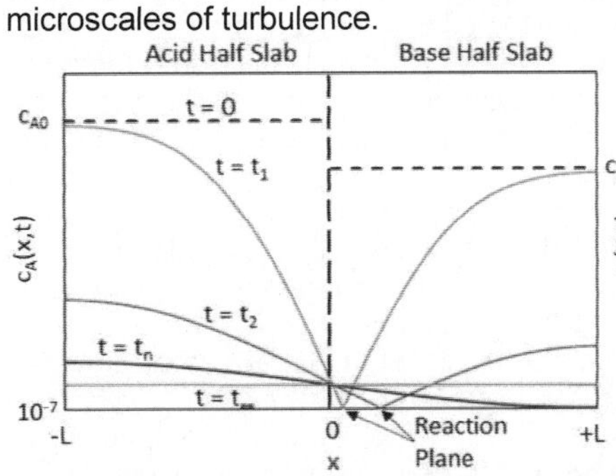

Figure 1. Concentration profiles in slabs of acid and base due to neutralization reaction. t_n is the neutralization time and t_∞ represents the concentration profile of residual acid at infinite time. Drawing is not to scale with initial concentrations used in the experiments.

https://doi.org/10.52202/074122-0025

Material and Methods

Equal single sheets of HCl and NaOH in stoichiometric ratios of 2:1 and 4:1 were impinged. The fluorophore was umbelliferone, which was added in equal concentrations to all solutions. Umbelliferone fluoresces brightly under basic conditions but is almost non-fluorescent at a pH less than 3. The impinging sheets were illuminated by a 365 nm UV light panel with dimensions of 8 cm x 12 cm. Since the UV power source was not constant over the entire mixed sheet, the relationship between fluorescence intensity and concentration was determined by impinging identical reference buffer solutions at one another, i.e. non-reactive solutions of known pH (as shown in the Graphical Illustration). Unlike the acid-base impingement, there were no concentration profiles in the mixed sheet of the buffer solutions and thus the measured fluorescence intensity (I_F) at any point in the mixed sheet applied to the pH of the buffer solution. For each experimental run, buffer solutions of pH 10, 8, 7, and 6 were typically self-impinged along with acid-acid impingement and base-base impingement.

The range of experimental conditions included impingement angles (2β) from 28° to 105°, single sheet velocities (u) from 3 m/s to 13.5 m/s, kinematic viscosities (ν) from $1x10^{-6}$ m²/s to $1.3x10^{-6}$ m²/s, and single sheet thicknesses at impingement (s_i) from 24 μm to 290 μm.

Some of the photographs from a typical experiment are shown in the Graphical Illustration. These photographs were imported into ImageJ as a stack so that a radial line, starting at the single sheet origin, could be drawn on all images at the exact same location. The I_F profiles of the radial lines for all solutions were measured using ImageJ and exported to a spreadsheet for further analysis. For the reference solutions, the pH was constant, so the profiles illustrate how I_F for a constant pH mixed sheet varies with UV light power and distance from the impingement zone. For the acid-base impingement, the pH varied along a radial line indicating the progress of the neutralization reaction.

Results and Discussion

Although an analytical solution exists for Eq. 1 with conditions given by Eqs. 3 and 4, the solution cannot be readily applied because the diffusion coefficients of acid and base are significantly different. Instead, a MatLab simulation of Eqs. 1 through 4 was constructed, which simulation was verified for accuracy by comparison with the analytical solution. The analytical solution and simulation provide the concentration gradients of hydroxide ion (or hydrogen ion) at any distance from the impingement zone along a radial line. Software code was separately added to the simulation to calculate the fluorescence intensity.

In order to calculate the fluorescence intensity (I_F) for comparison with experimental results, the concentration profile (Fig. 1) must be related to I_F. For the reference, constant-pH solutions, the coefficients of an I_F vs [OH^-] equation were determined at each pixel along a radial line. The resulting I_F vs [OH^-] equation produced a curve typical of an acid-base titration. A 4PL equation provided an excellent fit ($R^2 > 0.995$) at almost all pixel locations along the radial line. Exceptions were in the impingement zone and at the end of the mixed sheet where it breaks up into droplets. Typically, 1,000 to 2,000 I_F vs [OH^-] equations were determined: one for each pixel of distance from the impingement zone. These equations were then used by the simulation to calculate the associated I_F of the pH as it varied with time and distance as illustrated in Fig. 1.

The simulation divided each half slab into 100 slices. Since the velocity in the mixed sheets is constant, the time between pixels on the photographs is readily calculated. Typical time intervals were 2 to 30 microseconds with simulation incremental times about 1/10th of the pixel times. For each slice I_F was determined, and the total I_F was calculated as the sum of the I_F values of the 200 slices.

To start the simulation, a value of L was estimated and then adjusted as necessary to increase agreement between the experimental results and the simulation. Typically, L was increased or decreased by 0.5 microns, except at high ϵ where the increment was 0.25 microns. Figure 2 is a typical result of a comparison between the experimental I_F and the simulated I_F. Results for 103 radial lines from 36 experimental runs were obtained.

Figure 2. Comparison of experimental I_F profile with

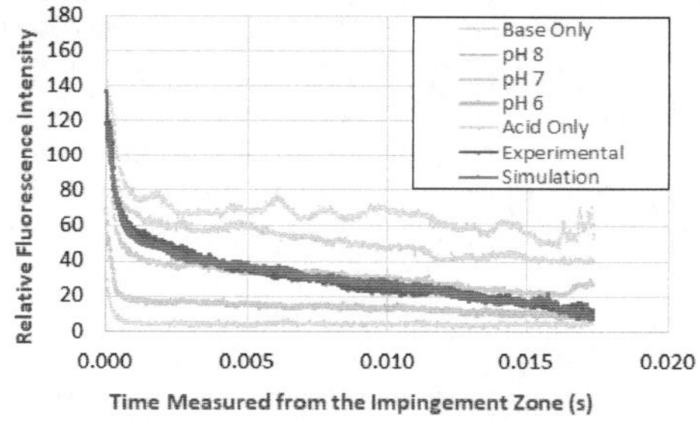

simulated I_F profile at L = 8 microns. Also shown are I_F profiles for reference mixed sheets at constant pH of 2.4 (acid), 6, 7, 8, and 11.2 (base). $c_{Ao}/c_{Bo} = 4$.

The experimental slab thicknesses ($2L$) were compared with turbulence microscales: isotropic Taylor microscale [$\lambda_g = (15\nu v'^2/\epsilon)^{0.5}$], Kolmogorov microscale [$\eta = (\nu^3/\epsilon)^{0.25}$] and an order of magnitude dissipative microscale [$\lambda = (\nu/\epsilon)^{0.5} v'$]. ϵ is the energy dissipation rate and ν is the kinematic viscosity. v' is the velocity associated with the energy given up at impingement, given by [3]:

$$v' = \sqrt{u^2(1 - COR^2)} \quad (6)$$

Figure 3 plots the mixing model slab thickness and these turbulence microscales divided by v' versus ν/ϵ. Figure 3 shows that $2L/v'$ lies between λ_g/v' and λ/v', but is significantly greater than η/v'. The experimental data in Figure 3 are well fit ($R^2 = 0.973$) by the following equation:

$$\frac{2L}{v'} = 0.1413 \left(\frac{\nu}{\epsilon}\right)^{0.3906} \quad (7)$$

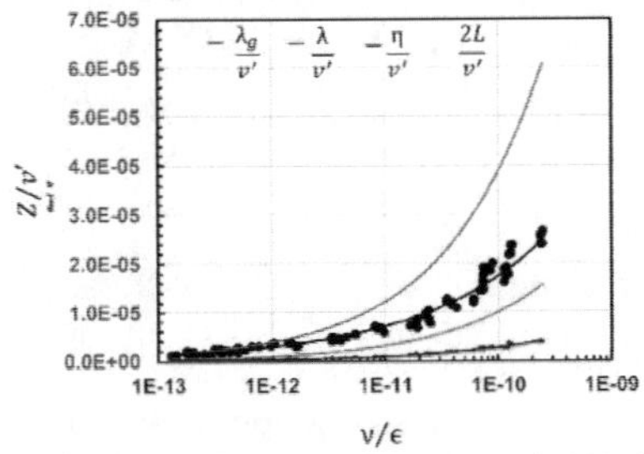

Figure 3. Comparison of various microscales (Z) divided by the fluctuating velocity, v', plotted against ν/ϵ.

However, the exponent in Eq. 7 may be masking the effects of other parameters since the experimental results lie between λ_g and λ, which are functions of $(\nu/\epsilon)^{0.5}$. Since the turbulence generated in the impingement zone is not fully developed, this additional parameter is likely to be the Reynolds number (Re).

In turbulent flows the range of eddy sizes is related to the Reynolds number of the largest eddy (size scale equal to Λ). For example, $\eta/\Lambda \sim (Re_\Lambda)^{-0.75}$ and $\lambda_g/\Lambda \sim (Re_\Lambda)^{-0.5}$. The study on the energy dissipation rate of impinging sheets [3] found that $t_\Lambda = C_{\epsilon A} D\Delta r/u$, where t_Λ is the large eddy turnover time, D is a geometric factor for the impingement zone volume, Δr is a length scale equal to $s_i/sin\beta$, and C_ϵ is the dimensionless dissipation rate coefficient, which for decaying turbulence has a minimum asymptotic value, $C_{\epsilon A}$, $\approx$ 0.5 [5]. Based on t_Λ , $\Lambda = C_{\epsilon A} D\Delta r$ and $Re_\Lambda = C_{\epsilon A} D\Delta r u/\nu$. The experimental results show that the parameter $2L/\Lambda$ is well fit (R^2 = 0.951) by Re_Λ multiplied by C_ϵ.

$$\frac{2L}{\Lambda} \approx 10(C_\epsilon Re_\Lambda)^{-0.75} \quad (8)$$

The energy dissipation rate (ϵ) for impinging sheets is given by [3]:

$$\epsilon = \frac{C_\epsilon}{C_{\epsilon A}}\epsilon_{IZ} = \frac{C_\epsilon v'^2 u}{C_{\epsilon A} D\Delta r} \quad (9)$$

where ϵ_{IZ} is a calculated value of ϵ based on the volume of dissipation equal to the impingement zone volume. When ϵ_{IZ} > 3x10^4 W/kg, $C_\epsilon \approx C_{\epsilon A}$. For the experimental results, ϵ varied from 5x10^3 W/kg to 7.7x10^6 W/kg. Rewriting Eq. 9 as $v'^2/\epsilon = C_{\epsilon A} D\Delta r\,/C_\epsilon u$ and substituting into the expression for the Taylor microscale yields:

$$\lambda_g = \left(\frac{15\nu C_{\epsilon A} D\Delta r}{C_\epsilon u}\right)^{0.5} \quad (10)$$

Dividing both sides by Λ results in,

$$\frac{\lambda_g}{\Lambda} = \left(\frac{15\nu C_{\epsilon A} D\Delta r}{C_\epsilon u C_{\epsilon A}^2 D^2 \Delta r^2}\right)^{0.5} = 3.873(C_\epsilon Re_\Lambda)^{-0.5} \quad (11)$$

Now, dividing Eq. 8 by Eq. 11 gives:

$$\frac{2L}{\lambda_g} \approx 2.58{C_\epsilon}^{-1/4}{Re_\Lambda}^{-1/4} \quad (12)$$

Substituting the definition for the Taylor microscale into Eq. 12 yields an expression for $2L/v'$ that includes the Reynolds number:

$$\frac{2L}{v'} \approx 10{C_\epsilon}^{-1/4}\left(\frac{\nu}{\epsilon}\right)^{1/2}{Re_\Lambda}^{-1/4} \quad (13)$$

Thus, $2L/v'$ varies with ν/ϵ in the same way as the Taylor microscale, and $2L/\Lambda$ is a function of Re_Λ in the same manner as the Kolmogorov microscale.

Figures 4a and 4b compare experimental values of $2L$ with calculated values of $2L$ using Eq. 13 and Eq. 7. The inclusion of Re_Λ provides improved estimates of $2L$ (R^2 = 0.935 in Fig. 4a while R^2 =0.889 in Fig. 4b). In Fig. 4a the best linear fit through the data is a closer fit to the line of equality than in Fig. 4b.

Also, the results from the earlier azo coupling study [1] were re-examined. Previously these experimental results were fit to the EDD model assuming a value of $2L = \eta$ for low-viscosity data and $2L = \sqrt{2}\lambda$ for high-viscosity data. $2L$ was recalculated using Eq. 13 for all of the azo-coupling data, yielding a satisfactory fit between experimental results and the model.

Finally, when $2L/\eta$ is plotted against Re_Λ, $2L/\eta$ tends towards an asymptotic value of about 6 at sufficiently high Re. Since the ratio of the turbulence scale of maximum dissipation (λ_d) to $\eta \approx 5$ [6], $2L \approx \lambda_d$ at sufficiently high Re and possibly at lower Re as well.

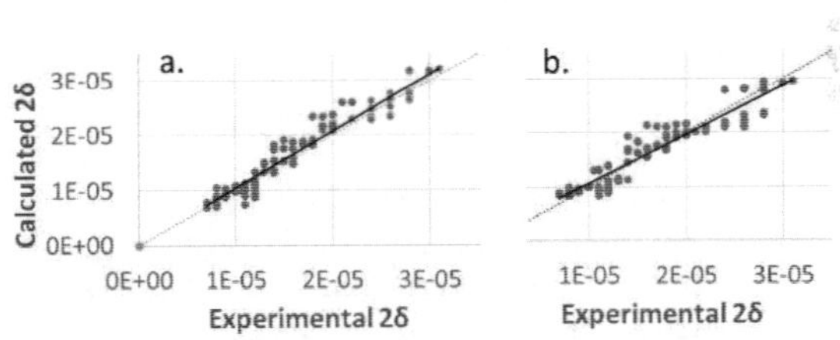

Figure 4. Comparison of $2L$ calculated by equation versus the experimental value of $2L$. 4a uses Eq. 13; 4b uses Eq. 7. ---- line of equality, — best linear fit to the data.

Conclusions

A simple micromixing model was used to determine the fixed slab thickness for diffusion of acid and base within the mixed sheet produced by equal impinging sheets. The diffusion slab thickness, $2L$, was found to lie between the Taylor microscale, λ_g, and a dissipative microscale, $\lambda = \lambda_g/(15)^{0.5}$. Similar to λ_g, $2L$ is a function of $(\nu/\epsilon)^{1/2}$ and is also a function of the Re of the large eddies in the impingement zone to the $-1/4$ power. The ratio of the slab thickness to the Kolmogorov microscale, $2L/\eta$, tends towards an asymptotic value of $\approx$ 6 with increasing Reynolds number. The turbulence microscale at which most of the dissipation occurs (λ_d) tends towards a value of λ_d/η = 5. Thus, the scalar diffusion thickness, $2L$, for impinging sheets is approximately equal to the turbulence length scale of maximum dissipation, λ_d.

References

[1] R.J. Demyanovich, J.R. Bourne, *Ind Eng Chem Res,* 28(6) (1989) 830.
[2] R.J. Demyanovich, *Chem Eng Sci X,* 12 (2021) 100113.
[3] R.J. Demyanovich, *Mixing 17*, Porto, Portugal, **2023.**
[4] J. Baldyga, J.R. Bourne, *Chem Eng Comm*, 28 (1984) 243.
[5] W.D. McComb, A. Berera, S.R. Yoffe, M.F. Linkmann, *Phys. Rev. E,* 91 (2015).
[6] *A First Course in Turbulence.* H. Tennekes, J. Lumley, Cambridge, MA, The MIT Press(1972)..

https://doi.org/10.52202/074122-0025

A dynamic compartment model approach for heterogeneous reactors MT1.6

Jérôme Morchain[1], *Carlos Mayorga*[1], *Alain Liné*[1], *Philippe Villedieu*[2], *(1) Toulouse University, INSA-TBI, 135 Avenue de Rangueil, 31077 Toulouse cedex, France. (2):Institute of Mathematics of Toulouse, Université de Toulouse, CNRS, INSA Toulouse, France*

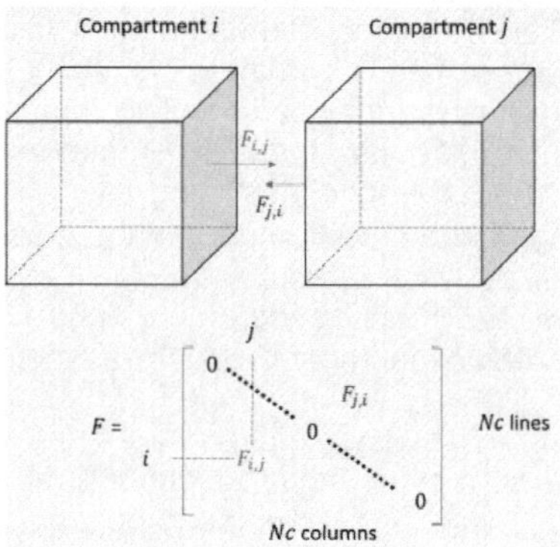

Multiple instantaneous flow fields simulated with CFD are stored and used to produce a dynamic description of fluid motion in the framework of compartment modelling. Periodic contributions due to impeller rotation are preserved and allow a good prediction of mixing dynamics at a minimal computational cost. Adding a stochastic particle transport algorithm allows computing local source terms via a Monte-Carlo approach. This eliminates the usual limitations regarding the number of properties attached to the suspended phase. Also it becomes straightforward to implement dynamic models for the particulate phase without any restriction in terms of complexity.

Introduction

The modelling and simulation of heterogeneous reactors, i.e. the coupling between mixing and reaction in multiphase reactor is a challenging task. It raises a number of theoretical and practical issues among which the formulation of closure terms for mass transfer between phases[1], the number of particle properties involved, the calculation of scalar and particle transport. The most rigorous approach is to combine probability density function to computational fluid dynamics. However, these demanding approaches are not affordable when the number of particle properties is large and/or the duration of the process is long. Indeed, solving multivariate population balance equation in the framework of a CFD code is practically out of reach if the dimension of particle property vector exceeds three. Also, running a CFD simulation over the total duration of a typical bioprocess (days) is not conceivable. An alternative is to build a compartment model through the spatial integration of the time averaged flow field obtained from a CFD simulation.

The originality of the proposed work is to use a series of instantaneous 3D flow fields to build an equal number of flow matrices between compartments leading to a dynamic compartment model approach. Coupled to a Monte-Carlo approach for the calculation of local source terms, this model reduction strategy allows the real-time simulation of mixing and reaction in a heterogeneous (bio) reactor.

Material and Methods

Up to now, the compartment model approach (CMA) was used to produce a time independent matrix of flowrate between compartments and an additional matrix of turbulent flux was added to account for the effect of turbulence on scalar transport[2]. In this work a series of 14 instantaneous 3D flow field are extracted from the unsteady CFD simulation of a 70 L stirred tank equipped with a Rushton turbine, during one impeller rotation. Each saved flow field is processed to produce a matrix of flow rates M_t, between the compartments. Hence, the contribution of periodic flow motion is embedded in the CMA which thus become dynamic. The scalar transport equation writes:

$$\frac{\partial CV}{\partial t} = M_t C + S.V$$

In order to compute the local source terms due to the reactions taking place in the suspended phase (either physical particles or living cells), a Monte-Carlo method is used. This approach is preferred to the integration of the density distribution in mass $m(\xi, \mathrm{x}, \mathrm{t})$. The advantage is the absence of limitation regarding the number of particle properties, ξ_p , one requirement is to use a sufficiently large number of virtual particles, $Np(t)$, to correctly estimate the integral source term.

$$S_j = \int m(\xi, \mathrm{x}, \mathrm{t})\, \mathrm{q}_j(\xi, t, x)\, d\xi$$
$$\cong \frac{1}{V} \sum_{p=1}^{Np(t)} w_p(t)\, \mathrm{q}_j\big(\xi_\mathrm{p}(t), t, x_p(t)\big)$$

Each numerical particle is known through its vector of internal properties $\xi_\mathrm{p}(t)$, its weight (i.e. the number of cells statistically represented by the particle) $w_p(t)$ and its location $x_p(t)$. The computation of the specific reaction rates $\mathrm{q}_j\big(\xi_\mathrm{p}(t), t, x_p(t)\big)$ depends on a concentration vector $C(x, t)$. It also refers to the particle properties and chemical reaction or a metabolic model. The change in particle properties requires a set of dynamic equations $\dot{\xi}$ while the change in particle location uses a stochastic particle tracking algorithm[3].

Results and Discussion

The dynamic compartment approach is first used to simulate the mixing of an inert scalar in a stirred tank. These results expressed in normalized

https://doi.org/10.52202/074122-0026

concentration are compared to the normalized particle density (number of particles per unit volume) divided by the expected density (total number of particles/ total volume). **Figure 1** show that the scalar transport predicts a mixing time between 15 and 20s similar to the mixing time obtained from correlations or in previously conducted LES simulation of the same reactor[4]. Also, the algorithm for particle transport is validated since identical results are found.

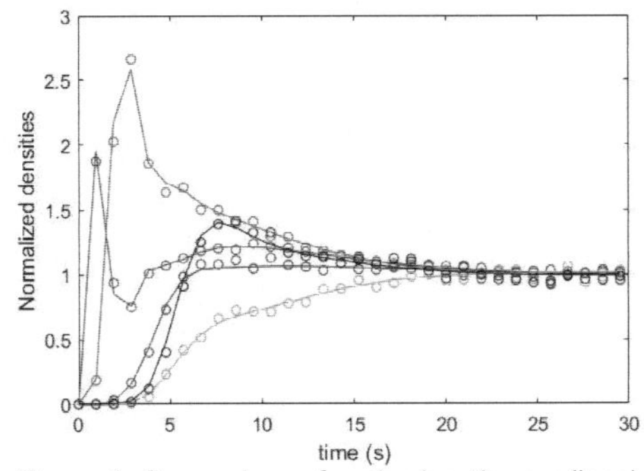

Figure 1. Comparison of scalar (continuous lines) and particle mixing (symbols) in a 70 L stirred tank.

It is observed that the variance of the scalar concentration and particle density fields are also very close to one another. Figure 2 further reveals that the difference decreases as the number of particles is increased. This observation is quite easy to understand from physical consideration but we propose here a theoretical demonstration which demonstrates the accurate implementation of the particle tracking algorithm and its consistent mathematical behaviour.

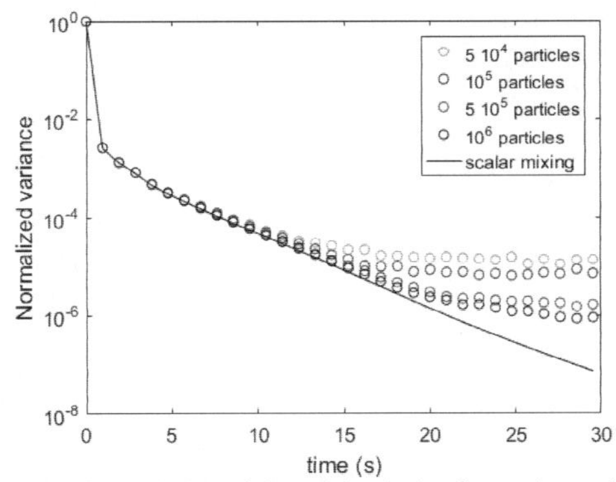

Figure 2. Evolution of the mixing index for scalar and virtual particle mixing.

It can be shown that the relative standard deviation in local particle number, the total number of particles and the volume of a compartment are linked by the relationship depicted in **figure 3**

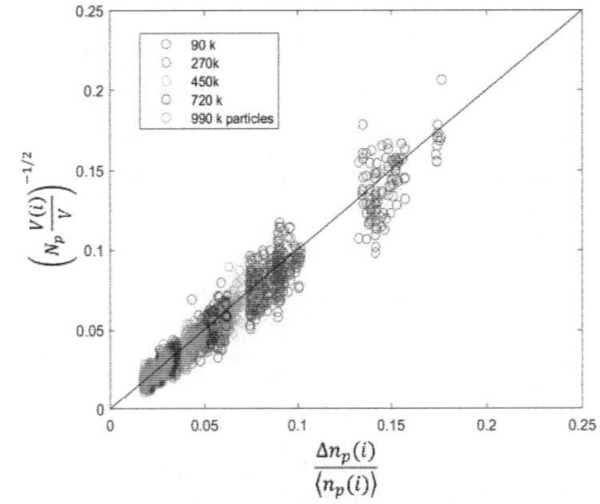

Figure 3. Statistical validation of the numerical algorithm for particle transport.

Finally, a reaction term is added to the simulation. Two approaches are considered: homogeneous reaction involving a dissolved scalar and a heterogeneous reaction where the dissolved scalar concentration is modified because of the consumption by virtual particles. The reactivity of virtual particle is adjusted so that the volumetric reaction term should be identical in both situations.

If the number of particle par compartment is large enough, fluctuations in local particle density (dashed lines) are moderate and the Monte-Carlo calculation of the local source term is identical to the homogeneous reaction rate, Figure 4. Obviously, the first order reaction described here is of little interest in itself but it is used to demonstrate the accuracy of the Monte-Carlo procedure implemented.

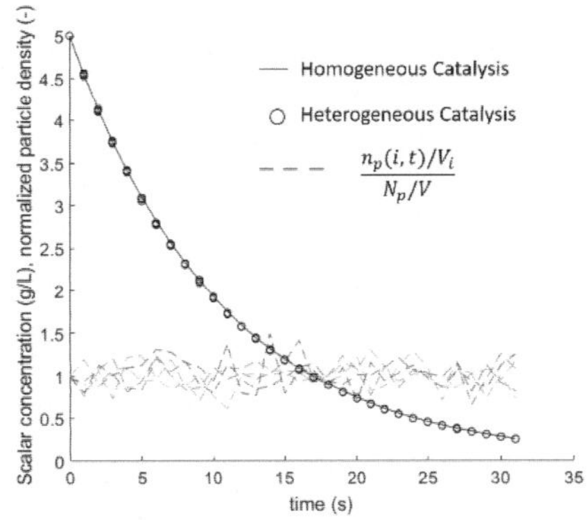

Figure 4. Homogeneous vs. heterogeneous calculation of the source term. Local particle density fluctuations in five randomly selected compartments (dashed lines)

Perspectives

Building upon this methodology, it is possible to perform the dynamic simulation of complex heterogeneous reactors, including bioreactors and to investigate the coupling between heterogeneities in both phases. This can be achieved in real-time while the methodology presents no restriction with respect to the number of properties attached to the suspended phase.

https://doi.org/10.52202/074122-0026

References

[1] Morchain, J.; Quedeville, V.; Fox, R. O.; Villedieu, P. The Closure Issue Related to Liquid–Cell Mass Transfer and Substrate Uptake Dynamics in Biological Systems. *Biotechnol. Bioeng.* **2021**, *118* (7), 2435–2447. https://doi.org/10.1002/bit.27752.

[2] Delafosse, A.; Collignon, M.-L.; Calvo, S.; Delvigne, F.; Crine, M.; Thonart, P.; Toye, D. CFD-Based Compartment Model for Description of Mixing in Bioreactors. *Chem. Eng. Sci.* **2014**, *106* (0), 76–85. https://doi.org/10.1016/j.ces.2013.11.033.

[3] Haringa, C.; Tang, W.; Noorman, H. J. Stochastic Parcel Tracking in an Euler–Lagrange Compartment Model for Fast Simulation of Fermentation Processes. *Biotechnol. Bioeng.* **2022**, *119* (7), 1849–1860. https://doi.org/10.1002/bit.28094.

[4] Delafosse, A. Analyse et Étude Numérique Des Effets de Mélange Dans Un Bioréacteur. PhD Thesis, Institut National des Sciences Appliquées, Toulouse, FRANCE, 2008.

Developing a soft sensor for viscosity measurement during batch mixing with evolving rheology and fill level

ER1.1

***Grace E. Cunningham*[1,2]**, *Mark J.H. Simmons*[1], *Jonathan J. O'Sullivan*[2]. *(1) Centre for Formulation Engineering, School of Chemical Engineering, University of Birmingham, Edgbaston, Birmingham B15 2TT, UK (2) Unilever Research and Development, Port Sunlight Laboratory, Quarry Road East, Bebington, Wirral, CH63 3JW, UK*

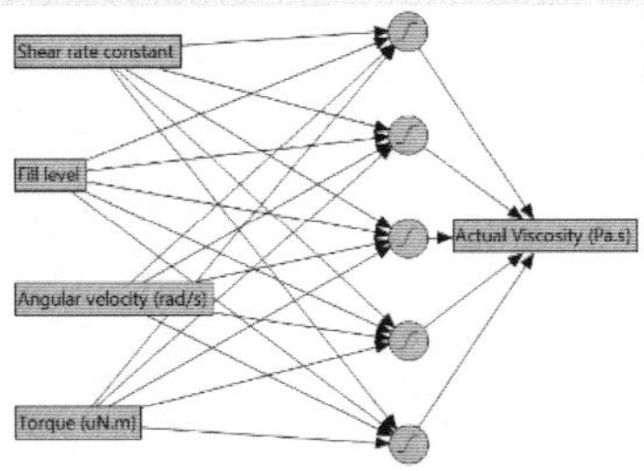

Formulated products are complex, multiphase materials which have been processed in a specific way to generate a structure, which imparts specific performance that the consumer desires. Manufacture of these materials often involves a batch mixing stage, where the rheology of the contents and the fill level change with sequential addition of materials into the vessel. Rheology is a good indication of the progression of the process and can be a critical process parameter. Torque-speed data have previously been used to determine process viscosity, but not when the batch fill level is less than 100% capacity. The purpose of this work is to collect torque-speed data for various rheology fluids at different vessel fill heights and generate a model for viscosity which can be used on-line to aid processing.

Introduction

Formulated products are complex, multiphase materials which have been processed in a precise way to generate a structure which imparts specific characteristics and performance that the consumer desires. Manufacture of these materials often involves a batch mixing stage, where the rheology of the contents and the fill level change with sequential addition of materials into the vessel. Understanding the mixing characteristics during this stage would aid product and process optimisation [1].

Measurement of torque and power during mixing is relatively simple, cost-effective, and suitable for online measurements in industrial applications [2]. Torque measurements enable process optimisation in many ways, for example in cases where a defined total energy input is required to control the product quality, enabling calculation of power per unit volume for use in scale-up, or to infer in-situ viscosity. The use of torque-speed data to determine process viscosity is not a new concept, but no existing techniques have been applied when the batch fill level is less than 100% capacity, which can often be the case during stages of formulated product manufacture.

The purpose of this work is, therefore, to investigate how torque measurements can enable process optimisation at different stages of formulated product manufacture when the batch fill level and rheology are evolving. To achieve this, torque-speed data has been collected using a rheometer for various rheology fluids at different vessel fill heights. Traditional mixer-viscometer techniques (torque curve method [3], matching viscosity method [4] and Couette analogy [5]) were initially applied to the data to investigate their suitability. Finally, a proof-of-concept study to develop an on-line soft sensor for viscosity during formulated product manufacture was achieved using a non-linear approach (neural network regression).

Material and Methods

A DHR-III Hybrid rheometer (TA Instruments, UK) was used to collect torque-speed data at three fill levels: 25%, 62.5%, 100% (height % of total fill height), at a range of speeds (0.125 m/s – 1.5 m/s, for four types of rheology fluids (Table 1). The fluids have complex rheological behaviour, but over the shear rates investigated can be simplified to fit the Ostwald-de-Waele power law model. The rheological properties of the fluids were measured using a cross-hatched parallel plate geometry (Table 1).

Table 1. Flow and consistency indices for investigated fluids [a]

Fluid	Notable rheological behaviour	Consistency index[b], K, Pa.s^n	Flow index, n, -
Glycerol	Newtonian	0.56	-
Model conditioner	Herschel-Bulkley, very shear-thinning	186	0.06
Xanthan gum	Viscoelastic	37	0.25
1.5% Carbopol		76	0.29

[a] Valid over a shear rate range of 10 s^{-1} to 500 s^{-1}
[b] Viscosity for Newtonian fluids if n =1

Three different close clearance geometries were investigated (Figure 1). The 3D printed geometries used were scaled down versions of pilot and factory scale agitators, designed and 3D printed to simplify future scale-up work. Details of the three mixer-viscometer techniques (torque curve method [3], matching viscosity method [4] and Couette analogy [5]) can be found in the literature. JMP statistical software (v 14.0) was used to develop the regression neural network model.

https://doi.org/10.52202/074122-0027

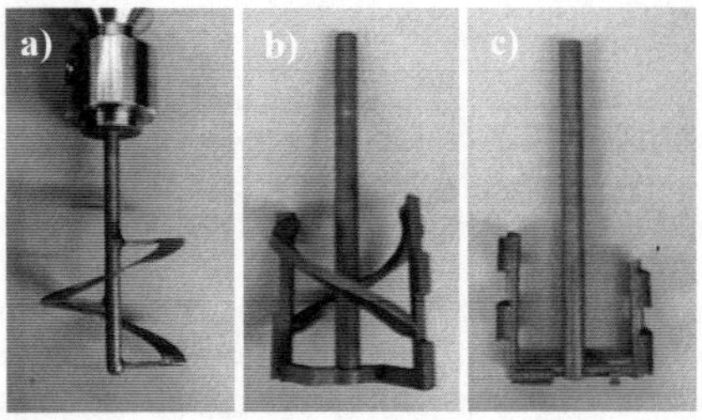

Figure 9. a) TA Instruments helical rotor (TA-HR), b) 3D printed helical ribbon (3D-HR), c) 3D printed anchor (3D-ANC)

Results and Discussion

The power curves for the three investigated geometries at different fill levels were generated using a Newtonian fluid (glycerol). Figure 2 shows an example for the HR geometry, and Figure 3 shows how the power constant varied with fill height for all geometries. At rheometer scale, (total batch size 34 g), the system remains in the laminar regime for all speeds (up to Re ~ 33) and fill levels. We can assume that the more viscous non-Newtonian fluids will also remain in the laminar regime, thus determining viscosity from torque-speed data using traditional methods should be viable.

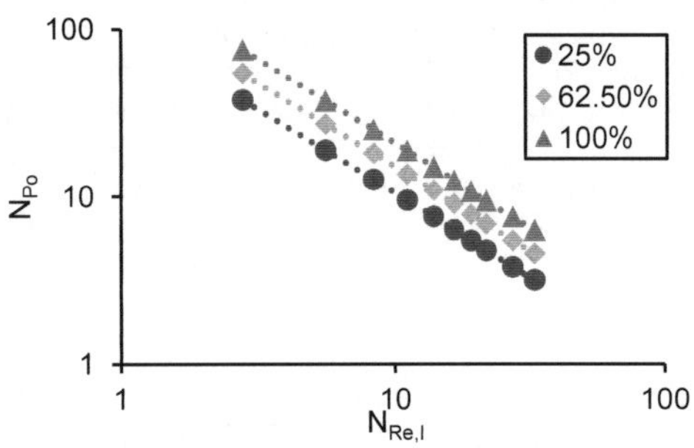

Figure 2. Power number vs. Reynolds number for the HR geometry at different fill levels, calculated using Glycerol

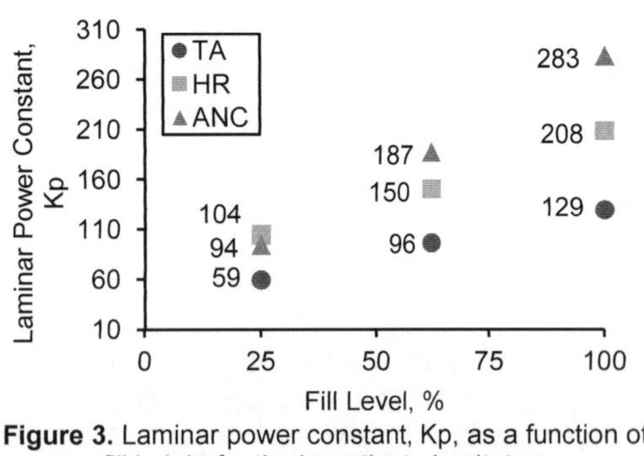

Figure 3. Laminar power constant, Kp, as a function of fill height for the investigated agitators

The authors initially investigated applicability of three different mixer-viscometer approaches (torque curve method, matching viscosity method and Couette analogy), however due to the low fill levels and complex rheology of the fluids affecting torque measurement, none of these approaches were suitable.

For the matching viscosity method and torque curve method, the aim is to characterise the mixer in terms of the mixer constant, k" (Eq. 1), and mixer coefficient (Metzner-Otto constant), k', (Eq. 2) calculated from torque measurements. For the model CD and Carbopol solution at 25% and 62.5%, there is not a linear relationship between the calculated shear rate (Eq. 3) and the angular velocity, thus a single value for k' cannot be applied (example shown in Figure 4).

$$k'' = \frac{\mu\Omega}{M} \qquad \text{(Eq. 1)}$$

$$\dot{\gamma}_a = k'\Omega \qquad \text{(Eq. 2)}$$

$$\dot{\gamma}_a = \left(\frac{Mk''}{K\Omega}\right)^{\frac{1}{n-1}} \qquad \text{(Eq. 3)}$$

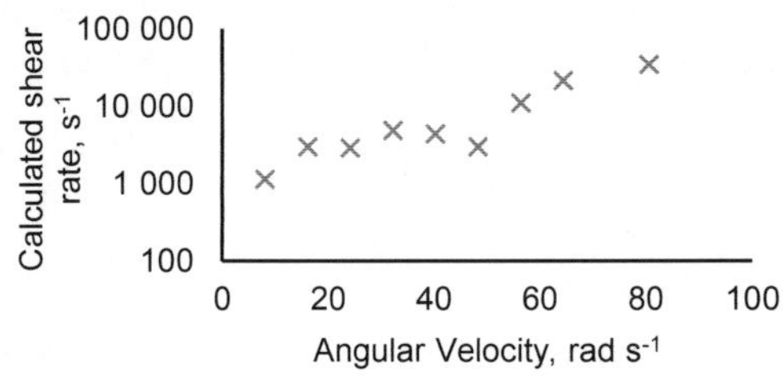

Figure 4. Plot of calculated shear rate (Eq. 3) vs angular velocity for model conditioner at 25% fill level, TA-HR geometry

Using the Couette analogy, the equivalent diameter of a virtual Couette analogue that exhibits the same power consumption as the impeller being investigated is calculated as a function of speed. It is established in the literature that the equivalent diameter is dependent on fluid rheology, and for highly shear-thinning fluids ($n < 0.3$) is proven to not be suitable [6], [7]. Thus, it was decided to explore the use of neural net regression models.

A neural network model was selected as it enables regression modelling of complex, non-linear data. The neural network consists of an input layer, hidden layer and an output layer. In the hidden layer, there are a number of neurons, which each have an associated activation function. Future work will include sensitivity analysis to optimise the model. As this was a proof-of-concept study, JMP software was used to generate the model. The inputs were

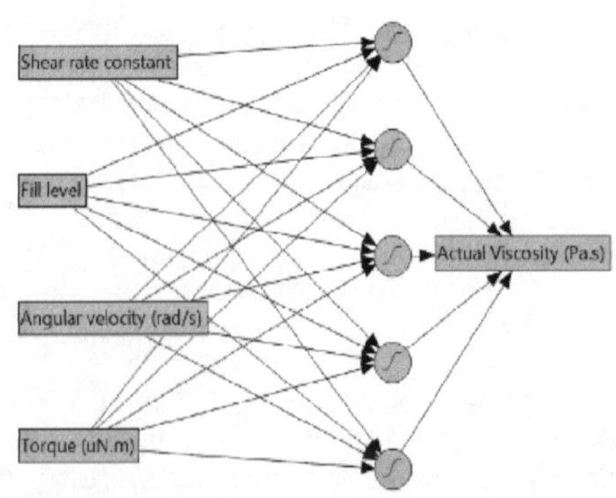

Figure 5. Schematic of neural network model for predicting viscosity of non-Newtonian fluids during manufacture

selected based on experimental observations (Figure 5). The number of neurons was varied between 3 and 6, with 5 neurons providing the highest R^2 (and smallest RMSE (Table 2). A k-fold cross validation technique using 5 folds was used.

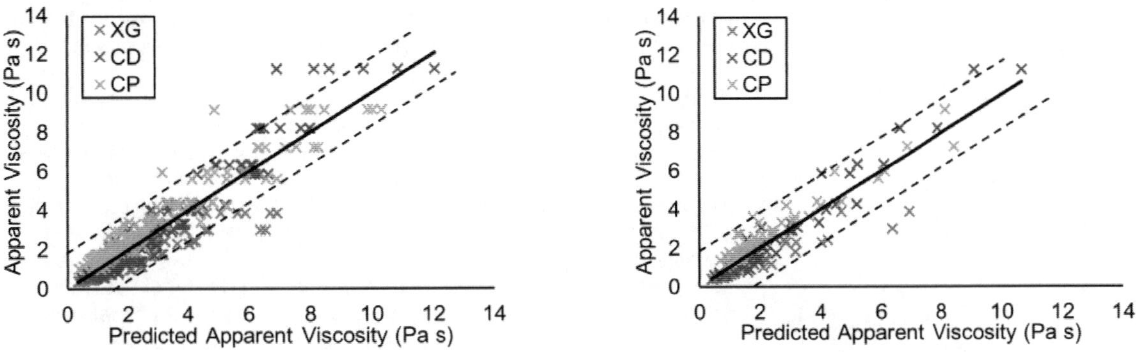

Figure 6. Plot of Actual vs. Predicted Values for Apparent Viscosity (Pa s) for a) training set and b) validation set

The model was improved by excluding the Newtonian fluid data, likely because the viscosity does not vary with shear rate as it does for non-Newtonian fluids. Figure 6 shows a plot of predicted vs actual value for apparent viscosity (Pa s).

Table 2. Statistics values for neural net regression model to predict viscosity

Value	Training	Validation
R^2	0.83	0.84
RMSE	0.48	0.47

Conclusions

A model has been developed which enables an adequate prediction of apparent viscosity for the three non-Newtonian fluids investigated over a range of speeds and fill levels, when the speed, fill level, geometry shear rate constant and torque measurements are known. This model could be applied to larger scale and used as a process optimisation tool during formulated product manufacture. The next steps of this work, to collect data and generate a model at pilot scale are currently underway.

This concept provides a novel approach to inferring viscosity from torque when the fill level of the batch is evolving and the rheology of the contents are unknown. This will enable improvement in scale-up and product quality for formulated non-Newtonian products, reducing waste and saving energy.

Acknowledgments

The authors would like to acknowledge Darren Lamb, Emily Robinson, Serena Todd, and Adam Rowatt from University of Strathclyde for their support designing the bespke mixing elements, and Alan Magowan from Laser Proto Europe Ltd for providing 3D printing services. Additionally, the authors acknowledge financial support received from the Centre for Doctoral Training in Formulation Engineering (EPSRC grant no.EP/S023070.1)

References

[1] M. Edwards, "Product Engineering - Some Challenges for Chemical Engineers," *Trans IChemeE,* vol. 84, no. A4, pp. 255-260, 2006.

[2] A. L. Bowler, S. Bakalis and N. J. Watson, "A review of in-line and on-line measurement techniques to monitor industrial mixing processes," *Chemical Engineering Research and Design,* vol. 153, pp. 463-495, 2020.

[3] M. E. Castell-Perez and J. F. Steffe, "Evaluating shear rates for power law fluids in mixer viscometry," *Journal of Texture Studies,* vol. 21, pp. 439-453, 1990.

[4] J. F. Steffe, Rheological Methods in Food Process Engineering, 2nd ed., East Lansing, Michigan: Freeman Press, 1996.

[5] A. Ait-Kadi, P. Marchal, L. Choplin, A.-S. Chrissemant and M. Bousmina, "Quantitative Analysis of Mixer-Type Rheometers using the Couette Analogy," *The Canadian Journal of Chemical Engineering,* vol. 80, 2002.

[6] P. Carreau, R. Chabra and J. Cheng, "Effect of rheological properties on power consumption with helical ribbon agitators," *Am. Inst. Chem. Eng. J.* , vol. 39, pp. 1421-1430, 1993.

[7] R. K. Thakur, C. Vial, G. Djelveh and M. Labbafi, "Mixing of complex fluids with flat-bladed impellers: effect of impeller geometry and highly shear-thinning behaviour," *Chemical Engineering and Processing,* vol. 43, pp. 1211-1222, 2004.

https://doi.org/10.52202/074122-0027

Population Balance Modelling of W/O/W Double Emulsions Preparation in Static Mixers

ER1.2

N. Lebaz[1], *K. Touma*[1], *N. Sheibat-Othman*[1]. *(1) University of Lyon, University Claude Bernard Lyon 1, CNRS, LAGEPP UMR 5007, F-69100, Villeurbanne, France, noureddine.lebaz@univ-lyon1.fr*

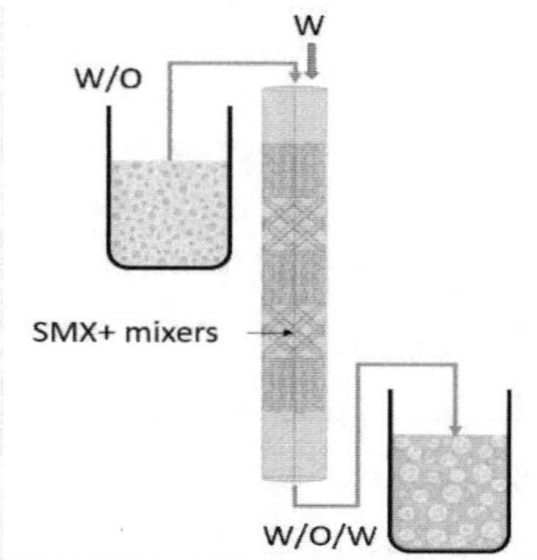

In this study, a novel experimental process is investigated for the preparation of double emulsions. The double emulsification is based on a two-step method: in the first step, a fine W/O emulsion is prepared using a high shear device and, in the second step, this primary emulsion is pumped into water through SMX+ static mixers for the formation of W/O/W double emulsions. Various physical chemical properties of the different phases and process parameters have been varied. It was demonstrated that static mixers are efficient for W/O/W double emulsification as a continuous preparation process. A population balance model was developed for the prediction of the droplet size distribution (DSD) of the outer emulsion at the downstream of the static mixers. The model was validated against experimental data and showed to be accurate for the prediction of droplet breakage occurring within the static mixers.

Introduction

Double emulsions (DEs), also known as liquid membrane systems, are small droplets immersed within larger ones. They are employed for instance for setting up fat-reduced food products or encapsulating different molecules in a single carrier for the formulation of pharmaceutics [1].

DEs are usually prepared using a two-step approach in which two successive emulsifications are performed. First, a primary emulsion is prepared where the inner phase is dispersed into the intermediate phase under high shear energy to form small inner droplets. Second, the primary emulsion is dispersed into a continuous external phase to form the outer droplets. The inner and outer droplets are stabilized using surfactants. During the second preparation step, different phenomena may occur, including outer droplet breakage and coalescence, and escape of the inner droplets. These phenomena are governed by physical chemical parameters as well as the energy dissipated by the emulsification device during the preparation. Very few modelling attempts regarding the formation of such systems are available in literature, which represents the main objective of this work.

Most of the reported DEs are prepared in batch mixing processes using impellers or rotor-stator devices. At the industrial scale, the development of continuous emulsification processes would be beneficial. The aim of this study is to investigate a novel double emulsification process based on the use of static mixers. Static mixers are motionless porous structures inserted into cylindrical pipes. They are largely employed at the industrial scale for miscible fluids mixing, mass and heat transfer enhancement and preparation of single emulsions. They present other advantages as recently reviewed by Valdés et al. (2022) [2]. A population balance model will be developed to describe droplet breakage within the static mixers and validated over a wide range of operating conditions. While modelling simple emulsification in static mixers was already addressed in the literature, the preparation of double emulsions fundamentally differs and requires particular treatment. First, as the inner droplets are themselves constituted of emulsions and the evolution of their viscosity (which greatly influences the breakage rate) depends not only on the oil phase but also on the encapsulated fraction of inner droplets, and possibly their size. Moreover, some leakage of inner droplets might occur, so affecting the volume of the outer droplets. Finally, salt is generally use to overcome the droplet Laplace pressure and increase the encapsulation efficiency, and it might have an effect on the surface tension.

Material and Methods

The inner W/O emulsion is prepared by mixing silicone oil (in which a lipophilic surfactant (Abil EM97S) is previously dissolved), with water (in which NaCl is previously dissolved at a concentration of 0.1 M) using a high shear device (Ultra-Turrax) at 12 000 rpm for 4 min. The viscosity of the silicone oil is varied from 5 mPa.s to 100 mPa.s and the dispersed phase fraction from 10 to 40 wt.%.

To prepare the W/O/W double emulsions, the inner emulsion is mixed with an aqueous phase (constituted of water, plus glycerol which acts as viscosity enhancer, and the hydrophilic surfactant Tween20) by pumping them through 20 SMX+ static mixing elements in series. The pressure drop in the system is monitored using a pressure gauge (Keller LEO1: 0-3 bar, ± 3 mbar).

The inner droplet size distribution is measured by dynamic light scattering (Malvern Zetasizer Nano ZS®) because of the sub-micron size of the water droplets, after dilution in oil. The viscosity of the W/O emulsions is measured using an MCR 302 rheometer (Anton Paar®). To assess the evolution of the DSD of outer droplets within the mixers, two samples are withdrawn systematically after 5 and 20 mixing elements. The DSD in this

https://doi.org/10.52202/074122-0028

case is monitored using laser diffraction (Mastersizer 3000, Malvern Instruments®). Finally, the quantification of the inner droplets release during preparation in the mixers is achieved by measuring the electrical conductivity of the samples. Indeed, as the aqueous inner phase contains NaCl, their release in the continuous external phase will increase the electrical conductivity. The investigated parameters include the viscosities of the three different phases, the inner droplets fraction and the flow rate during the second emulsification step (i.e. energy dissipation).

Regarding the population balance model, the breakage of the outer droplets is taken as the principle phenomenon occurring within the SMX+ mixers. Coalescence may be neglected because of the low external dispersed phase fraction (1 wt.%) and the saturation of the continuous phase by the hydrophilic surfactant [3]. Ostwald ripening can also be neglected due to the very low solubility of oil in water. Moreover, the spatial uniformity of mixing allows to formulate a homogeneous breakage population balance equation as [4]:

$$\frac{\partial n(v,t)}{\partial t} = \int_v^{\infty} b(v,v')g(v')n(v',t)\mathrm{d}v' - g(v)n(v,t) \quad (1)$$

where $n(v,t)$ is the number-based DSD density, v the outer droplet volume, $g(v)$ the breakage kernel for a droplet of size v and $b(v, v')$ is the daughter size distribution function giving the distribution of daughter droplets generated by the breakage of a droplet of volume v'. The first term on the right-hand side accounts for the birth of droplets of volume v resulting from the breakage of bigger droplets v'. The last term is the death or loss term of droplets of volume v, because of their breakage.

The formulation of the breakage kernel (equation 2) is based on the Coulaloglou and Tavlarides (1977) framework [5], extended first to consider droplets breakage within the inertial and dissipation subranges of turbulence and second to account for both surface and viscosity forces to the cohesion of the dispersed phase [3].

$$g(d) = \frac{C_1\sqrt{\overline{u^2}(d)}}{d}\exp\left(-\frac{C_2\sigma}{\rho_\mathrm{d}\, d\, \overline{u^2}(d)} - \frac{C_3\mu_\mathrm{d}}{\rho_\mathrm{d}^{3/2}\rho_\mathrm{c}^{-1/2}\, d\sqrt{\overline{u^2}(d)}}\right) \quad (2)$$

where d is the droplet diameter, $\bar{u}$ is the mean velocity of a turbulent eddy, σ and μ_d are the surface tension and dynamic viscosity of the outer dispersed phase respectively, ρ_d and ρ_c are the densities of the outer dispersed and external continuous phases respectively, and C_1, C_2, and C_3 are adjustable parameters.

Regarding the daughter size distribution function, the beta function (equation 3) is adopted which assumes that binary breakage of symmetric events has the highest probability [6].

$$b(v,v') = \frac{60}{v'}\left(\frac{v}{v'}\right)^2\left(1-\frac{v}{v'}\right)^2 \quad (3)$$

Equation (1) is solved numerically using the finite volume method [7] and the parameters of the breakage kernel are identified by minimizing the absolute difference between the experimental DSD (in volume fractions) and the corresponding model predictions.

Results and Discussion

The first remarkable result is the validation of the feasibility of preparing double emulsions using a two-step method in which the second step consists of a continuous process based on in line static mixers. At the best of our knowledge, this is the first time that such process is used for DEs preparation. Figure 1 shows an optical microscope image in which the internal water droplets are discernible inside the outer oil droplets confirming their effectiveness encapsulation.

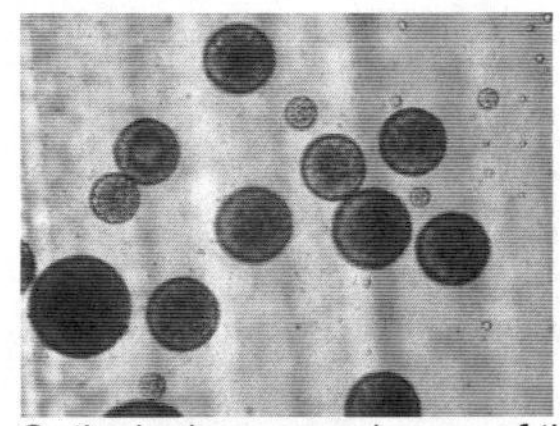

Figure 1. Optical microscope image of the W/O/W double emulsion droplets obtained at the outlet of the static mixers.

As expected, when increasing the inner droplets fraction from 10 to 40 %, the apparent viscosity of the inner emulsion is found to increase exponentially [8]. The increase in the viscosity of the inner emulsion (i.e. more cohesive inner dispersions) leads to a reduction in the breakage efficiency and so the DSDs of the generated DEs shift towards higher sizes as shown in Figure 2.

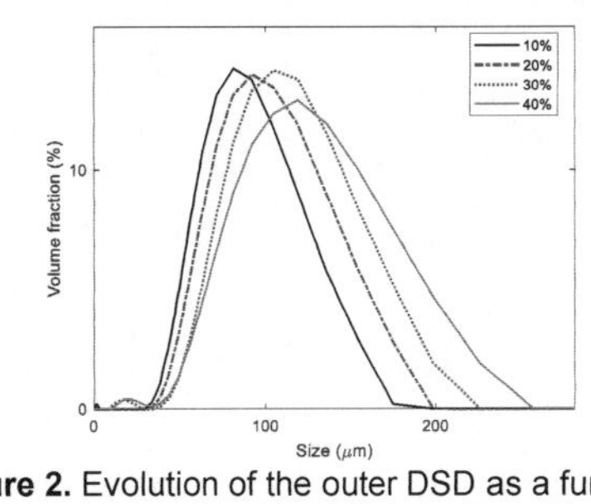

Figure 2. Evolution of the outer DSD as a function of the inner droplets fraction. Oil viscosity = 10 mPa.s.

Oppositely, when increasing the viscosity of the external aqueous phase from 1 to 10 mPa.s, at a constant Reynolds (Re) (in the case of an inner emulsion with 40 % of dispersed phase and an oil viscosity of 10 mPa.s), the DSD of the DE strongly shifts towards smaller sizes (Figure 3). To keep the same Re, the increase of the external phase viscosity is compensated by increasing the flow rate, and so the energy dissipation rate, leading to more breakage events.

https://doi.org/10.52202/074122-0028

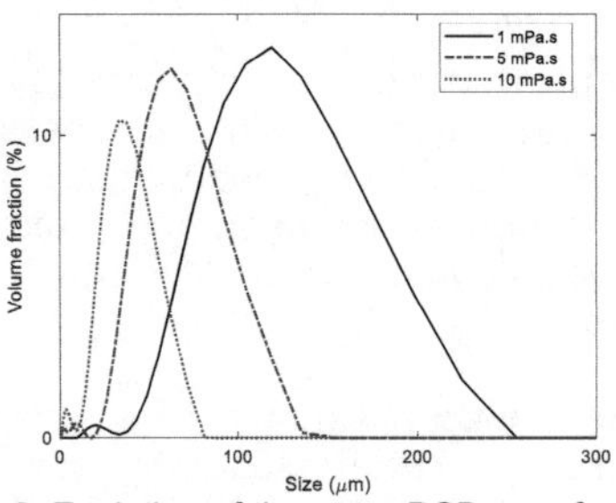

Figure 3. Evolution of the outer DSD as a function of the viscosity of the external phase (Re = 300).

Regarding modelling aspects, the outer DSD after 5 mixing elements is taken as the initial distribution (continuous line in Figures 4/5) and the aim is to predict the DSD at the outlet of the system (dotted black lines in Figures 4/5). As showed in Figures 4/5, the model predictions (dashed magenta lines) are very good with respect to the different parameters investigated such as the inner aqueous phase fraction (Figure 4) and the oil phase viscosity (Figure 5).

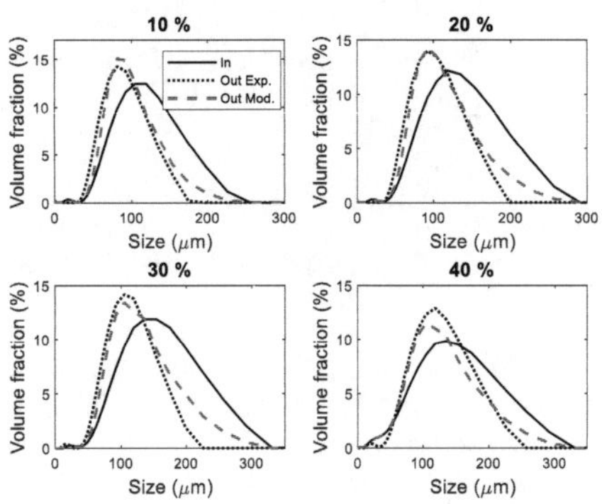

Figure 4. Effect of the inner dispersed phase fraction: experiments vs. model predictions

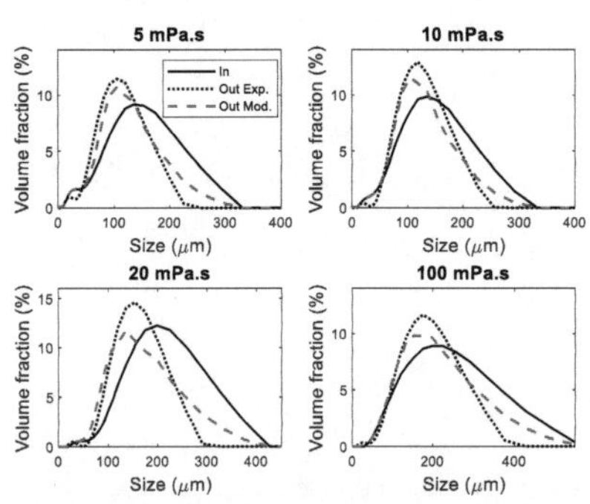

Figure 5. Effect of the oil phase viscosity: experiments vs. model predictions

To give a global look on the quality of the model predictions, Figure 5 compares the volume-based mean diameters (D_{43}) predicted by the model with the experimental measurements. As it can be seen, the predictions are excellent in the different cases despite the large domain of variation of the D_{43}. The deviation is less than 20 % for all the set of experiments.

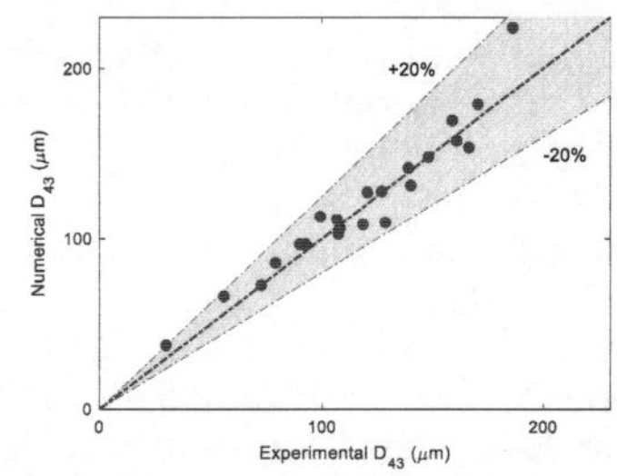

Figure 5. Comparison between the experimental volume-based mean diameters (D_{43}) and those predicted by the PB model for the different investigated cases.

Conclusions

In this study, the preparation of DEs using a two-step method is investigated. In the first step, a rotor-stator device is used for the preparation of W/O fine emulsions which are themselves dispersed in water thanks to the use of SMX+ static mixers. This is the first time that *in line* static mixers are used for the production of DEs. Different parameters are investigated including the viscosity of the different phases, the inner droplets fraction and the flow rate within the static mixers. It is showed that, despite the wide variation of these parameters, the preparation was successful. It was possible to prepare DEs with a sharp DSD with a small mean droplet diameter, around 20 μm. Furthermore, a population balance model is developed. It considers droplets breakage as a unique predominant mechanism in the system. The predictions of the model are excellent with regards to the experimental data, over a wide range of operating conditions.

Acknowledgments

This work was funded by Institut de Chimie de Lyon (ICL) and Agence Nationale de la Recherche under the grant agreement ANR-22-CE51-0006-01.

References

[1] S. Ding, C.A. Serra, T.F. Vandamme, W. Yu, and N. Anton, *Journal of Controlled Releas*, 31 (2019), 295.
[2] J.P. Valdés, L. Kahouadji, and O.K. Matar, *Chemical Engineering Research and Design*, 177 (2022), 694.
[3] N. Lebaz, F. Azizi, and N. Sheibat-Othman, *Industrial & Engineering Chemistry Research*, 61 (2022), 541
[4] F. Azizi, W. Abou-Hweij, N. Lebaz, and N. Sheibat-Othman, *Chemical Engineering Research and Design*, 178 (2022), 382.
[5] C.A. Coulaloglou and L.L. Tavlarides, *Chemical Engineering Science*, 32 (1977), 1289.
[6] M. Laakkonen, V. Alopaeus, and J. Aittamaa, *Chemical Engineering Science*, 61 (2006), 218.
[7] J. Kumar, G. Warnecke, M. Peglow, and S. Heinrich, *Powder Technology*, 189 (2009), 218.
[8] Y. Zhu, H. Gao, W. Liu, L. Zou, and D.J. McClements, *Journal of Texture Studies*, 51 (2020), 45.

Design and Scale up of a Mixing Process for the Manufacture of a Novel Automotive Coating

ER1.3

***H. Yu*[1,2]**, *S. Tsochataridou[1,3], K.Parsons[2], R. Kehn[4], R. Grenville[4], C. D. Rielly[1], N. G. Özcan-Taşkın[1], (1) Loughborough University School of AeroAuto, Chemical and Materials Eng. Department of Chemical Engineering Loughborough LE11 3RQ UK, N.Ozcan-Taskin@lboro.ac.uk; (2) MacDermid Alpha Electronics Solution MacDermid Autotype Ltd. Grove Road, Wantage, Oxfordshire, OX12 7BZ, UK, Herrick.Yu@macdermidenthone.com; (3) Now at: Mars Wrigley Slough SL1 4LG UK; (4)SPX FLOW, Inc. Mixing Solutions 135 Mount Read Boulevard Rochester, New York 14611, USA*

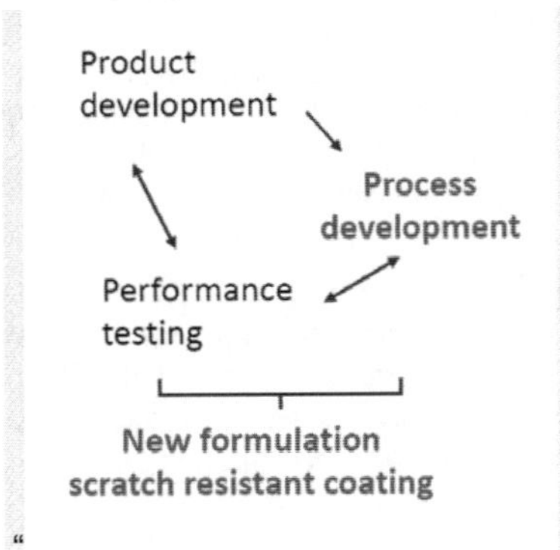

The study performed is part of a larger project aiming to develop high performance automotive coatings and associated processes to enable large scale manufacture of the new product. This part of the project relates to process development with the objective of establishing the comparative blending performance of two types of impeller, a sawtooth and a pitched blade turbine, at different scales. Comparable blending performance is noted in the turbulent and transitional regimes in tanks up to and including T= 0.30 m. In the large scale of T= 0.58 m diameter tank, mixing times were much longer with the sawtooth impeller. Measured and predicted mixing time values were in agreement for both impellers in T=0.30 m and for the PBT also in T= 0.58 m, but much longer for the sawtooth impeller at large scale. This deviation in the performance at large scale has highlighted the complexity of design and scale up using a sawtooth impeller.

Introduction

Improving fuel efficiency has been a major drive for the automotive original equipment manufacturers (OEMs). Consequently, the materials used for various automotive components have changed over the years to reduce vehicle weight and thereby fuel consumption: typical usage of plastics has increased from around 50 kg / vehicle in the mid 70's to > 150 kg / vehicle today. This trend has led to the need to develop high performance coatings for plastic surfaces as many plastics have limitations in terms of durability and physical robustness and are more prone to damage compared to metal. In addition, the range and complexity of functions that require driver interactions with the vehicle have increased with the introduction of touch screens for devices, such as integrated satellite navigation and control systems. It is highly desirable that such surfaces are sufficiently resistant to the effects of common consumer materials. These have brought up the need to develop high performance coatings to be used on interior surfaces to enhance properties such as hardness, scratch resistance, abrasion resistance as well as anti- fingerprinting whilst maintaining desirable optical properties, such as clarity and colour.

Typically, multi-layer coatings are employed to ensure both protection and the desired aesthetic appearance to the finish [1]. The appearance of the top clear coat is prone to degradation under harsh conditions. Scratch and mar, which result from mechanical damage, occur through two primary mechanisms: plastic and fracture flow, depending on the viscoelastic properties of the clear-coat and scratching condition. Two main strategies to produce highly scratch resistant clear coats are the optimisation of cross-linking behaviour of the clear-coat with appropriate components and introduction of reinforcing inorganic fillers into the clear-coat formulation [2]. It has been shown that abrasion-resistant coatings can be obtained incorporating metal oxide particles into formulations [for ex. 3,4]. Inorganic nanoparticles have hardness and elastic modulus greater than the organic polymers and hence introducing these particles into formulations enhances these properties of the coating.

This study has been performed as part of a larger project with the overall aim of developing a new formulation automotive coating with superior performance in quantities beyond lab scale. The three main strands of the project were therefore product development which was performed at small scale (1-2 litres); process development to ensure manufacture in large quantities with properties consistent with those at formulation scale and finally performance testing. Of these, the second is the main focus of this paper in particular the step at which a new additive in the form of a colloidal dispersion is blended with the resin. A few selected results from other parts of the study may also be included in the conference presentation.

The blending performance of a sawtooth impeller, which is typically used for dispersion processes, was investigated in comparison to that of the mixed flow pitched blade turbine (PBT) which is typically preferred for blending purposes. Due to the hazardous nature of the actual formulations, simulant liquids that replicate the rheological properties of these were used, covering a range of viscosities at different scales of operation.

Material and Methods

Experiments were conducted at Loughborough University using three cylindrical, fully baffled tanks. The smallest of these was of a diameter of T= 0.13 m and could be considered to be a replica of that used for formulations at MacDermid in volumes of around 2 litres. The largest tank of T= 0.58 m (~ 160

https://doi.org/10.52202/074122-0029

l) was close to that used at manufacturing scale, 200 litres. An intermediate scale tank of T= 0.30 m, (20 l) was also used. In addition, some results obtained at an independent laboratory, SPX FLOW, at a given scale of T=0.60 m are included for comparison. Drawings and detailed dimensions of the equipment used will be shown at the conference.

Two types of impeller used were a sawtooth impeller and a pitched blade turbine (PBT). There were some differences in the specific geometry of these from one scale to another, in addition to those inherent in the design of a sawtooth impeller. Overall, the sawtooth impellers were of Lightnin R500 type and the PBTs had 6 blades inclined at 45° and were of a diameter of D= T/2. The impellers used at SPX FLOW were of D= T/3, with the PBT having 4 blades.

Newtonian liquids, namely water and glycerol solutions, were used.

Power input was determined with a torque transducer at the smallest scale of T13, a torquemeter at T30 and shaft mounted strain gauge at the largest scale.

Observations of flow patterns, specifically how an added tracer propagates within the bulk, were made at all scales of operation using the colourisation-decolourisation technique, which is based on the colour change between iodine and sodium thiosulphate in the presence of starch indicator. This was performed over a wide range of viscosities and mixing time values were also determined based on visual observations. Mixing time measurements were also taken in T30 and T58 using conductivity probes, placed at 3 to 4 positions. Tracers were prepared using the bulk liquids so as to minimise the effects of physical property differences on mixing time as was shown in previous work [for ex 5,6]. Mixing time experiments were conducted several times: 4 times at SPX FLOW, and 10 times for each data set at Loughborough University.

Results and Discussion

Characteristic power number values were lower for the sawtooth impeller compared to the PBT as expected, also depending on the scale of operation. For example, at the largest scale, the power number for the sawtooth impeller was about 0.33 and 1.04 for the PBT in the turbulent regime. Characteristic power curves at three scales of operation in the Reynolds number range of around 10 to 10^5 will be presented at the conference.

The propagation of the added tracer and the last point of colour change were different with the PBT and sawtooth impellers, owing to the different flow patterns generated. These will be detailed at the conference.

Figures 1.a-c show the mixing time results obtained at three scales using the colourisation-decolourisation technique, Figure

1.d shows those obtained at intermediate and large scales with the conductivity technique also including those from SPX FLOW. At the smallest scale, transitional blending regime was covered; the dependency of mixing time on the volumetric power input was $\theta \propto \left(P/_V\right)^{-2/3}$, as expected, and the two types of impeller were of comparable performance in each of the four viscosity liquids (Figure 1.a). This also held at the intermediate scale over both the transitional and turbulent regimes, with a dependency of $\theta \propto \left(P/_V\right)^{-1/3}$ in the turbulent regime. On the other hand, mixing times recorded for the sawtooth impeller were much longer at the large scale (Figure 1.c).

Mixing time results obtained using the conductivity probes at the intermediate and large scale tanks were compared to predictions from correlations to estimate mixing time for 95% mixedness [7,8]:

$$Po^{1/3}\, Re\, Fo = 5.2 \pm 10\%$$

or

$$\theta_{95} = \frac{5.2\, T^{1.5}\, H^{0.5}}{Po^{1/3}\, N\, D^2}$$

Whilst good agreement is noted for both impellers in T30 and also for the PBT at the large scale, there is significant deviation with the sawtooth impeller at the large scale (Figure 1.d). Results from SPX FLOW also show that the sawtooth impeller is much slower in achieving homogeneity at a comparable scale. There is very limited previous work on the blending performance of sawtooth impellers and only in one case where the liquid height was higher than the tank diameter [9]. This deviation, which is scale dependent, would have significant implications in terms of design and scale up.

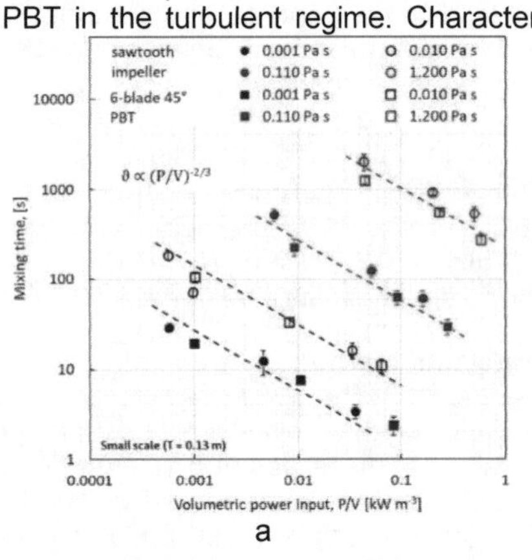

a

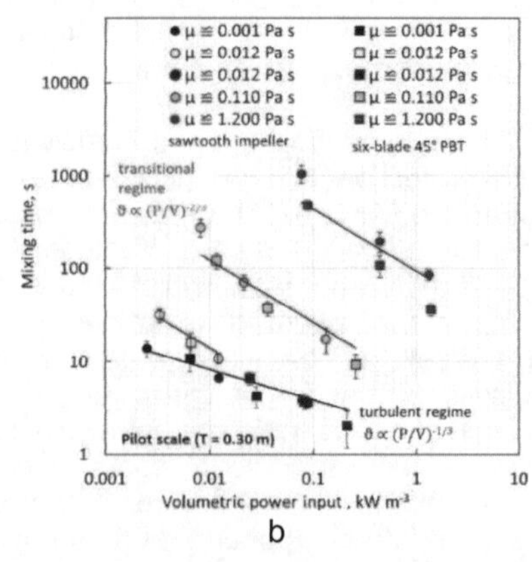

b

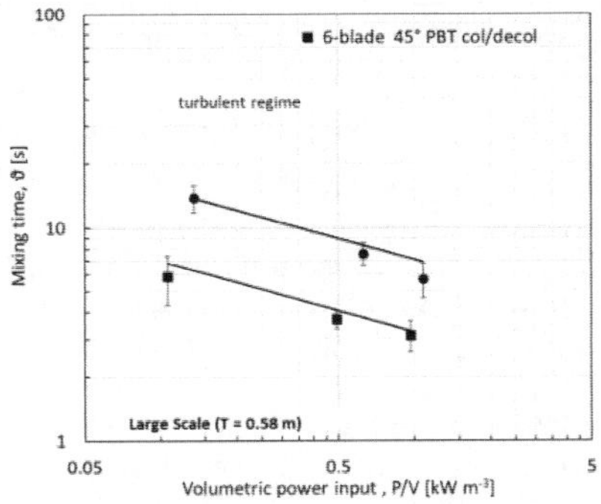

c

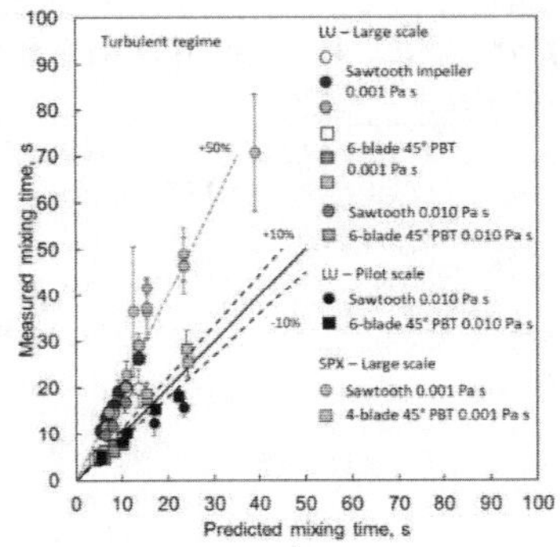

d

Figure 1 Mixing time results at three scales of operation: (a) in T=0.13 m; (b) T= 0.30 m; (c) T= 0.58 m tanks using the colourisation-decolourisation technique; (d) T=0.30, T= 0.58m tanks at LU and T= 0.60 m tank at SPX FLOW using the conductivity technique compared to predictions from the mixing time correlation.

Conclusions

The part of the study presented focuses the design and scale up of a process relevant to large scale production of a novel formulation, high performance automotive coating, developed with the addition of nanoparticles in the form of a colloidal dispersion. Experiments performed at three scales of operation from 2 to 160 litres correspond to those used for formulation and manufacture in industry with an additional intermediate size tank. The study with two types of impeller- a sawtooth and a mixed flow pitched blade turbine- have shown that regardless of the type and geometry of the impeller used, the blending performance is comparable in the turbulent and transitional regimes at a given volumetric power input when impeller-to-tank diameter ratio is also taken into account in tanks up to and including T= 0.30 m. In the large scale of T= 0.58 m diameter tank, the time to reach homogeneity is much longer with the sawtooth impeller as was demonstrated with both the colourisation- decolourisation or conductivity techniques. Mixing time values obtained through conductivity measurements were in agreement with those estimated using mixing time correlations for both impellers at the intermediate scale of T=0.30 m and only for the PBT at the large scale. This deviation in the performance of the sawtooth impeller at large scale, has highlighted the difficulty in making reliable estimates of blend time with this impeller when designing for larger volume manufacture. The poorer performance noted at larger scale would result in product properties being compromised. Further work to study the flow fields generated with a sawtooth impeller at different scales of operation would highlight this scale dependent performance difference.

Acknowledgments

A part of the study was performed during the InnovateUK funded project NEWNANO (contract nb 132726) which HY, ST, KP, CDR and NGOT gratefully acknowledge. HY also gratefully acknowledges the financial support from MacDermid for his PhD.

References

[1] Seubert, C.; Nichols, M.; Henderson, K.; Mechtel, M.; T. Klimmasch, T. Pohl (2010) The effect of weathering and thermal treatment on the scratch recovery characteristics of clearcoats. *J Coat Technol Res* **7**, 159–166 https://doi.org/10.1007/s11998- 009-9190-4

[2] Mohseni, M.; Ramezanzadeh, B.; Yari, H. and Gudarzi, M.M. (2012) The role of nanotechnology in automotive industries. In New Advances in Vehicular Technology and Automotive Engineering. Edited by Joao Paulo Carmo and Joao Eduardo Ribeiro http://dx.doi.org/10.5772/49939

[3] Scrinzi, E.; Rossi, S.; Kamarchik, P.; Deflorian F. (2011) Evaluation of durability of nano-silica containing clear coats for automotive applications. Progress in Organic Coatings, p: 384-390 https://doi.org/10.1016/j.porgcoat.2011.04.009

[4] Bautista, Y.; Gomez, M.P.; Ribes, C.; Sanz, V. (2011) Correlation between the wear resistance and the scratch resistance for nanocomposite coatings Prog. Org. Coat, 70, 4, p: 178- 185 https://doi.org/10.1016/j.porgcoat.2010.09.022

[5] Burmester, S.S.H., Rielly, C.D., Edwards, M.F. (1992). The mixing of miscible liquids with large differences in density and viscosity. In: King, R. (eds) Fluid Mechanics of Mixing. Fluid Mechanics and Its Applications, vol 10. Springer, Dordrecht. https://link.springer.com/chapter/10.1007/978-94-015-7973-5_10

[6] Jones, P.N.; Özcan- Taşkin, G. N. (2005) Effect of physical property differences on blending Chem. Eng and Tech. Vol 28, 8, p: 908- 914. https://doi.org/10.1002/ceat.200500012

[7] Ruszkowski, S. (1994) A rational method for measuring blending performance and comparison of different impeller types. Proc. 8th European Conference on Mixing, pp. 283-291. Inst. Chem. Engrs. Rugby,U.K.

[8] Grenville, R. K., Ruszkowski, S. and Garred, E. (1995) Blending of miscible liquids in the turbulent and transional regimes. Paper 1.5, 15th NAMF Mixing Conference, Banff, Canada.

[9] Rodgers, T.L.; Cooke, M.; Siperstein, F.R.; Kowalski, A. (2009) Mixing and dissolution times for a cowles disk agitator in large-scale emulsion preparation. *Ind. Eng. Chem.*, 48, 14, 6859–6868 https://doi.org/10.1021/ie900286s

https://doi.org/10.52202/074122-0029

Mixing of non-Newtonian fluid in a stirred tank: experimental characterization

ER1.4

Ernest Simon[1,2], *Jérôme Morchain*[1], *Frédéric Augier*[2], *Marion Servel*[2], *Alain Liné*[1] *(1) Toulouse University, INSA-TBI, 135 Avenue de Rangueil, 31077 Toulouse cedex, France. (2) IFP Energies nouvelles, 69360 Solaize, France*

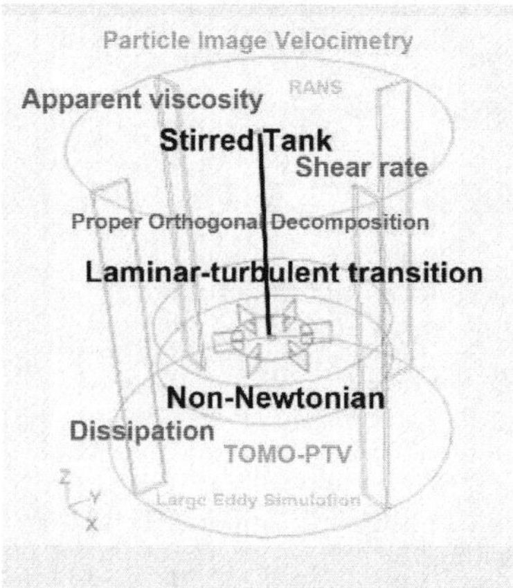

To increase our understanding and to resolve the data gap for CFD validation, we present a detailed particle image velocimetry study of hydrodynamics of a stirred tank reactor with a shear thinning fluid.
Bulk flow visualizations show that the flow field is highly influenced by the rheological behavior of the fluid. Moreover, an analysis of the shear rate and viscosity revealed important spatial heterogeneities. An estimation of the Reynolds number classified the rotational speeds as the onset of the transitional regime and in the transitional regime. The data set consists of local mean and fluctuating velocities at different locations in the tank, which are available for validation and further study.
The objectives of the general study are twofold: (i) analyze hydrodynamics at different flow regimes produced by two different impellers in non-Newtonian fluid (CMC) and (ii) provide a reliable data set to evaluate the strengths and weaknesses of CFD models.

Introduction

The production of ethanol from biomass is recognized as one of the possible ways to reduce greenhouse gases and partially replace fossil fuels. The production process requires the mixing of biomass suspensions exhibiting a strong shear thinning behavior. The dependence of the apparent viscosity on the shear rate implies that several flow regimes coexist in the vessel. Although there has been recent progress in the literature, much remains to be learned about the fundamental phenomena controlling flow and mixing of non-Newtonian fluids.

The present work deals with the experimental study of hydrodynamics of a stirred tank with a non-Newtonian fluid. A particular interest is given to the phenomena taking place in a laminar-turbulent transition regime.

The main concern is the evaluation of the shear rate $\dot{\gamma}$ from experimental data in order to estimate locally the rheological characteristics of the fluid.

The most common approach is to use Metzner-Otto's relationship ($\dot{\gamma}_{avg} = k_S N$) which is only valid for Newtonian fluid in the laminar regime. Pérez et al. ([4]) extended this relation to non-Newtonian fluids for turbulent regime. However, in both cases, the evaluation of the shear rate is based on macroscopic variables, while for a non-Newtonian fluid, the apparent viscosity μ_a depends on the local shear rate defined as $\dot{\gamma} = \sqrt{2tr(S^2)} = \sqrt{\frac{1}{2}\sum_{i=1}^{3}\sum_{j=1}^{3}\frac{\partial u_i}{\partial x_j}\left(\frac{\partial u_i}{\partial x_j}+\frac{\partial u_j}{\partial x_i}\right)}$. Hence, it is necessary to evaluate locally $\dot{\gamma}$ from experimental measurements.

In the next step related to CFD, our PIV data base will be compared to numerical simulations. The main question is how to simulate the apparent viscosity. In DNS, it would be sufficient to use local and instantaneous shear rate. Using Large Eddy Simulation, the local shear rate is not solved, as the smallest scales of turbulence are filtered out. With the URANS approach, although legitimate in industrial applications, the question arises as to the validity of the transport equation for the viscous dissipation rate of the turbulent kinetic energy on the one hand, and therefore the estimation of the apparent viscosity from the shear rate on the other.
In our study, URANS and LES simulations will be performed. From comparisons of simulations with experiments, it will be possible to estimate the modelling refinement necessary and sufficient to estimate the average velocity fields, the kinetic energy fields and even the shear rate fields.

Material and Methods

In our case of interest, the biomass suspensions are opaque. To characterize them with optical methods, it is necessary to work with a transparent model fluid. The first issue is thus to find a model fluid, transparent and able to mimic the rheology of complex media. CMC was chosen to perform PIV experiments. CMC 7HF (Carboxymethyl Cellulose, Ashland) is mixed with high purity water. A transparent solution that mimics the rheological characteristics of biomass fluids is obtained. The CMC solution is modelled with a Carreau-Yasuda law (C-Y; equation (1)), which captures the shear thinning behaviour.

$$\mu = \mu_\infty + (\mu_0 - \mu_\infty)(1 + (\lambda\dot{\gamma})^a)^{(n-1)/a}, \quad (1)$$

where n is power law index, μ_0 and μ_∞ the viscosities at zero and infinite shear rate, λ the characteristic time, a a dimensionless parameter and $\dot{\gamma}$ the magnitude of the shear rate. The CMC 1.5 wt% flow curves (Figure 1 and 2) are obtained from a Thermo Scientific Mars III rheometer with a smooth stainless-steel Cone and Plate (6°) geometry. The time-dependant behaviour of the solution is described Figure 1 (frequency sweeps test) and the rheogram is plotted on Figure 2.

The tank used in this study is a standard cylindrical tank (28.7L), equipped with four equally spaced

https://doi.org/10.52202/074122-0030

removable baffles (width B=T/10). The tank is made of glass and has a diameter and a liquid height of T = H = 0.3 m. The tank and the cubic tank in which it is placed (to minimize optical refraction) share the same flat bottom.

Two impellers are tested in this study, a standard Rushton turbine (RT) and a HTPG4, each with a diameter D=T/3=0.1m.

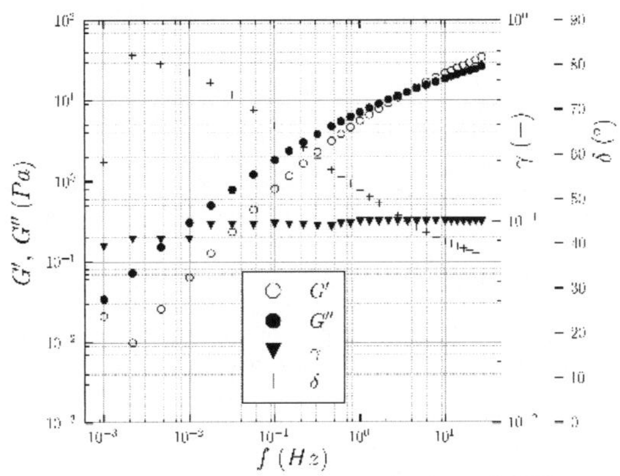

Figure 1. Time-dependant behaviour of CMC.

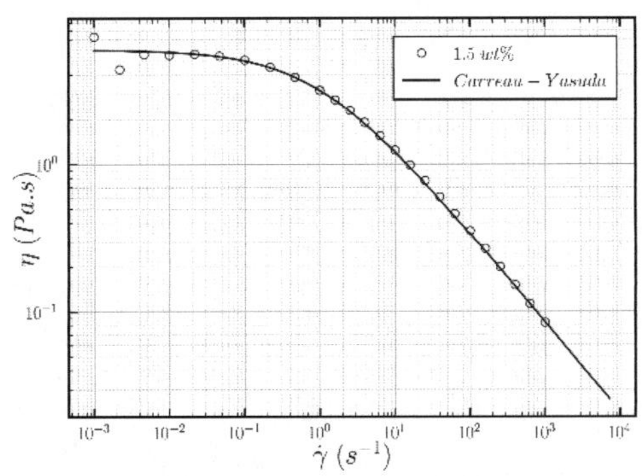

Figure 2. Rheological profile of CMC 1.5wt% with a calibrated Carreau-Yasuda (C-Y) model ($\mu_\infty = 0,\ \mu_0 = 5.9,\ \lambda = 1.05,\ a = 0.68,\ n = 0.39$).

The experimental method used in that study is 2D2C PIV. The PIV system is composed of a Nd-Yag laser and an image acquisition system provided by Dantec. Both vertical and horizontal (below the impeller) planes are obtained. Each PIV acquisition is based on 5000 instantaneous velocity fields. The baffles can be rotated to take measurements in different positions relative to the baffles. Since only two components of the velocity are available within the 2D fields, working vertically and horizontally allows to retrieve three velocity components in a 2D plane at the intersection. The 2D2C PIV not being resolved in time, Proper Orthogonal Decomposition (POD) technique is used in postprocessing to assess the different contributions to the instantaneous velocity.

From the power number N_p of the Rushton turbine, at a fixed rotation speed N, it is possible to estimate the power dissipated per unit mass <ε> in the reactor based on global quantities.

$$< \varepsilon >= \frac{P}{\rho V},$$

where $P = N_P \rho N^3 D^5$ ([3]) is the power input, V the tank volume. From that value, the Taylor micro scale λ and the Kolmogorov scale η can be evaluated. Both values are compared to the interrogation area (IA) of the PIV. The former must be at least of the order of the IA so that the velocity gradients are well resolved. If the latter is smaller than the IA then the whole energy spectrum is solved and the local dissipation can be precisely evaluated. Thus, special attention is paid to the initial estimation of these scales.

The acquisitions are realized at different Reynolds number Re, in the range $100 < Re < 6000$, to study the laminar-turbulent transition.

Results and Discussion

The first study is performed in the laminar regime. The instantaneous velocity depends only on the mean and fluctuant periodic velocities.

$$u(t) = \overline{u} + \tilde{u}(t).$$

Both contributions are obtained with POD. The first objective is to determine the contribution of each component to the total shear rate $\dot{\gamma}_{tot}$. The authors main interrogation lays in the expression of the mean apparent viscosity $\overline{\mu_a}(\dot{\gamma})$. It can be obtained with (a) the averaged instantaneous shear rate $\overline{\mu_a}(\dot{\gamma})$ and (b) using the time-averaged shear rate $\mu_a(\overline{\dot{\gamma}})$ ([1]). The instantaneous shear rate will be estimated from instantaneous 2D2C PIV data. Thus, the second objective is to compare the two fields of viscosity based respectively on $\overline{\mu_a}(\dot{\gamma})$ or $\mu_a(\overline{\dot{\gamma}})$.

The second part of this study is dedicated to the laminar-turbulent transition. The instantaneous velocity depends now also on a third contribution, the fluctuant turbulent velocity.

$$u(t) = \overline{u} + \tilde{u}(t) + u'(t).$$

Here again, POD is required to determine the different contributions to u. Special care is given to the PIV resolution; it is necessary for the PIV IA to be comparable to Kolmogorov scale. As for the laminar case, the different contribution to $\dot{\gamma}_{tot}$ are assessed and the two statistical methods are compared.

The main advantage of POD technique is that, the modes being orthogonal and thus uncorrelated, the total shear rate can be decomposed into two or three contributions (laminar or turbulent) ([2])

$$\dot{\gamma} = \sqrt{\overline{\dot{\gamma}}^2 + \tilde{\dot{\gamma}}^2 + \langle \dot{\gamma}'^2 \rangle}.$$

The instantaneous shear rate will be estimated from instantaneous 2D2C PIV data. Thus, the two fields of viscosity based respectively on $\overline{\mu_a}(\dot{\gamma})$ or $\mu_a(\overline{\dot{\gamma}})$ will be compared.

https://doi.org/10.52202/074122-0030

(a) (b)

Figure 3. (a) standard Rushton turbine and (b) HTPG4 impeller coated with paint to minimize the laser refraction.

Perspectives

In this work, an experimental mixing analysis using a non-Newtonian fluid (CMC) stirred with two different impellers has been carried out. Local shear rates and local apparent viscosities as a function of different impeller frequencies N were determined (covering the onset of transitional and turbulent regime). Moreover, the spatial heterogeneity of both shear rates and viscosities is shown to be greater when the rotational speed decreases.

This experimental work presented here, based on 2D2C PIV, is part of a more general study. The research project aims at evaluating and modifying/adapting existing CFD turbulence models so that the local shear distribution generated in a reactor accounts for the spatial and temporal evolution of the rheology of a non-Newtonian fluid.

The work presented here is the first step of the experimental study which will include a 2D-3C PIV study and a study with TOMO-PTV in the frame of collaboration with Fluid Mechanics Institute in Toulouse. The goal is to access, *in fine*, the set of components of the strain rate tensor S_{ij} on which $\dot{\gamma}$ depends.

A CFD study is conducted in parallel with a first study of the stirred tank in URANS and Large Eddy Simulation (LES); crucial questions arise there too in terms of definition of local shear rate and expression of the apparent viscosity. The final objective is to evaluate the strengths and weaknesses of CFD models in the light of the experimental PIV/SPIV/TOMO-PTV databases.

Acknowledgments

This work was supported by IFP Energies Nouvelles (IFPEN).

References

[1] Fernandes del Pozo, D., Liné, A., Van Geem, K. M., Le Men, C., & Nopens, I. (2020). Hydrodynamic analysis of an axial impeller in a non-Newtonian fluid through particle image velocimetry. AIChE Journal, 66(6), e16939.

[2] Gabelle, J. C., Morchain, J., Anne-Archard, D., Augier, F., & Liné, A. (2013). Experimental determination of the shear rate in a stirred tank with a non-newtonian fluid: Carbopol. AIChE Journal, 59(6), 2251-2266.

[3] *Bujalski, W. A. L. D. E. M. A. R., Nienow, A. W., Chatwin, S., & Cooke, M. (1987). The dependency on scale of power numbers of Rushton disc turbines. Chemical Engineering Science, 42(2), 317-326.*

[4] Pérez, J. S., Porcel, E. R., López, J. C., Sevilla, J. F., & Chisti, Y. (2006). Shear rate in stirred tank and bubble column bioreactors. Chemical Engineering Journal, 124(1-3), 1-5.

[5] Zhang, Y., Gao, Z., Li, Z., & Derksen, J. J. (2017). Transitional flow in a Rushton turbine stirred tank. AIChE Journal, 63(8), 3610-3623.

[6] Paul LE, Victor AAO, Suzanne MK. Handbook of Industrial Mixing: Science and Practice. New York, NY: John Wiley and Sons; 2003

[7] Montante G, Paglianti A. Fluid dynamics characterization of a stirred model bio-methanation digester. Chem Eng Res Des. 2014;3:38-47.

[8] Saeed S, Ein-mozaffari F, Upreti SR. Using computational fluid dynamics to study the dynamic behavior of the continuous mixing of Herschel - Bulkley fluids. Ind Eng Chem Res. 2008;47:7465-7475

[9] Arratia PE, Arratia PE, Kukura J, Lacombe J, Muzzio FJ. Mixing of shear-thinning fluids with yield stress in stirred tanks mixing of shear- thinning fluids with yield stress. AIChE J. 2006;52(7):2310-2322

[10] Venneker BCH, Derksen JJ, Van den Akker HEA. Turbulent flow of shear-thinning liquids in stirred tanks-the effects of Reynolds number and flow index. Chem Eng Res Des. 2010;88(7):827-843

[11] Metzner A, Otto RE. Agitation of non-Newtonian fluids. AIChE J. 1957;3(1):3-10

[12] Zhang, Y., Wang, L., Huo, H., Li, Z., & Gao, Z. (2022). Computational and experimental investigation of flow fields in a Rushton turbine stirred tank with shear-thinning fluid. Asia-Pacific Journal of Chemical Engineering, 17(1), e2735.

https://doi.org/10.52202/074122-0030

Viscoelastic fluid flow in a helical static mixer – a comprehensive numerical study.

ER1.5

Thomas P John[1], Robert J Poole[2], Adam J Kowalski[3], Cláudio P Fonte[1]. (1) Department of Chemical Engineering, The University of Manchester, Manchester, U.K., thomas.john@manchester.ac.uk. (2) School of Engineering, The University of Liverpool, Liverpool, U.K., robpoole@liverpool.ac.uk. (3) Unilever R&D, Port Sunlight Laboratory, Bebington, U.K.

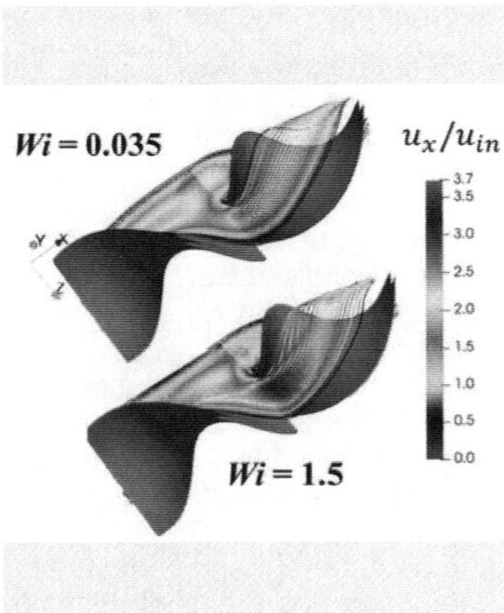

Static mixers are commonly used devices for the mixing of complex formulated products. For the first time, in this study, we simulate viscoelastic fluid flow through a helical static mixer using Computational Fluid Dynamics (CFD). The results show that increasing elasticity of the fluid leads to asymmetric flow bifurcations around the mixing elements (see graphical abstract), resulting in significantly poorer mixing performances. These results are in agreement with the experimental observations reported in the literature. With the CFD results in this study, we have quantified the degree of asymmetry at bifurcation points in the mixer and correlated this with the mixing performance for a range of Reynolds (*Re*) and Weissenberg numbers (*Wi*). We also investigate the effect of increasing the polymer extensibility (at fixed *Re* and *Wi*). Increasing extensibility caused the solution to become transient rather than steady-state and led to more pronounced asymmetry in the flow. Both findings also match previous experimental observations.

Introduction

Static mixers are commonly used during the processing of formulated products, which often exhibit complex rheological properties such as viscoelasticity. Viscoelasticity can lead to interesting, counter-intuitive, and often unexpected, phenomena such as die swell and the rod-climbing effect. In helical static mixers, it has been found previously [1, 2], using Planar Laser Induced Fluorescence (PLIF), that increasing viscoelasticity causes a reduction in the mixing performance. Without access to the flow variables in the fluid domain, the PLIF results can only be used to infer the changes in the flow pattern which cause the reduction in mixing performance.

In this study, we use Computational Fluid Dynamics (CFD) to simulate viscoelastic fluid flows through a helical static mixer with 8 mixing elements. The results from this study greatly help us to understand the effect of viscoelasticity on mixing in helical static mixers, and also, more generally, viscoelastic fluid flows in geometries which involve bifurcations of the fluid.

Material and Methods

We use RheoTool (version 5.0) [3], an open source CFD code based on OpenFOAM®, to run all simulations. In the momentum equation, the total deviatoric stress is split into the sum of a polymeric stress $\boldsymbol{\tau}_p$ and a solvent stress which is modelled as a Newtonian fluid with viscosity η_s. The viscoelastic constitutive model describing the transport of $\boldsymbol{\tau}_p$ was the FENE-CR model, which is given by

$$\boldsymbol{\tau}_p + \lambda\left(\frac{\overset{\nabla}{\boldsymbol{\tau}_p}}{f}\right) = \eta_p(\nabla\boldsymbol{u} + \nabla\boldsymbol{u}^{\mathrm{T}}) \qquad [1]$$

where $\boldsymbol{u}$ the velocity, λ the relaxation time of the polymer chains, η_p the polymeric viscosity, and f is given by

$$f = \frac{L^2 + (\lambda/\eta_p)\mathrm{Tr}(\boldsymbol{\tau})}{L^2 - 3} \qquad [2]$$

The parameter L^2 limits the extensibility of the polymeric chains. The term $\overset{\nabla}{(\cdot)}$ represents the upper convected derivative, which is the time derivative written in a co-ordinate system which translates, rotates, and deforms with the continuum, and is required to ensure the tensorial model is objective. We define the Weissenburg and Reynolds numbers as $Wi = \lambda U_{in}/D$ and $Re = \rho U_{in} D/(\eta_s + \eta_p)$ respectively, where D is the pipe diameter. We investigate a range of Wi and Re with $L^2 = 50$, and we also investigate $L^2 = 500$ and $L^2 = 5000$ for $Re = 0.49$ and $Wi = 0.43$. The FENE-CR model, first derived by Chilcott and Rallison [4], is a modification of the original FENE-P model, and has a constant shear viscosity in steady simple shear flow, making it suitable for studying Boger fluids. We use η_p = 8.16 Pa s and η_s = 0.01 for all simulations.

All simulations were performed on the University of Manchester's Computational Shared Facility. A hexahedral mesh with ~5 million cells was used. The no slip boundary condition was used at all solid walls. The inlet to the mixer was set as a fixed uniform velocity (with velocity u_{in}) and zero gradient for pressure, whilst the outlet was set at atmospheric pressure and with a zero gradient for velocity. The log-conformation method was used to solve the constitutive equation [5] in order to avoid the high

https://doi.org/10.52202/074122-0031

Weissenberg number problem. The high-order CUBISTA scheme [6] was used to discretise the convective terms in both the momentum and constitutive equation. Laplacian terms were discretised using the Gauss linear scheme with the correction for surface normals. The implicit Euler scheme was used for time discretisation. The time step was auto-adjusted during the simulations using the maximum Courant number at each time step. The maximum Courant number was varied between roughly 0.05 for higher Wi cases and 0.4 for lower Wi cases. For all $L^2 = 50$ cases, the simulations were run until steady-state was reached. The $L^2 = 500$ and $L^2 = 5000$ cases, which did not reach a steady-state, were run for approximately $t = 13\lambda$.

Results and Discussion

Figure 1 shows the contours of u_x/u_{in} at the end of the 5th mixing element for various Wi at $Re = 0.04$. As the elasticity (quantified by *Wi*) increases to 0.12, the flow preferentially passes below the upcoming element (the horizontal grey bar). Since the twisting direction of 5th element is clockwise, the flow is preferentially bifurcating against the twisting direction (ie. downwards instead of upwards). As *Wi* is increased further to 1.5, a strong asymmetry is observed at the element intersection, and the flow now passes over the top of the up-coming element.

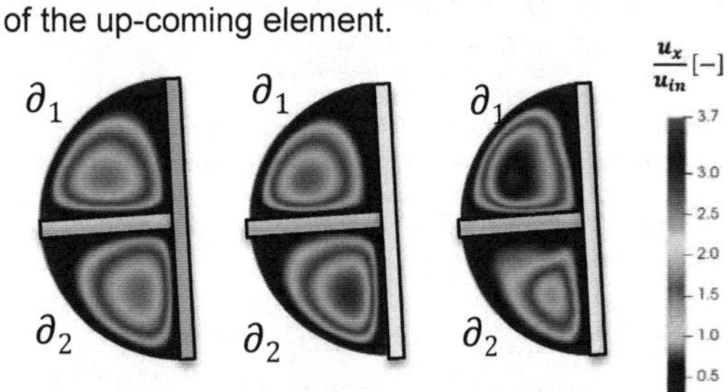

(a)Wi = 0.01 **(b)**Wi = 0.12 **(C)**Wi = 1.5

Figure 1. Contours of u_x/u_{in} at the end of the 5th element for Re = 0.04 and L^2=50.

We quantify the asymmetry using the following asymmetry parameter

$$\Delta = \frac{\max(u_x)_{\partial_1} - \max(u_x)_{\partial_2}}{\max(u_x)_{\partial_1} + \max(u_x)_{\partial_2}} \quad [3]$$

where ∂_1 and ∂_2 represent the quadrants of the circular (normal in the axial direction) slice of the fluid domain, which can be seen in Figure 1. For positive values of Δ, the fluid is bending with the twisting direction of the previous element, and for negative Δ, the fluid is bending against the twisting direction of the previous element. For $\Delta = 0$, the bifurcation over the up-coming mixing element is essentially symmetric. The asymmetry at the bifurcation points, caused by the viscoelasticity, drastically alters the mixing performance. Figure 2 shows the mixing patterns predicted by the CFD results at the end of the 6th mixing element for Re=0.04. For Wi=0.01, the mixing pattern is

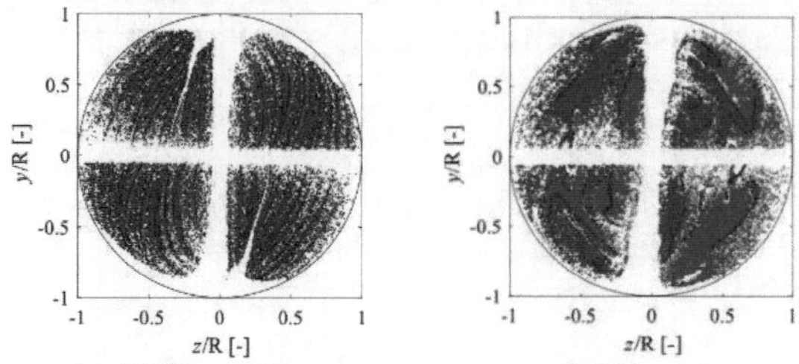

(a)Wi = 0.01 **(b)**Wi = 1.5

Figure 2. Mixing patterns at the end of the 6th element for Re = 0.04 and L^2=50.

similar to that which would be expected for a Newtonian fluid in helical static mixer. A good degree of mixing has been achieved, with a large number of small striations.

For Wi=1.5, the mixing performance is significantly worse, as the striations are thicker and there is a larger distribution of striation thickness, caused by the asymmetries in the flow bifurcations.

Figure 3a shows the contour of the asymmetry parameter in the *Re*-*Wi* space at the end of the 4th mixing element. It can be seen that for *Re* below approximately 1, increasing *Wi* increases the magnitude of Δ but its sign is negative. For *Wi* above approximately 0.1, the sign of Δ changes abruptly, and the magnitude keeps increasing with increasing *Wi*. For the highest values of *Re* studied,

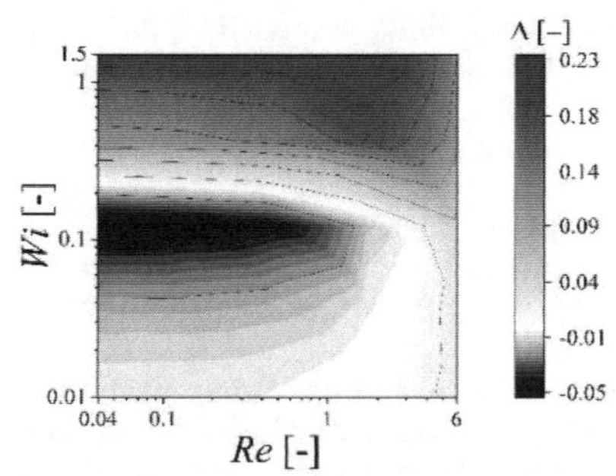

Figure 3. Contour of Δ at the end of the 4th element for the range of *Re* and *Wi* investigated for L^2=50.

The sign of Δ is positive for all *Wi* studied. However, for high *Re* and high *Wi*, the magnitude of Δ is lower than for low *Re* and high *Wi*, indicating that increased inertia acts to promote symmetry at element intersections.

Viscoelastic asymmetries and instabilities have been observed, both experimentally and numerically, in a number of geometries including the cross-slot [7, 8] and channel flow around a cylinder [9]. In these symmetrical geometries, the asymmetry clearly behaves as a symmetry breaking instability. In the case of the helical static mixer, the geometry is not symmetric due to the twisting of the elements. Therefore, is it not absolutely clear whether the asymmetry in this geometry is the result of an elastic instability, or an elastic stress modifying the flow field such as, for example, via a hoop stress.

There are at least two possible scenarios which might explain the results shown in Figure 3. One is that the asymmetry is caused solely by an elastic instability, and the change from negative to positive Δ is a change from one solution branch of a pitchfork bifurcation to the other. In this scenario, there should be two stable solutions (one positive and one negative) for most of the points in the *Re-Wi* space beyond the critical *Wi*. The other, likely more plausible scenario, is that for relatively low *Wi,* the change in Δ with increasing *Wi* is the result of an elastic stress pushing the fluid. The sudden change in the sign of Δ as *Wi* Is increased further is then likely the onset of an elastic instability.

Figure 4 shows the mixing index at the outlet (normalized by the mixing index at the inlet, I_0) of the mixer as a function of the magnitude of Δ. It can clearly be seen that there is a proportionality between the degree of asymmetry and mixing index, indicating that asymmetry at the element intersections is the cause for the change in mixing performance when the elasticity is changed.

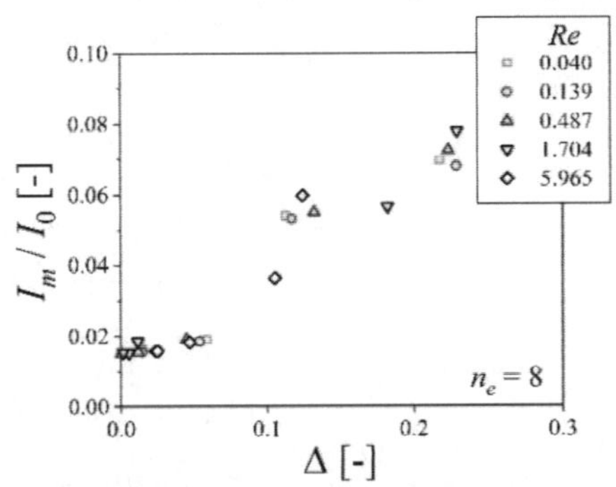

Figure 4. Mixing index at mixer outlet *vs* Δ at the end of the 4th mixer element for L^2=50.

Figure 5 shows contours of u_x/u_{in} for Re = 0.487 and Wi = 0.429 with (a) L^2=500 and (b) L^2=5000. These simulations did not achieve true steady-state and the contours are for $t \approx 12.5\lambda$. Increasing the value of L^2 greatly increases the asymmetry of the bifurcation of the flow. (Note: the FENE-CR model approaches the Oldroyd-B model in the limit L^2 tends to infinity). If the cause of the asymmetry is a viscoelastic instability or an elastic stress, it would be expected that both of these effects are exacerbated by increasing L^2, since increasing L^2 causes both higher normal stresses and higher extensional viscosities.

Conclusions

We have, for the first time, investigated numerically the effect of viscoelasticity on flow

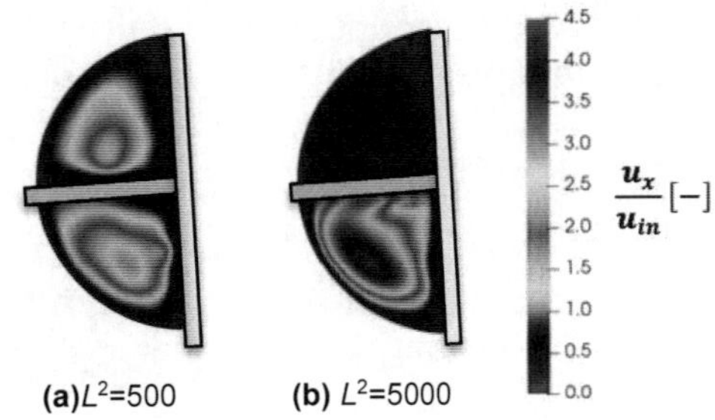

Figure 5. Contours of u_x/u_{in} at the end of the 5th element for *Re* = 0.487 and *Wi* = 0.429 and for L^2=500 (a) and L^2=5000 (b). $t \approx 12.5\lambda$.

asymmetries and mixing performance in a helical static mixer. Previously, experimental results showed that increasing elasticity caused the mixing performance to drop. Our CFD results show that the change in mixing performance is caused by asymmetric bifurcations of the flow at the intersections of the mixer elements. We have quantified the degree of asymmetry at these points and correlated this with the mixing index. The asymmetry was found to exhibit complex behaviour with varying elasticity. As *Wi* is increased, there is a sudden change in the direction of the asymmetry, indicating the possible onset of an instability. These results provide much needed insight into the role that elasticity can play in the performance of mixing processes.

References

[1] .Ramsay, J., et al., *Mixing performance of viscoelastic fluids in a Kenics KM in-line static mixer.* Chemical Engineering Research and Design, 2016. **115**: p. 310-324.

[2]. Migliozzi, S., L. Mazzei, and P. Angeli, *Viscoelastic flow instabilities in static mixers: Onset and effect on the mixing efficiency.* Physics of Fluids, 2021. **33**(1): p. 013104.

[3]. Pimenta, F. and M.A. Alves. *rheoTool.* 2016; Available from: https://github.com/fppimenta/rheoTool.

[4]. Chilcott, M.D. and J.M. Rallison, *Creeping flow of dilute polymer solutions past cylinders and spheres.* Journal of Non-Newtonian Fluid Mechanics, 1988. **29**: p. 381-432.

[5]. Fattal, R. and R. Kupferman, *Constitutive laws for the matrix-logarithm of the conformation tensor.* Journal of Non-Newtonian Fluid Mechanics, 2004. **123**(2): p. 281-285.

[6]. Alves, M.A., P.J. Oliveira, and F.T. Pinho, *A convergent and universally bounded interpolation scheme for the treatment of advection.* International Journal for Numerical Methods in Fluids, 2003. **41**(1): p. 47-75.

[7]. Poole, R.J., M.A. Alves, and P.J. Oliveira, *Purely Elastic Flow Asymmetries.* Physical Review Letters, 2007. **99**(16): p. 164503.

[8]. Rocha, G.N., et al., *On extensibility effects in the cross-slot flow bifurcation.* Journal of Non-Newtonian Fluid Mechanics, 2009. **156**(1): p. 58-69.

[9]. Varchanis, S., et al., *Asymmetric flows of complex fluids past confined cylinders: A comprehensive numerical study with experimental validation.* Physics of Fluids, 2020. **32**(5): p. 053103.

https://doi.org/10.52202/074122-0031

Dispersed phase viscosity and drop size distributions: Experiments, modeling, and a discussion of Population balance fit parameters. ER1.6

***L. Hohl*[1]**, *S. Röhl*[1], *M. Kraume*[1]. *(1) Technische Universität Berlin, Ackerstraße 76, 13355 Berlin, Germany, Lena.Hohl@tu-berlin.de*

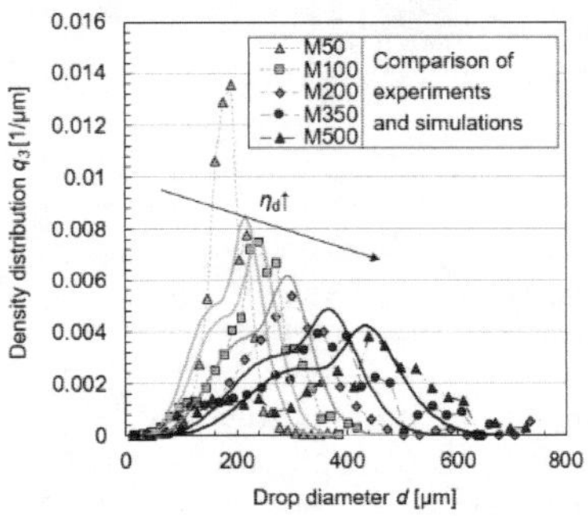

Measurements of dynamic and steady state drop size distributions were performed in liquid multiphase systems with water and different silicone oils in a discontinuously operated stirred tank. The oils used as dispersed phase significantly vary in dynamic viscosity, while density and interfacial tension are nearly constant. Population balance models for partially mobile interfaces are applied and their ability to predict drop size distributions at different process conditions is evaluated. The impact of different daughter drop size distribution shapes on simulation results is quantified. Furthermore, a discussion about the determination of Population balance fit parameters and their impact on the simulation results is performed.

Introduction

A reliable prediction of drop size distributions in liquid multiphase systems often is crucial for process design and optimization. In Population balance equations (PBEs), the change of a drop size distribution over time is calculated based on the balance between drop breakage and coalescence rates [1]. In contrast to complex, spatial-resolved Computational fluid dynamic simulations (CFD), the numerical effort to solve PBEs is small. Hence, they are a valuable tool for a first description of liquid/liquid systems. Although a combination of PBEs with CFD can further improve simulation accuracy, the PBE performance should be investigated independently to identify its strengths and weaknesses beforehand.

The dispersed phase viscosity is one of several physical properties affecting drop breakage and coalescence rates [2]. Although the general impact of dispersed phase viscosity on drop size distributions has been quantified experimentally before, this physical property is not included in many PBE models available in literature. Based on very detailed own experiments, different PBE models are compared and evaluated. Furthermore, their optimization potential regarding the fit procedure and the used daughter drop size distribution is discussed [12].

Material and Methods

Ultrapure water and different silicone oils M50, M100, M200, M350 and M500 (Carl Roth GmbH & Co. KG) were used. Densities were measured using the oscillating u-tube principle (DSA 5000M, Anton Paar GmbH). Interfacial tensions between water and silicone oils were determined with the Pendant drop method after reaching steady-state conditions (OCA20, Dataphysics). Dynamic viscosities of the Newtonian liquids were determined with a cone-and-plate rheometer (MCR 302, Anton Paar GmbH). The value of dynamic viscosity corresponds to the oil name and is the only oil property that varies significantly (see Table 1).

Table 1. Physical properties of used liquids (T = 20°C) [12]

liquid	dynamic viscosity [mPa s]	density [kg m^{-3}]	Interfacial tension [mN m^{-1}]
water	1	998.20	- - -
M50	53		
M100	103		
M200	211	969.06 ± 3.58	38.62 ± 1.61
M300	366		
M500	506		

A double-walled stirred tank (V = 0.7L, H/D = 1.6, d/D = 0.45) equipped with baffles and a Rushton turbine was used. Experiments were performed at fully turbulent conditions using different agitation speeds (n = 500, 600 and 700 rpm) and dispersed phase fractions (φ = 0.05 and 0.1). All experiments were performed at T = 20°C. An endoscope measurement technique and subsequent automated image analysis (SOPAT GmbH) were applied to determine the drop size distributions of the agitated systems in-situ. Every drop size distribution was calculated using at least 5000 measured drop diameters from a series of 200 images.

The software Parsival (Computing in Technology GmbH) was used to solve the PBEs. Each PBE model contains certain parameters that can be fitted to experimental data, in this case with an implemented parameter estimation routine minimizing the residual between experiments and simulation. Only one set of fit parameters for each PBE model is used to describe all different process conditions and oil/water combinations. Population balance models by Alopaeus et al. [3,4], Podgórska & Bałdyga [5,6], Zhou et al. [7,8] and Chesters [9] we compared. For further details concerning the models we refer to the respective original publications. The considered drop size

https://doi.org/10.52202/074122-0032

range was 1-1000 µm while assuming binary breakage. A Gaussian distribution with a mean of 500 µm and a standard deviation of 25 µm was used as the initial drop size distribution. Different types of daughter drop size distributions were applied: Gaussian, beta, and bimodal distribution types [1,10], a m-shaped distribution developed by observing breakage phenomena in a pump-mixer by Zhou et al. [7,8] and a u-shaped distribution suggested by Tsouris and Tavlarides [11]. The parameter fitting procedure was performed as an iterative process using two experiments. Thereby, the distribution shape and the dynamic changes of Sauter mean diameter over time were taken into account [12].

Results and Discussion

In a first step to compare the different PBE models, the breakage rate was calculated using the model by Alopaeus [3,4] and the coalescence rate was calculated using models by different authors with a Gaussian daughter drop size distribution. Exemplary results for steady state Sauter mean diameters including different oil types and agitation speeds are depicted in Figure 1. The best agreement between experiments and simulations is achieved with the Alopaeus model and, therefore, further simulations shown in this abstract were performed with this approach. However, we would like to point out that effects such as the choice of daughter drop size distribution and PBE fit parameters (which will also be discussed in the following) severely affect simulation results. Therefore, this evaluation of models is just preliminary and will be investigated further.

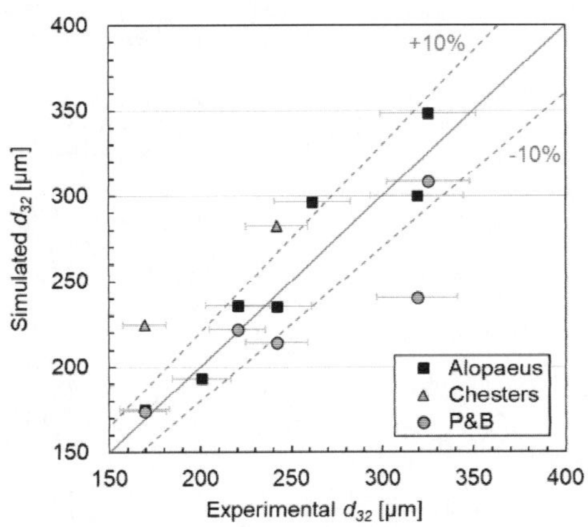

Figure 1. Results using the Alopaeus model for breakage rate [3,4,12] and coalescence rate models by different authors [3-6,9] and a Gaussian daughter drop size distribution.

Experimental and simulated steady state Sauter mean diameters for the complete Alopaeus model [3,4] using different daughter drop size distributions are shown in Figure 2. The impact is significant and best results were achieved using the daughter drop size distributions proposed by Zhou et al. [7,8] and by Tsouris and Tavlarides [11] in particular. Results include different oil types, agitation speeds and dispersed phase fractions [12]. It should be noted that only one set of fit parameters used for the simulations and that a Gaussian daughter drop size distribution was used for the fitting procedure. Interestingly, the daughter drop size distributions leading to the best agreement with experimental data significantly differ from the Gaussian bell shape (m-shape and u-shape, respectively) [12]. To study if a determination of new fit parameters for each daughter drop size distribution will improve simulation results further is part of ongoing investigations. The impact of dispersed phase fraction is not resolved satisfactorily (largest deviations in the parity plot) and should be subject of model optimization. The impact of different energy dissipation rates (stirrer speeds, respectively) was represented well, but occurring small deviations might be reduced further by using PBE compartment models or combining PBE with CFD to describe spatial inhomogeneities.

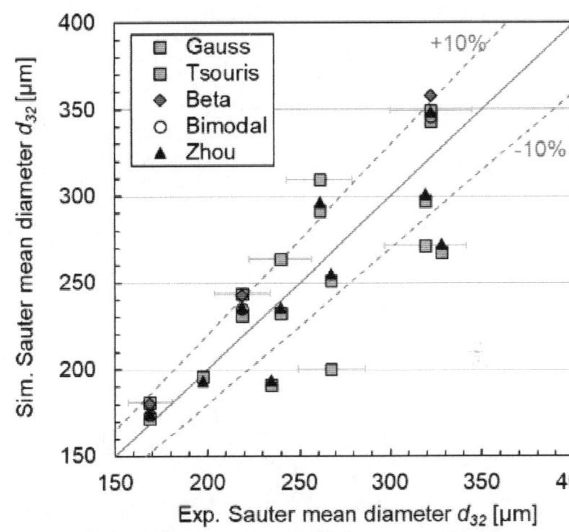

Figure 2. Results using the Alopaeus model for breakage and coalescence rate [3,4] and different daughter drop size distributions [1,7,8,10-12]. Experimental error bars are only shown exemplary for better graph clarity.

The aim of this work is not only to analyse and predict the impact on mean diameters in steady state, but also the complete drop size distributions and dynamic behaviour. Corresponding results with the Alopaeus model [3,4] using the daughter drop size distribution by Tsouris & Tavlarides [11] are shown in Figure 3 [12]. Agitation started at t = 0 s with n = 700 rpm and the energy dissipation was reduced to n = 500 rpm after t = 1800 s. Image series were taken every 30s to analyse the dynamic response of the system. The balance between breakage and coalescence clearly is shifted towards coalescence with rising dispersed phase viscosity and it takes longer to reach a new steady state. For M350 and M500 at n = 500 rpm partial segregation occurred, preventing a reliable image analysis. The steady state drop size distributions at t = 1800s are shown in Figure 4. A

https://doi.org/10.52202/074122-0032

deviation between simulated and experimental results clearly is visible, but the results are considered as surprisingly good considering that only one set of fit parameters was used to describe different oil types and the dispersed phase viscosity varies by an order of magnitude [12].

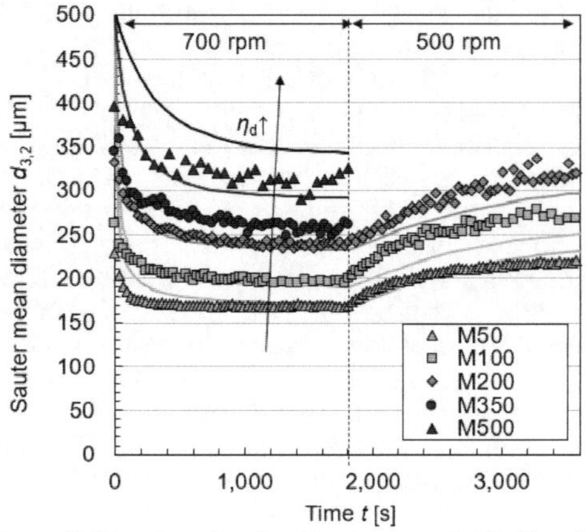

Figure 3. Results using the Alopaeus model [3,4] and the daughter drop size distributions by Tsouris & Tavlarides [11] (φ = 0.05) [12].

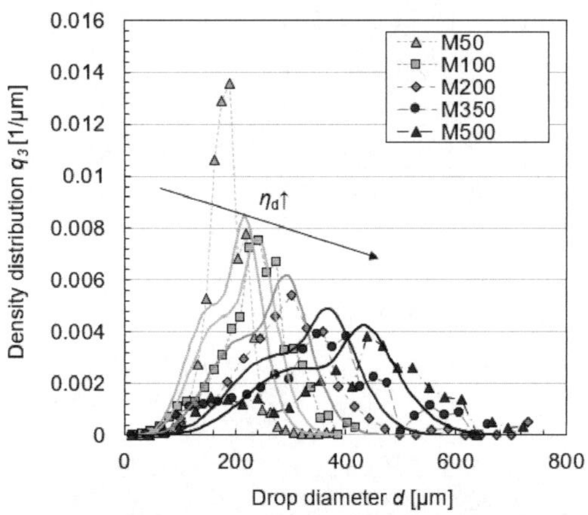

Figure 4. Results using the Alopaeus model [3,4] and the daughter drop size distributions by Tsouris & Tavlarides [11]. Steady state drop size distributions at n = 700 rpm, t = 1800s (compare Figure 3) (φ = 0.05) [12].

Conclusions

Results of steady-state and dynamic experiments with rising dispersed phase viscosity, dispersed phase fractions and stirrer speeds agree with expectations and literature. An iterative PBE fit procedure to steady state drop size distributions and dynamic Sauter mean diameters extracted from two experiments led to suitable fit parameters for the investigated PBE models [12].

The impact of rising dispersed phase viscosity on DSD shape and on Sauter mean diameters is best represented with the DDSD by Tsouris and Tavlarides [11, 12]. However, the agreement between experimental and simulated drop size distributions could be improved after an evaluation of other daughter drop size distribution shapes and might even be optimized using new fit parameters for the most promising daughter drop size distributions. The determination of the interdependent PBE fit parameters is crucial for simulation performance, but often only briefly discussed in publications. In future work, the fit procedure for all PBE models will be further evaluated, optimized and compared to literature.

Acknowledgments

Financial support by the Collaborative Research Centre (TRR 63 – subproject B8 - 56091768) funded by the German research foundation and the Berliner Chancengleichheitsprogramm (BCP) funded by the Berlin Senate Administration is gratefully acknowledged.

References

[1] L. Coulaloglou, C. Tavlarides, *Chem Eng Sci.,* 32 (1977), 1289–1297.
[2] R. V. Calabrese, T. P. K. Chang, P. T. Dang, *AIChE Journal*, 32 (1986), 657-666.
[3] V. Alopaeus, J. Koskinen, K. I. Keskinen, *Chemical Engineering Sci.* 54 (1999), 5887–5899.
[4] V. Alopaeus, J. Koskinen, K. I. Keskinen, J. Majander, *Chem Eng Sci.* 57 (2002), 1815–1825.
[5] W. Podgórska, *Chem. Eng. Res. Des.* 85 (2007), 721-729.
[6] W. Podgórska, J. Bałdyga, *10th European Conference on Mixing,* Delft, The Netherlands, 2000, 141–148.
[7] H. Zhou, J. Yang, S. Jing, W. Lan, Q. Zheng, S. Li, *Ind. Eng. Chem. Res.* 58 (2019), 23458–23467.
[8] H. Zhou, X. Yu, S. Jing, H. Zhou, W. Lan, S. Li, *Chem Eng Sci.* 195 (2019), 23–38.
[9] A. K. Chesters, *Chem. Eng. Res. Des.* 69 (1991), 259–270.
[10] S. Maaß, S. Buscher, S. Hermann, M. Kraume, *Biotechnol J.* 6 (2011), 979–992.
[11] C. Tsouris, L. Tavlarides, *AIChE Journal* 40 (1994), 395–406.
[12] L. Hohl, S. Röhl, M. Kraume, *Chem. Eng. Tech.* 46 (2023), 1-12 (early view version, DOI:10.1002 /ceat.202200589).

Quantifying mixing processes towards sustainable manufacture of bioinspired nanomaterials

PP1.1

***Richard Hodgkinson**[1]*, *Yahaya Baba[1], Mauro Chiacchia[1] and Siddharth Patwardhan[1]*. (1) Green Nanomaterials Research Group, Department of Chemical and Biological Engineering, University of Sheffield, S1 3JD, U. K.*
**S.Patwardhan@sheffield.ac.uk.*

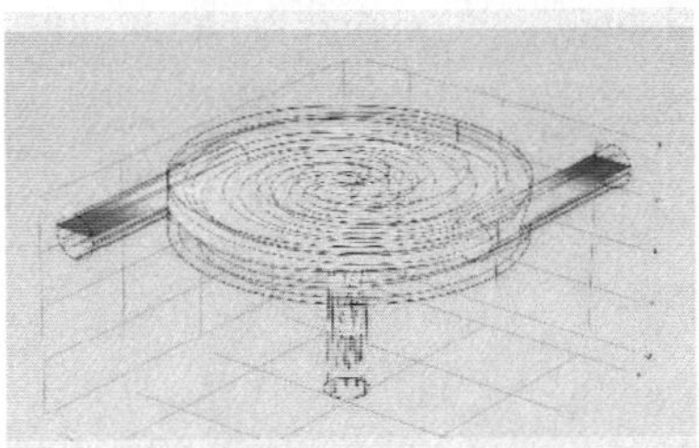

The choice of a stirred tank or continuous mixer places different technical challenges on process scale-up, but more importantly, they inherently achieve vastly different mixing timescales – a feature likely to affect final product properties in complex synthesis situations. In this talk, we report on our use of experimental mixing analysis techniques to explore these processes in the context of nanomaterials production.

Introduction

Nanomaterials have a vast potential in applications such as food, coatings, cosmetics, textile, transport, healthcare, and electronics, with a multibillion market. While many new nanomaterials are being discovered, their manufacturing is unsustainable and therefore their potential is not realised. Emulating natural designs, we have developed sustainable approaches to produce high value nanomaterials, in particular bioinspired silica (BIS) [1]. BIS has received an unmatched attention in recent times owing to its green synthesis, which offers a scalable, sustainable and economical method to produce high value silica for wide ranging applications including catalysis, environmental remediation, biomedical and energy storage [2]. To enable BIS to reach market, it is crucial to develop a strategy to scale-up manufacturing. BIS is one of many nanomaterials which we believe is strongly affected by synthesis conditions, although at the present time the complexities of the reaction pathway are not fully understood. As a result, exploring the effects of varying precursor mixing conditions – particularly mixing time – and the development of associated scale-up methodologies which maintain equivalent conditions is of critical importance.

Results and Discussion

Firstly, we will briefly introduce our recently published work [3] where we characterised mixing timescales within a stirred tank using a non-invasive colorimetric method coupled with novel image analysis (a workflow is shown in Figure 1 and full details are published in ref. [3]). The degree of mixing and mixing time were determined from this analysis for different impeller speeds and feed injection locations. The results from mixing analysis and materials characterisation (Figure 1 and ref. [3]) suggested that the bioinspired silica formation is controlled by meso- and, to a lesser extent, micro-mixing. Based on the new data from this investigation, a mixing time correlation is developed as function of the Reynolds number (see details in ref. [3]).

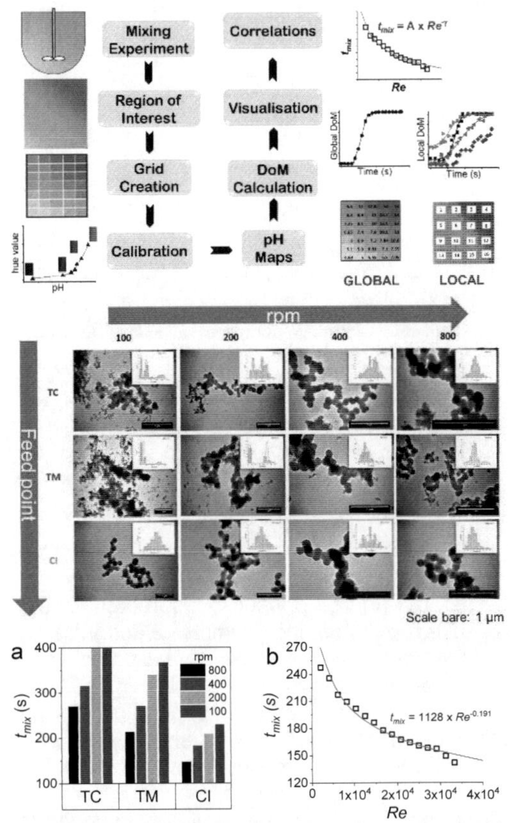

Figure 1. *(top) Schematic representation of the workflow and associated key results. (middle) TEM micrographs and particle size distribution plots (insets) for BIS produced under varying conditions of impeller speed and feed location. (bottom left) Experimental mixing times obtained from different feed locations (TC – top corner, TM – top middle and CI – close to impeller) and impeller speeds (100-800 rpm). (bottom right) The experimental mixing time as a function of Re (open symbols), the solid line shows a power model fit (R^2 = 0.95). Images reproduced with permission from ref. [3].*

While these results provide valuable insights on mixing occurring during BIS synthesis, the stirred tank geometry can be limiting in terms of energy dissipation and hence mapping a wider range is not possible. In order to address this issue, further work was performed using multi-inlet vortex mixers

https://doi.org/10.52202/074122-0033

(MIVM). MIVMs provide a wide range of flow conditions (Re = 10-10,000) and energy dissipation ($\epsilon = 10^{-1} - 10^3$ W/kg), covering the lab-scales as well as plant scales; hence they are very useful in this research. We used a custom designed MIVM as shown in the inset of Figure 2.

We will then move onto recent work applying MIVMs to BIS synthesis. Since such continuous mixers are not particularly amenable to image analysis techniques, we applied the well-known Villermaux-Dushman competing chemical reaction system to experimentally quantify mixing timescales against a range of flowrates. The correlations in ref. [4] were used to convert measured optical density to mixing times as shown in Figure 2.

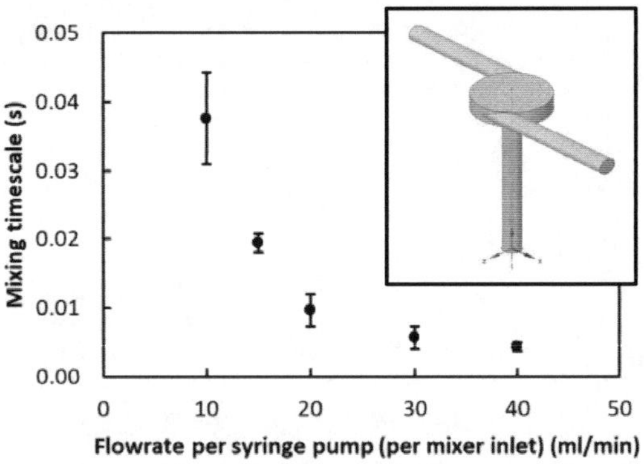

Figure 2. *Mixing times determined from a manually sampled Villermaux-Dushman experiment for a MIVM similar to the CAD image shown. A raw plot of optical absorbance is qualitatively identical with similar error bars. The MIVM has a mixing chamber diameter of 6mm and height of 1mm.*

In order to streamline the process and improve experimental accuracy, we developed a dedicated low-cost DIY in-line optical measurement system. This used 3D printing, simple electronics and LabVIEW code which was readily validated against a commercial spectrometer and can be easily recreated. The construction and operation of this will be briefly described: Figure 3 shows part of the setup. Not only does it easily allow a sweep of flow conditions, but also provided important insights on future general experimental protocols in continuous mixer studies and guiding our on-going work.

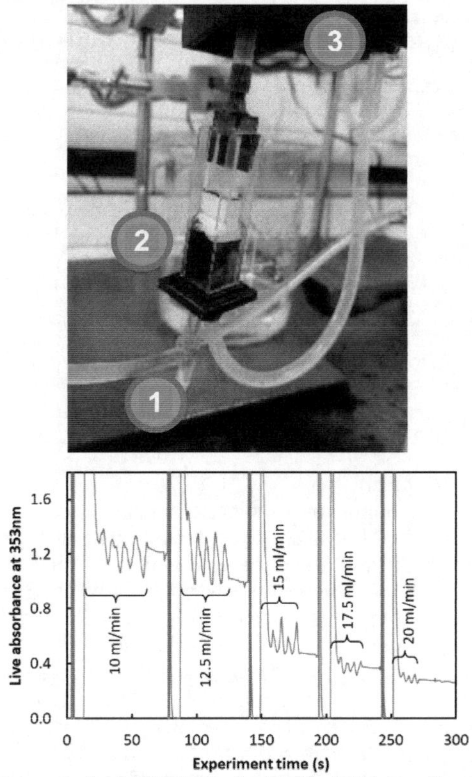

Figure 3*: (top) MIVM (1) attached to DIY in-line flow cell (2) (shown removed from DIY optical beam path enclosure (3)) for low cost automated Villermaux-Dushman characterisation of continuous mixers. (bottom) Example time series absorbance data from the setup showing experimental variability live during an automated run.*

BIS synthesis was then performed for a range of mixing times and the presentation will discuss the results from measuring precursor conversion, the BIS yield and BIS morphologies. Briefly, consistent with the observations from the reactions performed in the CSTR [3], the precursor conversion was not affected by varying flowrates/mixing times.

Using the mixing characterisation reported above, present and future work consists of investigating the relationship between flow properties, the BIS formation process and the particle properties. These results from stirred tank reactor and MIVM (V-D and BIS synthesis) will be compared and analysed to develop mixing knowledge and scale-up rules.

References
[1] Robert Pilling and Siddharth V. Patwardhan, "Recent Advances in Enabling Green Manufacture of Functional Nanomaterials: A Case Study of Bioinspired Silica". *ACS Sustainable Chemistry & Engineering* 2022 10 (37), 12048-12064.
[2] Siddharth V. Patwardhan, and Sarah S. Staniland, *Green Nanomaterials: From Bioinspired Synthesis to Sustainable Manufacturing of Inorganic Nanomaterials*; IOP Publishing, 2019.
[3] Yahaya D. Baba, Mauro Chiacchia, and Siddharth V. Patwardhan, "A Novel Method for Understanding the Mixing Mechanisms to Enable Sustainable Manufacturing of Bioinspired Silica". *ACS Engineering Au*, DOI: 10.1021/acsengineeringau.2c00028
[4] Jean-Marc Commenge and Laurent Falk, "Villermaux–Dushman protocol for experimental characterization of micromixers". *Chemical Engineering and Processing: Process Intensification* 2011 50 (10), 979-990.

Advanced control of microstructure in ice cream by agitation and mixing operations during freezing process

PP1.2

Hayato Masuda[1,2], *Tomohiro Ryuzaki*[2], *Hiroyuki Iyota*[1,2]. *(1) Osaka Metropolitan University, 3-3-138 Sugimoto Sumiyoshi-ku, Osaka, Japan, hayato-masuda@omu.ac.jp. (2) Osaka City University, 3-3-138 Sugimoto Sumiyoshi-ku, Osaka, Japan*

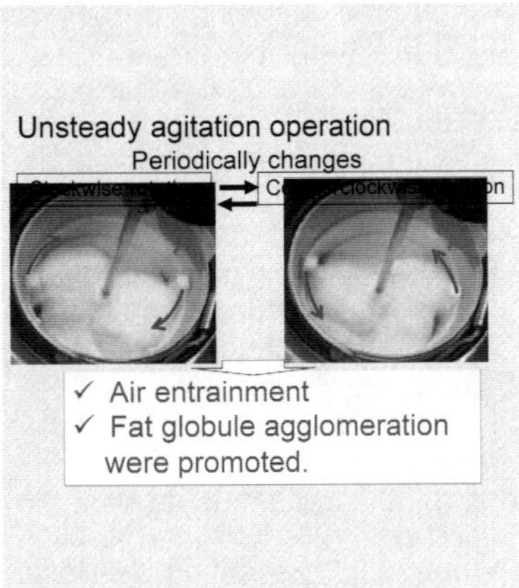

In the ice cream manufacturing, agitation and mixing operations plays an important role to sufficiently disperse ice crystals, bubbles, and fat globules during a freezing process. To achieve the advanced control of microstructure in ice cream, in this study, the effect of agitation operation on the air entrainment and fat globule agglomeration was investigated using a batch-type freezer with an anchor impeller. In addition to the steady agitation, an unsteady agitation with the periodical change in rotational direction was utilized. From the temperature measurement and flow observation, compared with the steady agitation, more effective scraping the frozen phase from the cooling surface was confirmed in the unsteady agitation case. Furthermore, the air entrainment and far globule during the freezing process were promoted by the unsteady agitation. Thus, the unsteady agitation is considered an effective tool to control the formation process of microstructure in ice cream.

Introduction

Ice cream, which consists of ice crystals, bubbles, fat globules, and unfrozen serum phase, is one of the most popular desserts in the world, and the consumption continuously increases. In the ice cream manufacturing, there are several steps, e.g., preparation of ice cream mix, freezing, and hardening. Generally, the freezing process is the most important because the ice cream structure, which determines texture, creaminess, and melt in mouth, is basically formed during freezing. It is noted that the freezing process in the ice cream manufacturing means not only freezing of water phase but also air incorporation and fat globule agglomeration. In industrial processes, a scraped surface heat exchanger with the rotating dasher is utilized to continuously mix the ice cream mix with air introduced from the inlet. In many cases, the ice cream is introduced at 4°C, and the frozen products are drawn at –4°C [1]. The products at this stage have still the fluidity. In the freezing process, three phenomena mainly occur: (i) nucleation and growth of ice crystals, (ii) breakup and coalescence of air bubbles, and (iii) fat globule agglomeration. In each phenomenon, the most dominant driving force is the shear force caused by the rotation of dasher [2, 3]. Besides, dispersion of each constitutive element (ice crystal, air cell, and fat globule) is important for the formation of ice cream structure during the freezing process. Thus, mixing/agitation plays an important role in the freezing process. Nevertheless, the freezing process has not been investigated so far from a perspective of mixing/agitation.

Previously, using a batch-type freezer, Masuda et al. [4, 5] investigated the effect of agitation speed during the freezing process on the size of air cells in the ice cream. They reported that smaller air bubbles with its narrower size distribution in the case of higher agitation speed. Because smaller bubbles with its narrower size distribution improve the creaminess [6], higher agitation operation is considered preferable for manufacturing creamy ice cream. Nevertheless, the kinetics in each phenomenon of constitutive elements during the freezing process remains unclear because only the final frozen product was analyzed in the previous study [4, 5]. Besides, the freezing time was restricted by a rated torque of motor. Thus, in this study, the effect of agitation and mixing operation on the change in bubbles and fat globules sizes during the freezing process was investigated in detail by developing an agitation system with higher rated torque enables to agitate for longer freezing time. Furthermore, an unsteady agitation which periodically changes the rotational direction was also performed aiming the improvement of the agitation and mixing performance.

Material and Methods

The experimental setup is shown in Fig. 1. The setup comprises the following devices: batch freezer (Ice-100, Cuisinart), DC motor, driver, torque meter (TPS-A-10NM, Kyowa Electric Co., Ltd.), and decelerator. The anchor-type impeller [diameter (D): 0.144 m, height (H): 0.08 m, thickness (w): 0.01 m] was utilized to scrape the frozen phase from the cooling surface. The vessel diameter was 0.147 m. Thus, the clearance between the impeller and vessel wall (cooling surface) was 0.0015 m. Compared with our previous study [4, 5], in this system, longer agitation during freezing was possible due to higher rated torque (2.0 N·m). The agitation/cooling were conducted for 2400 s. After 600 mL of commercial ice cream mix (Yamada Milk Product Co., Ltd.), which consists of 8.0 % fat, 10.0 % nonfat milk solids and unspecified

https://doi.org/10.52202/074122-0034

amount of emulsifier, stabilizer and sugar, was poured into the freezer, agitation and cooling were started. The rotational speed of impeller, N, was set at $N = 60$ or 90 rpm. In addition, the unsteady agitation with the periodical change of rotational direction was conducted. As shown in Fig. 2, the rotational direction was periodically varied with the constant agitation speed using a microcomputer control circuit. The impeller speed for the unsteady agitation, N_{us}, was set at $N_{us} = 60$ rpm. The period was set at 5 s in this study.

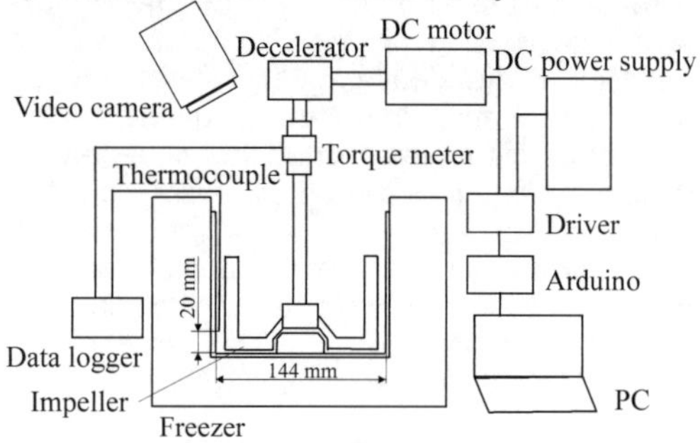

Figure 1. Experimental setup.

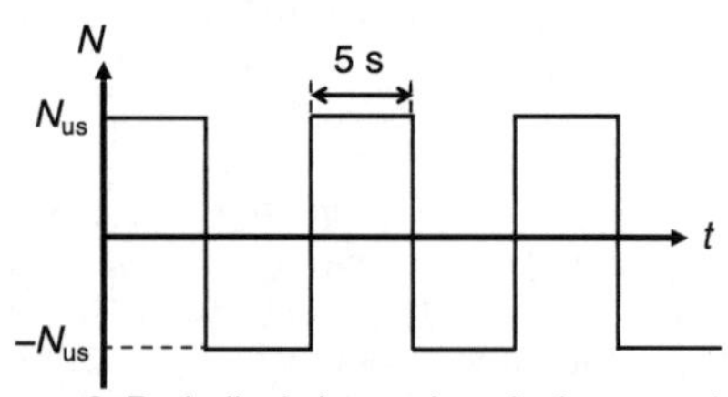

Figure 2. Periodical change in agitation speed with time.

The temperature of sample was measured during freezing using a K-type thermocouple with a diameter of 1 mm at the cooling wall 20 mm from the bottom surface. The agitation/cooling time, t, was regarded as $t = 0$ s when the measured temperature was 5°C. A small amount of sample was taken from the central part of freezer at t = 1200, 1800, and 2400 s, and immediately analyzed using a digital microscope (Leica DMS 1000, Leica Microsystems GmbH) for the measurement of air bubbles in the sample and a laser diffraction particle size distribution analyzer (LS230, Beckman Coulter Inc.) for the measurement of fat globule size distribution. From the image (1942 × 2590 pixels^2) obtained by the microscope, the number of air bubbles was counted using an ImageJ software. As an example, Figure 3 shows the image of bubbles within ice cream at t = 1200 and 1800 s. Besides, the area ratio of air cells in the image was calculated. This ratio can be approximately regarded as an overrun value.

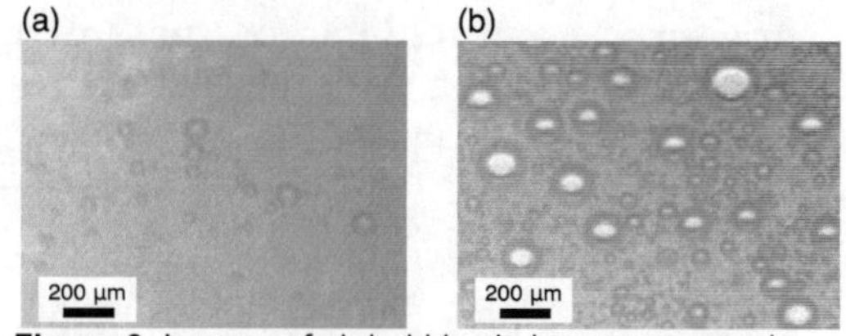

Figure 3. Images of air bubbles in ice cream sample prepared at $N = 90$ rpm: t = (a) 1200 s, (b) 1800 s.

Results and Discussion

Figure 4 indicates the change in temperature with time at N = (a) 60, (b) 90 rpm, and (c) N_{un} = 60 rpm. As shown in Fig. 4 (a), the temperature constant phase between approximately t = 1000 and 1400 s was observed. It is considered that, in this phase, freezing of water phase, namely the growth of ice crystals, locally proceeds near the cooling surface. On the contrary, no clear constant phase was observed in the case of higher agitation (Fig. 4 (b)). This is because ice crystals were dispersed well from the cooling surface into the bulk fluid preventing the local growth due to more efficient scraping. It is particularly worth noting that, as shown in Fig. 4 (c), the similar tendency was found in the case of unsteady agitation at $N_{us} = 60$ rpm in spite of lower absolute value of agitation speed than $N = 90$ rpm. From the recorded movie, the high efficiency scraping was confirmed when the rotational direction was switched. Nevertheless, there was no clear difference in the agitation torque during the freezing process between $N = 60$ rpm and N_{us} = 60 rpm. Thus, the unsteady agitation was effective for dispersion of ice crystals preventing the local freezing at the cooling surface.

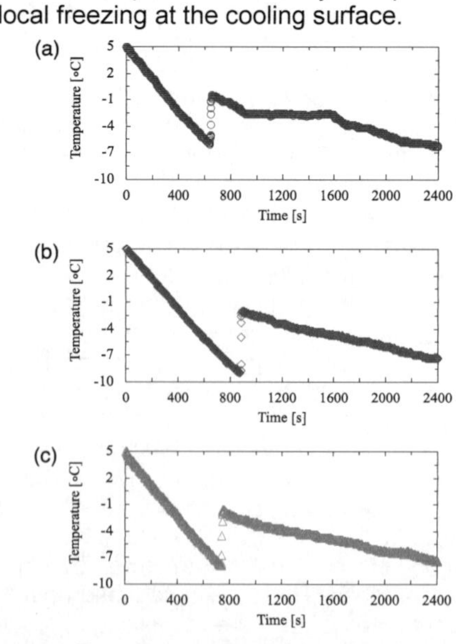

Figure 4. Temperature change in agitation speed with time: at (a) N = 60, (b) 90 rpm, and (c) N_{us} = 60 rpm.

Figure 5 shows the change in bubble area ratio within the ice cream with time. According to our previous study [4], taking bubbles in the ice cream mix occurs after freezing of water phase is advanced to a certain degree. Actually, as shown in Fig. 5, the bubble area ratio drastically increased between 1200 s and 1800 s. In this study, bubbles were taken from the free surface disturbed by the agitation because of no aeration. Compared between N = 60 and 90 rpm, it was confirmed that higher agitation speed leads to larger increase rate of bubble area ratio. Thus, more intense disturbance of the free surface by higher agitation speed is desirable for the enhancement of bubble entrainment. Remarkably, at t = 1200 s, the unsteady agitation (N_{us} = 60 rpm) was more effective for the bubble entrainment than steady agitation conditions (N = 60 and 90 rpm). Judging from the recorded movie during the freezing process, the air entrainment mainly occurred cavity regions where is generated behind the rotating impellers, as shown in Fig. 6. In the case of steady agitation, cavity regions steadily existed moving with the rotational impeller. On the contrary, the regions (behind the impeller) periodically changed the position due to the periodical change of rotational direction. It is considered that this non-localization of the regions enhanced the air entrainment at the initial stage (around t = 1200 s).

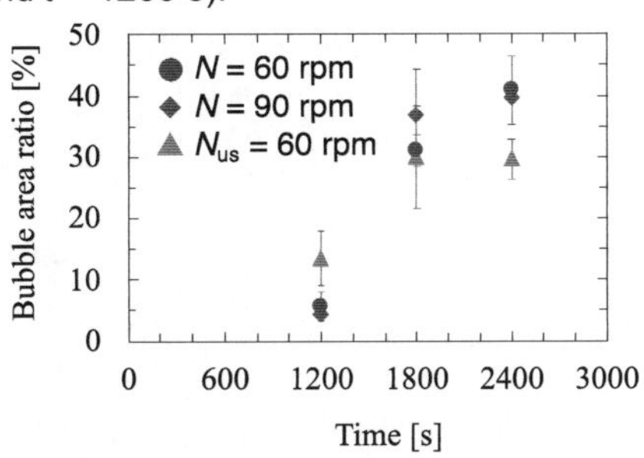

Figure 5. Change in bubble area ratio with time.

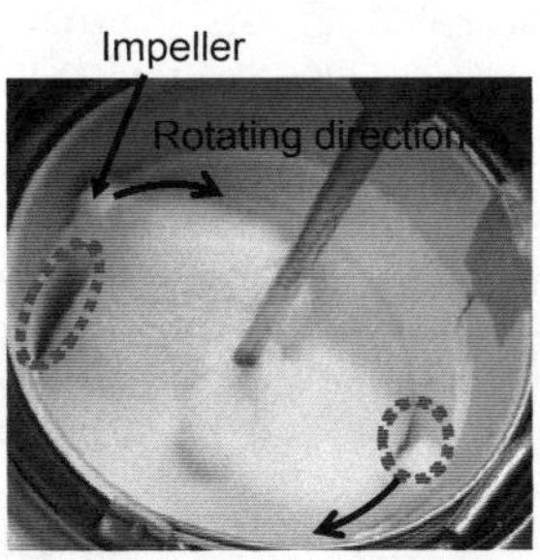

Figure 6. Snapshot during freezing at N = 60 rpm. The air entrainment mainly occurs behind the rotating impeller within red dashed line.

Enhancement of air entrainment is important for not only providing light sensation in mouth but also promoting fat globule agglomeration that occurs at the interface of bubbles [6]. Figure 7 indicates that the effect of agitation operation on the fat globule size distribution in samples at t = 1200 s. The size of primary fat globule is considered about 0.4 μm. Although there was no clear difference in the amount of bubble between N = 60 and 90 rpm (Fig. 5), agglomerate having about 2 μm diameter was detected only at N = 90 rpm as shown in Fig. 7. This result suggests higher shear force by higher agitation speed promotes agglomeration of fat globules. In the case of N_{us} = 60 rpm, however, more agglomerates were found than the steady agitation cases due to the presence of more bubbles as described above. Thus, the unsteady agitation has a possibility to promote fat globule agglomeration with lower agitation speed.

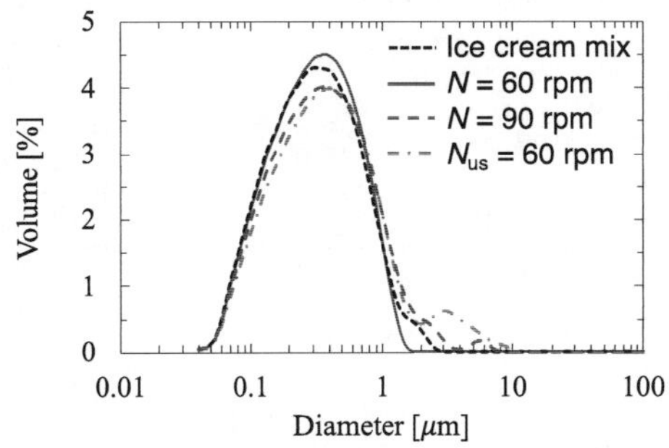

Figure 7. Fat globule size distribution in ice cream sampled at t = 1200 s. As a reference, the fat globule size distribution in ice cream mix is also plotted.

Conclusions

The effect of agitation and mixing operations on the change in bubbles and fat globules sizes during the freezing process was investigated using the batch-type freezer. In addition to the steady agitation, the unsteady agitation with the periodical change in the rotational direction was utilized. With respect to the steady agitation, higher agitation was effective for the air entrainment from the free surface and agglomeration of fat globules. Compared with the steady agitation, the unsteady agitation promoted scraping frozen water phase from the cooling surface. In addition, the air entrainment at t = 1200 s was enhanced due to the non-localization of the regions where the air entrainment mainly occurs. As a result, agglomeration of fat globules was promoted by the presence of more bubbles which is regarded as the medium for the agglomeration. Therefore, the unsteady agitation during the freezing process is considered an effective tool for the advanced control of the formation process of microstructure in ice cream.

Acknowledgments

This research was partially supported by JSPS KAKENHI grant numbers 20K21110 and 21KK0261.

References

[1] H. D. Goff, R. W. Hartel, in *Ice cream* (7th Eds.), New York, Springer, 2013.
[2] Y. Chang, R. W. Hartel, *Journal of Food Engineering*, 55 (2007) 71.
[3] E. M. Drewett, R. W. Hartel, *Journal of Food Engineering*, 78 (2007) 1060.
[4] H. Masuda, M. Sawano, K. Ishihara, M. Shimoyamada, *Journal of Food Process Engineering*, 43 (2020) e13369.
[5] M. Sawano, H. Masuda, H. Iyota, M. Shimoyamada, *Chemical Engineering Transactions*, 87 (2021) 337.
[6] H. Wildmoser, J. Scheiwiller, E. J. Windhab, *LWT-Food Science and Technology*, 37 (2004) 881.

https://doi.org/10.52202/074122-0034

Reactive Co-Precipitation in a Continuous Stirred Tank Reactor: Comparison Between CFD-PBM and Compartment Model Approaches

PP1.3

***A. Querio*[1]**, *M. Shiea[2], A. Buffo[1], G. Boccardo[1], D. Marchisio[1]. (1) Politecnico di Torino, Department of Applied Science and Technology, Institute of Chemical Engineering. Corso Duca degli Abruzzi, 24, 10129, Torino, Italy. (2) Centre on Sustainable Energy, Fondazione Bruno Kessler, Via Sommarive 18, 38123, Trento, Italy. andrea.querio@polito.it*

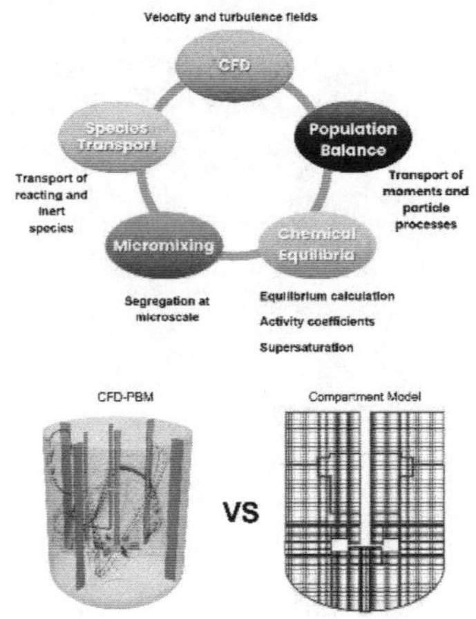

In this work, a modelling framework to simulate a reactive co-precipitation of Ni-Mn-Co hydroxide in a continuous stirred tank reactor is described. It couples computational fluid dynamics with population balance equation to describe the evolution of particle size due to nucleation, growth, aggregation and breakage. To solve the population balance equation, a quadrature method of moments approximation is employed. Moreover, a rigorous solver for the chemical equilibria is implemented to determine the supersaturation, i.e., driving force of the process, and a micromixing model to consider the segregation at molecular scale. In the end, a different less demanding method, the compartment model, is presented. The comparison between the results obtained by the different approaches, i.e., full CFD-PBM and compartment model simulations, are showed.

Introduction

Reactive co-precipitation is a very common process used to produce solid particles in solution. Particularly, the co-precipitation of nickel, manganese and cobalt hydroxide is really interesting because it is the precursor of cathodes for Li-ion batteries that are growing exponentially in diverse applications, like electric vehicles. In order to obtain batteries with desired characteristics, e.g., high energy density and long lifetime, is of fundamental importance produce cathode precursor materials with specific properties and morphologies that are strictly influenced by the co-precipitation process. Therefore, a modelling framework can be very useful to perform different simulations in order to optimize the operating conditions of the process, reducing the need of large and costly experimental campaigns.

In the modelling framework are included different features, like a micromixing model that is implemented to consider the segregation at molecular scale that, in this kind of system, can occur due to the fast injections of the reactants; the model considered is a multi-environment micromixing model. Moreover, a module to solve the chemical equilibria of the process was developed; the co-precipitation occurs in presence of ammonia that is a chelating agent that enhance the precipitation of Ni-Mn-Co hydroxide. In the expression used to determine the supersaturation are included the activity coefficients to take in consideration the deviation of the reacting medium from the ideal solution. In the end, to evaluate the evolution of the particle population, the quadrature method of moments (QMOM) is employed to solve the population balance equation which describe the size of the precipitated particles; the size evolution depends on different particle processes like nucleation, growth, aggregation and breakage.

In this work, two different approaches are presented: a full CFD-PBM simulation and a compartment model. The latter consists in a compartmentalization of the reactor using an agglomerative clustering algorithm, starting from information obtained by a CFD simulation followed by the resolution of balance equations for compartments. The co-precipitation process is performed in system that are in turbulent regime and so, for the CFD simulations, a turbulent model of those available in the Reynolds-Averaged Navier Stokes equation approach is employed, i.e., the *k*-*ε* standard model.

Material and Methods

The synthesis of Ni-Mn-Co hydroxide is performed in an aqueous solution in presence of ammonia; all the reactions and kinetic constants can be found in the work of Van Bommel and Dahn [1].

In **Figure 10**, a sketch of the simulated system is shown.

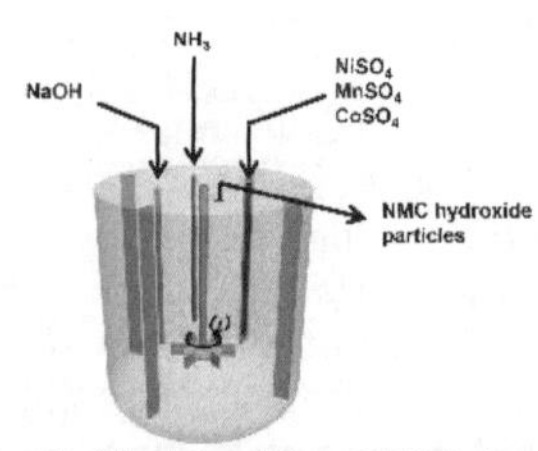

Figure 10. Sketch of the CSTR simulated

As mentioned before, a multi-environment micromixing model is implemented. The segregation at molecular scale has to be considered because it slows down the chemical

https://doi.org/10.52202/074122-0035

reactions that characterize the synthesis. Each feed is considered as an environment that mix with the other and it is described by a probability (or volume fraction) and local concentrations of the species of the specific environment. These are considered constant, while the probabilities decrease to the detriment of the reacting environment; the probabilities of the feed environments are tracked by the following equation:

$$\frac{\partial p_i}{\partial t} + \mathbf{u}_\mathrm{f} \cdot \frac{\partial p_i}{\partial \mathbf{x}} = \frac{\partial}{\partial \mathbf{x}} \cdot \left(\Gamma_t \frac{\partial p_i}{\partial \mathbf{x}} \right) - r_i \ ,$$

where p_i is the probability (or volume fraction) of the *i*-th environment, $\mathbf{u}_\mathrm{f}$ is the fluid velocity, identical to that of the particles because they have a negligible inertia, Γ_t is the turbulent diffusivity and r_i is the is probability flux from the *i*-th environment to the reacting one. The probability of the reacting environment is the complement to one of the summations of the first three environment probabilities. The flux r_i depends on the probability p_i and the local turbulence properties of the system.

As mentioned before, the driving force of the co-precipitation process is the supersaturation, which lead to the formation and growth of particles. In order to determine it, it is necessary to solve the chemical equilibrium of the process, so the transport equations for the concentrations of nickel, manganese, cobalt, ammonia, sodium and sulphate are solved. The volume-weighted total concentration of the chemical species cited before are tracked by the following equation:

$$\frac{\partial s_\alpha^{(4)}}{\partial t} + \mathbf{u}_\mathrm{f} \cdot \frac{\partial s_\alpha^{(4)}}{\partial \mathbf{x}} = \frac{\partial}{\partial \mathbf{x}} \cdot \left(\mathrm{D}_{eff} \frac{\partial s_\alpha^{(4)}}{\partial \mathbf{x}} \right) + h_\alpha^{(4)} \ ,$$

where D_{eff} is the effective diffusivity, summation of the molecular and turbulent diffusivities, and $h_\alpha^{(4)}$ is the source term for the chemical species α, summation of two different contributions: first, the flux of species α from the environment *i* to the reacting one and, second, the sink term for the species that react during the co-precipitation. The chemical equilibrium is solved by a Newton-Raphson method; as result, the equilibrium concentration of ammonia and nickel, manganese, cobalt and hydroxyl ions are obtained, which are fundamental to evaluate the supersaturation of the process. This is determined by the ratio between the chemical activities of the involved ions, i.e., nickel, manganese, cobalt and hydroxyl, and the solubility product of Ni-Mn-Co hydroxide.

As explained before, the increase of supersaturation leads to the formation of particles and their growth. The evolution of particle population is very important in this kind of process, which can be described by the particle size distribution (PSD). This express the expected number density of particles and its evolution in time can be evaluated from the population balance equation (PBE). In order to solve the PBE, a quadrature method of moments (QMOM) is adopted; this approximation consists in defining the PSD by a summation of *N* weighted Dirac delta functions. In this way, the weights (w_i) and abscissa (L_i) of the quadrature approximation can be specified from the lower-order moments of the PSD. After evaluating w_i and L_i by an adaptive quadrature technique, all the particle processes that describe the evolution of moments in time, can be described by weights and abscissas.

The evolution of moments in time and in the spatial domain are tracked by the following expression:

$$\frac{\partial \langle m_k^{(4)} \rangle}{\partial t} + \mathbf{u}_\mathrm{f} \cdot \frac{\partial \langle m_k^{(4)} \rangle}{\partial \mathbf{x}} = \frac{\partial}{\partial \mathbf{x}} \cdot \left(\Gamma_t \frac{\partial \langle m_k^{(4)} \rangle}{\partial \mathbf{x}} \right) - p_4 h_k^{(4)} \ ,$$

where $\langle m_k^{(4)} \rangle$ is the volume-weighted moment of order *k* in the reacting environment. The source term of the *k*-th moment, $h_k^{(4)}$, is the given by the contributions of nucleation, growth, aggregation and breakage. A more detailed description of the expression used to model the particle processes mentioned can be found in our previous works [2], [3]; all the parameters used in the framework were identified by fitting the experimental data via an optimization algorithm [4].

The full CFD-PBM simulations requires very long computational time to reach the stationary results and so, in order to reduce the computational cost and make the solver more robust, an operator-splitting approach and an hybrid parallelization were implemented in the model: the former consists in dividing the calculations of the transport equations of variables in two steps, interspersed with the integration of source terms, while the latter is a parallelization performed by MPI and OpenMP that allows us to use MPI for the decomposition of the system and OpenMP for the source integration.

Despite the implementation of the operator-splitting approach and the hybrid parallelization, the solver requires more than two weeks to complete a full CFD-PBM simulation, so a less demanding model has been developed: the Compartment Model. This approach consists in dividing the reactor in a finite number of small and completely mixed compartments that exchange mass between them, using the agglomerative clustering algorithm of the module of python scikit. The division of the reactor is based on the results obtained by a preliminary CFD simulation without PBE. From this short simulation, the turbulent dissipation rate and probability fields are taken as criterion at the base of the compartmentalization: the latter represents the zones of the system where the supersaturation is higher, while the first describe the turbulent conditions of the reactor. The governing equations solved by the compartment model are similar to the ones tracked in the full CFD-PBM; the source terms are the same, while the diffusive terms are not considered, and the convective ones are

https://doi.org/10.52202/074122-0035

represented by the mass exchange between one compartment and its neighbours.

Results and Discussion

First, in **Figure 11** some typical results of the full CFD-PBM simulations are shown; in this case only the nucleation and growth of the particles are considered. As can be seen, the solver is able to reach a convergence solution predicting a reasonable profile of the Sauter mean diameter (SMD), determined by the ratio between the area-weighted moment m_3 and the moment m_2.

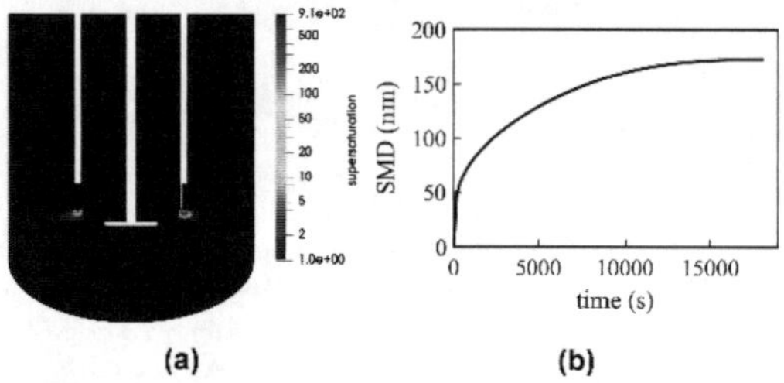

Figure 11. Contour plot of the supersaturation (a) and profile of SMD at the outlet of the reactor (b)

Furthermore, also with the compartment model can be obtained similar results, as can be seen in **Figure 12**. It is interesting to note that is possible to reduce the error between the compartment model and the full CFD-PBM increasing the number of subdivisions, however this approach reaches a final solution that no longer changes, even if we increase the number of compartments.

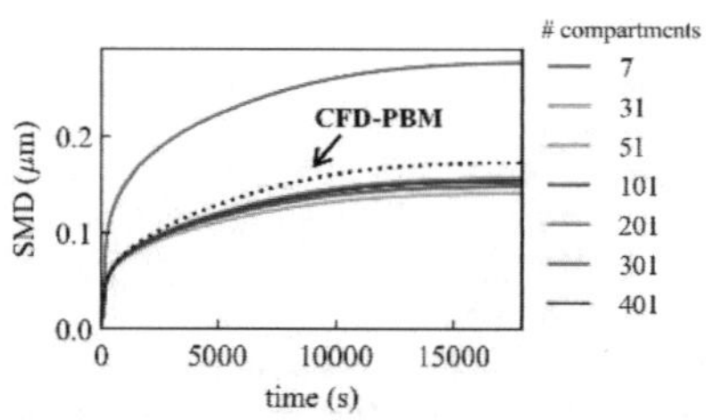

Figure 12. Comparison between SMD predictions of the full CFD-PBM model and compartment model, changing the number of compartments.

In **Figure 13** is showed the profile of moments m_0 obtained by the compartment model. In this case, is possible to see that there is a greater difference in the results obtained by the two approaches. This can be attributed to the loss of information relating to the diffusive transport terms.

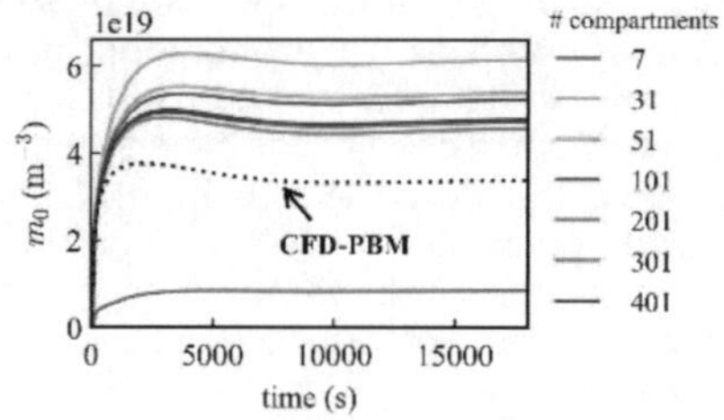

Figure 13. Comparison between m_0 predictions of the full CFD-PBM model and compartment model, changing the number of compartments.

However, it is noteworthy to highlight that the compartment model, using 51 compartment and 26 cores, reaches the steady-state solution in less than 5 hrs, while the full CFD-PBM approach needs more than 2 weeks with 42 cores.

Conclusions

An innovative framework has been developed for simulate reactive co-precipitation in continuous system, e.g., continuous stirred tank reactor. Thanks to this model, a full CFD-PBM simulations can be performed, predicting good results like the particle size distribution and the particle population evolution. Moreover, a less demanding model, i.e., the Compartment Model, has been developed. Adopting this approach, it is possible to obtain a good approximation of the real solution in much less time than the full simulations. In the end, increasing the number of compartments, the results converge to a solution that do not depends on the choice of the number of compartments.

The code can be downloaded from the following link: https://github.com/mulmopro/wet-synthesis-route.

Acknowledgments

The research reported in this paper was funded by European Union, Horizon 2020 Programme, SimDOME Project, Grant Agreement No 814492. The views and opinions expressed in this publication are the sole responsibility of the authors and do not necessarily reflect the views of the European Commission/Research Executive Agency.

Computational resources were provided by HPC@POLITO, a project of Academic Computing within the Department of Control and Computer Engineering at the Politecnico di Torino (http://www.hpc.polito.it).

References

[1] A. van Bommel and J.R. Dahn, *Journal of The Electrochemical Society*, 156 (2009) 5.

[2] M. Shiea, A. Querio, A. Buffo, G. Boccardo. D.L. Marchisio, *Chemical Engineering Research and Design*, 177 (2022).

[3] D.L. Marchisio, R.D. Vigil, R.O. Fox, *Journal of Colloid and Interface Science*, 258 (2003) 2.

[4] M.L. Para, M. Alidoost, M. Shiea, G. Boccardo, A. Buffo, A.A. Barresi, D.L. Marchisio, *Chemical Engineering Science*, 254 (2022).

Mixing Influence on Magnesium Hydroxide Precipitation: Computational Modelling and Kinetics Identification

PP1.4

A. Raponi[1], R. Achermann[2], S. Trespi[2], S. Romano[3], G. Battaglia[3], A. Cipollina[3], M. Mazzotti[2], A. Buffo[1], M. Vanni[1], D. Marchisio[1]. (1) Politecnico di Torino, Corso Duca degli Abruzzi, Turin, Italy, antonello.raponi@polito.it (2) Swiss Federal Institute of Technology, Leonhardstrasse 21, Zurich, Switzerland, ramona.achermann@ipe.mavt.ethz.ch (3) Università degli Studi di Palermo, Viale delle Scienze Ed. 6, Palermo, Italy salvatore.romano06@unipa.it

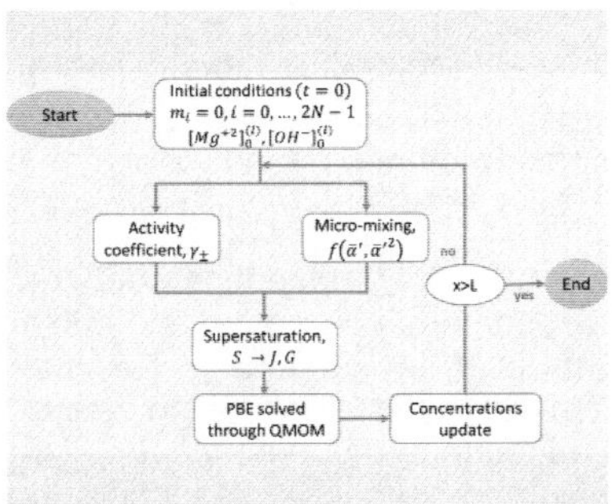

Magnesium is a raw material, whose recovery can be operated from waste-brines via precipitation, as magnesium hydroxide. The design, optimization and industrial scale-up of the precipitation process involves the development of computational tools to predict the complex interplay of mixing, hydrochemistry, primary nucleation, molecular growth, and aggregation. In this contribution, a simplified model, informed by computational fluid dynamics (CFD) simulations and an integrated population balance model (PBM), validated against experimental data from a T-mixer is presented. Once the unknown kinetics parameters are identified, the computational framework's scalability and transferability is tested.

Introduction

Mixing, at all scales, plays a paramount role in crystallization and precipitation processes [1]. In fact, the reactants mixing influences the supersaturation build-up which in turn affects both primary nucleation and molecular growth, as well as indirectly aggregation [2]. Therefore, it is crucial to account for the mixing influence on precipitation, especially when identifying unknown kinetics parameters. In this contribution, the importance of mixing on magnesium hydroxide is investigated and a computational model to describe the interplay between the phenomena involved is presented. The work is organized as follows: (i) a mono-dimensional framework informed by computational fluid dynamics (CFD) simulations is developed, (ii) unknown kinetic parameters are determined using multivariate optimization tools to match experimental data obtained through a circular cross-section static T-mixer and, to conclude, (iii) the computational framework is used to make predictions on new reactor geometries. The model aims at designing and optimizing a magnesium hydroxide first-generation prototype to control granulometry starting from brines, generally disposed as waste.

Material and Methods

Experimental tests

As described in the introduction, the experimental setup is composed of a circular cross-section T-mixer of constant diameter of 2 mm and total arms and mixing channel lengths of 4 cm. Ad-hoc experiments using $MgCl_2$ at different initial concentrations of [0.125, 0.25, 0.5, 0.75, 1] M, were performed, to mimic real brines:

$$MgCl_2 + 2NaOH \rightarrow Mg(OH)_2 + 2NaCl \quad (1)$$

The flow rates at each inlet were kept constant and equal to 1160 ml/min.

Computational modelling

As aforementioned, mixing strongly affects precipitation and it turns out to be, generally, one of the controlling phenomenon. This complicates the identification of the unknown kinetic parameters for magnesium hydroxide precipitation. These are inferred through a mono-dimensional model which accounts for a wide variety of physical contributions as shown in the graphical abstract: (i) micro-mixing, (ii) activity coefficients and (iii) well-tuned local turbulence from CFD simulations. In this section, a brief description of the model is provided to reinforce the importance of the micro-mixing to obtain system-independent kinetics.

Marchisio et al. [3] proposed to model extremely fast, irreversible reactions involving only two equations: the Reynolds-Averaged mixture fraction, $\bar{\alpha}$, and its variance, $\bar{\alpha}'^2$. The mixture fraction, in fact, accounts for the macro-mixing whereas its variance accounts for the micro-mixing. Moreover, they link the variance dissipation rate to the turbulent variables, $k - \varepsilon$, as follows:

$$\mathbf{u}\frac{\mathrm{d}\bar{\alpha}'^2}{\mathrm{dy}} = -\frac{C_\Phi}{2}\frac{\varepsilon}{k}\bar{\alpha}'^2 \quad (2)$$

where C_Φ is a tuning parameter set to 2 in the simulations and $\mathbf{u}$ is the mean fluid velocity, constant along the mixing channel coordinate "y" (since the system is assumed to behave like a plug flow reactor). Once the variance equation is solved, it is used to evaluate the supersaturation level along the mixing channel coordinate using the β-PDF approach. The supersaturation is, then, used to calculate the source terms in the Population Balance Equation (PBE) solved exploiting the Quadrature Method of Moments (QMOM) [4,5]. Moments equations (in this case three nodes, i.e., six moments) are solved accounting for primary nucleation (J), both

https://doi.org/10.52202/074122-0036

homogeneous and heterogeneous, molecular growth (G) and aggregation ($\overline{B_k}-\overline{D_k}$) as follows:

$$\mathbf{u}\frac{\mathrm{d}m_k}{\mathrm{d}y} = L_c^k J + \int_0^\infty kL^{k-1}Gn\,dL + \overline{B_k} - \overline{D_k} \quad (3)$$

with k = 0…5. Moreover, regarding the primary nucleation and molecular growth, these expressions are employed:

$$J = \boldsymbol{A}_1 \mathrm{e}^{\left(-\frac{\boldsymbol{B}_1}{\ln(S+1)^2}\right)} + \boldsymbol{A}_2 \mathrm{e}^{\left(-\frac{\boldsymbol{B}_2}{\ln(S+1)^2}\right)} \quad (4)$$

$$G = \boldsymbol{k}_g S^{g} \quad (5)$$

where $\boldsymbol{A}_1, \boldsymbol{A}_2, \boldsymbol{B}_1, \boldsymbol{B}_2, \boldsymbol{k}_g, \boldsymbol{g}$ are the parameters undergoing the optimization process. Concerning the aggregation, both the Brownian and turbulent contributions are accounted for, and their mutual influence is neglected. Therefore, the two contributions are simply summed. The parameters arising from aggregation contribution are $\boldsymbol{C}_1$ and $\boldsymbol{A}_\mathrm{p}$, respectively related to the collision frequency and particles bridge strength [6].

$$\beta_{\mathrm{agg}} = 10^{C_1}(\beta_{\mathrm{Br}} + \beta_{\mathrm{tr}})\eta_{\mathrm{agg}}\left(\boldsymbol{A}_\mathbf{p}^{1/2}\right) \quad (6)$$

The eight parameters are, first, tuned exploiting the T-mixer experimental tests. For each run (i.e., for each concentration), four characteristic moments ratios are extracted for a final amount of 20 points to fit. Then the framework is validated on a different geometry: a combination of a Y-mixer, two diverging channels and a final tube of constant diameter.

Results and Discussion

In this section the constrained optimization results, reported in Table 1, model prediction and extension to other geometries are presented and discussed.

Table 1. Ranges and final values of the constrained optimization between the mono-dimensional model and the T-mixer experimental tests

Parameters	Range	Value
A₁, particles/(m³s)	$10^{19} - 10^{29}$	$1.486 \cdot 10^{26}$
B₁, /	$250 - 350$	301.44
A₂, particles/(m³s)	$10^{10} - 10^{18}$	$7.41 \cdot 10^{14}$
B₂, /	$10 - 10^2$	30.34
k_g, m/s	$10^{-13} - 10^{-9}$	$2.51 \cdot 10^{-10}$
g, /	$1 - 2$	1.001
C₁, /	$0 - 3$	0.835
A_p, N/m²	$10^0 - 10^7$	5.897

Although, as introduced in the previous section, the used data points were twenty, for the sake of brevity and clarity only d_{10} (ratio between the first and zero[th] order moment) are reported.

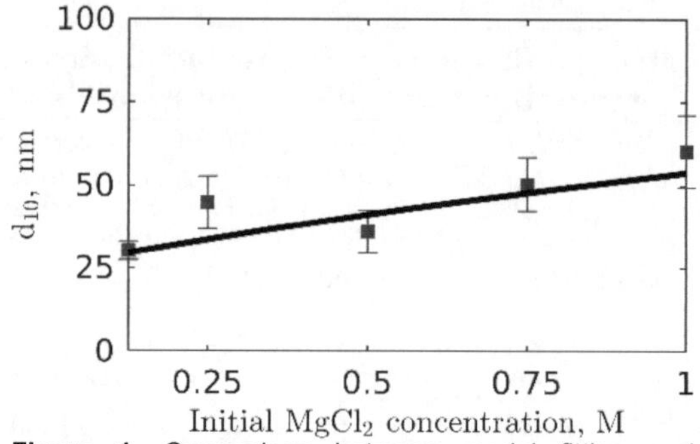

Figure 1. Comparison between model fitting and experimental data in the T-mixer.

As shown in Figure 1, the agreement between the model and the experiments is accurate and within the standard deviation which means that the parameters set in Table 1 is a good candidate to be tested for validation.

Therefore, a new geometry composed of a Y-mixer, two diverging channels and a final tube was again both experimentally and numerically tested. New experimental tests, including the investigation at lower concentrations to be sure the model could properly predict, were executed and CFD simulations of the geometry were run to account for the new local turbulence.

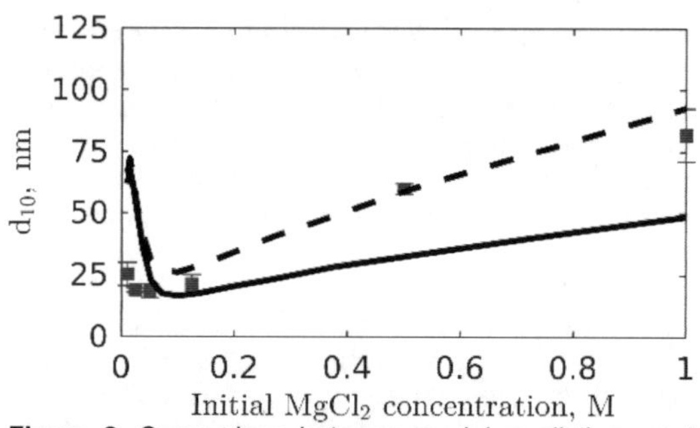

Figure 2. Comparison between model predictions and experimental data in the Y-mixer. $\boldsymbol{A_1} = 1.486 \cdot 10^{26}$ (*solid line*), $\boldsymbol{A_1} = 1.486 \cdot 10^{25}$ (*dashed line*).

As shown in Figure 2, the model can both predict the numerical trend and the minimum detected experimentally. Figure 2, moreover, presents the model trend when the primary nucleation pre-exponential factor is decreased by one order of magnitude. Although it might seem a big difference, one order of magnitude for that parameter is more than acceptable because it represents, by definition, the homogeneous nucleation rate when the supersaturation goes to infinite which makes it extremely difficult to be estimated. It is noteworthy, that the minimum presence is of strong conceptual interest. In fact, being the model able to predict that value and the numerical trend, authors can, not only validate the model, but also interpret and understand the main interfering phenomena, both at molecular level (micro-mixing dependent) and at the meso-scale, such as the secondary processes.

https://doi.org/10.52202/074122-0036

Moreover, a new correction factor for the collision frequency is accounted for, tested, and validated. From eq. (6), authors tried to provide a physical interpretation of the C_1 parameter linking it to the third-order moment (also related to the system magma density) as follows:

$$\beta_{\text{agg}} = 10^{C_1' m_3}(\beta_{\text{Br}} + \beta_{\text{tr}})\eta_{\text{agg}}\left(A_{\text{p}}^{1/2}\right) \quad (7)$$

The multivariate optimization was run again starting from the T-mixer experimental data and predicting, then, the Y-mixer ones. The new parameters set, as one can see in Table 2, is very close to the first one found (Table 1), but with a much better agreement in terms of framework predictions (Figure 3).

Table 2. Ranges and final values of the constrained optimization between the mono-dimensional model and the T-mixer experimental tests involving the novel correction factor

Parameters	Range	Value
A_1, particles/(m^{3}s)	$10^{19} - 10^{29}$	$10^{24.6}$
B_1, /	$250 - 350$	301.44
A_2, particles/(m^{3}s)	$10^{10} - 10^{18}$	10^{14}
B_2, /	$10 - 10^2$	30.34
k_g, m/s	$10^{-13} - 10^{-9}$	$2.51 \cdot 10^{-10}$
g, /	$1 - 2$	1.001
C_1', /	$0 - 3$	20.4
A_p, N/m^2	$10^0 - 10^7$	5.75

Figure 3 shows the model predictions for the Y-mixer dataset when the novel correction factor is used.

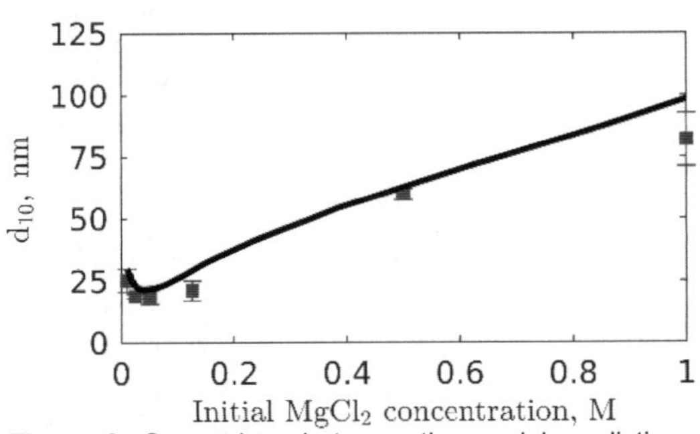

Figure 3. Comparison between the model predictions and the experimental data using the novel correction factor for the aggregation kernel.

Conclusions

Magnesium hydroxide is rising new interest since its wide application range. Therefore, authors implemented a code to describe and determine its precipitation kinetics. The computational framework is tuned on experimental data run using a circular cross-section T-mixer and validated on a geometry composed of a Y-mixer, two diverging conducts and a constant diameter pipe. The numerical predictions turned out to be in good agreement with the new experimental data points. The validated mono-dimensional framework, therefore, can be used to understand complex interacting phenomena and, thanks to its predictive feature, it sets itself as a powerful tool for the design and optimization of a magnesium hydroxide first-generation prototype.

Acknowledgments

This project has received funding from the European Union's Horizon 2020 research and innovation programme under Grant Agreement No. 869467 (SEArcularMINE). This output reflects only the author's view. The Executive Agency for Small and Medium-sized Enterprises (EASME) and the European Commission cannot be held responsible for any use that may be made of the information contained therein.

References

[1] D.L. Marchisio, A.A. Barresi, CFD simulation of mixing and reaction: The relevance of the micro- mixing model, Chem Eng Sci. 58 (2003) 3579–3587.

[2] P.M. Orlewski, M. Mazzotti, Modeling of Mixing-Precipitation Processes: Agglomeration, Chem Eng Technol. 43 (2020) 1029–1039.

[3] Marchisio D.L., and Fox R.O. (2016) Reacting Flows and the Interaction between Turbulence and Chemistry. In: Reedijk, J. (Ed.) Elsevier Reference Module in Chemistry, Molecular Sciences and Chemical Engineering. Waltham, MA: Elsevier. 21-Jan-2016

[4] D.L. Marchisio, J.T. Pikturna, R.O. Fox, R. Dennis Vigil, A.A. Barresi, Quadrature Method of Moments for Population-Balance Equations, AIChE Journal. 49 (2003) 1266–1276.

[5] D.L. Marchisio, R.D. Vigil, R.O. Fox, Quadrature method of moments for aggregation-breakage processes, J Colloid Interface Sci. 258 (2003) 322–334.

[6] R. David, P. Marchal, J.P. Klein, J. Villermaux, Crystallization and precipitation engineering-III. A discrete formulation of the agglomeration rate of crystals in a crystallization process, Chem Eng Sci. 46 (1991) 205–213.

https://doi.org/10.52202/074122-0036

Static Mixer Performance Enhancement using Measurements from a Novel Method for Continuous PLIF

MT2.1

Tom Robinson[1,2], *Peter Droegemueller*[2], *Mark J. H. Simmons*[1], *(1) University of Birmingham, United Kingdom, txr438@student.bham.ac.uk/tom.robinson@calgavin.com. (2) CALGAVIN LTD, Station Road, Alcester, United Kingdom.*

Planar Laser-Induced Fluorescence (PLIF) is a key experimental technique in the assessment of static mixers and is frequently the process by which Computational Fluid Dynamics (CFD) is validated. A prominent downside in continued and extensive PLIF research is the cost of often expensive test fluids that are disposed of after testing. This research presents a 'Continuous PLIF', methodology where fluid components are actively regenerated. The Rhodamine 6G tracer has been removed by adsorption thereby eliminating the test fluid as a consumable and enables PLIF in a novel continuous temperature-controlled loop. The PLIF images are high resolution, exhibit broad dynamic range, and have been captured using automated testing. Post-processing the images and comparing with measured pressure drop generated quantified performance curves that compare market available static mixer inserts against a novel design.

Introduction

Mixing is a critical step in numerous chemical processes, particularly when handling viscous fluids. Components within products are required to be in a mixed state (or at a controlled level of segregation). A lack of adequate mixing during processing can create yield loss through reaction inefficiencies and reduced product performance. These aspects require control and understanding, but knowledge of the mixedness of a product is as complex as mixing it. There is no thermocouple equivalent for mixing, limited design equations from first principles, and the measurement techniques are expensive and complicated. When is something truly mixed? Or more realistically, when is it mixed enough? Historically, mixing has been assessed using mostly experimental methods but improvements in numerical simulations have facilitated a shift from solely experimental research [1-3]. However, validation of in silico models remains key and laboratory experimentation is essential. In this research, the Planar Laser-Induced Fluorescence (PLIF) technique is developed to improve both resolution and efficiency of measurement from which mixing statistics are extracted to compare static mixers. Conventionally, this involves mixing multiple fluids and disposing into a waste tank once mixed [4-7]. This is costly due to the wastage of potentially expensive fluids – once mixed, they cannot be un-mixed. This work describes a novel methodology for continuous PLIF measurements which allows rapid, high-quality, repeatable, and quantifiable mixing analysis of static mixers for laminar flow.

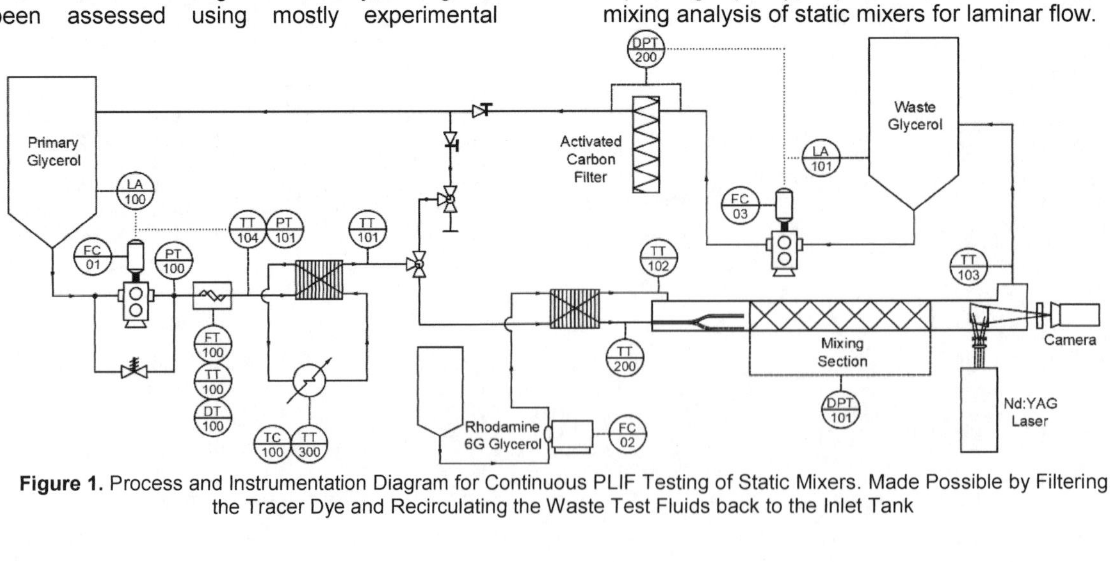

Figure 1. Process and Instrumentation Diagram for Continuous PLIF Testing of Static Mixers. Made Possible by Filtering the Tracer Dye and Recirculating the Waste Test Fluids back to the Inlet Tank

https://doi.org/10.52202/074122-0037

Material and Methods

Figure 1 shows a Process and Instrumentation Diagram of the test rig. Rhodamine 6G dye is dissolved into a 96 wt.% aqueous glycerol mixture at 0.5 mg rhodamine / L glycerol. This dosed glycerol is injected though a 6 mm tube into a velocity-matched primary flow of 96 wt.% aqueous glycerol in a 23 mm ID tube. The injection position is adjustable between either the wall or the centre of the primary stream. The injection is at the inlet of an interchangeable mixing section that houses a static mixer up to 300 mm long. After the tracer dye is blended into the primary flow, a thin sheet of 532 nm laser light from a Nd:YAG pulsed laser illuminates a cross-section of the pipe. This light causes the dye to fluoresce, and the distribution and concentration of this tracer is mapped by capturing its fluorescence with a 4 MP 12-bit high-speed camera situated normal to the flow.

Images are quantified by correlating the light intensity to the dye concentration at a 12 µm spatial resolution. Figure 2 contains the calibration verification when equating dye concentration to pixel intensity. 50 images are taken over 25 s for each test and averaged, calibrated, and quantified using a Python post-processing script. The script outputs a measured Coefficient of Variance (CoV) that is used in conjunction with pressure drop derived Fanning Friction factor, f_F to compare the performance of various static mixers.

$$f_F = \frac{\Delta P\, r}{L\rho u^2} \qquad CoV = \frac{\sqrt{\frac{1}{n-1}\sum_{i=1}^{n}(C_i - \bar{C})^2}}{\left(\frac{C_i - C_0}{C_\infty - C_0}\right)}$$

Unlike conventional static mixer PLIF, the novelty lies in the introduction of a continuous and temperature-controlled flow loop. By introducing an inline 5 µm activated carbon filter to adsorb the tracer dye, the glycerol can be regenerated; waste fluid material is regenerated in a separate flow loop and returns to the inlet tank. Concurrently to the image capture tests, a differential pressure transducer measures the equally important pressure drop induced by the static mixer.

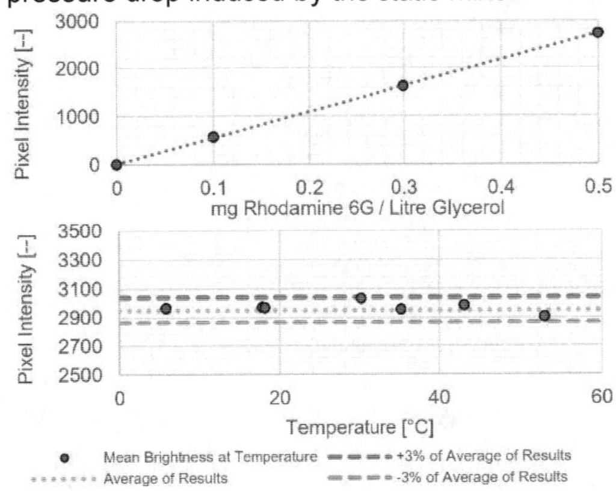

This PLIF facility leverages the temperature-viscosity curve of glycerol to control the fluid to temperatures between 5 °C and 60 °C, thereby enabling a viscosity range between 40 - 2600 cP.

Figure 2. Top - Shows the Linear Relationship between Dye Concentration and Light Intensity. Bottom - Sho Temperature and Light Intensity Independence

By introducing a variable flow rate, a three order of magnitude range of Reynolds (0.01 – 16) is realised with a single fluid. The test rig is run autonomously, finding steady states, and automatically stepping through test parameters. This is made possible through extensive data acquisition of digital and analogue measurements combined with I/O controls that communicate with a logic control system built in DASYLab. The mixing test section also functions as an in-line pipe rheometer by measuring pressure drop when empty, ensuring accurate fluid property data.

To test this setup, a series of experiments were run on Sulzer SMX™ and SMX™ plus mixers as well as a novel, first-generation design from CALGAVIN. The SMX elements both consist of 23 mm repeats that are sequentially inserted at 90° to one another. The CALGAVIN insert is a 32 mm repeating insert that tessellates without rotation.

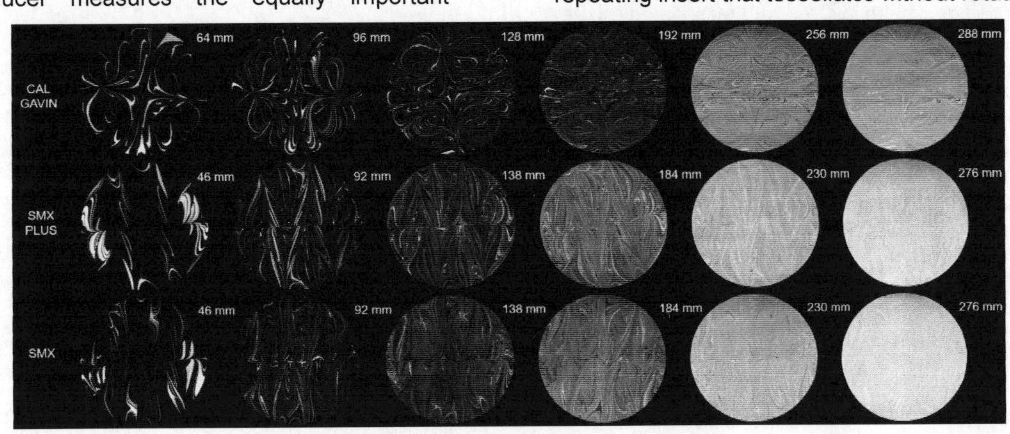

Figure 3. PLIF Images Comparing Mixing Performance and Demonstrating Image Clarity. (Contrast Enhanced for Viewability and are Therefore No Longer Quantifiable) Re: 2, Viscosity: 400 cP, Temp.: 23 °C, Flowrate: 0.015 kg.s[-1].

https://doi.org/10.52202/074122-0037

Results and Discussion

PLIF tests were conducted every 46 mm (two repeats) for the SMX and SMX plus and every 32 mm (one repeat) for the CALGAVIN insert up 300 mm. Each of these repeats were tested at twenty levels between values of Reynolds numbers from 0.01 to 16. The images for every experiment were recorded and post-processed to produce a final calibrated and averaged image for each experiment. Figure 3 contains a small selection of images from the tests completed. Each row is a different static mixer design, indicated on the LHS and each column an increasing static mixer length. This Figure showcases the clarity of the mixing images taken with the test facility. White pixels represent the injected dye stream, the black is the primary flow. As the dye is blended into the fluid the levels of grey depict the quantifiable concentration of the dye at that location.

Figure 4 displays the Fanning Friction factor computed from pressure drop. The tests span different mixer lengths and are normalised over the maximum 300 mm. These measurements span the full temperature range from 5 – 60 °C with a minimum power regression of 0.9986. This demonstrates the reliability of temperature, viscosity, mass flow and pressure measurement.

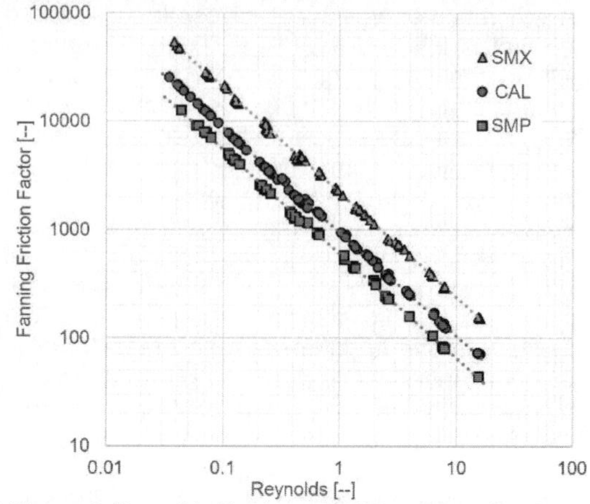

Figure 4. Reynolds Number vs Friction Factor for all repeat tests normalised over maximum 300mm length.

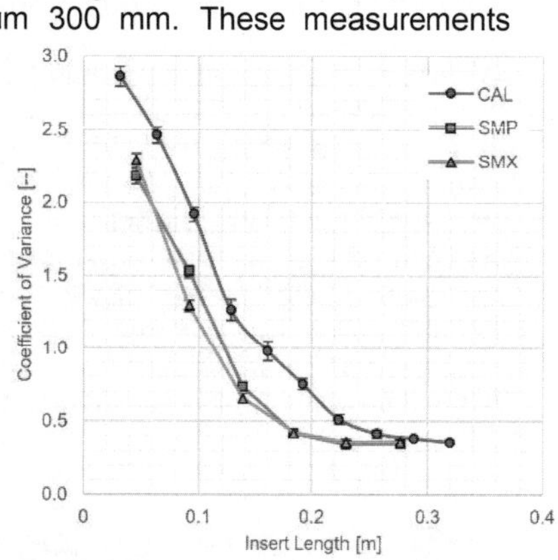

Figure 5. Mixer Length vs Coefficient of Variance

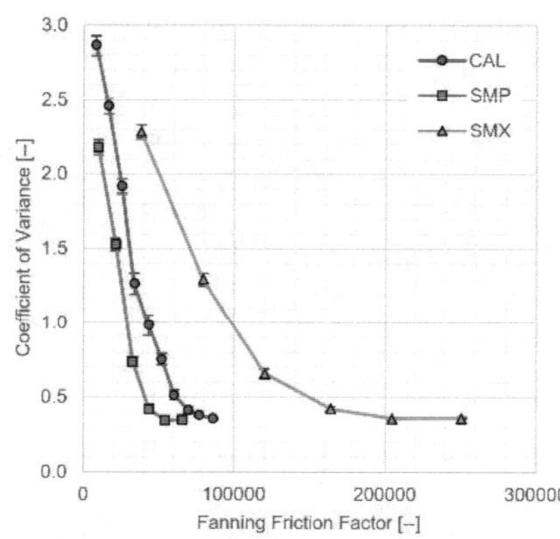

Figure 6. Friction Factor vs Coefficient of Variance

Computing the CoV of every image captured and graphing it against the insert length for the Fanning Friction factor generates performance curves comparing static mixer inserts. Figure 5 shows the decrease in CoV with increasing mixer length, with Figure 6 showing the corresponding energy penalty. Each point is the mean of twenty tests with error bars indicating one standard deviation – demonstrating the repeatability of the testing.

Conclusions

The development of this methodology enabled a detailed analysis of the tested static mixers. For instance, Figure 6 quantifiably demonstrated that the CALGAVIN mixer exhibits higher mixing / pressure efficiency than the Sulzer SMX™ and lower than the Sulzer SMX™ plus for the range of Reynolds numbers tested. These conclusions are made possible by highly detailed PLIF imagery and consistent pressure drop measurement with results shown to be repeatable.

PLIF, typically a destructive testing technique, is utilised in a novel continuous manner which enabled rapid, automated testing. This resulted in an improved methodology that limited experiment cost and encouraged more comprehensive testing and will allow faster static mixer prototyping.

Acknowledgments

TR wishes to acknowledge the support of CALGAVIN LTD and the EPSRC in funding the Centre for Doctoral Training in Formulation Engineering: EP/S023070/1.

References

[1] A. Lehwald, D. Thévenin, K. Zähringer, *Experiments in Fluids,* 48(5), (2010), 823-836.
[2] Y. Kim, O. Nematollahi, H. Dong Kim, K. Chun Kim. *J Vis*, 24, (2021), 671–681
[3] S. Migliozzi, L. Mazzei, P. Angeli, *Physics of Fluids*, 33(013104), (2021), 1-19.
[4] D. Hobbs, F. Muzzio, *Chemical Engineering Science*, 53(18), (1998), 3199-3213.
[5] F. Alberini, M. Simmons, A. Ingram, *AIChE Journal*, 60(1), (2014), 332-342.
[6] J. Aubin, D. Fletcher, C. Xuereb, *Chemical Engineering Science*, 60, (2005), 2503-2516.
[7] B. Nyande K. Thomas, R. Lakerveld, *Ind. Eng. Chem. Res*, 60(14), (2021), 5264-5277.

Numerical and experimental investigation of the effect of baffles and probes on the flow in small-scale reactors

MT2.2

A. D. Charalambidou[1], T.Wyrobnik[1], A. Ducci[2], M. Micheletti[1] (1)University College London, Department of Biochemical Engineering, Gower Street, London WC1E 6BT, a.charalambidou@ucl.ac.uk, m.micheletti@ucl.ac.uk. (2)University College London, Department of Mechanical Engineering, Torrington Place, London, WC1E 7JE, a.ducci@ucl.ac.uk.

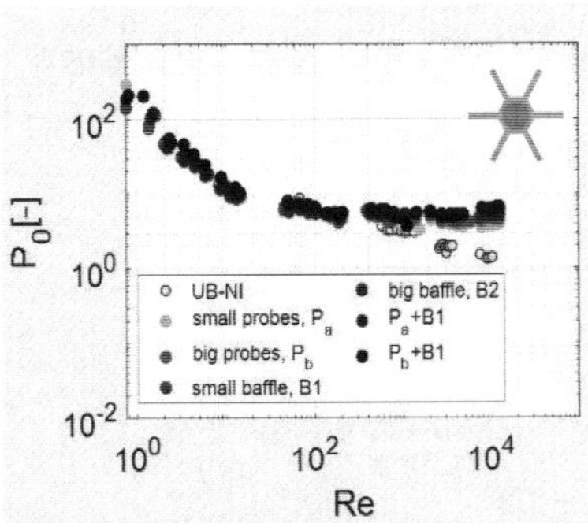

Power number curve for 250 mL small-scale bioreactor equipped with a radial Flat Blade Turbine (FBT).

The biopharmaceutical industry faces continuous pressure to reduce process development costs and decrease time to market. Therefore, the development of small-scale devices capable of multivariate process optimisation is essential. The aim of this work is to assess the power characteristics in several commercially available small-scale bioreactors, used for the production of biopharmaceuticals, equipped with axial and radial flow impellers. Results show that the presence of probes has a significant impact on the flow and the power consumption in a way similar to baffles and leads to a Power number (P_0) increase of 1.5-2.5 times compared to the unbaffled (UB) scenarios for axial and radial flows respectively. Detailed flow characterisation around the probes indicates the influence of probe configuration on power consumption and highlights that probe geometry optimisation can lead to at least 10% reduction in P_0 values.

Introduction

The design of economically feasible and profitable production processes has driven the biopharmaceutical industry towards the intensification of upstream experimentation and investigation of robust and predictive models for systems biology, following quality by design principles (QbD). In this context miniature bioreactors (i.e. mL-scale) have been widely adopted both as screening tools and Scale-Down Models (SDMs) operating under GMP manufacturing processes [1], [2]. For SDMs to be reliable and efficient, equivalence of the critical quality attributes (CQAs) and performance parameters should be demonstrated and preserved across different scales. Nevertheless, scaling among bioreactors with different volumes and geometries can be challenging due to the resulting differences in mixing and flow that may significantly impact reactor performance and process efficiency (i.e. lead to poor mixing and/or high shear environments) [3].

To date the majority of scaling up/down studies in stirred tank reactors (STRs) have focused on pilot and large scale systems (10-70 L), though it is crucial to extend those studies to smaller reactors, where probes and baffle dimensions relative to the reactor size might greatly affect the overall flow and mixing performance and thus the robustness of scaling equations. In this framework, a systematic characterisation of engineering features in scale down bioreactors is particularly important and a thorough understanding of the flow field in small-scale STRs, where probes cumulative volume can reach nearly 10-15% of the total working volume, is essential to evaluate the way critical design parameters affect the fluid dynamics underpinning upstream operations [4].

The aim of this work is to assess the power characteristics in axial and radial flow patterns in several commercial SDMs used for the production of biopharmaceuticals. Moreover, the flow patterns developed in the presence of baffles and probes at intermediate *Re*, where SDMs tend to operate, are assessed both computationally and experimentally in order to evaluate the impact of design parameters on reactor performance and assist in the optimisation of bioreactor design.

Material and Methods

The ungassed power inputs across all vessels at different rotational speeds were measured with an air bearing system using a digital force gauge, mounted at a distance l from the bioreactor axis. Power number (P_0) measurements were performed in SDMs with working volumes of 60 mL – 1 L equipped with and without internals (including baffles and probes) across the whole *Re* range (see Graphical abstract: results obtained for a 250 mL SDM equipped with a radial six-blade Flat Blade Turbine (FBT)). For Power number measurements the working fluid is water-glycerol mixtures with density 998.15 kg/ms$<\rho<$1,260 kg/m^3 and dynamic viscosity $9.9\cdot10^{-1}$ kg/ms$<\mu<1.02\cdot10^{-3}$ kg/ms with higher and lower values corresponding to 100% v/v glycerol and water concentration respectively, and *Re* varying in the range 1$<Re<$25,000 [5].

Thorough investigation of the flow dynamics was conducted for a 250 mL small-scale bioreactor via experimentally validated Computational Fluid Dynamics (CFD) simulations, as described in [6]. Briefly for the CFD studies in the tank, Large Eddy Simulations (LES) were used with a grid density of 3 million elements corresponding to a spatial resolution of 7 μm$<\Delta x<$1.8 mm, matching Taylor microscale. The time step size was dt=$3.3\cdot10^4$ s corresponding to an angular displacement of $d\varphi$~0.5° and 60 impeller revolutions were simulated in total. The bioreactor is equipped with three

https://doi.org/10.52202/074122-0038

equally spaced baffles, two cylindrical probes for pH and oxygen monitoring and a radial FBT. For flow investigation the working fluid is water with density ρ=998.2 kg/m^3 and viscosity μ=0.001003 kg/ms, while the rotational speed is N=250 rpm corresponding to Re=3,732.

Results and Discussion

Impact of probes on power

The impact of internal type, including probes and baffles, on Power number is presented in **Figure 1** according to results based on two SDMs: a 250 mL small-scale vessel equipped with a radial FBT and a 60 mL commercial scale-down bioreactor (EasyMax™, Mettler-Toledo) equipped with an axial Pitched Blade Turbine (PBT). The results are presented at Re>5,000 where changes in P_0 values due to the presence of internals are more evident and, in the presence of probes, data are presented in terms of probe blockage ratio, V(%), defined as the ratio of the probe volume over the total fluid working volume. For both flow types it is evident that at Re>5,000, P_0 becomes independent of Re in the presence of probes, irrespectively of the presence of baffles. This phenomenon appears more pronounced when the radial impeller is used as after the first probe addition P_0 becomes constant while for the axial impeller it continues to slightly decrease as illustrated in **Figure 1**(a) and (b) respectively.

The impact of probes and baffles on P_0 was investigated in more detail for both radial and axial flows. **Figure 1** shows a gradual rise of P_0 with increasing probe V(%) for both FBT and PBT, with P_0 being amplified up to 3 and 2.5 times respectively, at the highest probe V(%), when compared to the unbaffled-no internal (UB-NI) scenario. Similarly assessment of the sole impact of baffles indicated a further 20-25% and 6% increase of the P_0 for the radial and axial flow impellers respectively, highlighting the higher sensitivity of radial flow patterns on internal presence. P_0 results for all baffled SDMs studied were in good agreement with literature data corresponding to vessels with standardised ratios [7], [8], but discrepancies arose in the configurations where both probes and baffles were present. In the latter case P_0 values seem to be underestimated by ~30% in the literature. Those findings underline the importance of accurate P_0 measurements in small-scale vessels especially when power per unit volume (P/V) is used as a scaling parameter. In such cases underestimated P_0 values will lead to underestimated shear levels and dissimilarities in flow across scales.

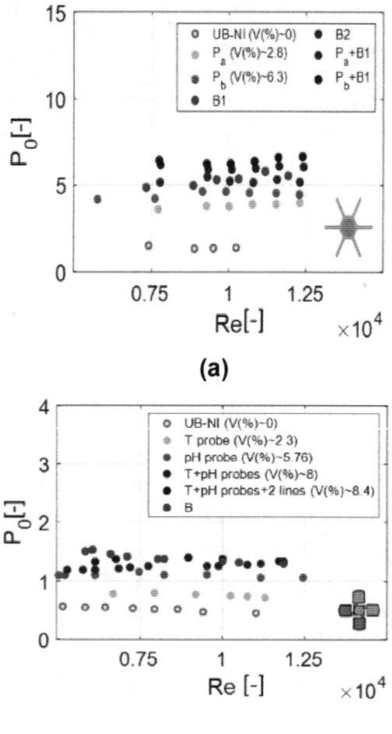

Figure 1. Impact of probes and baffles on the experimentally measured P_0 at Re>5,000 for: (a) radially agitated 250 mL SDM equipped with a six-blade FBT and (b) axially agitated 60 mL EasyMax™ (Mettler-Toledo) equipped with a PBT.

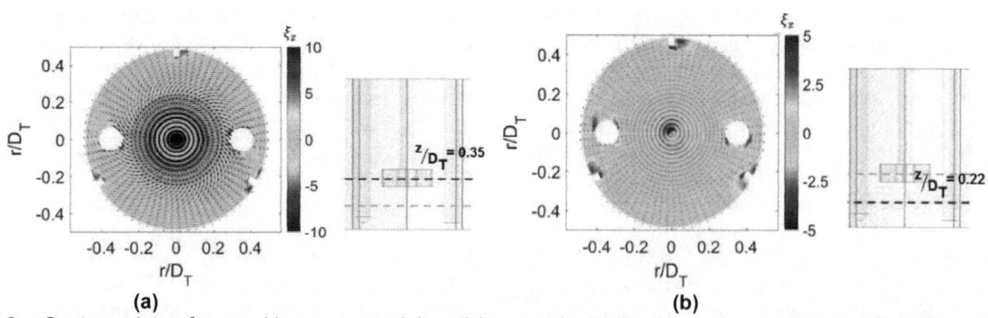

Figure 2. Contour plots of ensemble average axial vorticity magnitude ($\xi_z = D_T\omega_z/u_{tip}$, where D_T, ω_z and u_{tip} the tank internal diameter (D_T=67 mm), the axial vorticity and the impeller tip speed respectively) extracted from CFD dataset for two horizontal planes at two different elevations in the 250 mL SDM equipped with baffles sized $B1{\sim}D_T/16$, two probes of equal diameter $D_{Pa} = 0.12D_T$ for pH and oxygen monitoring (P_a) and a radial FBT: (a) horizontal plane at the impeller centreline (z/D_T ~0.35) and (b) horizontal plane below the impeller (z/D_T ~0.22). The different elevations of the horizontal planes investigated are indicated with the red dotted line on the bioreactor schematic.

l

https://doi.org/10.52202/074122-0038

mpact of probes on flow

The impact of the presence of probes on vorticity distribution was studied in multiple bioreactor planes as a direct way to assess the size and shape of the probe wake and an indirect way to obtain an approximate estimate of the shear in proximity to the probe walls. Indicative results are presented in **Figure 2** where ensemble average of the axial vorticity magnitude is presented for two horizontal planes corresponding to two different elevations in the 250 mL radially agitated SDM. Maximum vorticity values ($|\xi_z|$ ~5.5) are denoted at the corners of the baffles where small counter-clockwise rotating vortices are formed, thus increasing the total resistance of the system. In the presence of probes the wake generated by the impeller stream impinging on the probe surface, is evident at both bioreactor elevations, but appears to change directions along the probe height. At the impeller level, as shown in **Figure 2**(a), the flow impingement on the probes occurs in proximity of the impeller blades, where the radial velocity component is high, thus leading to the generation of a wake around the probes pointing outwards towards the vessel wall; for the horizontal plane below the impeller the intensity of the radial velocity component drops and tangential motion becomes more pronounced, changing the direction of the wakes formed around the probes which in **Figure 2**(b) appear parallel to the vessel wall.

It can be postulated that the formation of the wake occurring in the presence of probes is linked to an increase in resistance and form drag in the tank interior, which increases turbulence levels and could explain the rise in P_0 between configurations with and without probes. The replacement of cylindrical probes with novel probe configurations, for example probes having aerodynamic shapes, looks promising to reduce the form drag.

Conclusions

In this work, thorough characterisation of power requirements in commercial SDMs with different working volumes and configurations is conducted both experimentally and computationally. Investigation of the sole impact of probes and baffles on P_0 showed that the presence of probes led to a rise of turbulent P_0 equivalent to that produced by baffles, while at high *Re* results indicated that P_0 tends to flatten with increasing V(%).

Flow studies in the presence of probes revealed the formation of a wake developed around the cylindrical probes with changing direction depending on the tank elevation and interaction with the impeller jet. This wake was proven to be responsible for the P_0 increase observed in the presence of probes.

Comparison of the measured P_0 in the presence of both baffles and probes indicated an increase in Power number in small-scale reactors in the presence of internals which seems to deviate significantly from well-established empirical data that are commonly used for scaling studies. When scaling occurs based on constant P/V, discrepancies in P_0 between small and large vessel volumes will lead to variations in local energy dissipation rates affecting the flow patterns and mixing performance. Those disparities are expected to grow as reactor scales become smaller and highlight that a detailed engineering characterisation of small-scale vessels is essential for scaling optimisation and the development of efficient SDMs operating equivalently to larger scale bioreactors.

Acknowledgments

This work was supported by the Centre of Doctoral Training (CDT) in Emerging Macromolecular Therapies, Engineering and Physical Science Research Council (EPSRC); ER/L015218/1.

References

[1] V. Sandner, L. P. Pybus, G. McCreath, and J. Glassey, *Biotechnol. J.*, 14 (2019), 1700766.

[2] R. Bareither and D. Pollard, *Biotechnol. Prog.*, 27 (2011), 2.

[3] F. Li, Y. Hashimura, R. Pendleton, J. Harms, E. Collins, and B. Lee, *Biotechnol. Prog.*, 22 (2002^), 696.

[4] A. W. Nienow *et al.*, *Biochem. Eng. J.*, 76 (2013), 25.

[5] T. A. Wyrobnik, S. Oh, A. Ducci, and M. Micheletti, *Chem. Eng. Sci.*, 2021, 117263

[6] A.-D. Charalambidou, M. Micheletti, and A. Ducci, *AIChE J.*, 252 (2022).

[7] D. Chapple, S. M. Kresta, A. Wall, and A. Afacan, *Chem. Eng. Res. Des.*, 80 (2002), 364.

[8] R. L. Bates, P. L. Fondy, and R. R. Corpstein, *Ind. Eng. Chem. Process Des. Dev.*, 2 (1963), 310.

https://doi.org/10.52202/074122-0038

Macro- and Micromixing and Synthesis of Oxide Particles in Microreactors with Intensively Swirling Flows

MT2.3

Rufat Sh. Abiev[1,2], *Irina V. Makusheva*[1,2]. *(1) St. Petersburg State Institute of Technology (Technical University), Moskovskiy pr. 26, St. Petersburg, Russia. (2) Ioffe Physical Technical Institute, Russian Academy of Sciences, Politekhnicheskaya ul. 26, St. Petersburg, Russia. abiev.rufat@gmail.com*

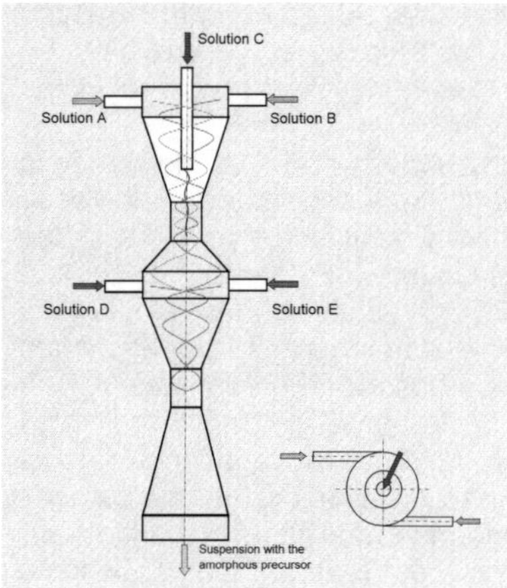

The influence of the hydrodynamics (flow rates Q, specific energy dissipation rate ε) on the micromixing in a two-step microreactor with intensively swirled flows (MRISF-2, see Graphical Illustration) was studied experimentally. Three methods of liquid input into the reactor were compared: (i) through the upper tangential and axial nozzles (TU1, Ax); (ii) through two upper tangential nozzles (TU1, TU2); (iii) through the upper and lower tangential nozzles (TU1, TL2). Segregation index X_s used as a measure of micromixing level was determined by means of iodide iodate reaction method. High efficiency is caused by the kinetic energy concentrated in a small volume of liquid near the neck. The best behavior in terms of micromixing was achieved for the (TU1, Ax) connection scheme: the level of $X_s \approx 0.01$ for $\varepsilon \approx 30$ W/kg, and comes down with growing ε to $X_s \approx 0.002$ for $\varepsilon \approx 30000$ W/kg.

Introduction

The growing interest in the production of nanosized particles over the past 2-3 decades is due to their special properties and wide areas of application, namely in order to obtain materials with special magnetic properties, catalysts, contrast agents for MRI, and many other applications. The most commonly used synthesis methods are the sol-gel method (and its variants - direct or reverse co-precipitation), the relatively new hydrothermal method and glycine-nitrate combustion.

Both at the macro- and microscale levels, the quality of mixing in laboratory reactors with magnetic stirrers is extremely low, which in many cases explains the difficulties in obtaining high-purity particles in solution synthesis, as well as the reasons for the formation of rather large agglomerates.

One of the effective methods for improving the quality of mixing is the use of intensively swirling flows. The combination of intensive swirling with a small volume (smaller than 0.5 ml) allows us to achieve high values of specific energy dissipation rate.

The objective of this work is to study the effect of the specific energy dissipation rate on the quality of micromixing in a two-stage microreactor with intensively swirling flows (MRISF-2, see Graphical Illustration) and to determine the most efficient operating mode and method for supplying reagent solutions.

Material and Methods

The main dimensions of the studied MRISF-2 were: the inner diameters of input pipes 5 mm, the diameters of output pipes 21 mm, the first neck diameter is 6.5 mm, the second neck is 8.0 mm. The total height of the device was 195 mm (including connecting nipples).

Studies to determine the quality of micromixing using the iodite-iodate technique were based on the recommendations given in [1, 2]. In this work, sets of concentrations are given in [1, 2] corresponding to the optimal optical densities that can be measured by means of UV-VIS spectrophotometer SF-2000 (Russia).

The local velocities were varied along with the flow rates, according to continuity equation, and for the range of flow rate from 1.0 to 3.0 L/min in each feeding branch pipe (having diameters of 5 mm) velocities were in the range $v_{1a} = v_{1b} =$ 0.87…2.53 m/s, whereas in the neck of the MRISF-2 the velocity was in the range $v_n =$ 4.83…13.58 m/s, see [3] for further details.

The specific energy dissipation rate ε (W/kg) was calculated by use of equation:

$$\varepsilon = \frac{N}{\rho V} \qquad (1)$$

where ρ is the density of solution, kg/m^3; V is the volume of the micromixing zone, m^3; N is the energy dissipation rate (spent on mixing), W.
The volume of the micromixing zone was considered as a sum of the neck volume (V_1 = 0.15 mL) and the zone between the end of the nozzle and the entrance into the neck (V_2 = 0.20 mL), $V_0 = V_1 + V_2 = 0.35$ mL.

To determine the value of N for each case, the Bernoulli equation for a system with two inputs and one output was derived [3].

$$\rho g Q_{1a}\left(z_{1a}+\frac{p_{1a}}{\rho g}\right)+\rho g Q_{1b}\left(z_{1b}+\frac{p_{1b}}{\rho g}\right)+\rho Q_{1a}\alpha_{1a}\frac{v_{1a}^2}{2}+$$

$$+\rho Q_{1b}\alpha_{1b}\frac{v_{1b}^2}{2}=\rho Q_2\alpha_2\frac{v_2^2}{2}+\rho g Q_2\left(z_2+\frac{p_2}{\rho g}\right)+\rho g Q_{1a}h_{w1a}+\rho g Q_{1b}h_{w1b}$$

where Q_i – liquid flow rate, m^3/s; z_i is the height from the branch pipe to the immersion height of

https://doi.org/10.52202/074122-0039

the apparatus, m; p_i – pressure, Pa; v_i is the velocity in the section, m/s; g is the gravity acceleration, m/s^2; *i* takes the following values: 1a - first inlet, 1b - second inlet, 2 - outlet.

The iodide-iodate method [1, 2] is based on the use of a system of competing parallel reactions was used to determine the quality of micromixing.

Three methods of supplying solutions to the MRISF-2 (Table 2) were studied to determine their effect on energy consumption and the quality of micromixing. In all cases, the volume of the zone of active mixing was taken equal to the sum of the volumes of the neck and the zone in front of it: into the upper tangential and axial branch pipes (TU1, Ax); into two upper tangential branch pipes (TU1, TU2); into the upper tangential and lower tangential branch pipes (TU1, TL2).

Results and Discussion

The plot shown in Fig. 1 allows to compare the specific energy dissipation rate for various methods of liquids supply into the MRISF-2. In all cases the correlation was described by the equation (where n was very close to 3.0):

$$\varepsilon = A_1 Q^n . \qquad (1)$$

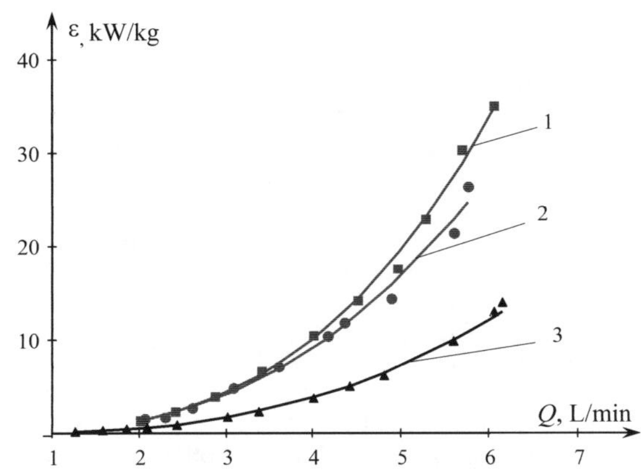

Figure 1. Dependence of the specific energy dissipation rate on the total flow rate ($Q = Q_2$) for various methods of liquids supply into the MRISF-2: 1 – (TU1, Ax); 2 – (TU1, TU2); 3 – (TU1, TL2).

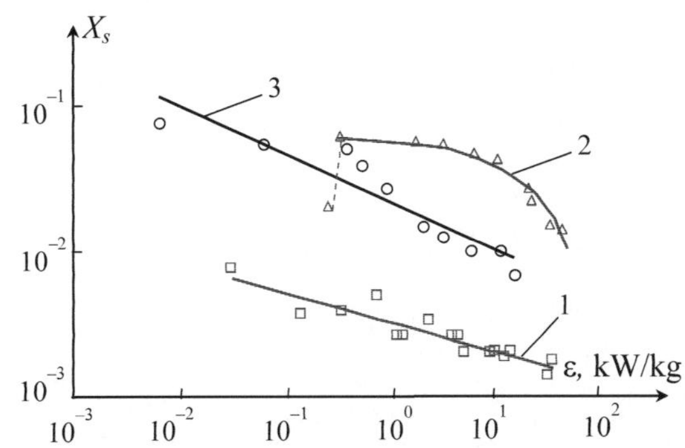

Figure 2. Dependence of the segregation index X_s on the specific energy dissipation rate ε (kW/kg) for various liquids supply methods:

1 – (TU1, Ax), 2 – (TU1, TU2), 3 – (TU1, TL2)

Fig. 2 demonstrates the influence of the specific energy rate and methods of liquids supply into the microreactor MRISF-2. The specific energy consumption when supplied through pipes (TU1, Ax) differs a bit from the supply through pipes (TU1, TU2). This is apparently due to the fact that in both cases the total flow Q_2 passes through both the first and the second necks.

The fact of the two tangential flows supplied through tangential branch pipes TU1 and TU2 simultaneously more intensive centrifugal forces was observed. As a result there are either the empty volumes in both upper and lower chambers filled by an air or some bubbles in the liquid caused by more intensive swirled flow.

The reason of the worse micromixing for the case (TU1, TL2) at the same flow rates compared to the case (TU1, Ax) is that the main (micro)mixing takes place in the second chamber, where the liquid from the branch pipe TU1 arrives through the first neck, mixing with the second flow which, in turn comes from the branch pipe TL2.

Conclusions

The micromixing in a two-step microreactor with intensively swirled flows (MRISF-2) was studied experimentally by use of iodide iodate reaction method. Three methods of two liquids supply were studied: (i) through the upper tangential and axial nozzles (TU1, Ax); (ii) through two upper tangential nozzles (TU1, TU2); (iii) through the upper and lower tangential nozzles (TU1, TL2).

The dependence of the segregation index X_s on the specific energy dissipation rate ε was revealed to be in good coincidence with the exponential function

$$X_s = A_2 \varepsilon^m . \qquad (2)$$

The best results were obtained for the connection (TU1, Ax) (X_s was within 0.01 and 0.002, i.e. very close to the ideal micromixing).

Falk and Commenge [1] on the base of 8 micromixers studies comparison get $\varepsilon \approx 10^1$–10^4 W/kg, and corresponding mixing time was identified as being in the range of t_m = 0.1 to 0.003 s. These values of t_m, on the other hand, are related to segregation index X_s in the range from 0.01 to 0.0004.

Some examples of oxide particles in MRISF-2 are discussed as well.

Acknowledgments

The work was supported by the Russian Scientific Foundation (project no. 20-63-47016).

References

[1] L.Falk, J.-M. Commenge, *Chem. Eng. Sci.*, *65* (2010) 405-411.

[2] P. Guichardon, L. Falk, *Chem. Eng. Sci.* **2000**, *55*, 4233–4243.

https://doi.org/10.52202/074122-0039

Observing the Interaction of Kolmogorov Length Scales in Lab and Industrial Scale Continuous Flow Loops

MT2.4

***Suzanne Kresta*[1]**, *William A Campbell*[1], *(1) University of Saskatchewan Dept. of Chemical and Biological Engineering, 57 Campus Dr., Saskatoon, Saskatchewan, Canada.*

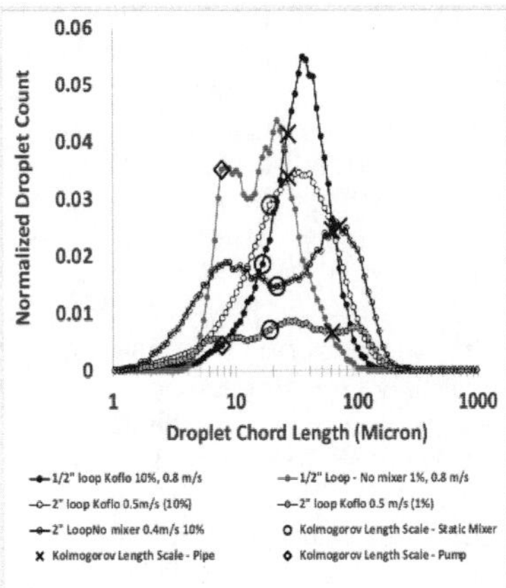

The Kolmogorov microscales describe the dimensions of the smallest turbulent eddies. In the case of continuously flowing liquid-liquid dispersions, the Kolmogorov scales of the system components (pipe, static mixer, pump) are are known to interact with the droplets of the dispersed phase and result in drop size distributions that line up with the Kolmogorov scales. In order to observe this phenomenon directly, liquid-liquid dispersion experiments were conducted at the bench scale and industrial scale, where an oil-in-water dispersion was continuously pumped through a pipe flow loop including a static mixer for up to 300 cycles, while data was collected on the pressure, temperature and drop size distribution. By calculating the respective Kolmogorov scales for pump, static mixer and pipeline at various flowrates, the shape and peak of the DSD can be observed to interact with the Kolmogorov scales which are calculated to be in the range of 7 to 120 microns.

Introduction

In the specific case of liquid-liquid dispersions, the Kolmogorov length scale describes the size of the smallest turbulent eddies[1], and thus is a useful scaling factor for predicting the shape and peaks found in drop size distributions.

While stirred tanks can be suitable for small scale turbulence experiments such as L-L dispersions, the rate of energy dispersion is very focused near the impeller and thus the time scale to achieve consistent dispersion throughout the vessel can be quite long. On the other hand, static or in-line mixers can achieve a relatively consistent dispersion throughout the pipe diameter in a single pass when the mixer is sized and selected appropriately for the desired outcome. It was with this understanding that this research was undertaken to combine static mixers with the continuous flow loop. Both a PVM and FBRM were used to continuously monitor the chord length distribution, with the goal of gaining further understanding of the interactions between Kolmogorov scales in the development of a liquid-liquid dispersion.

Material and Methods

A series of liquid-liquid dispersion experiments were carried out at 2 scales: a bench scale 12 mm flow loop and an industrial scale 50.8 mm flow loop. The bench flow loop was constructed with a 2.5 m length and 0.5 litre volume (Figure 1), and the 50.8 mm continuous flow loop was assembled with 24 meters in total length and 53 litres total volume (Figure 2). These experiments were carried out in such a manner that the data collected could be used to calculate the Kolmogorov length scales for the system components. In both cases, Koflo static mixers were used as the mixing element, Conosol 260

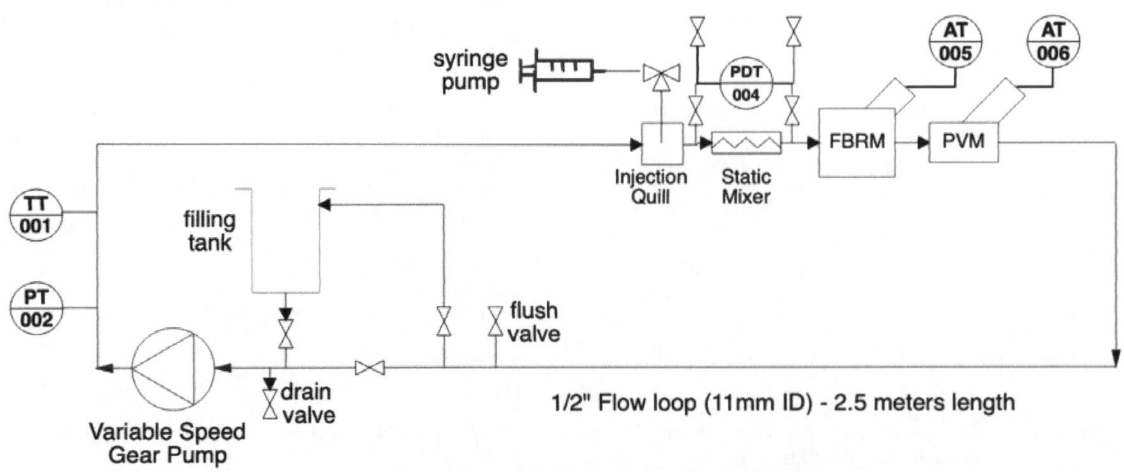

Figure 1. Bench Scale Flow Loop Schematic (1/2" diameter)

https://doi.org/10.52202/074122-0040

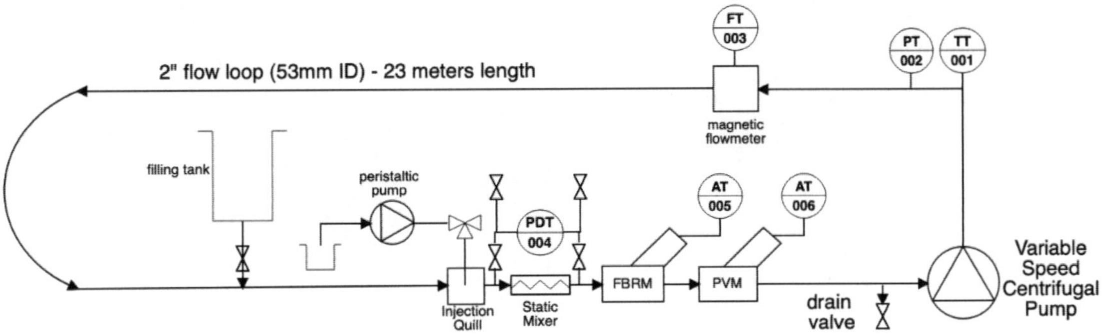

Figure 2. SRC Flow Loop Schematic (2" diameter)

(petroleum distillate, density = 0.839, viscosity = 3.74 cSt[2]) was used as the dispersed phase fluid, and water was the continuous phase (RO water was primarily used at the bench scale, while tap water was used in the 50.8 mm flow loop). Similar instrumentation was used at both scales; temperature and pressure measurements were similar for both and the same analytical instruments were used at both scales as well – the Mettler Toledo FBRM (G400) and PVM instruments (Particleview v19).

Table 1. L-L Dispersion Experimental Design Table

Variable	**1**	**2**	3	4
12.7 mm Flow Loop				
Flowrate	3 lpm	3.5 lpm	4 lpm	4.5 lpm
Holdup	1%	10%		
Cont. Phase	RO Water	Tap Water		
Disp. Phase	Conosol 260			
Mixer	Koflo	None		
Surfactant	Tween20	None		
50.8 mm Flow Loop				
Flowrate	40 lpm	53 lpm	66 lpm	76 lpm
Holdup	0.1%	1%	10%	
Cont. Phase	Tap Water			
Disp. Phase	Conosol 260			
Mixer	Koflo	Sulzer SMV	None	
Surfactant	Tween20			

The experimental design that was developed for each of these experimental scales identified key operational constraints for the mixers, while targeting the turbulent-high transitional regime between Reynolds numbers of 5000 to 20000 for the same static mixer at both scales. For the bench scale loop, the gear pump and the pipe diameter constrained the flowrate to the range of 0-4.5 lpm, which constrained the velocity range to 0.8 m/s and Reynolds number to 8700 and below.

Results and Discussion

The data collected in this work provides up to 15 million data points per experiment. The Focused Beam Reflectance Measurement (FBRM) instrument measures and counts the chord lengths of droplets passing by its focal length using a high-speed rotating laser in 100 length 'bins' in the range of 1 to 1000 microns, measuring and counting hundreds of chord lengths per second. The FBRM was mounted such that its focal face lies at the centre point of flow at a 45 degree angle, face-in-flow.

The other key measurement in these experiments was the pressure, and more specifically the pressure loss occurring over each pipe length and the static mixer.

The CLD was collected over a single flow loop cycle time and then analysed as an evolving drop size distribution from cycle 1 up to cycle 300. Cycle 100 was found to represent a well-evolved distribution and by superimposing the Kolmogorov scales for the mixer, pipe, and pump (where available) the interaction between the Kolmogorov scales and the CLD can be extracted (see Figures 3 and 4).

By normalizing each experiment's CLD count against the total number of droplets counted at cycle 100 for that experiment, many different experiments can be directly compared. Of note in these results is how the Kolmogorov scale of the pipe and the mixer together seem to interact with the shape of the CLD for a holdup of 10%, while for 1% holdup the pump apparently contributes to a greater degree in drop breakup.

The pump is observed to have a Kolmogorov scale between 5 and 10 microns for both scales of equipment and a peak in this scale range can be observed in the CLD for scenarios without a static mixer, and for 1% holdup.

Where no static mixer was used a bimodalCLD is observed at 100 cycles, with the two modes aligning with the Kolmogorov scale of the pump and pipeline, while the addition of a static mixer seems to limit the production of fine drops due to shear forces in the pump, suggesting that coalescence is also an important mechanism in the development of the CLD.

https://doi.org/10.52202/074122-0040

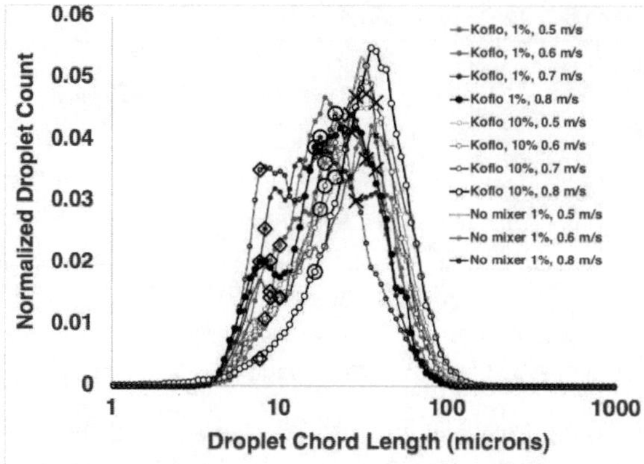

Figure 3. Drop Size Distribution for 12.7mm Flow Loop (O's represent Kolmogorov scales for static mixer, Xs for the pipeline, diamonds for the pump)

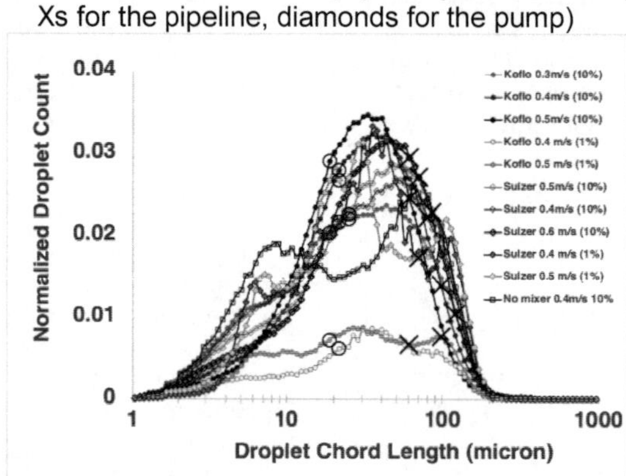

Figure 4. Chord Length Distribution for 50.8mm Flow Loop (O's represent Kolmogorov scales for static mixer, Xs for the pipeline)

Conclusions

In this work, large scale systems with hold-up of dispersed phase near 1%, showed repeated cycling and a wide, multimodal distribution. At 10% holdup the CLD becomes unimodal and more Gaussian. The Kolmogorov scales of static mixers and pipelines are shown to interact to influence the peak drop size. When a static mixer is not used, the Kolmogorov scale associated with the pump promotes drop breakup and the presence of a static mixer seems to counteract this by promoting coalescence. [2]

Acknowledgments

We would like to recognize the primary sponsor of this project, the National Research Council of Canada, as well as the Saskatchewan Research Council Pipe Flow Technology Centre who supported and assisted in the execution of this project. We would also acknowledge our research partners in this project, Alexandra Komrakova (University of Alberta), Federico Alberini (University of Bologna, University of Birmingham) and our colleagues at 4t2 Instruments (United Kingdom).

References

[1] A. E. Komrakova, D. Eskin, and J. J. Derksen, "Numerical study of turbulent liquid-liquid dispersions," *AIChE Journal*, vol. 61, no. 8, pp. 2618–2633, Aug. 2015, doi: 10.1002/aic.14821.

[2] Calumet Specialty Products Partners L.P., "Aliphatic Solvents - Typical Properties." Accessed: Dec. 19, 2022. [Online]. Available: https://calumet.com/wp-content/uploads/2022/09/Typical_Properties_one_page_HR.pdf

Effect of the kinematic viscosity on liquid flow hydrodynamics in vortex mixers

MT2.5

Gozde Gecim[1], *Ertugrul Erkoc*[2] *(1) Department of Chemical Engineering, Faculty of Engineering and Natural Sciences, Bursa Technical University, Bursa, Turkey, gozde.salkic@btu.edu.tr (2) Department of Chemical Engineering, Faculty of Engineering and Natural Sciences, Bursa Technical University, Bursa, Turkey, ertugrul.erkoc@btu.edu.tr.*

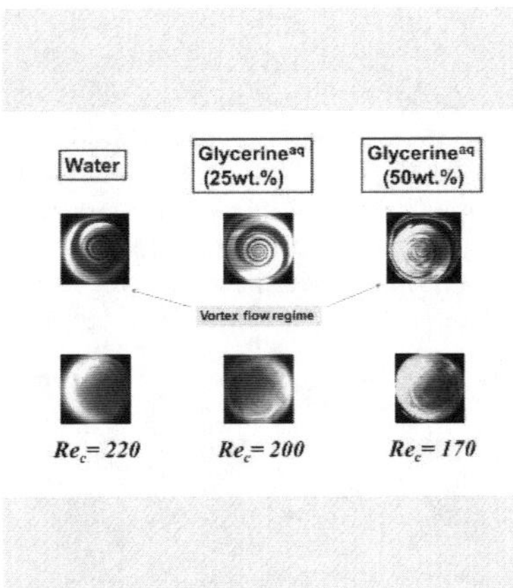

The effect of the kinematic viscosity of similar fluids on flow hydrodynamics and the onset of the instability was investigated to acquire information on mixing behaviour of vortex mixers. To assess the influence of the kinematic viscosity, water and two glycerine aqueous solutions with varying weight percentages were used and compared with our earlier study of gas flow hydrodynamics. The critical Reynolds number of the flow regime transitions was found using PLIF and PIV flow-field techniques. It was observed that the kinematic viscosity has a linear relation with the critical Reynolds number; however, the coefficient is different for liquids and gases. Additionally, kinematic viscosity also dictates the minimum Reynolds number where the fluids have a full turn inside the chamber. It was observed that fluids having higher kinematic viscosities have a full turn at lower Reynolds numbers, and the critical Reynolds number value decreases with increasing kinematic viscosity.

Introduction

Jet mixers are widely used as fast mixing devices because of their high mass and heat transfer rates and mixing efficiency [1]. Confined impinging jets (CIJs) and T-jets, which have opposed cylindrical/square jets and an impingement region to offer increased mixing, are the most commonly employed. The hydrodynamics of impinging jets are dependent on the momentum ratio of the jets. This can constrain the design of the geometry due to the differences or competition between the jets. Thus, vortex mixers with tangentially opposed jets where the equal momentum of the jets is not critical can be advantageous for applications requiring fast mixing [2,3].

Although the liquid flow hydrodynamics in CIJs or T-jets have been extensively studied, there is still a lack of knowledge about vortex mixers [4,5]. Flow regime characterization, when the flow changes from steady to chaotic, is crucial for understanding mixing hydrodynamics. Gas flow hydrodynamics in vortex mixers was previously revealed by our group [6]. Smoke-seeded flow is concentrated in one part of the chamber at $Re < 70$, according to experiments on gas flow in vortex mixers. This flow regime is called stratified flow as a certain segregation of fluid streams cannot be observed. At $Re = 70$, seeded flow completes a full rotation of the chamber, which is known as the vortex flow regime. As the Re number increases, radial and axial oscillations occur in the flow. The critical Re number is defined as the point at which the balance between these oscillations breaks and the flow becomes irregular. This is also known as the onset of the unsteady flow regime, and as the Re number increases, flow evolves to the chaotic regime [6].

In a recent work, our group presented the effect of kinematic viscosity on the critical Re number for gas flow hydrodynamics when employing the vortex mixer design provided. It was found that gases with higher kinematic viscosities will have a single defined vortex at lower Reynolds numbers and the formation of a full vortex in the channel is governed by the momentum rate of the jets, which is the function of gas property (density) and velocity [7]. The present study aims to reveal the liquid flow hydrodynamics and determine the effect of kinematic viscosity for the presented vortex mixer geometry. The liquid flow hydrodynamics were investigated by PLIF and PIV experiments to characterize the flow field and turbulent parameters in the mixing chamber with respect to liquid physical properties.

Material and Methods

The detailed schematic view of the vortex mixer geometry ($D_c/D_{inj} = 2$) and the experimental system was given in our recent works where D_c is the diameter of the chamber and D_{inj} is the diameter of the injectors (Figure 1) [5,6]. In this study, liquids were supplied from the tanks using the pumps, and the flow was controlled by flow controllers (Coriflow).

Three sets of experiments were performed when using water and aqueous glycerine solutions with weight percentages of 25 % and 50 % at equal Re numbers of jets. The physical properties of glycerine aqueous solutions at room conditions were determined using the general expressions for density and dynamic viscosity [5]. Kinematic viscosity values were calculated from the dynamic viscosity over density.

The evolution of liquid flow regime transitions and liquid mixing was examined based on the Re number of the jets. The critical Re numbers for liquids with different physical properties were established.

https://doi.org/10.52202/074122-0041

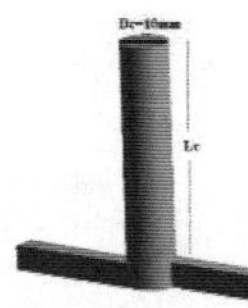

Figure 1. 3D view of the vortex reactor

For both PLIF and PIV experiments, a CMOS camera (Speedsense 1040) with 2330x1723 megapixels resolution and frame rate of 30 Hz was used to capture the frames lighted by each laser flash. The laser source used is a double laser head from Dantec Dynamics™ (Nd-YAG laser (532 nm, 50 mj) at a frequency of 96 Hz.

Table 1. Fluid properties at experimental conditions.

	ρ (kg/m³)	μ (kg/m.s)	ϑ (m²/s)
Water	996.0	9.59E-4	9.63E-07
Glycerine aq (25.0 wt.%)	1062.3	2.12E-3	2E-06
Glycerine aq (50.0 wt.%)	1128.4	5.65E-3	5.49E-06

To visualize the liquid flow and determine the flow regime transitions in PLIF experiments, rhodamine B, a fluorescent substance, was introduced to one of the incoming flows. A bandpass filter was used to ensure visibility by the laser source. Radial flow visualization experiments were performed at the chamber height of 30 mm, which is defined as plane-3.

For the PIV experiments, to obtain the liquid flow field and turbulent characteristics, one of the incoming liquid flows was seeded with polyamide seeding particles. The velocity vectors were calculated from the displacement of seeding particles between two subsequent frames. The Adaptive PIV method, which was thoroughly detailed in our most recent work, was used to disclose the liquid mixing in the presented vortex mixer geometry [5].

Results and Discussion

Figure 1 demonstrates the time-averaged flow visualization images from radial flow visualization experiments performed at chamber height of 30 mm. For pure water and glycerine aqueous solution of 25 wt.%, the vortex flow behaviour was gained at $Re = 40$, and the flow developed similarly until $Re = 130$. However, for glycerine aqueous solution of 50wt.%, the formation of vortex flow behaviour was observed at $Re = 30$. Moreover, the flow looks like it evolves differently than the others, as the fluid streamlines are not observable even at low Re numbers. This can be attributed to the high viscous forces of the glycerine aqueous solution of 50 wt.%.

To determine the flow regime transitions, it is important to analyze both instantaneous and time-averaged flow images. From Figure 2a, it can be seen that liquid flow streamlines disappeared beyond $Re \geq 220$, and the vortex behavior of the flow became almost unnoticeable for water flow. After this point, shortened fluid streams occur and the thickness of the boundary layer decreases. Thus, during this regime, convection becomes more prominent than diffusion resulting in enhanced mass and heat transfer rates. Although the flow hydrodynamics of Glycerine^{aq} (25 wt.%) evolves in the same way with water flow at lower Re numbers, the flow regime changes from steady to unsteady start at $Re = 200$ (Figure 2b). Additionally, the critical Re number of Glycerine^{aq} (50 wt.%) is defined at around $Re = 170$ from the flow visualization experiments. As the weight of the glycerine increases in the aqueous solution, the average kinematic viscosity of the solution increases as shown in Table 1. It can be concluded from both time-averaged and instantaneous results, flow regime changes from steady to unsteady at lower Re numbers for the liquids with higher kinematic viscosities.

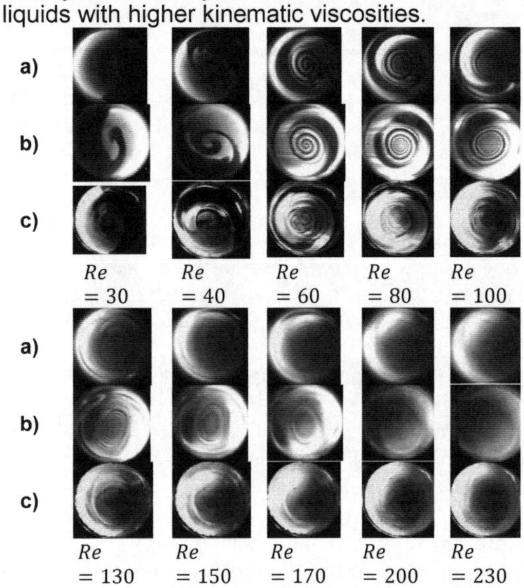

Figure 1. Time-averaged flow images at plane-3 a) Water b) Glycerine ^{aq} (25wt.%) c) Glycerine ^{aq} (50wt.%).

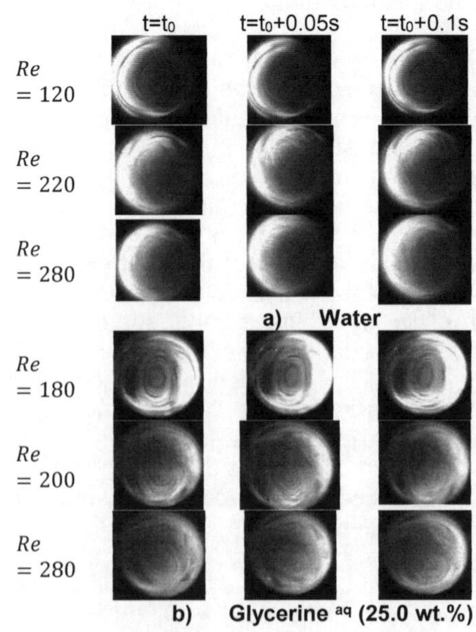

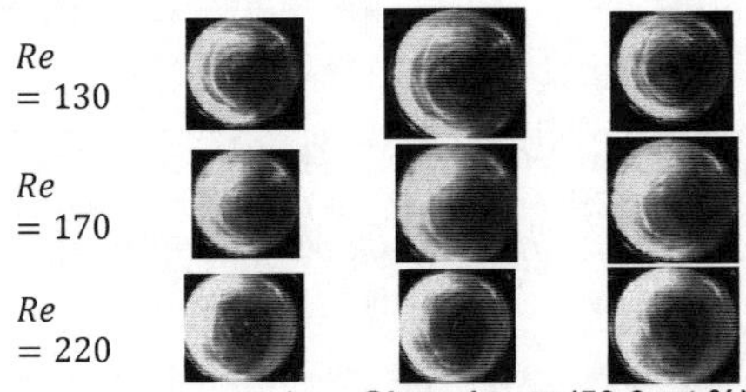

c) Glycerine [aq] (50.0 wt.%)

Figure 2. Instantaneous flow images at plane-3 a) Water b) Glycerine [aq] (25wt.%) c) Glycerine [aq] (50wt.%).

Table 2 summarizes the three liquid flow regime regions that were identified: the stratified flow regime, the vortex flow regime, and the unsteady engulfment regime. While the mixing is mainly dependent on the diffusion at stratified flow, it becomes dependent on the convection due to the axial instabilities created by the tangential characteristics of the jets.

Table 2. Flow regime transitions.

	Stratified Flow Regime	Vortex Flow Regime	Unsteady flow regime
Water	$Re < 40$	$40 \leq Re < 220$	$Re \geq 220$
Glycerine[aq] (25.0 wt.%)	$Re < 40$	$70 \leq Re < 200$	$Re \geq 200$
Glycerine[aq] (50.0 wt.%)	$Re < 30$	$30 \leq Re < 170$	$Re \geq 170$

In Figure 3, the critical Re number of the liquid and gas flows in the presented vortex mixer geometry is demonstrated. The decreasing trend of the Re_c based on the increasing kinematic viscosity is found to be similar for liquid and gas flows. Re_c was found to be very close for three gases, namely methane, argon, and nitrogen, due to their close kinematic viscosity values [6,7]. Thus, the linear relationship between Re_c and kinematic viscosity is more pronounced for liquids. Additionally, the calculated coefficients of trendlines for liquids and gases were shown to be different. The operating Re number of the proposed vortex mixer design to ensure efficient mixing may be precisely established for the liquids and gases with the range of kinematic viscosity that is supplied.

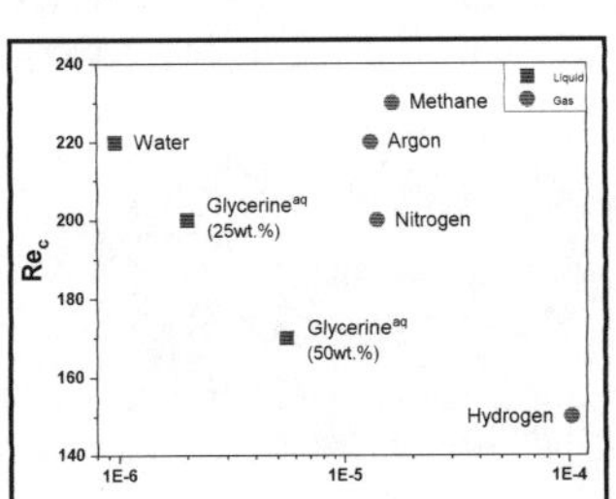

Figure 3. The effect of kinematic viscosity on the critical Re number.

From the time-averaged flow field in Figure 4, it can be deduced that the tangential flow of coming jets through the chamber walls is the main characteristic of the liquid flow. As the Re number increases, vortex wandering motion of the flow around the center axis is clearly noticeable. Similarly to gases, liquids of higher kinematic viscosities have a full vortex rotation at lower Re numbers. Furthermore, the balance between centrifugal force and the pressure gradient breaks earlier.

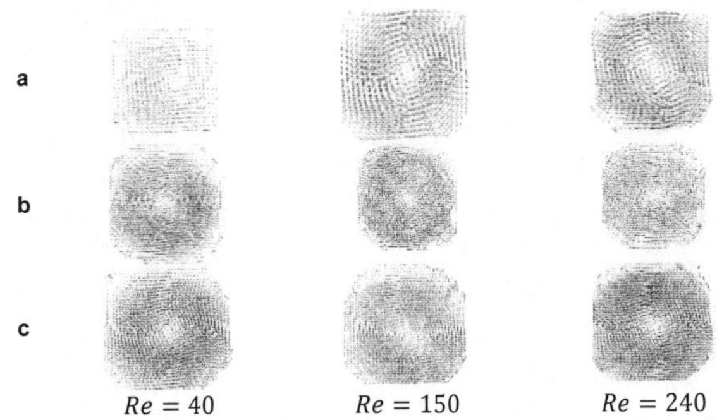

Figure 4. Time-averaged velocity vectors of determined flow regime regions a) Water b) Glycerine [aq] (25wt.%) c) Glycerine [aq] (50wt.%).

Conclusions

To successfully reveal the design characteristics of vortex mixers, it is important to examine both liquid flow hydrodynamics and gas flow hydrodynamics. In the presented study, three liquids with different physical properties were injected from the tangential jets. Flow visualization and PIV measurements were performed. Under the first turning point of the flow, from segregated to vortex flow, mixing was dominated by diffusion. As Re number increased, shear stress between fluid streams started to rule the flow. The evolution of the liquid flow hydrodynamics in the vortex mixer chamber was found to be very similar to the gas flow hydrodynamics. The linearly decreasing trend of the critical Re number based on the kinematic viscosity was found for liquids. Kinematic viscosity was proven to be one of the most important factors affecting liquid flow hydrodynamics.

References

[1] E.F. Alboiu, S. Rus, N.I. Albouiu, M. Degeratu, Energy Procedia, 112 (2017), 3-10.

[2] J. C. Cheng, M. G. Olsen, R. O. Fox, *Applied Physics Letters,* 94 (2009), 204104.

[3] Y. Shi, R. O. Fox, M. G. Olsen, *Applied Physics Letters*, 99 (2011) 204103.

[4] W. F. Li, K. J. Du, G. S. Yu, H. F. Liu, F. C. Wang, *AIChE Journal*, 60 (2014) 3033-3045.

[5] R. J. Santos, E. Erkoç, M.M. Dias, A. M. Teixeira, J. C. B. Lopes, *AIChE journal*, 54 (2008) 1153-1163.K.

[6] G. Gecim, E. Erkoc, *Industrial & Engineering Chemistry Research*, 60 (2021) 5674-5687.

[7] G. Gecim, E. Erkoc, *Industrial & Engineering Chemistry Research,* 61 (2022) 1192-1206.

https://doi.org/10.52202/074122-0041

Local interpretation of micro mixing through the Villermaux-Dushman reaction and the imaging UV-Vis spectroscopy

MT2.6

Torben Frey[1], *Markus Grabellus*[2,3], *Xuan My Le*[1], *Marko Hoffmann*[1], *Frank Herbstritt*[3], *Marcus Grünewald*[2], *Michael Schlüter*[1]. *(1) Institute of Multiphase Flows, Eissendorfer Straße 38, Hamburg, Germany, info@ims-tuhh.de. (2) Laboratory of Fluid Separations, Universitätsstraße 50, Bochum, Germany, sekretariat@fluidvt.ruhr-uni-bochum.de. (3) Ehrfeld Mikrotechnik GmbH, Mikro-Forum-Ring 1, Wendelsheim, Germany, info@ehrfeld.com.*

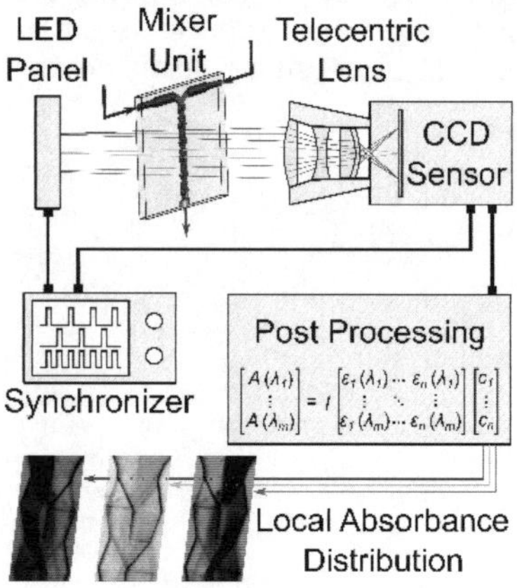

Global UV-Vis spectroscopy together with the Villermaux-Dushman test reaction is an established tool to characterise the overall mixing performance in process equipment, without giving local information concerning micro mixing. The novel imaging UV-Vis spectroscopy is developed in order to uncover local micro mixing phenomena with the Villermaux-Dushman reaction in a transparent split-and-recombine milli mixer manufactured from selective laser-induced chemical etching (SLE). The absorbance of backlight with multiple discrete wavelengths is recorded with a high local resolution through a telecentric lens (as shown in the Graphical Illustration) and post-processed into 2D-concentration fields of reaction components. The method presented in this work broadens conventional UV-Vis spectroscopy to creating a 2D-map of micro mixing within the split-and-recombine mixer to allow for insights into reactive micro mixing mechanisms and optimisation of mixer geometries.

Introduction

Milli- and micro mixers are used in continuous polymerisation processes to prevent or at least reduce fouling as shown by Bayer et al. [1], caused by the enhanced micro mixing and the reduction of dead zones. The field of micro fluidics is rapidly expanding with the emergence of new manufacturing procedures, high computational resources and new analysis systems.

Optical analysis systems in reactive systems, such as shadowgraphs, Laser-Induced Fluorescence (LIF), Near Infrared (NIR) or UV-Vis spectroscopy, are often used to quantify micro mixing characteristics [2]. From the known chemical timescale of a limiting reagent the time scale of micro mixing can be estimated [3,4,5]. Drawbacks of these methods can either lie in the limitation of only one optically detectable species [4] or in the lack of local information due to the necessity of micro mixing models, e.g., with the Villermaux-Dushman reaction [5].

The novel imaging UV-Vis spectroscopy [6,7] presented in this work is able to simultaneously detect the absorbance of multiple species at different wavelengths locally and consequently creates a 2D-map of the selectivity and micro mixing of the mixer.

Material and Methods

The Villermaux-Dushman system consists two parallel reactions with different reaction rates compete for a limiting reagent, shifting the selectivity towards the slow reaction when micro mixing is incomplete. The first acid base reaction

$$TRIS + H^+ \rightarrow TRISH^+ \quad (1)$$

converts fully dissociated perchloric acid using tris(hydroxymethyl)aminomethane (TRIS) buffer solution, which is shown effective for the global UV-Vis method [8]. The second slow reaction (Dushman reaction)

$$5\,I^- + IO_3^- + 6\,H^+ \rightarrow 3\,I_2 + 3\,H_2O \quad (2)$$

is known as iodide–iodate reaction and occurs only in areas of incomplete micro mixing.

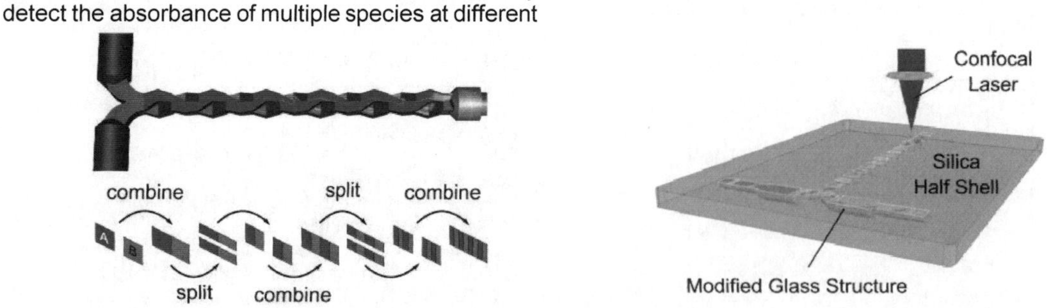

Figure 1. (a) Top view of the split-and-recombine mixer geometry and mixing mechanism, (b) Manufacturing of the silica mixer half shell by means of SLE.

While global UV-Vis spectroscopy detects the absorbance of triiodide formed from the equilibrium

https://doi.org/10.52202/074122-0042

$$I_2 + I^- \rightleftharpoons I_3^- \quad (3)$$

after a residence time section, the imaging UV-Vis spectroscopy captures the local triiodide concentration as well as the iodine formed from equation (2).

Imaging UV/Vis spectroscopy requires an optical access, which is achieved by means of Selective Laser-induced Chemical Etching (SLE) of the split-and-recombine structure displayed in Figure 1 (a) into fused silica half shells (Figure 1 (b)).

The mixer unit (model: Kaskadenmischer 15 by Ehrfeld Mikrotechnik GmbH) is backlighted by a custom printed LED panel. The LEDs used have dominant wavelengths of $\lambda_1 = 278\,\mathrm{nm}$, $\lambda_2 = 395\,\mathrm{nm}$, and $\lambda_3 = 465\,\mathrm{nm}$. The PCO Sensicam qe records light down to wavelengths of $200\,\mathrm{nm}$ at $60\,\mathrm{fps}$, invisible to the human eye. Using the Beer-Lambert law

$$\begin{bmatrix} c_1 \\ \vdots \\ c_n \end{bmatrix} = \frac{1}{l} \begin{bmatrix} \varepsilon_1(\lambda_1) & \cdots & \varepsilon_n(\lambda_1) \\ \vdots & \ddots & \vdots \\ \varepsilon_1(\lambda_m) & \cdots & \varepsilon_n(\lambda_m) \end{bmatrix}^{-1} \begin{bmatrix} A(\lambda_1) \\ \vdots \\ A(\lambda_m) \end{bmatrix} \quad (4)$$

for multiple recorded absorbances A at wavelengths λ_1 to λ_m, the local depth-averaged concentration distributions of species c_1 to c_n are calculated at each pixel. The local species' absorptivities ε_1 to ε_n are obtained from calibration solutions in the split-and-recombine mixer. The spectra of triiodide and iodine solutions (Figure 2) are used to calibrate the setup.

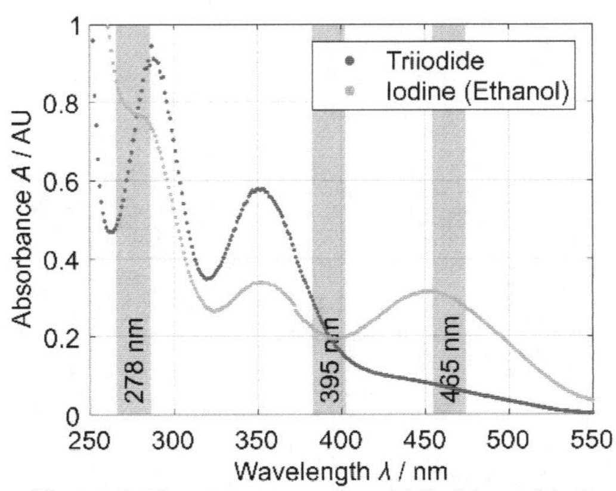

Figure 2. Absorbance spectra of triiodide and iodine solutions and dominant wavelengths of imaging UV-Vis.

The locally resolved 2D-concentration field is validated by global UV-Vis spectroscopy through the segregation index X_S

$$X_S = Y_{I_3^-} / Y_{I_3^-,\mathrm{max}} \quad (5)$$

based on the triiodide yield $Y_{I_3^-}$ and compared to the Incorporation Model [9] in order to compute the mixing times. The new non-toxic TRIS buffer system is compared to conventional systems using global UV-Vis spectroscopy.

Results and Discussion

The depth-averaged concentration distribution of iodine and triiodide is recorded for $10 < Re < 350$ in symmetric mixing conditions. The mixing channel has depths between 1.5 and $2.2\,\mathrm{mm}$. Since iodine is hardly soluble in water, it reacts according to equation (3) to form polyiodide ions.

The 2D-concentration distribution of triiodide is displayed in Figure 3 for a high acid concentration, leading to a visible yield of triiodide in the first section of the split-and-recombine mixer. The concentration map shows an increased yield of triiodide (and incomplete micro mixing) at the contact points of the buffer and acid solutions. The split-and-recombine pattern can be observed as a serial multi-lamination of the formed lamella.

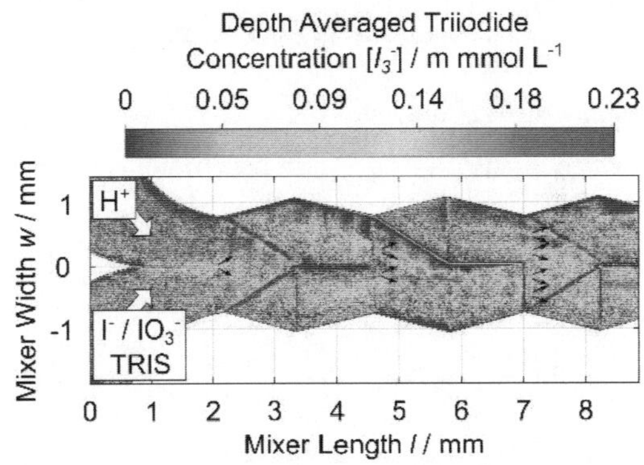

Figure 3. Depth averaged triiodide concentration in the split-and-recombine mixer at $Re = 77$ with $[HClO_4] = 0.04\,\mathrm{M}$, $[KIO_3] = 2.5\,\mathrm{mM}$, $[KI] = 1.3\,\mathrm{M}$ at room temperature.

By tweaking the acid concentrations, the reaction timescale can be modified towards the characteristic micro mixing time of the mixer unit. The concentration map shows not only if micro mixing is incomplete before the chemical reaction (2) occurs, but also the location within the mixer.

Conclusions

A novel method for utilising the Villermaux-Dushman reaction in micro- and milli systems is validated to visualise micro mixing maps. The method is compared to established the UV-Vis spectroscopy.

Imaging UV-Vis spectroscopy gives local insight into the interaction of residence time and micro mixing in reactive systems. Together with global UV-Vis analytics, it is a powerful tool for design and optimisation studies of milli- and micro-systems with high importance for research and industry.

Acknowledgments

The authors gratefully acknowledge the support by the Federal Ministry for Economic Affairs and Climate Action (BMWK) within the ENPRO 2.0 initiative, project "Continuous polymerisation in modular, intelligent reactors resistant to the formation of deposits (KoPPonA 2.0); sub-project: CFD modelling of deposit formation processes", funding code 03EN2004H.

https://doi.org/10.52202/074122-0042

References

[1] T. Bayer, D. Pysall, O. Wachsen, in *Microreaction Technology: Industrial Prospects*. W. Ehrfeld (Eds.), Berlin, Heidelberg, Springer, 2000.
[2] J. Aubin, M. Ferrando, V. Jiricny, *Chem. Eng. Sci.*, 65 (2010) 2065-2093.
[3] K. Rodermund, F. Schael, F. Herbstritt, J. Heck, M. Grünewald, Chem. Ing. Tech., 83 (2011) 1036–1043.
[4] M. Hoffmann, M. Schlüter, N. Räbiger, *Chem. Eng. Sci.*, 61 (2006) 2968-2976.
[5] J. Reckamp, A. Bindels, S. Duffield, Y. Liu, E. Bradford, E. Ricci, F. Susanne, A. Rutter, *Org. Process Res. Dev.*, 21 (2017) 816–820.
[6] F. Kexel, A. von Kameke, J. Tenhaus, M. Hoffmann, M. Schlüter, *Chem. Ing. Tech.*, 93 (2021) 830-837.
[7] T. Frey, F. Kexel, K. Dittmer, S. Bohne, M. Hoffmann, H. Trieu, M. Schlüter, *Front. Chem. Eng.*, 4 (2022) 874019.
[8] E. Arian, W. Pauer, *Ind. Eng. Chem. Res.*, 61 (2022) 9192–9205.
[9] E. Arian, W. Pauer, *Chem. Eng. Res. Des.*, 175 (2021) 296-308.

https://doi.org/10.52202/074122-0042

Incipient Air Entrainment in Unbaffled Laminar Stirred Vessels: Its Origins and Detection

MF1.1

B. C. Watson[1,2], *M. J. H. Simmons*[1], *E. H. Stitt*[2], *C. R. Tipton*[2]. *(1) School of Chemical Engineering, University of Birmingham, Birmingham, B15 2TT, United Kingdom, BCW962@student.bham.ac.uk. (2) Johnson Matthey, Belasis Avenue, Billingham, TS23 1LB, United Kingdom, Ben.Watson@matthey.com.*

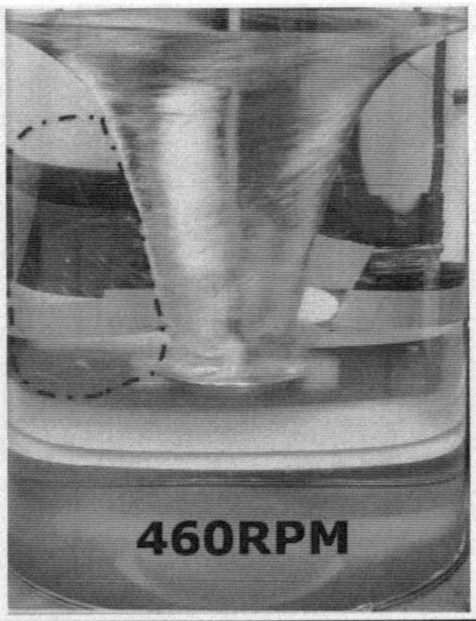

The avoidance of air entrainment within process unit operations, such as stirred tanks, is important in many formulated liquid products, since air bubbles can remain entrapped in the product and are deleterious to its final properties. Optical methods show that air entrainment begins at 460 RPM in a 290 mm unbaffled stirred tank using a 100mm Rushton Turbine with a clearance of 100mm. This method shows that entrainment begins when the vortex intercepts the impeller blades and in turn breaks the surface, not when the vortex hits the disc of the impeller. A Rankine vortex model was used to compute the position of the vortex free surface within the tank and further highlights the cause of incipient entrainment within the system. To detect the point of incipience and in turn control to a set condition, an acoustic sensor was placed on the outside of the tank and the variation in signal produced at different impeller speeds was investigated. This technique has shown acoustics to be a good indicator of entrainment.

Introduction

Stirred tanks are ubiquitous in the production of formulated liquid products by industry. The avoidance of air entrainment is important in many of these formulated products such as inks to eliminate pinhole imperfections on the surface of the fuel cell membranes. [1]

However, unbaffled stirred tanks naturally entrain air when the free surface of the vortex intercepts the blades of the impeller and thus breaks the free surface. This occurs above a critical impeller speed which is a function of the impeller design, tank design and rheological properties of the fluid.

There are many methods for predicting vortex depth within a stirred tank but none that are common for predicting the position of the free surface. [2] The Rankine vortex model can however be used for this purpose.

The Rankine vortex is made of two key components: a rigid inner core and an irrotational vortex in the outer region of the system. Both of these components are a function of angular velocity, ω, tank radius, r and the core radius, a as seen in Equation 1 and Figure 1. [3] The core radius is a system dependent variable and is found experimentally.

$$u_\theta(r) = \begin{cases} \omega r, & 0 < r \leq a \\ \dfrac{\omega a^2}{r}, & a < r \leq R \end{cases} \quad (1)$$

Modifying Equation 1 with respect to Bernoulii's principle can allow the height of the free surface with respect to the tip of the vortex to be found, as seen in Equation 2. The tip of the vortex z^* can be found by integrating under the free surface, which due to the conservation of volume, can give the position of the vortex peak.

$$z(r) - z^* = \begin{cases} \dfrac{\omega^2 r^2}{2g}, & 0 < r \leq a \\ \dfrac{\omega^2 r^2}{g}\left(1 - \dfrac{a^2}{2r^2}\right), & a < r \leq R \end{cases} \quad (2)$$

This model can be used to predict the impeller speed region where entrainment should be expected to begin as seen in Figure 2.

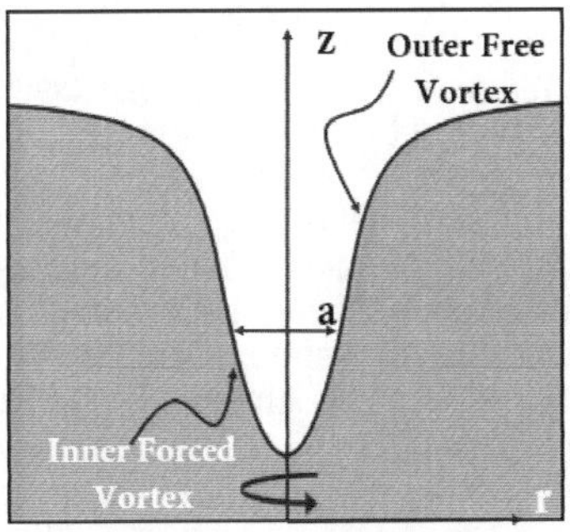

Figure 1. Rankine vortex Free Surface co-ordinate system assuming radial symmetry.

Of critical importance industrially is detection of the previously mentioned air entrainment. One possible solution to this problem is the development of a sensor that could detect the onset of entrainment. This sensor would then be used to control the impeller speed to give a desired amount of air entrainment.

A torque sensor would not be able to be used for this application as it is insufficiently sensitive at low levels of entrainment, and it does not offer a unique gas hold up at high levels of entrainment. Optical methods could be utilised for translucent tanks, but this is not feasible for industrial applications. Electrical resistance tomography (ERT) has also

https://doi.org/10.52202/074122-0043

been used in gas-liquid systems, but for this application it would be insufficiently sensitive at low levels of entrainment and impractical for industrial deployment.

Entrained stirred tank systems create noise via three main mechanisms: turbulent noise, bubble break up and bubble resonances. Turbulent noise can be filtered out, leaving the remaining two as the main acoustic events. As both of these events are directly related to the bubbles within the tank it means that an acoustic method is appropriate for this application. [4]

Acoustics emissions is a non-invasive technique and can be applied as either an on-line or in-line measurement technique, leading to lower scale-up costs. It is relatively inexpensive compared to ERT or a torque meter set up and torque cannot be used for industrial applications. Thus, acoustics is a scalable technique in comparison to the alternative techniques. Acoustic systems have simple set-ups, with very few parts, and has no moving parts leading to high reliability. [5]

This research is thus concerned with the evaluation of an acoustic sensor for the detection of incipient air entrainment within a stirred tank. The overall aim is to find any underlying characteristic of the acoustic data to detect incipient air entrainment, when the first few air bubbles are drawn into the bulk liquid.

Material and Methods

The agitated system was a T = 290 mm diameter, cylindrical flat-bottomed Perspex tank with no baffles. The impeller used was a stainless steel six blade Rushton disc turbine [6] with diameter D = 0.1 m *($D/T \approx 1/3$)*, the thickness of the stainless-steel plate was 2 mm. The bulk liquid was distilled water and its level was set to H = 290 mm *(H/T = 1)*.

The other regime of interest, as well as incipient entrainment, is no entrainment, defined as no air trapped within the bulk liquid. The two regimes were initially found by experimentally using an optical method with incipient entrainment being found to begin at 460 RPM. This enables validation against the acoustic results.

A Physical Acoustics R.45I low frequency acoustic sensor was used in this experiment with a resonant frequency of 18.55 kHz and diameter of 28.6 mm. It was attached to the tank using silicon grease which was recommended by the manufacture for the best acoustic coupling to the tank. [7] The R.45I was connected to a Vallen Systeme GmbH decoupling box which provided a 28 V ghost voltage to the system. This means that it supplied the required voltage to the sensor then decoupled it from the acoustic signal before it reached the oscilloscope [8]. The digital oscilloscope used was a Pico Technologies PicoScope 5243A with a resolution of *16 bits* and bandwidth of 100 MHz. [9] The experiment consisted of recording an acoustic signal at each impeller speed ranging from 40 RPM to 600 RPM in steps of 20 RPM. The acoustic signal was recorded for 20 s at a sampling frequency of 100 kHz. The acoustic data files can resolve a maximum frequency of 50 kHz with a resolution of 0.05 Hz per bin.

A Hilbert Transform and Savitsky-Golay filter were utilised to remove much of the noise within the time series and highlight the peaks within the data [10][11]. A thresholding technique is used to create Figure 3 where the peaks above a threshold, in this instance 6.44 V, are counted for each impeller speed applied.

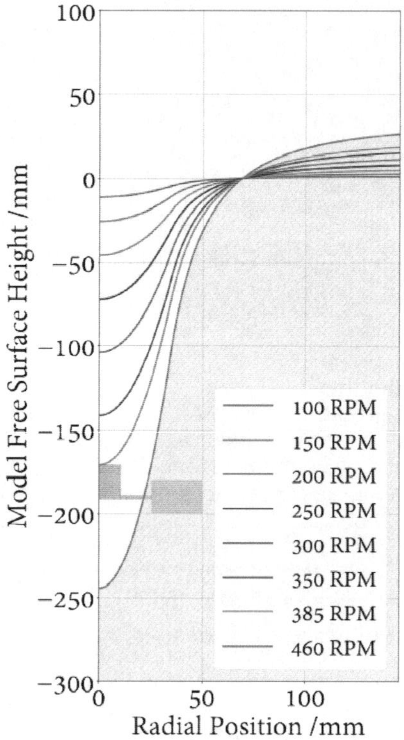

Figure 2. Rankine vortex model of a vortex in a 290mm diameter stirred tank.

Results and Discussion

The Rankine vortex analysis seen in figure 2 shows that it can be used to provide a reasonable estimate for this geometry. At 460 RPM the surface of the vortex will intercept the blade thus showing that entrainment will begin on or before this rotational speed.

In figure 3, the data points before 460 RPM have very small errors and rest on the *x*-axis therefore, we can reasonably assume that no threshold events occur before this speed. Beyond it, we see an initial increase in threshold events, in coincidence with the optically evaluated point of incipient entrainment. The sharp increase in threshold events shows that it can be used as a detection method for incipient air entrainment. Additionally, the sharp increase in threshold events means it would be well equipped within a control system to react quickly to changes within the stirred tank.

There is then a sharp increase in threshold events until 500 RPM, where the change in threshold events with respect to impeller speed change begins to

https://doi.org/10.52202/074122-0043

settle out. This shows that the amount of air being entrained in the tank could be controlled between 460 RPM and 500 RPM. Thus, if the process can handle a small amount of air within the system and is more reliant on impeller speed, this could be controlled within this region. However, after 500 RPM it would be unknown how much additional air is in the system and thus not be feasible to control the exact amount of air beyond this speed.

It should be noted that when the number of threshold events moves off the x-axis at 460 RPM a substantial increase in error in the datapoints are seen. It is understandable that the error is increased in this region due to the sampling time frame. The longer the time frame the more likely the acoustic signal will be fully representative of tank thus reducing the error.

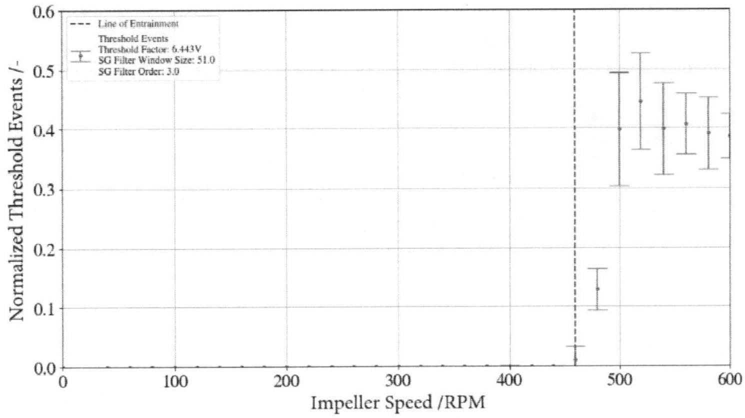

Figure 3. Normalized threshold events with respect to impeller speed showing an increase in threshold events at 460RPM which is optically the point of incipience.

Conclusions

The Rankine Vortex model can be effectively used to predict the region where entrainment will begin in an unbaffled stirred tank. An acoustic emissions method has been developed to detect the point of incipient air entrainment within a stirred tank. This was conducted by placing an R.45I acoustic sensor on the side of the stirred tank and recording and analysing the signal produced to detect threshold events due to air entrainment. At the point of entrainment, there was sharp rise in the number of threshold events. This showed that the threshold method is effective in detecting incipient air entrainment in a stirred tank.

Acknowledgments

B. Watson acknowledges funding received from the Centre for Doctoral Training in Formulation Engineering (EPSRC grant no. EP/S023070/1) and Johnson Matthey.

References

[1] Phillips, A., Ulsh, M., Mackay, J., Harris, T., Shrivastava, N., Chatterjee, A., Bender, G. (2020). The Effect of Membrane Casting Irregularities on Initial Fuel Cell Performance. Fuel Cells, 20(1), 60-69.

[2] Busciglio, A., Grisafi, F., Scargiali, F., Daví, M.L., Brucato, A. Vortex shape in unbaffled stirred vessels: Experimental study via digital image analysis (2011) Chemical Engineering Transactions, 24, pp. 1387-1392.

[3] W. Rankine, A Manual of Applied Mechanics, ninth ed. (C. Griffin and Co., London, UK,1958).

[4] *M. Minnaert. On musical air-bubbles and the sound of running water. Philosophical Magazine, 16(104):235–248, 1993.*

[5] A.P. Wade, Acoustic emission: is industry listening? Chemom. Intell. Lab. Syst. 8 (1990) 305–310.

[6] *Rushton, J. H., Costich, E. W., & Everett, H. J. (1950). Power characteristics of mixing impellers. Chemical Engineering Progress, 46(8), 395-404.*

[7] Physical Acoustics Corporation, "R.45I Sensor: Very Low Frequency Sensor," Tech. Rep. (Mistras, New Jersey, 2020).

[8] Vallen Systeme. GmbH, "Accessories for Acoustic Emission Systems: Specification," Tech. Rep. (Wolfratshausen, Germany, 2021).

[9] PicoTechnology, "PicoScope 5000 Series," Tech. Rep. (Pico Technology, St Neots, Cambridgeshire, 2020).

[10] Hilbert, E. Schmidt, and A. Pietsch, eds., Integralgleichungen und Gleichungen mit unendlich vielen Unbekannten, nachdr. ed., Teubner-Archiv zur Mathematik No. 11 (Teubner, Leipzig, 1989).

[11] Savitzky and M. J. E. Golay, "Smoothing and Differentiation of Data by Simplified Least Squares Procedures." Analytical Chemistry 36, 1627–1639 (1964).

https://doi.org/10.52202/074122-0043

Dissimilar gas flow field characterization in vortex mixers MF1.2

Gozde Gecim[1], Ertugrul Erkoc[2] (1) Department of Chemical Engineering, Faculty of Engineering and Natural Sciences, Bursa Technical University, Bursa, Turkey, gozde.salkic@btu.edu.tr (2) Department of Chemical Engineering, Faculty of Engineering and Natural Sciences, Bursa Technical University, Bursa, Turkey, ertugrul.erkoc@btu.edu.tr.

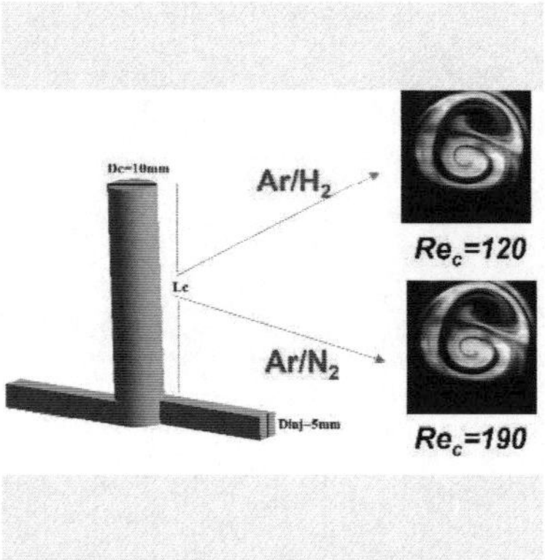

The characterization of flow hydrodynamics in vortex mixers plays an important role in gas mixing applications. This study examines the mixing of two different gases, hydrogen (H_2) and nitrogen (N_2), with argon (Ar) at varying Reynolds (Re) numbers between 30 and 300 by using the flow visualization and PIV techniques. The influence of gas physical properties such as density, viscosity and mass diffusivity on the flow field characteristics was investigated. While the first critical point of the flow was observed nearly at the same Re number ($Re \cong 60$) for Ar/H_2 and Ar/N_2 flows, the second critical point was found to be at $Re \cong 120$ for Ar/H_2 and at $Re \cong 210$ for Ar/N_2. Since hydrogen has higher gas velocity and thus a higher shear stress than that of nitrogen at the same Re number, the critical Re number of the Ar/H_2 flow was determined to be lower than that of Ar/N_2 flow.

Introduction

Vortex mixers made of a cylindrical chamber connected by two or more tangential opposed jets are of great importance for various reactions requiring fast mixing [1]. These mixers provide a swirling vortex pattern near the jets which is the type of flow rotating about an axis like tornado [1, 2]. Differently from Confined Impinging Jet (CIJ) mixers, equal momentum of the jets is not vital for enhanced mixing due to the absence of an impingement region in the vortex mixers [3].

Aside from the liquid mixing studies, the enhancement of mixing problems for gas mixture flows is of great importance in engineering. To ensure adequate reaction rates with perfect reactant distributions and obtain homogeneous mixtures for highly efficient combustion systems is the challenge of chemical industries related with gasification and oxidation processes, combustors, or fluidized bed systems [4]. Previous gas mixing studies on the jet mixers mostly in micrometers ranges are focused on the numerical approaches like Computational Fluid Dynamics (CFD) or Direct numerical Simulation Monte Carlo (DSMC) [5, 6]. A limited number of experimental gas mixing studies are concerned on either the measurement of gas concentrations at the reactor outlets [4] or flow visualization experiments like PIV [7] and Pressure sensitive paint (PSP) [8] methods.

Recently, gas flow hydrodynamics- including methane (CH_4), argon (Ar), nitrogen (N_2), and hydrogen (H_2) in a vortex mixer having tangential jets was studied by our group [9,10]. The critical Re numbers of the similar gas flows from the jets (CH_4/CH_4, Ar/Ar, N_2/N_2, H_2/H_2) where steady flow dynamics turn into an unsteady flow regime were determined. The effect of the gas physical properties on flow-field hydrodynamics inside the vortex chamber was presented. The flow hydrodynamics for hydrogen, having the highest kinematic viscosity of all, was found to develop at lower Re numbers compared to others [9,10].

Investigation of mixing of dissimilar gases relevant to industrial processes, such as hydrogenation reactions are important. Thus, the aim of the current study is to investigate the flow hydrodynamics of dissimilar gases for the presented vortex mixer geometry. In this regard, two sets of experiments were performed by mixing argon gas with hydrogen and nitrogen gases, respectively. While argon and nitrogen gases have similar physical properties like density, viscosity, and diffusion coefficients, these values are quite different for argon and hydrogen. The effects of this similarity and dissimilarity between gas properties on the flow structures and flow regime transitions were pointed out by flow visualization and PIV experiments inside the mixing chamber.

Material and Methods

The detailed schematic view of the vortex mixer system is shown in Figure 1. Gases were supplied from gas cylinders controlled by flow controllers (Bronkhorst F201-AV). Vortex mixer, dimensions are shown in Figure 1, was made of PMMA material to enable the transparency of the flow.

Two sets of experiments including argon/hydrogen and argon/nitrogen gases at equal Re numbers of jets were performed to reveal the flow hydrodynamics of dissimilar gases inside the vortex chamber. For both scenarios, the incoming argon flow was seeded with smoke by being passed through a smoke reservoir which is filled by a smoke generator (Safex Fog Generator F2010).

For imaging the instantaneous flow behaviour, a PIV system was installed in the vortex mixer setup as shown in Figure 1. A CMOS camera (Speedsense 1040) with 2330x1723 megapixels resolution and frame rate of 30Hz was used to capture the frames lighted by each laser flash. The laser source used is a double laser head from

https://doi.org/10.52202/074122-0044

Dantec Dynamics™ (Nd-YAG laser (532nm, 50mj) at a frequency of 96 Hz. To visualize the gas flow and compute the velocity fields in the frames captured by camera, argon flow was seeded with a white smoke generated by a smoke generator (Safex Fog Generator F2010).

To obtain the average behavior of the flow, the root means square (rms) method which calculates the standard deviation of corresponding pixels from their mean value was applied. Then, the particle displacement between two consecutive frames is computed by Adaptive PIV method (with width x height size of 32x32 interrogation areas) using Dantec Dynamics Software™. Afterwards, the average filter and vector statistics methods were utilized consecutively in order to obtain the instantaneous velocity vector map and mean velocity vector of the flow, respectively. Additionally, the evolution of dimensionless turbulent intensity values obtained from velocity fluctuations with Re number was studied in order to determine the critical Re number region.

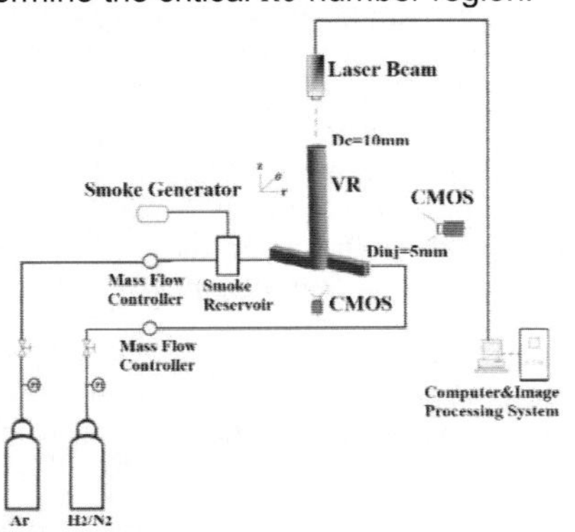

Figure 1. Schematics of the experimental setup.

Results and Discussion

Radial flow visualization experiments were performed for chamber heights ranging from 10 mm to 200 mm (Figure 2). For both gas combinations, the vortex behaviour of the flow was gained at $Re = 60$ and develops similarly until $Re = 100$ at a chamber height of 40 mm. Flow streamlines of argon, however, are thinner in the Ar/N_2 flow than they are in the Ar/H_2 flow. This can be attributed to the fact that the shear stress between the fluid streams of hydrogen is higher than that of nitrogen [4]. From $Re = 100$ to $Re = 110$, there is a second swirl flow that is similar to the vortex structure for the Ar/H_2 flow (in Figure 2).

Beyond $Re \geq 120$, gas flow streamlines disappeared, and the vortex behavior of the flow became almost unnoticeable. The flow hydrodynamics of Ar/H_2 flow evolves in the same way for higher Re numbers. On the other hand, flow streamlines of Ar/N_2 flow are visible until $Re = 200$. Besides, the development of flow hydrodynamics is quite different than Ar/H_2 flow. The vortex structure in Ar/N_2 flow has more than one turn, and elliptic regions in the upper and lower sides of the vortex center are formed between $Re = 100$ and $Re = 200$. The development of flow hydrodynamics for Ar/N_2 is similar between $Re = 210$ and $Re = 280$.

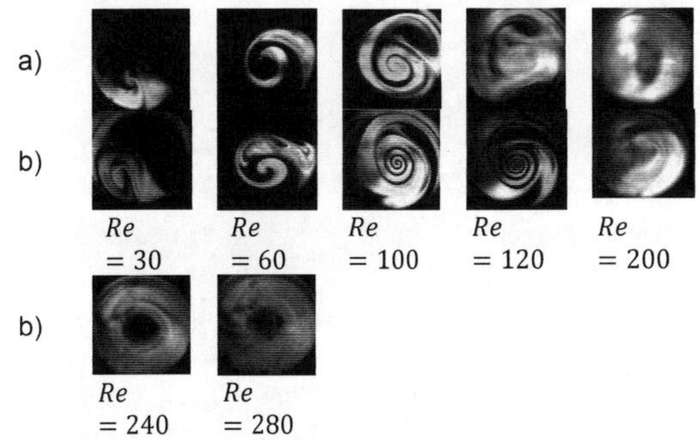

Figure 2. Time-averaged flow visualization images (rms) between $Re = 30$ and $Re = 280$ at 40 mm of chamber height a) Ar/H_2 b) Ar/N_2.

Flow regime regions determined for Ar/H_2 and Ar/N_2 flows was presented in Table 1 including stratified flow regime, vortex flow regime and unsteady engulfment regime. At stratified flow regime and vortex flow regimes, mixing is mainly dependent on the diffusion. As the velocity increases after the critical Re number of the flow, axial instabilities arise. This is due to the tangential character of the jets, which causes chaotic flows. Overall, mixing becomes independent of diffusion and is solely dependent on convection. [7, 11].

Table 1. Flow regime transitions.

	Stratified Flow Regime	**Vortex Flow Regime**	**Unsteady flow regime**
Ar/H_2	$Re < 70$	$70 \leq Re < 120$	$Re \geq 110$
Ar/N_2	$Re < 70$	$70 \leq Re < 190$	$Re \geq 190$

From the instantaneous flow-field in Figure 3, velocity vector maps of $Re = 70$ and $Re = 110$ remained unchanged over time for Ar/H_2 gas flow. However, the displacement of vortex center is noticeable after $Re = 120$ with an increasing trend of axis until $Re = 200$. Although the same overall behavior of the flow has been observed for Ar/N_2 flow, the flow has not gained its vortex structure at $Re = 70$. However, the stable vortex formation has been observed at $Re = 120$. By further increasing Re number, vortex wandering motion of the flow changes with time. From Figure 3 at $Re = 240$, vortex center seems to be broken by formation of small vortices around the vortex center indicating the radial and axial oscillations of the flow.

https://doi.org/10.52202/074122-0044

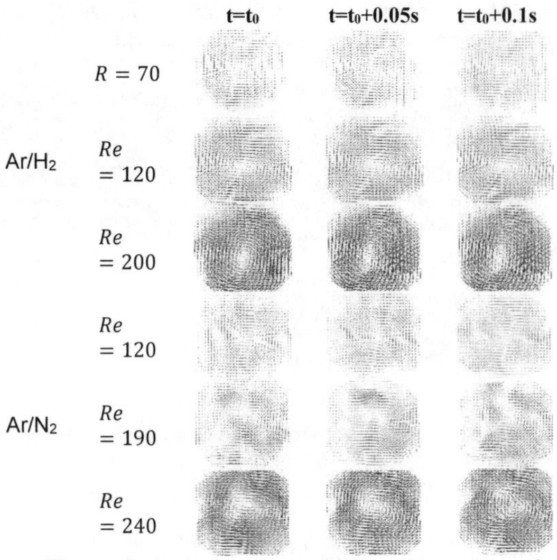

Figure 3. Instantaneous velocity vector maps of determined critical points of the flow from PIV.

Additionally, to validate the critical Re numbers of the flow visualization experiment, the critical Re number region where the turbulent intensity starts to increase dramatically has been determined. The results of dimensionless turbulent intensity values based on the Re numbers were demonstrated in Figure 4. From these plots, the increasing trend of turbulent intensity has started at $Re = 120$ for Ar/H_2 flow and at $Re = 210$ for Ar/N_2 flow. The dimensionless turbulent intensity values are almost constant until these Re numbers. Following these points, they increase with a higher trend. This results in increased turbulence quantities in the flow as the Re numbers of the jets increase. The flow visualization results were compatible with the critical Re number of the flows at which the flow changed from steady to unsteady.

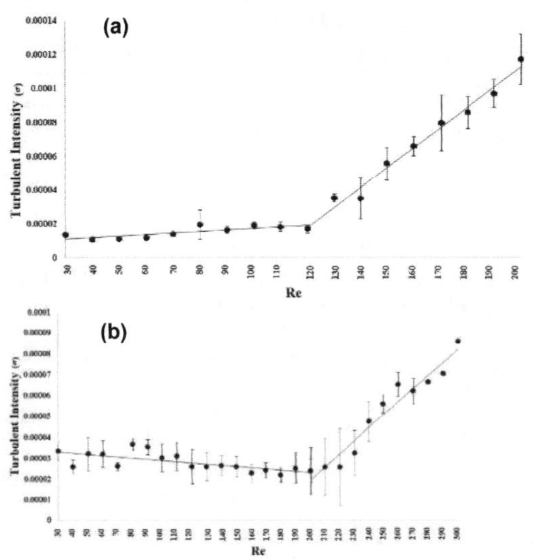

Figure 4. Plot of the variation of average turbulent intensity values with respect to Re number at rz plane for a) Ar/H_2 b) Ar/N_2.

Conclusions

In the presented study, the characterization of flow regimes for vortex mixers where different gases were injected from tangential jets was performed by flow visualization and PIV measurements. Flow fields for the selected gases were qualitatively and quantitatively visualized in detail. The transition from segregated to vortex flow regime was observed at $Re = 60$ for both gas flows. They were dominated by diffusion under this Re number. As Re number increased, shear stress between fluid sub streams started to dominate the flow. The second transition region which is called as critical Re number of the flow was determined at lower Re number for Ar/H_2 due to the higher gas flow rates and injection momentum of hydrogen. For Ar/H_2 flow, hydrogen encounters with argon which has lower velocity and inertial force. The incoming hydrogen flow overcomes the argon easier than hydrogen. This results in lower critical Re number for Ar/H_2 flow $(Re = 120)$ than that of H_2/H_2 flow $(Re = 150)$ [9]. For Ar/N_2 flow, however, argon and nitrogen have similar physical properties resulting in similar gas velocities and inertial forces. That is why the critical Re number for Ar/N_2 flow $(Re = 210)$ is nearly the same with N_2/N_2 flow $(Re = 200)$ or Ar/Ar flow $(Re = 220)$ [9].

Acknowledgments

This work was supported by the Scientific and Technological Research Council of Turkey (TÜBİTAK) [118M391].

References

[1] J. C. Cheng, M. G. Olsen, R. O. Fox, *Applied Physics Letters,* 94 (2009), 204104.
[2] S. T. Chan, J. T. Ault, S. J. Haward, E. Meiburg, A. Q. Shen, *Physical Review Fluids*, 4 (2019) 084701.
[3] Y. Shi, R. O. Fox, M. G. Olsen, *Applied Physics Letters*, 99 (2011) 204103.
[4] K. Haas-Santo, P. Pfeifer, K. Schubert, T. Zech, D. Hönicke, *Chemical engineering science*, 60 (2005) 2955-2962.
[5] D. Gobby, P. Angeli, A. Gavriilidis, *Journal of Micromechanics and microengineering,* 11 (2001) 126.
[6] F. Yan, B. Farouk, *Microscale Thermophysical Engineering*, 6 (2002) 235-251.
[7] W. F. Li, K. J. Du, G. S. Yu, H. F. Liu, F. C. Wang, *AIChE Journal*, 60 (2014) 3033-3045.
[8] C.-Y. Huang, S.-A. Wan, Y.-H. Hu, *International Journal of Heat and Mass Transfer*, 111 (2017) 520-531.
[9] G. Gecim, E. Erkoc, *Industrial & Engineering Chemistry Research,* 61 (2022) 1192-1206.
[10] G. Gecim, E. Erkoc, *Industrial & Engineering Chemistry Research*, 60 (2021) 5674-5687.
[11] R. J. Santos, E. Erkoç, M.M. Dias, A. M. Teixeira, J. C. B. Lopes, *AIChE journal*, 54 (2008) 1153-1163.

Extending Bubble Breakage and Coalescence Models for Presence of Solids in Stirred Tanks

MF1.3

Ville Alopaeus[1], *Yiting Xiao*[2,3], *Xiangyang Li*[4], *Haoliang Wang*[2,3], *Chao Yang*[2,3,5]
(1) Aalto University, School of Chemical Engineering, 00076 Aalto, Finland. Ville.Alopaeus@Aalto.fi; (2) CAS Key Laboratory of Green Process and Engineering, Institute of Process Engineering, Chinese Academy of Sciences, Beijing 100190, China; (3) School of Chemical Engineering, University of Chinese Academy of Sciences, Beijing 100049, China; (4) School of Chemical Engineering, Sichuan University, Chengdu 610065, China; (5) Green Manufacture Institute, Chinese Academy of Sciences, Beijing 100190, China

The intrusive image-based method was used to investigate bubble size and holdup distributions in a gas-liquid-solid stirred tank reactor. The presence of solid increased the drag force between gas and liquid, thus affecting gas holdups. The parameters of bubble breakup and coalescence model were optimized with the measured bubble size distribution and local holdup data, using a population balance model. Parameter optimization was done using Matlab's optimization algorithm, in which a compartmental model for the fluid flow for all three phases was implemented. Based on the analysis, it was found that both breakup and coalescence frequencies increased due to the presence of solids. The developed model of bubble behaviour provides a basis for numerical simulation design and scale-up of industrial stirred reactors.

Introduction

Gas-liquid-solid systems are used extensively in various fields of chemical industries. In these systems, bubble size distributions as well as gas holdups are essential for accurate prediction of fluid flow and mass transfer, and ultimately reactor performance.

Although population balance models are relatively well established in pure gas-liquid systems, the effect of solids on bubble dynamics is not so clear. However, recent development of experimental techniques and population balance models allows us to improve this deficiency.

In the present work, newly acquired experimental data is first measured with two-phase systems (gas-liquid and solid-liquid) in order to calibrate models with these systems, and finally three-phase experimental data for bubble size distributions and local hold-ups in the presence of solids is used to optimize bubble breakage and coalescence parameters.

Material and Methods

The experiments were conducted in a cylindrical vessel (T = 0.282 m) with a flat bottom stirred by a Rushton turbine (D = T/3) located at T/3 from the vessel bottom. Four baffles of width T/10 were mounted perpendicular to the vessel sidewall. Static liquid height H is equal to T. Gas holdup and bubble size of gas-liquid system was measured in three positions (r/R, z/H) = (0.645, 0.071), (0.645, 0.248), and (0.645, 0.886). Solid holdup of the solid-liquid system was measured at the same positions and additionally at (0.645, 0.39). The air was supplied through a ring sparger of inner diameter 0.126 mm with 30 holes of 1 mm diameter. The liquid phase is tap water. Solid particles are uniformly spherical and 1 mm in size, with a density of 1350 kg/m^3. The impeller speed is set to 405 rpm. The average fraction of solid phase is 5 vol% and the volumetric gassing rate is 0.5 m3/h. Bubbles and particles were identified using vision probe and image analysis explained in detail in [1].

Fluid flow was modelled using a compartmental approach [2]. Compartment division is shown in Figure 1. Gas feed was assumed to enter compartment 9.

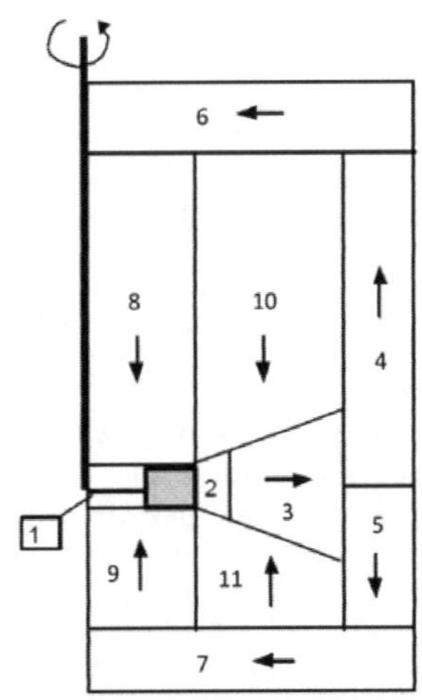

Figure 1. Compartmentalization of the standard stirred tank

Gas bubble and solid particle movement with respect to main flow was modelled using standard drag models, augmented with turbulence correction and effect of the other dispersed phase. Bubble breakage and coalescence models were taken from the work of Laakkonen et al. [3]. Solution of the population balance equations was done with high order moment conserving method of classes [4].

Results and Discussion

The parameters of the Laakkonen et al. breakage

https://doi.org/10.52202/074122-0045

and coalescence models were first calibrated against pure gas-liquid data using Levenberg–Marquardt algorithm implemented in Matlab. Objective function was based on local values (experimental and predicted) of bubble size distribution and gas holdup. Observations have shown that finding models that predict both properties correctly at the same time is absolutely critical. If bubble size distribution would be correct with erroneous gas holdup, breakage and coalescence model parameters would be erroneous as well, as gas holdup is affecting breakage rate with first order dependency but coalescence rate with second order dependency. It is a two-way coupling: when bubble size distribution is incorrect, bubble rise velocities will be incorrect, resulting in errors in gas holdups.

With original parameters, bubble sizes were fairly well predicted, but in order to make comparison between systems with and without solids more meaningful, the parameters were re-fitted.

In addition, various drag models in the solid-liquid system were tested and those predicting the spatial variations in solids volume fraction distribution were selected for further analysis.

After calibrating the model with two-phase systems, model parameters for gas-liquid-solid systems were optimized. It was found that the drag models of Brucato [5] and Khopkar [6] were most appropriate in this case. Figure 2 shows the results with the best drag models along with the optimized bubble breakage and coalescence rate parameters.

Based on the optimized parameters it can be concluded that presence of solids 1) increases both bubble breakage and coalescence rates in this system (this may depend on mixing conditions, physical properties and particle characteristics, 2) affect bubble drag and therefore local gas holdups.

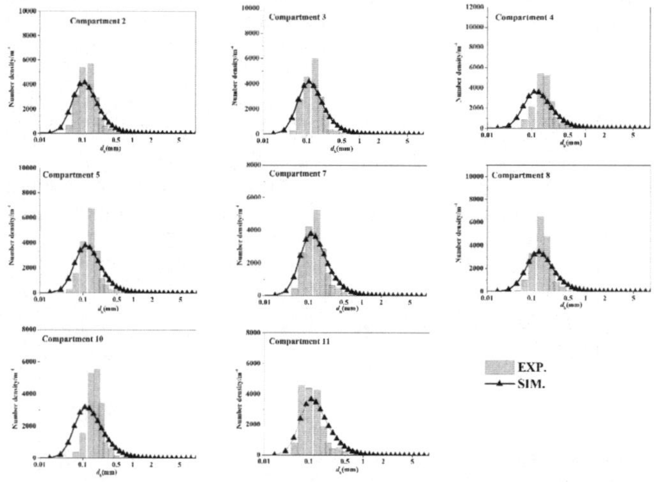

Figure 2. Comparison of experimental and predicted bubble size distributions corresponding to various compartments

Conclusions

Bubble breakage and coalescence are affected by presence of solid phase. This complicated system with multiple inter-phase couplings can be modelled by population balances and proper fluid flow description. Although detailed computational fluid dynamics is the preferred option for detailed studies, it is not a tool of choice when optimizing multiple parameters against experimental data. In order to overcome this problem, a compartmental model was used here. The parameter optimization showed that solid phase affects both bubble breakage and coalescence as well as bubble rise velocities. It was also observed that it is essential to use both local bubble size distributions as well as local gas holdups in the optimization algorithm.

References

[1] Y. Xiao, X. Li, S. Ren, Z.S. Mao, C. Yang, Hydrodynamics of gas phase under typical industrial gassing rates in a gas-liquid stirred tank using intrusive image-based method, Chem. Eng. Sci. 227 (2020) 115923.

[2] V. Alopaeus, J. Koskinen, K.I. Keskinen, J. Majander, Simulation of the population balances for liquid-liquid systems in a nonideal stirred tank. Part 2-parameter fitting and the use of the multiblock model for dense dispersions, Chem. Eng. Sci. 57 (2002) 1815–1825

[3] M. Laakkonen, P. Moilanen, V. Alopaeus, J. Aittamaa, Modelling local bubble size distributions in agitated vessels, Chem. Eng. Sci. 62 (2007) 721–740.

[4] V. Alopaeus, M. Laakkonen, J. Aittamaa, Solution of population balances with breakage and agglomeration by high-order moment-conserving method of classes, Chem. Eng. Sci. 61 (2006) 6732–6752.

[5] A. Brucato, F. Grisafi, G. Montante, Particle drag coefficients in turbulent fluids, Chem. Eng. Sci. 53 (1998) 3295–3314

[6] A. R. Khopkar and V. V. Ranade, CFD simulation of gas-liquid stirred vessel: VC, w regimesS33, and L33 flo, AIChE J. 52 (2006) 1654–1672

Influence of temperature on the mixing and separation of water-oil dispersions in a continuous gravity settler

MF1.4

Song Ye[1], *Lena Hohl*[1], *Matthias Kraume*[1]. *(1) Technische Universität Berlin, 10587 Berlin, Germany, song.ye@campus.tu-berlin.de.*

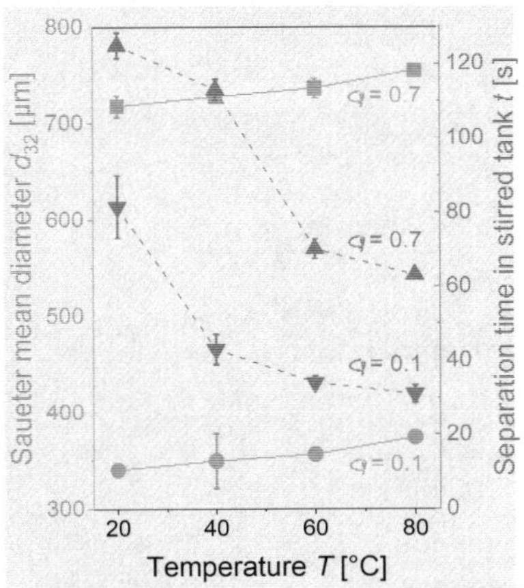

Liquid-liquid mixing and separation is a fundamental process in many industries. However, the operating temperature in these industries is hard to keep constant or even never noticed. Moreover, there need to be more studies concerning the effect of temperature on droplet size and phase separation. This work studied the influence of temperature on the mixing and separation in both batch-stirred tank and continuous gravity settler, trying to reduce this knowledge gap. Results show that in both systems, the varied temperature has a tiny influence on the droplet size but enhances phase separation significantly.

Introduction

Liquid-liquid mixing and separation are fundamental for increasing the interfacial area and mass transfer between different phases in the chemical, oil, pharma, and food industries. Many parameters affect the mixing and separating performance, such as liquid properties, mixing intensity, oil phase ratio, temperature, etc. In this work, the temperature is the main parameter of interest.

Material and Methods

In this work, the influence of temperature on the mixing and separation of water and oil was first tested in a batch-stirred tank and then applied to a continuous gravity settler. Deionized water and paraffin oil were used for both systems, where water was the continuous phase and paraffin oil was the dispersed phase. The dispersed phase fraction varied from 0.1 to 0.7 with an increment of 0.2, and the temperature varied between 20 °C and 80 °C. Table 1 summarizes the liquid properties under different temperatures.

Table 1: Liquid characteristics at different temperatures

	Temp. [°C]	Density [kg/m³]	Viscosity [mPas]	Interfacial tension [mN/m]
water	20	998.29	1.00	
	40	992.25	0.65	
	60	983.13	0.47	
	80	971.60	0.35	36.70
Paraffin oil	20	819.72	9.73	
	40	806.50	5.30	
	60	793.23	3.33	
	80	777.06	2.35	

In the batch-stirred tank, a Rushton turbine with a diameter of 100 mm was used to generate the dispersions, and the tank was fully baffled. The schematic drawing of the stirred tank can be seen in Figure 1, and the detailed dimension is provided in Table 2. An in-situ endoscope system was utilized to determine the droplet size in the stirred tank. An external camera was used to record the dynamic phase separation. The power input was kept constant at 317.5 W/m³ for different dispersed phase fractions. The stirring speed for the dispersed phase fraction at 0.1 was first determined to be 200 rpm. Due to the density differences, the stirring speed for different dispersed phase fractions could finally be determined from 200 rpm to 212 rpm.

Table 2: detailed dimensions for the stirred tank

H [mm]	T [mm]	C [mm]	H_B [mm]	t_B [mm]
200	160	50	225	2.5

In the continuous gravity settler, as shown in Figure 2, a stirred tank with two Rushton turbines was applied to generate the water-oil dispersions. The total volume of the settler is around 10 L, and 5 L for the stirred tank. A transparent glass vessel is the main settler with a diameter of 150 mm, where water-oil dispersion is separated. The droplet size was also determined with the in-situ endoscope system. The liquid interface was controlled at half the height of the separator during the experiment. The separation efficiency was determined by collecting the unseparated oil at the water outlet of the main separator. The inlet conditions for the continuous gravity settler are dispersed fraction φ = 0.1 and 0.2, temperature T = 20, 50, and 80 °C, total inlet flow rate Q = 200, 240, and 280 L/h, inlet droplet size d_{32} = 180, 200 and 250 μm.

Results and Discussion

In this section, the results will be presented and discussed in two parts: batch stirred tank and continuous gravity settler.

Figure 3 shows the droplet diameter changing with different temperatures and dispersed phase fractions. As it can be seen, with the rising temperature, the droplet diameter only increases in the range of 1.90% to 9.92% for different dispersed

https://doi.org/10.52202/074122-0046

phase fractions, which results from the increase of coalescence together with the decrease of breakage in the stirred tank [1, 2, 3]. However, the separation time was sharply decreased in the range of 49.60% to 67.39%, as shown in Figure 4. Temperature significantly shortened the separation because not only the viscosity of the continuous phase is reduced dramatically, leading to the higher sedimentation velocity, but also the coalescence was enhanced under higher temperatures. The dynamic separation process was carefully recorded and studied. Figure 5 shows an example separation curve where the dispersed phase fraction is 0.7 with the varied temperature from 20 °C to 80 °C. It is evident that elevated temperatures significantly impact sedimentation and coalescence processes, as indicated by their respective slopes.

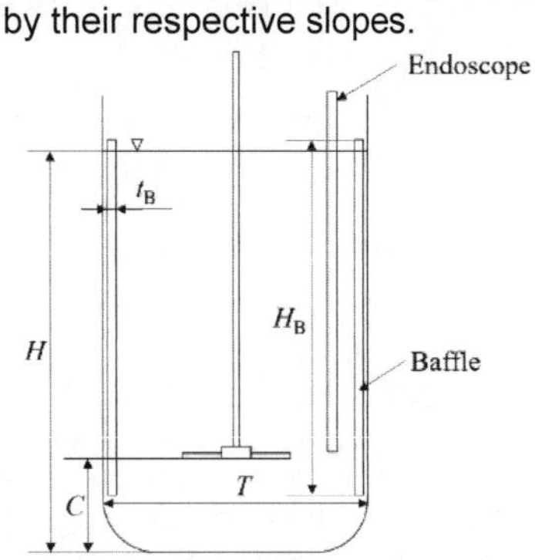

Figure 1. Schematic drawing of the stirred tank.

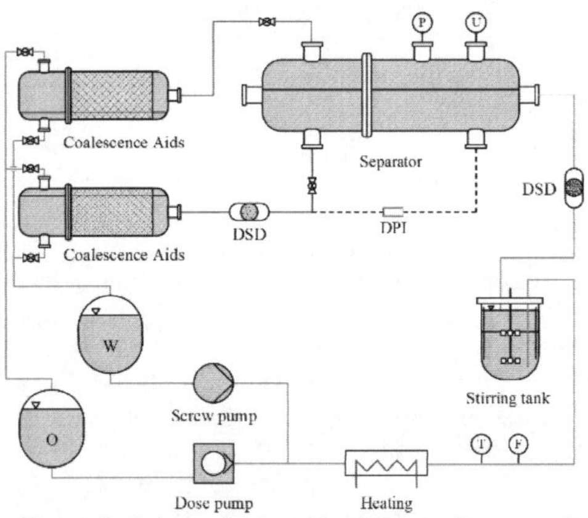

Figure 2. Schematic drawing of the continuous gravity settler.

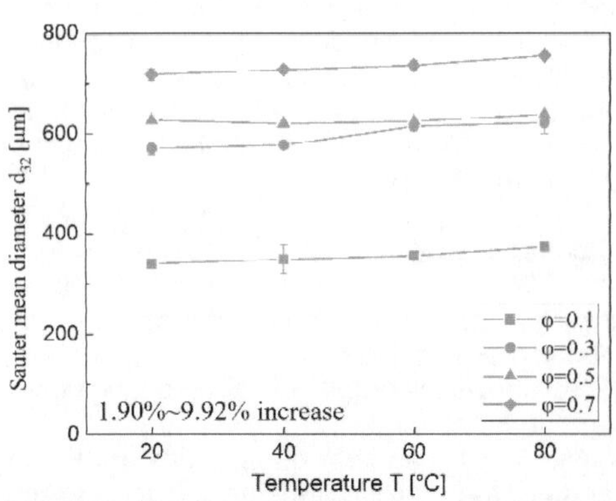

Figure 3. Sauter mean diameter vs. temperatures in the stirred tank.

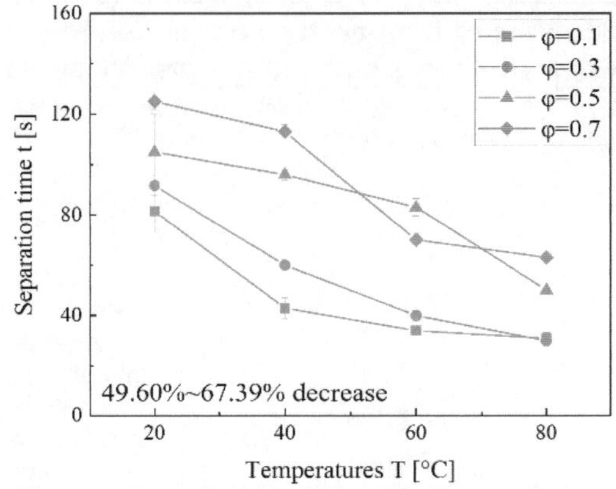

Figure 4. Separation time vs. temperatures in the stirred tank.

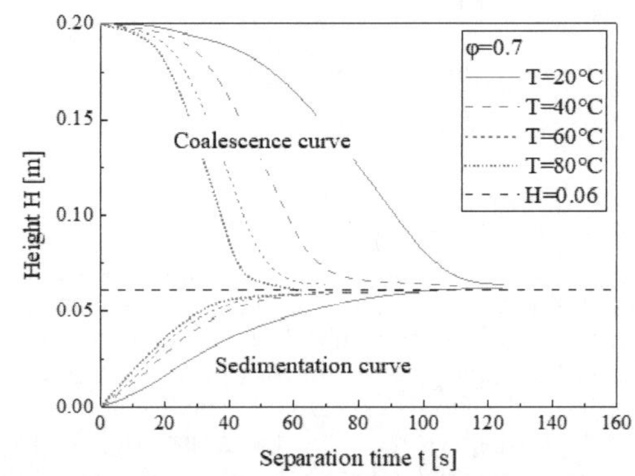

Figure 5. Separation curve for dispersed phase fractions at 0.7 for different temperatures in the stirred tank.

Figure 6 shows the inlet droplet diameter over stirring speed for different dispersed phase fractions and temperatures. The curve for different temperatures nearly collapsed together showing a similar conclusion as we got in batch stirred tank. With faster sedimentation and coalescence processes, the separation efficiency increases for higher temperatures, as expected, as seen in Figure 7.

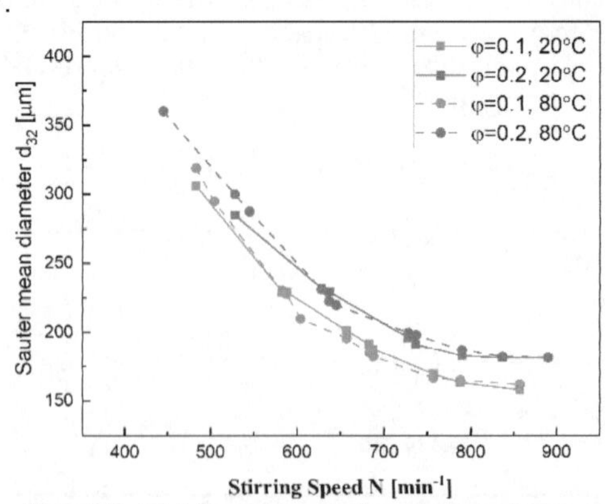

Figure 6. Inlet Sauter mean diameter vs. stirring speed under 20 and 80 °C in the continuous settler system.

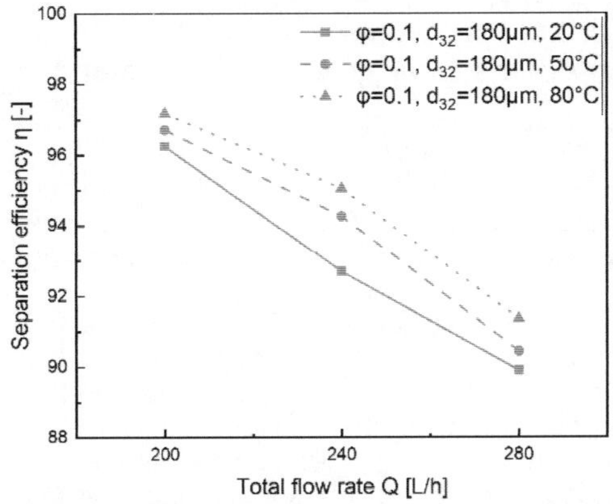

Figure 7. Separation efficiency vs. total inlet flow rate under 20, 50, and 80 °C in the continuous settler system.

Conclusions

The influence of temperature on the mixing and separation was studied in both batch stirred tank and continuous gravity settler. Results show that the varied temperature has a tiny influence on the droplet diameter, but the separation time is reduced significantly. Similar results were obtained in continuous gravity settler that the inlet droplet diameter nearly remains constant, but the separation efficiency increases with rising temperatures.

Acknowledgments

This study was partly supported by the China Scholarship Council.

References

[1] G.V. Jeffreys and G.A. Davies. Coalescence of liquid droplets and liquid dispersion. In Recent Advances in Liquid–Liquid Extraction, pages 495–584. Elsevier, 1971.

[2] B.J. Carroll. The stability of emulsions and mechanisms of emulsion breakdown. Surface and Colloid Science, 9(1):1–65, 1976.

[3] K. Arai, M. Konno, Y. Matunaga, and S. Saito. Effect of dispersed-phase viscosity on the maximum stable drop size for breakup in turbulent flow. Journal of chemical engineering of Japan, 10(4):325–330, 1977.

.

 https://doi.org/10.52202/074122-0046

High Energy Dissipation Rates from the Impingement of Free Paper-Thin Sheets of Liquids: Determination of the Volume of the Energy Dissipation Zone

Poster.1

Robert J. Demyanovich. *RJD Technologies, Inc., Seattle, WA, USA, bob@rjdtech.com.*

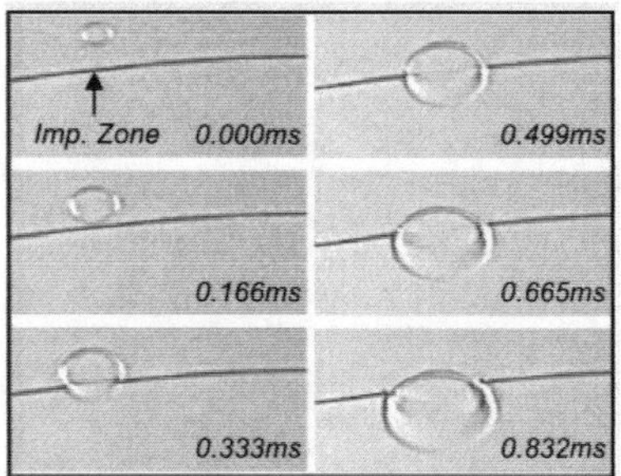

Movement of a hole formed in a single sheet. After ~ 0.5ms the hole center has passed through the impingement (imp.) zone into the mixed sheet.

The micromixing time of impinging thin liquid sheets depends upon the energy dissipation rate (ϵ). Earlier work studied the coefficient of restitution of the collision, which sets the amount of energy released [1]. The purpose of this study was to determine the volume of the energy dissipation zone (EDZ). High-speed video was used to measure the velocity of features, generated in the single sheet, as they passed through the impingement zone (IZ) and into the mixed sheet. At $\epsilon_{IZ} > 3\text{x}10^4$ W/kg the time required for the velocity to be reduced from the single sheet velocity to the mixed sheet velocity was equal to the liquid residence time in the IZ, indicating that the EDZ is equal to the IZ ($\epsilon \approx \epsilon_{IZ}$). At $\epsilon_{IZ} < 3\text{x}10^4$ W/kg, the volume of the EDZ is smaller than the IZ and a correction factor ($C_\epsilon/C_{\epsilon A}$) can be applied to ϵ_{IZ} to calculate ϵ. These results agree with the notion from turbulence energy cascade theory that large eddies (produced in the IZ) lose their energy within one turnover time (which is $\approx$ to the IZ residence time).

Introduction

When thin liquid sheets are impinged, energy is released as a result of the collision [1]. High energy dissipation rates appear to result from the dissipa-tion of kinetic energy in a very small amount of liquid (typically of order 0.001 gm). Estimated energy dissipation rates of order 10^6 - 10^7 W/kg are readily attainable. Rapid micromixing (1 to several ms) of relatively large flow rates of liquids (1 to 10 L/min) has been observed at low viscosities [2].

Fig. 1 is a photograph of the impingement of equal single sheets at an angle (2β) of 45^0.

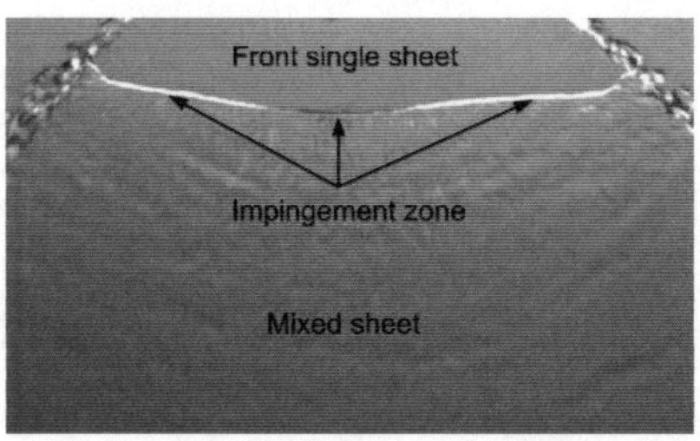

Figure 1. Front view of the impinge-ment of two, equal water sheets at an impinge-ment angle of 45°.

For impinging sheets, ϵ is defined as follows:

$$\epsilon = P/\rho V \qquad (1)$$

where P is the kinetic energy per unit time that is lost from the single sheets as a result of the collision, ρ is the liquid density, and V is the volume within which the kinetic energy is dissipated. The coefficient of restitution (COR) of equal impinging sheets has been studied ([1] - subsequently referred to as the COR study). The coefficient of restitution is defined as:

$$COR = v_m/u \qquad (2)$$

where u is the velocity of the single sheet and v_m is the velocity of the mixed sheet after impingement.

The COR study confirmed earlier results that flow within single sheets proceeds at constant velocity along radial lines emanating from the single sheet origin [3,4]. The velocity of the mixed sheet (not including the impingement zone) was also shown to be constant with liquid flowing along radial lines similar to single sheets.

The flow equation for liquid sheets is given by:

$$m = \rho\varphi R s u \qquad (3)$$

where m is the single sheet mass flowrate, R is the radial distance from the sheet origin, s is the sheet thickness at R and φ is the spread angle of the sheet. Thus, the power given up into the mixed sheet at impingement (i) is:

$$P = 2m(u^2/2 - v_m^2/2) = \rho\varphi R_i s_i u^3(1 - COR^2) \qquad (4)$$

The velocity (v') associated with the kinetic energy released as a result of the collision is given by,

$$v' = u(1 - COR^2)^{1/2} \qquad (5)$$

The COR study set the amount of energy released as a result of the impingement (the numerator in Eq. 1); the focus of this work was to more precisely deter-mine the volume or mass of the zone in which the kinetic energy is dissipated (denominator in Eq. 1).

Material and Methods

Velocities were measured from high-speed video of the displacement of features, which form in a liquid sheet when a soluble or emulsified oil is added to the liquid [5,1]. In the COR study, sheet velocities were measured from double exposure photographs of the movement of holes over a fixed time interval. However, features were seldom observed to form prior to impingement. To generate features prior to impingement commercially available bleach, containing emulsified lavender oil, was used.

Sheet velocities of features created in 48.5% w/w (6 cP) and 65.8% w/w (26 cP) aqueous glycerine were measured. Higher viscosities

https://doi.org/10.52202/074122-0047

dampen impact waves yielding features in the mixed sheet with less distortion, and thereby, increasing the accuracy of measurements. Viscosities up to 70 cP have been found to have no effect on the velocity [3]. Further, the COR study demonstrated that the COR is not a function of surface tension. Standard values of the liquid physical properties were taken from the literature. The temperature for all experiments was 20°C ± 2°C. Methods for creating and impinging sheets are provided elsewhere [6].

Two to three second videos were taken at speeds of 24,047 frames per second. Short clips of a few milliseconds in duration were extracted for post-processing. Post-processing involved increasing exposure and sharpening the video the maximum amount possible using video editing software.

Results and Discussion

The Graphical Illustration contains extracted frames from a video sequence of the movement of a hole through the impingement zone (IZ). As shown in Fig. 2 the center of each instance of the hole, as it moved, was determined by fitting the hole with an ellipse drawn using ImageJ software. ImageJ provided the coordinates of each ellipse center, which were exported to a spreadsheet. These coordinates were measured in pixels and frame-to-frame pixel displacement was converted to mm by a reference measurement from a ruler (with 0.5 mm divisions) on each video clip.

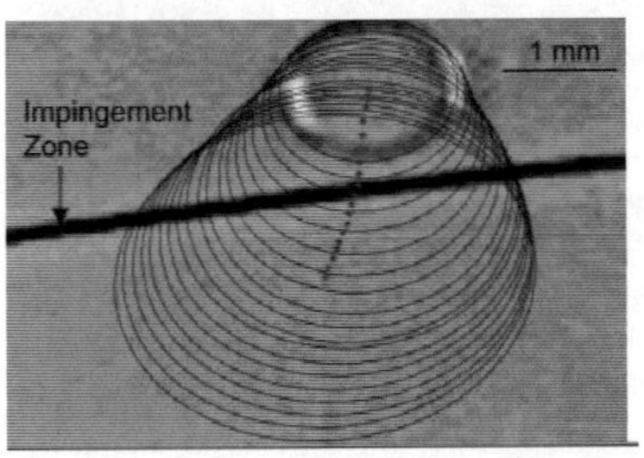

Figure 2. Sequence of movement of a hole. Each instance of the hole is fit with an ellipse using Image J. The ellipse centers are shown in red.

From a two to three second video, multiple different hole sequences were extracted for measur-ing movement. Within each hole sequence between 12 to 25 displacements of the hole were measured (only six are shown in the Graphical Illustration whereas all 19 are shown in Fig. 2).

Measurement of hole displacement was separated as movement in the single sheet, movement in the impingement zone and movement in the mixed sheet. Only the impingement zone and mixed sheet were considered to be perpendicular to the camera.

The experimental parameters and results are listed in Table 1. For each run, the mean mixed sheet velocity (v_m) is an average of the velocities measured in the mixed sheet for the number of feature sequences. The single sheet velocity was calculated based on the measured pressure drop [1]. The measured COR was calculated using Eq. 2.

A comparison of columns 6 and 7 shows good agreement between the COR measured in this study and the calculated COR based on an inelastic collision ($cos\beta$). As found in the COR study, however, the experimental COR was somewhat lower for an inelastic collision at higher viscosity/density when compared with water.

Column 8 in Table 1 lists the number of displacement vectors that could be measured in the impingement zone. This number is a function of 2β, the single sheet thickness at impingement (s_i), the velocity in the impingement zone and the camera frames per second. s_i ranged from 140 to 340 μm.

Table 1. Results summary for the inelastic collision of thin liquid sheets.

#	2β (°)	Nom. ΔP (kPa)	Mean u (m/s)	Mean v_m (m/s)	Ave. exp. COR	cosβ	Nbr. of vec-tors meas. in IZ	Δr (μm)	Δm (μm)	IZ res. time (ms)	Δw (μm)
1	48	10.0	2.08	1.76	0.85	0.91	6	742	590	0.328	373
2	48	14.5	2.54	2.02	0.80	0.91	6	695	473	0.233	283
3	46	12.0	3.53	3.23	0.91	0.92	3	355	323	0.097	222
4	46	19.5	4.57	4.09	0.89	0.92	2	412	361	0.084	367
5	75	15.0	3.37	2.47	0.73	0.79	3	500	333	0.118	301
6	75	10.0	2.75	2.02	0.74	0.79	4	564	379	0.171	278
7	67	10.4	3.10	2.63	0.85	0.83	3	426	367	0.128	275
8	67	12.0	3.25	2.61	0.80	0.83	3	593	460	0.167	385
9	67	17.8	3.94	3.12	0.79	0.83	2	339	255	0.083	246

In the impingement zone the measured displacements (S) of the features were plotted on a position versus time (t) chart where the velocity is equal to the slope at any point. All charts were well fit by a rational model ($R^2 \geq 0.9999$) of the form:

$$S = (a + bt)/(1 + ct + et^2) \tag{6}$$

In Table 1, Δw is the width of the energy dissipation zone and is equal to the distance from the beginning of the impingement zone to the point at which the sheet velocity has decreased from u to v_m. Since $\Delta w \leq \Delta m$ for every run (within experimental error), the results indicate that u is reduced to v_m by the time liquid reaches the end of the impingement zone. As a result, the previous diagram of a side view of the impingement of equal single sheets shown in Fig. 4 of the COR study is replaced with the new diagram shown in Fig. 3.

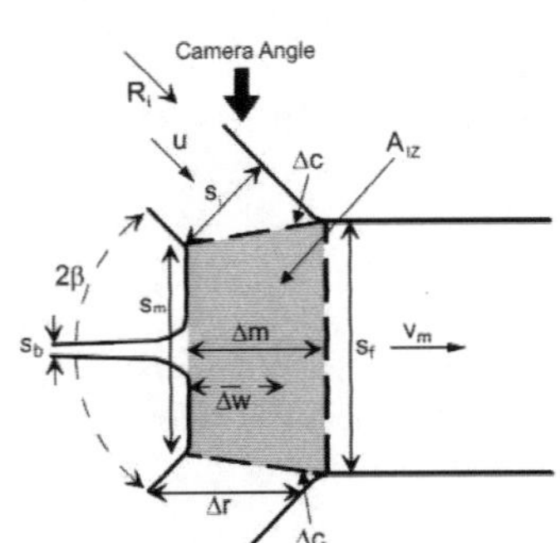

Figure 3. Side view of collision of equal impinging sheets. A_{IZ} – cross section-al area of IZ; Δc – projection of single sheet on mixed sheet; Δr - $s_i/sin\beta$; s_b – backward mixed sheet rim thickness; s_m, s_f – thickness at begin-ning and end of IZ.

https://doi.org/10.52202/074122-0047

Since the velocity has decreased to v_m by the time liquid reaches the end of the projection width, Δc, the impingement zone will now be defined as the mass of liquid between the two projections. While the actual volume in which the energy is released may be different, calculation of the impingement zone volume based on this assumption provides a reference measurement of the energy dissipation volume (the projection of the single sheet onto the mixed sheet is the main visible clue of impingement). A method, based on the results of the experiments, will be provided for adjusting the reference energy dissipation rate as necessary.

s_f [1], s_m, and Δm are given by the following equations:

$$s_f = s_i[(COR + cos\beta)/COR^2] = Bs_i \qquad (7)$$

$$s_m = s_i(1 + cos\beta) \qquad (8)$$

$$\Delta m = \Delta r\left(1 - \frac{Bcos\beta}{2} + \frac{cos\beta}{2} + \frac{cos^2\beta}{2}\right) = F\Delta r \qquad (9)$$

Thus, A_{IZ} is equal to:

$$A_{IZ} = s_i F\Delta r[(1 + cos\beta + B)/2] = Ds_i\Delta r \qquad (10)$$

The impingement zone mass is equal to $\rho\varphi R_i s_i D\Delta r$. Dividing Eq. 4 by the IZ mass yields an energy dissipation rate based on the IZ volume:

$$\epsilon_{IZ} = u^3(1 - COR^2)/D\Delta r = v'^2 u/D\,\Delta r \qquad (11)$$

Eq. 11 can be compared to ϵ derived from turbulence energy cascade theory:

$$\epsilon = C_\epsilon U_\Lambda^2/t_\Lambda \qquad (12)$$

where t_Λ is the turnover time of the large, energy-containing eddies (size equal to Λ) and U_Λ is the characteristic velocity of the large eddies. C_ϵ is the dimensionless dissipation rate coefficient, which for decaying turbulence has a minimum asymptotic value, $C_{\epsilon A}$, $\approx$ 0.5 [7]. C_ϵ should be independent of Reynolds number (Re) [8], but is often found to be a function of Re [9]. The experimental results showed that kinetic energy is dissipated within the IZ, therefore, ϵ_{IZ} is the minimum value of ϵ and when making comparisons with cascade theory, $C_\epsilon = C_{\epsilon A}$. Setting Eq. 12 equal to Eq. 11 yields,

$$C_{\epsilon A}U_\Lambda^2/t_\Lambda = v'^2 u/D\,\Delta r \qquad (13)$$

indicating that $U_\Lambda = v'$ and $t_\Lambda = C_{\epsilon A}D\Delta r/u$. For the experimental results, the ratio of the large eddy turnover time (t_Λ) to the IZ residence time (col. 11 in Table 1) varies from 0.85 to 1.05. The experimental results then are in good agreement with the notion from turbulence energy cascade theory that large eddies lose their energy within one turnover time.

For the experimental results:

$$\frac{\epsilon}{\epsilon_{IZ}} = \frac{C_\epsilon}{C_{\epsilon A}} = \frac{\Delta m}{\Delta w} \qquad (14)$$

Eq. 14 allows for the calculation of the actual energy dissipation rate, ϵ, from ϵ_{IZ}, the reference rate based on a dissipation volume equal to the IZ volume.

Fig. 4 plots $C_\epsilon/C_{\epsilon A}$ vs ϵ_{IZ} for the experimental data.

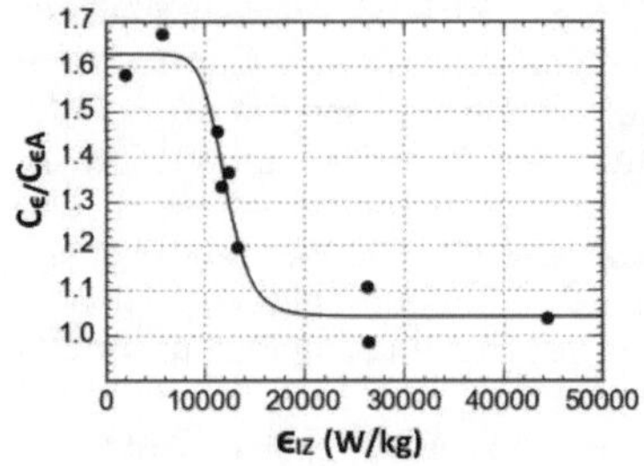

Figure 4. Variation in $C_\epsilon/C_{\epsilon A}$ as a function of ϵ_{IZ}. — Hill function (see Eq. 15).

Curve fitting software was used to fit the data to 85 nonlinear models. The best fit was a Hill type function (R^2 = 0.96), which has the following equation:

$$\frac{C_\epsilon}{C_{\epsilon A}} = a + \frac{b\epsilon_{IZ}^c}{d^c + \epsilon_{IZ}^c} = 1.627 - \frac{0.585\epsilon_{IZ}^{9.71}}{12217^{9.71} + \epsilon_{IZ}^{9.71}} \qquad (15)$$

where the units for ϵ_{IZ} are W/kg. The minimum asymptotic value, $C_\epsilon/C_{\epsilon A}$, is 1.04 (a-b), indicating that C_ϵ approaches $C_{\epsilon A}$ to within 4%.

To calculate the energy dissipation rate of impinging sheets, ϵ_{IZ} should first be calculated from Eq. 11. If $\epsilon_{IZ} \geq 3 \times 10^4$ W/kg, then $\epsilon \approx \epsilon_{IZ}$. If $\epsilon_{IZ} < 3 \times 10^4$ W/kg, $C_\epsilon/C_{\epsilon A}$ can be determined from Eq. 15 and ϵ can subsequently be calculated using Eq. 14.

C_ϵ values significantly higher than $C_{\epsilon A}$ have been published in the literature [7,8,9]. C_ϵ has been found to be a function of Re and reaches $C_{\epsilon A}$ at high Re.

The Taylor-microscale Reynolds number ($Re_T = v'^2\sqrt{15/\nu\epsilon}$) is often used to determine the Re of a turbulent flow (fully developed turbulence is achieved when $Re_T > 100 - 140$ [10]). In this study, there was no correlation of C_ϵ with Re_T. According to Vassilicos [9], the dependence on Re is a result of a significant non-equilibrium region existing in various turbulent flows causing C_ϵ to be a function of the ratio of Re at the inlet of the flow and the local turbulence Re. This disparity in Re does not exist for the impingement zone of impinging sheets. Instead, C_ϵ for impinging sheets is a function of the amount of energy to be injected relative to the large eddy turnover time.

Conclusions

The experimental results demonstrate that for impinging sheets, the release of energy resulting from the collision occurs within the impingement zone. At $\epsilon_{IZ} < 3 \times 10^4$ W/kg, the energy dissipation zone is significantly smaller than the impingement zone and $C_\epsilon > C_{\epsilon A}$. At values of $\epsilon_{IZ} > 3 \times 10^4$ W/kg, the energy dissipation zone is approximately equal to the impingement zone such that $C_\epsilon \approx C_{\epsilon A}$ and $\epsilon \approx \epsilon_{IZ}$.

These results are in good agreement with the notion from turbulence energy cascade theory that large eddies lose their energy within one turnover time. For impinging sheets, the largest eddies are produced in the impingement zone and have a turnover time that is approximately equal to the residence time of the liquid in the impingement zone.

The value of C_ϵ did not correlate with Re as other researchers have found. This is most likely because there are no differences in Re within the impingement zone. Instead, C_ϵ for impinging sheets is a function of the amount of energy to be injected relative to the large eddy turnover time.

References

[1] R.J. Demyanovich, *Chem Eng Sci X,* 12 (2021) 100113.
[2] R.J. Demyanovich, J.R. Bourne, *Ind Eng Chem Res,* 28(6) (1989) 830.
[3] N. Dombrowski, D. Hasson, D.E. Ward, *Chem Eng Sci*, 12 (1960) 35.
[4] C. Clanet, E. Villermaux, *J. Fluid Mech.* 462 (2002) 307.
[5] N. Dombrowski, R.P. Fraser, G.T. Peck, *J Sci Instrum,* 32(9) (1955) 329.
[6] R.J. Demyanovich, *U.S. Pat. 4,735,359* (1988).
[7] W.D. McComb, A. Berera, S.R. Yoffe, M.F. Linkmann, *Phys. Rev. E,* 91 (2015).
[8] B.R. Pearson, T.A. Yousef, N.E.L. Haugen, A. Brandenburg, P.Å. Krogstad, *Phys. Rev. E,* 70 (2004) 6.
[9] J.C. Vassilicos, *Annu. Rev. Fluid Mech.* 47 (2015) 95.
[10] P.E. Dimotakis, J. Fluid Mech. 409 (2000) 69.

https://doi.org/10.52202/074122-0047

Enhancement of fluid mixing in an agitated thin film evaporator with multistage blades

Poster.2

***Koichi Inata*[1]**, *Kazutaka Takata*[1]*.(1) Department of Mechanical Engineering, National Institute of Technology, Kure college, 2-2-11 Agaminami Kure, Hiroshima 737-2206 Japan, Corresponding email:k-takata@kure-nct.ac.jp*

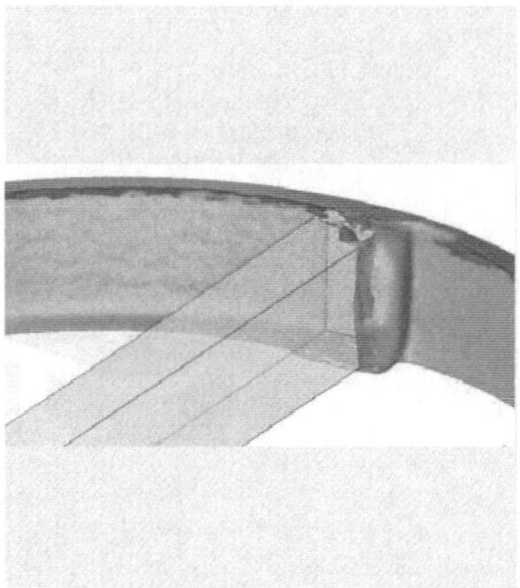

An agitated thin-film evaporator forms a liquid thin film on the inner surface scraped by the blade and fillet at the corner of the blade tip and inner wall, and the former contributes to evaporation and the latter to fluid transport and surface renewal of the film. In this study, flow structure with free surface in an equipment using high viscosity fluid is tried to clarify by means of CFD, and discussed the enhancement of fluid exchange between the fillet and the film. The predicted results showed that the majority of feed flow is downstream at the fillet with little fluid exchange with the film. It was suggested that material exchange between the film and the fillet is necessary to improve evaporation performance. A double-stage blade configuration that mixes the film and the fillet was proposed. In this study the flow structure of double-stage configuration was also predicted, and the mechanism of fluid exchange was discussed.

Introduction

An agitated thin-film evaporators are used in the chemical, food, pharmaceutical, and other process industries to remove residual monomers and impurities to improve product purity. It is operated under vacuum condition so that it makes suitable for processing highly thermosensitive materials. Figure 1 shows a schematic flow diagram in the equipment. By agitating the flowing liquid film by the blade with a small clearance (distance between the blade tip and inner surface wall), a thin film is formed in the inner surface and a fillet is formed at the corner between the blade tip and the tank wall, where the thin film contributes to evaporation and the fillet contributes to surface renewal of the film.

In this study, flow structure with complicated three dimensional free surface and mass transfer in the equipment has tried to clarify when a highly viscous fluid is used. Both methods of CFD and physical experiment are used to clarify the flow structure and material transport in the equipment. And discussed the factors that may enhance the evaporation performance.

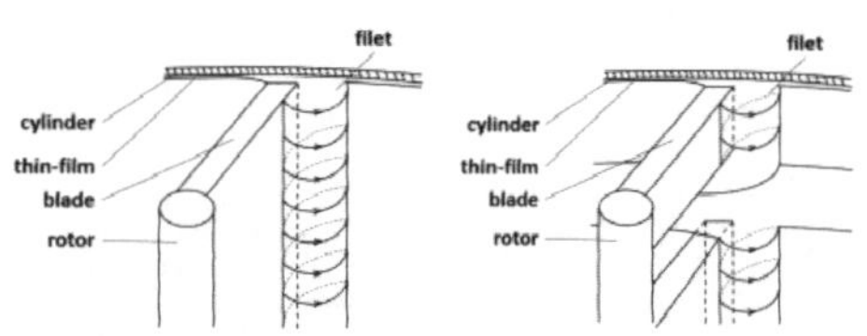

Figure 1. Schematic diagram of flow pattern with single and double blades configurations

Experiments

Figure 2 shows the schematic diagram of experimental equipment. It consists of double stage of blades inside a transparent acrylic cylinder with an inner diameter of 200 mm. Fluid was supplied to the dispersion plate inside the cylinder by a gear pump and flowed uniformly down to the top of the upper blades. The blades rotated at 200 rpm and the volumetric flow rate was 15 litters/hour. A laser displacement meter was used to measure the film thickness. To confirm the presence of fluid exchange between the fillet and the liquid film in double-stage blades, the conductivity method was applied to measure the concentration at the fillet to determine the fluid exchange rate [2].

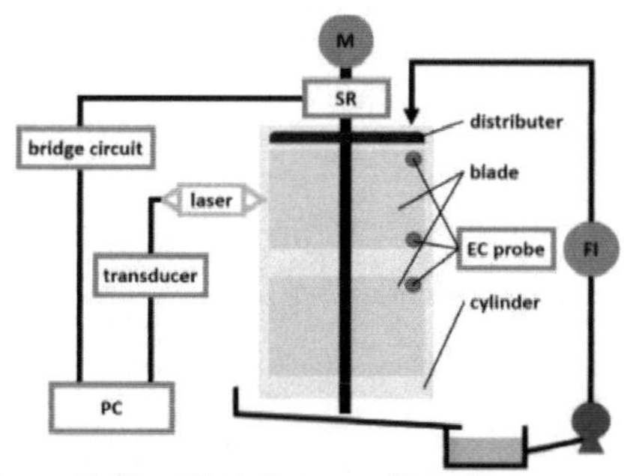

Figure 2. Simplified diagram of thin-film evaporator

Numerical Calculation（CFD）

CFD studies in previous research have been conducted in regions of relatively low viscosity [3]. Here, CFD was performed using a high viscous fluid and taking into account the free surface. The free surface model is based on the VOF method, and the relative effects of interfacial tension, gravity, and centrifugal force are considered. The analysis conditions are shown in Table 1. In this calculation, we first focused on to clarify the flow mechanism of the liquid film and fillet with single stage blade. Next, a multi-stage blade model was introduced to enhance fluid exchange between the fillet and film, and the flow mechanism that contributes to fluid mixing between the two was elucidated. A schematic of the flow of double stage blades was already shown in Fig.1. Analysis code scFLOW [3] was used as the solver.

https://doi.org/10.52202/074122-0048

Table 1. Analysis and experimental conditions

	CFD	Experiment
Inner Diameter(m)	0.2	
Blade Dimensions(m)	0.197×0.016×0.028	0.197×0.016×0.15
Clearance(m)	0.0015	
Viscosity(Pas)	10	
Rotations numbers(rpm)	200	

Results and Discussion

(1) Flow characteristic of single stage

Figure 4 shows the formation of the fillet and thin film on a single-stage blade (bird's-eye view) and a cross-sectional view focusing on the vicinity of the blade. When focusing on the area near the blade, a puddle of liquid (fillet) is generated in the direction of rotation at the corner of blade tip and inner surface wall, and a thin film is formed after the blade. From the analysis results, macroscopic flow balance can be determined by Eq.(1).

$$Q=2\pi D\cdot\delta_m\cdot U_m+A_f\cdot U_f \quad (1)$$

Here, Q : supply flow rate [m^3/s], D : inner tank diameter [m], δ_m : thin film thickness [m], Um : downward mean velocity of film [m/s], A_f : cross sectional area of the fillet [m^2], U_f: downward mean velocity of the fillet [m/s])

Here the predicted film thickness is compared with the measured one, the thickness was about 55% of the clearance both in measured and CFD, which was a little larger than that of the laminar Couette flow. Under the conditions of this calculation, the flow rates of thin film and fillet were found to be about 33:67, and about 70% of the total flow rate was found to be flowing down in the fillet with approximately 30times larger velocity of the thin film.

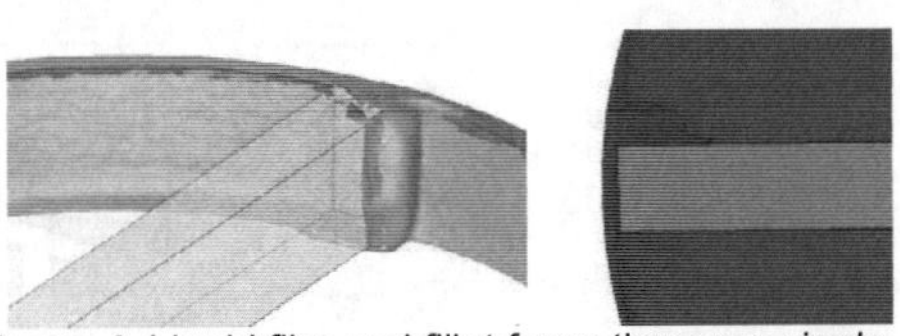

Figure 4. Liquid film and fillet formation on a single-blade configuration (CFD)

(2) Flow and mixing characteristics of double stage

Figure 5 shows the formation of the fillet, and thin film on the double-stage configuration. The fillet and thin film merge between the blades, and here the film thickness equals to the Nusselt's falling film thickness. And then, the thicker liquid film than the clearance is disturbed by the top of the lower blade edges, causing strong disturbance to the merged liquid, suggesting that the liquid is actively mixed here. Based on this hypothesis, the concentration at the fillet section was measured by the electrical conductivity method to evaluate the magnitude of mass exchange, *M,* between the fillet and the film. Tracer with KCl was carefully injected into the liquid film region from the tank wall, and the rate of mass exchange between the single-stage and double-stage were evaluated. Table-2 shows the results and suggests that little mass exchange was observed in the single-stage, but the improvement of mass exchange was observed in the double-stage. Further CFD analysis is required how the material is exchanged between the fillet and the film in the merge region.

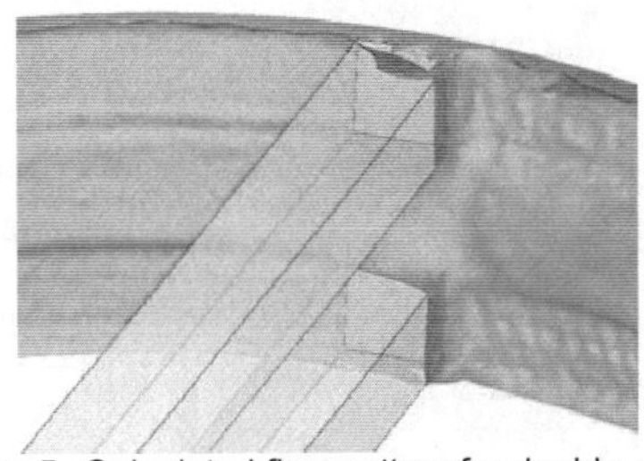

Figure 5. Calculated flow pattern for double-stage blades

Table 2. Measured M in single and double stages

	Single	Double
M	0.01	0.076
$M = (C_{f2} - C_{f1})/(C_{ave} - C_{f2})^{1)}$		

Conclusions

In this study, flow and mixing characteristics and the method of mixing enhancement in the equipment were examined, and the results obtained in this study are summarised as follows.

(1) The results of CFD showed that the fraction of flow in the fillet is approximately 70% of total flow rate with over 30times larger flow velocity of the thin film.

(2) In single stage configuration, the mixing between the fillet and film is suppressed, however, the double stage can enhance the mass exchange by merging the fluid of film and fillet between the blades.

Acknowledgments

Authors would like to express their gratitude to Kobelco Eco-Solutions Co.,Ltd. for its cooperation in conducting this research.

References

[1] Komori,S.,K.Takata, Y.Murakami, Flow and Mixing characteristics in an agitated thin film evapotrator with vertically aliened blades, J.Chem.Eng.Jpn. 22,346-351

[2] Sanjay,B.P.,A.S.Mujumdar, B.N.Thorat, CFD analysis of flow pattern in the agitated thin film evaporator, Chem. Eng. Res. Des., 90(2012) 757-765

[3] Cradle CFD scFLOW | Hexagon

https://doi.org/10.52202/074122-0048

Development of a high-efficiency mixing equipment with a large projected area

Poster.3

Kai Ogata[1], *Kazutaka Takata*[1]. *(1) Department of Mechanical Engineering, National Institute of Technology, Kure college, 2-2-11 Agaminami Kure, Hiroshima 737-8506 Japan, Corresponding email:k-takata@kure-nct.ac.jp*

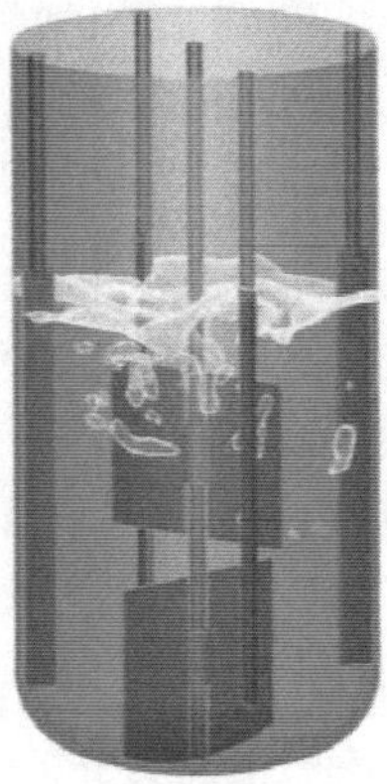

Present study investigates the effect of impeller projected area of rotation on the mixing performance with the aim of developing a high-performance mixing equipment. A 2-stage paddle impeller was selected, and physical experimental and numerical methods were used to evaluate its performance by changing the projected area. A rotating torque, mixing performance, heat transfer characteristics, solid suspension performance, and discharge flow rate from the mixing blade in a wide viscosity range were evaluated. The results showed that the 2-stage wide paddle impeller with a larger projected area has advantages in all evaluation items. In terms of mixing performance, the highest discharge efficiency ($Nq/Np^{1/3}$) was achieved in comparison with existing mixing impellers. These results indicate that the increase of projected area allows the mixing impeller to be both versatile and highly functional.

Introduction

Mixing equipment is widely used in various industries, and its performance leads to significant impact on quality and process efficiency. Therefore, it is extremely important to select the correct mixing equipment that suit for the process. The final objective of this study is to develop a high-efficiency mixing equipment. Based on the fact that the projected area of rotation makes an impact on mixing performance [1]. In this study rotation torque, mixing time and its process, discharge performance, solid suspension and heat transfer characteristics were evaluated by changing rotation projected area, and try to clarify the optimum dimension of the impeller.

Experiments

Figure 1 shows a schematic diagram of the experimental setup. Table 1 also shows the dimensions of the impellers used in this study. Three types of 2-stage paddle impeller[2] were prepared to vary the projected area of rotation, and 4-paddle impeller was prepared as an existing impeller[3]. Water (0.001 Pas) and corn syrup (0.7~10 Pas) were used as fluids to evaluate mixing performance in a wide viscosity range. In order to compare the performance of impellers, the power per unit fluid volume was set to the same.

The rotating torque was measured using a non-contact torque transducer, and the idle torque was subtracted to obtain the torque required to mix the fluid only. The mixing process and mixing time were measured using a colorimetric reaction with iodine and sodium thiosulfate. An aluminium alloy with an inner diameter of 386 mm was used to measure heat transfer performance. Thin precision thermocouples were attached to the inner and outer wall surfaces of the tank to determine surface temperature and evaluate the local heat transfer coefficient. In heat transfer experiments with highly viscous liquids, it is necessary to evaluate the viscosity coefficient at the wall surface. In this study, the temperature dependence on the viscosity coefficient of corn syrup solution used was preliminary evaluated and corrected using the following equation (applicable viscosity coefficient > 0.1Pas).

$$log\mu = \frac{A * 10^3}{T} - B$$
$$A = 1.288 * log\mu_{20} + 1.500$$
$$B = 3.393 * log\mu_{20} + 5.116 \qquad (1)$$

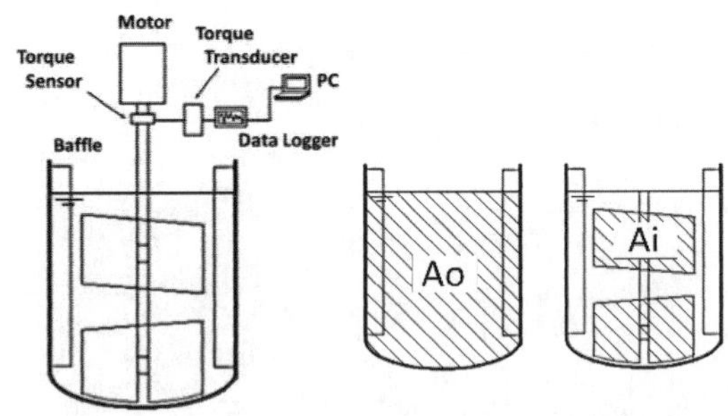

Figure 1. Schematic diagram of experimental equipment

Numerical Calculation (CFD)

In this study, a method of CFD was applied to elucidate the flow structure in the tank. The rotating impeller was formulated using the boundary-jointed lattice method, and an unsteady analysis was performed. Detached Eddy Simulation (DES) was applied to the turbulent to the transition regions to consider the scales of various eddy sizes that may affect mixing[4]. For the calculation of mass and heat transfer, the coupled analysis was performed with the momentum transport equation. For the effect of

https://doi.org/10.52202/074122-0049

Table 1. Dimensions of the impellers

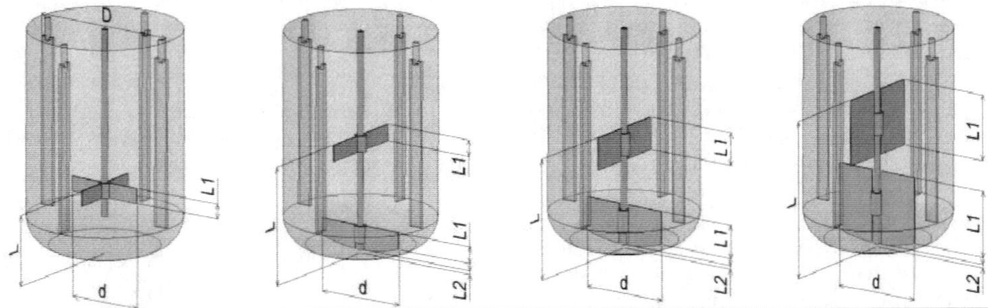

Name	4-padlle	2-stage small padlle	2-stage middle padlle	2-stage wide padlle
D (mm)	400			
d (mm)	200	240	240	240
L (mm)	190	320	330	430
L1 (mm)	40	45	90	180
L2 (mm)	-	10	10	10

temperature on the viscosity coefficient during the analysis of heat transfer, the temperature calculated for each time step was corrected using equation (1). The VOF method was also employed for the liquid surface to account for its shape and the effect of the shape of the liquid surface on mixing performance. Analysis code"scFLOW" was used as the solver. Figure 2 shows an example of an analytical model.

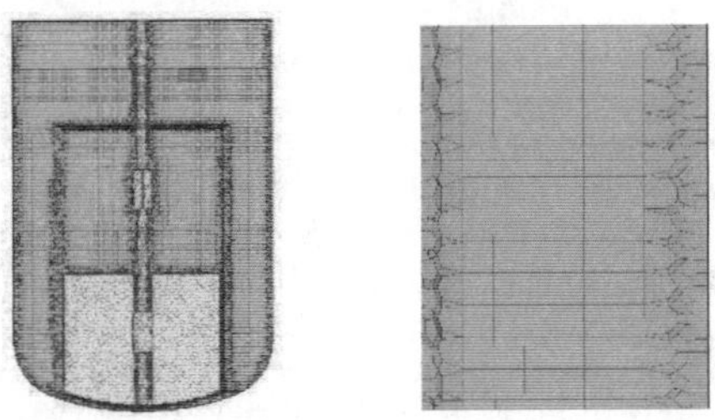

Figure 2. Mesh model (left: whole, right: near wall)

Results and Discussion

(1) Correlation of rotation torque

The measured and calculated rotation torque was evaluated as power number and plotted against the mixing Reynolds number in Figure 3. It is found that the measured and calculated values are in good agreement, indicating that the prediction of the flow by CFD is sufficiently accurate.

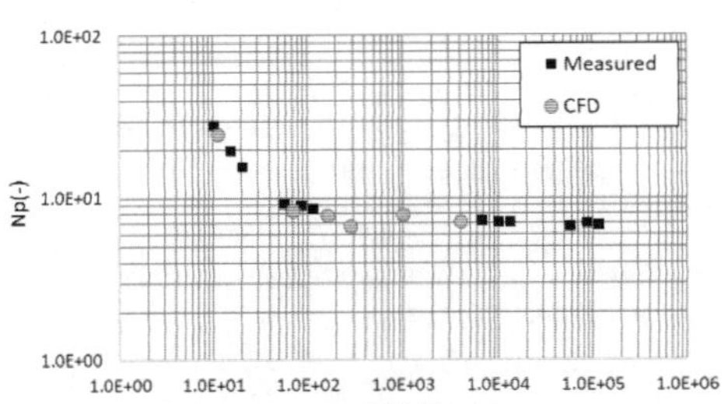

Figure 3. Correlation between rotating torque and mixing Reynolds number

(2) Effect of projected area on mixing characteristics

The measured mixing times are shown in Figure 4. The difference in mixing time by projected area is small for viscosity of 0.02 Pas or less. On the contrary, it is shown that the mixing time reduces when using a larger projected area. This shows that the impeller with large projected area showed the versatility in the wide range of viscosity.

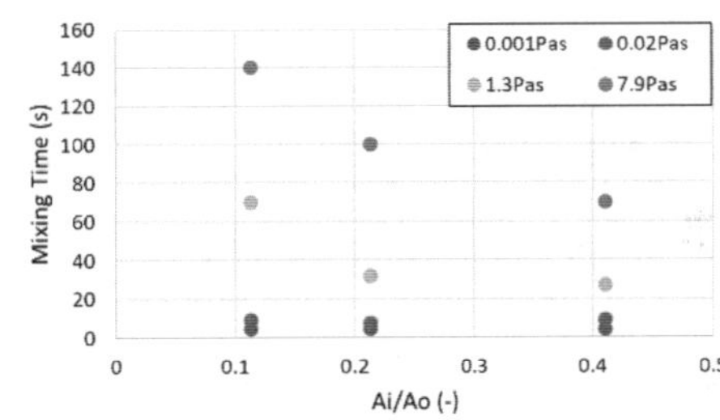

Figure 4. Influence of mixing time on projected

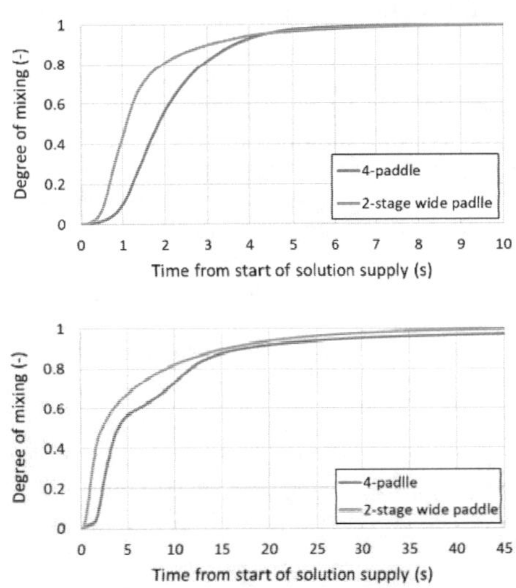

Figure 5. Concentration diffusion (Upper figure: water (0.001 Pas), Downer figure: corn syrup (1 Pas))

https://doi.org/10.52202/074122-0049

Regarding mixing process, the concentration of the passive tracer was evaluated in two type's impellers. The concentration change over the time is shown in Figure 5. When the fluid is water, there is no significant difference in performance between 4-paddle and large projected area. However, as the viscosity increases, the difference in mixing performance becomes more pronounced, resulting in a time reduction of about 32% for the large projected area.

(3) Solid-liquid mixing (Particle dispersion)

Figure 6 shows the percentage of particles present per unit volume over the mixing elapsed time, and compared two types of impellers. In this calculation, 3000 particles with a density of 1300 kg/m3 and an average particle diameter of 0.1 mm were placed at the bottom of the mixing tank, and the process of dispersion in the tank with mixing flow was determined. It can be seen that the impeller with large projected area completes dispersion in about 3 seconds, whereas the 4-paddle impeller takes about 10 seconds. This difference is thought to be that the flow by the large projection area reaches the entire area of the tank. In fact, the calculated average flow velocity in the tank is about twice that of 4-paddle impeller, which is also a factor that greatly affect performance.

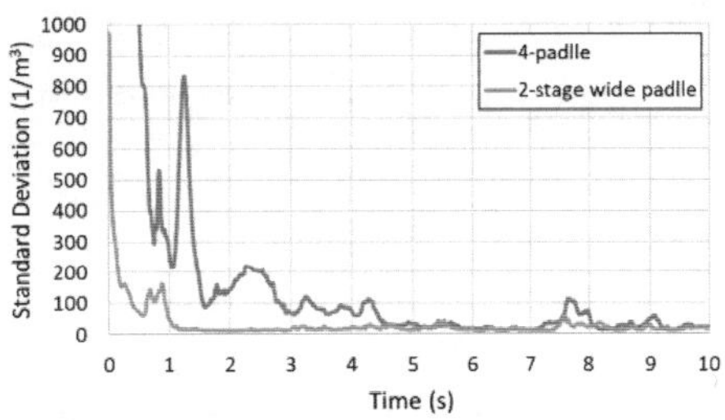

Figure 6. Comparisons of particle dispersion in two impellers

(4) Heat transfer evaluation

The local heat transfer coefficient at the inner wall of the mixing tank was evaluated and irrustrated against the axial direction[5]. It is found that there is a distribution in the heat transfer coefficient depending on the location of the mixing impellers. This correlates with the magnitude of the flow velocity discharged from the impellers. Although the peak of heat transfer coefficient was higher for the 4-paddle with higher rotational speed, the average value for the tank wall was about 20% higher for the larger projected area.

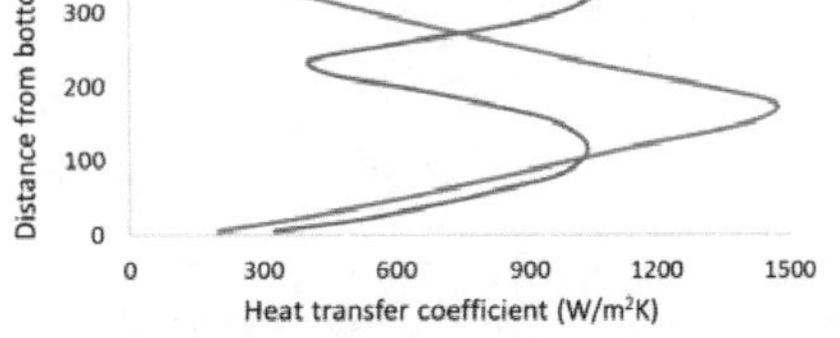

Figure 7. Distribution of local heat transfer coefficients

(5) Comparison with existing impellers

The dimensionless value $Nq/Np^{1/3}$ was calculated from the calculated discharge flow rate of each impeller and power. The results are shown in Figure 8 by comparing the numbers with another existing impellers. It is found that impeller with a large projected area is the most effective in discharge performance.

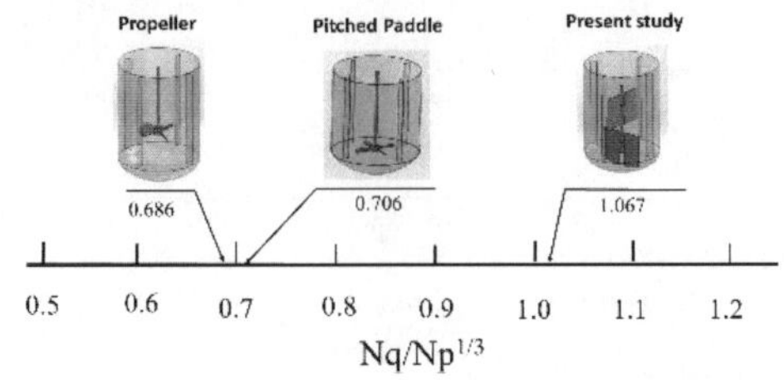

Figure 8. Comparisons of discharge efficiency

Conclusions

In this study, the influence of projected area of the impeller on the mixing performance was investigated for the purpose of developing a highly efficient mixing both using the methods of experimental and CFD. The results obtained in this study are summarized as follows.

1) Based on the experimental values and CFD, the application of DES is effective for the prediction of mixing flow in turbulent and transient regions.
2) The results of the mixing characteristics such as discharge, mixing, solid-liquid dispersion, and heat transfer indicated that the impeller with the larger projected area showed advantage in performances. The results also indicate that the impeller has the aspect of a multifunctional and versatile mixing impeller.
3) Mixing performance can be improved by influencing the flow by the impeller to the entire tank, and one of the methods is to increase the projected area. In case of two stage paddle impellers used in this study, the optimal shape can be determined by carefully examining the discharge balance of the upper and lower paddle impellers.

Acknowledgments

Authors would like to express our thanks to JKA foundation for its support in conducting the present research (2020M-142,2023M-303).

References

[1] Hasegawa, S., K.Takata, Prediction of Performance in an Agitated Tank with Double-stage 2-Bladed Wide Paddle impeller,Proceedings of 16th Europian Conference on Mixing No.204(2018)

[2] *Komori S, Y.Murakami, Turbulent mixing in baffled stirred tanks with vertical-blade impellers, AIChE Journal(1988), 34, 6, 932-937*

[3] Nagata S.、Mixing, Kodansha(1970)

[4] Gimbun,J, C.D.Rielly, Z.K.Nagy,JJ.Derksen, Detached Eddy Simulation on the Turbulent Flow in a Stirred Tank、AIChE J(2011), 58,10,3224-3241

[5] Critical Data Tables, Corn Sweeteners,Starches and Other Corn Carbonhydrate Products, Corn Refiners Association, Inc., Revised 2018.

https://doi.org/10.52202/074122-0049

Particle Suspension in Spherical Vessels

Poster.4

Adam Krupica[1], *Tomáš Jirout*[2]. *Czech Technical University in Prague, Prague, Technická 4, Prague, Czech Republic, (1) adam.krupica@fs.cvut.cz. (2) tomas.jirout@fs.cvut.cz.*

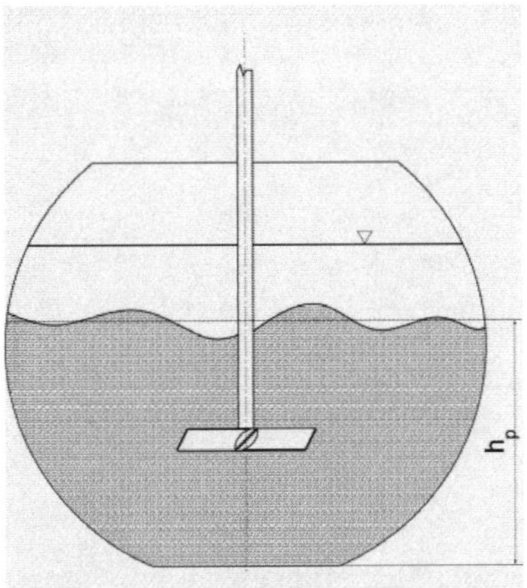

Performance of particle suspension using various agitators and jet configurations was tested in a spherical vessel. Although uncommon, spherical vessels can be found throughout the industry, most commonly as storage tanks or anaerobic digesters, however, there seems to be no data regarding how to efficiently facilitate particle suspension in this geometry.

Therefore, assessment of the necessary power input for particle suspension and the corresponding level of homogenisation was performed utilizing video analysis. Of the devices used, most were capable of achieving just-suspended state, however achieving homogeneity in the whole volume was problematic. The best performing from the tested methods were the one which least enhanced tangential flow, such as Rushton turbine or PB6 agitator. Due to high difference in densities between the particles and liquid, jets were performing poorly. However, when applied to suspensions of comparable densities, their efficiency was one of the best.

Introduction

Although uncommon in the industry, spherical vessels are often used as anaerobic digesters. For such application, homogenised batch is often desired for better nutrition delivery to the microorganisms. The general design of such reactors is predominantly assessed based on their biogas production [1]. There is however little amount of information available regarding the underlying process of particle suspension in such geometry. It is therefore the goal of this work to assess the performance of the most used method for particle suspension, agitator mixing, based on its power consumption.

Material and Methods

The ability of the various agitators (Table 1.) and jet configurations (axial, radial and tangential) to suspend particles was tested in a spherical vessel with an inner diameter of 320 mm. The vessel was filled with 13 litres of water. To obtain comparable results, the experiments were performed in sets with constant particle concentration inside the batch. Between each round of experiments 850 grams of 1 ~ 1.3 mm glass beads (density 2600 kg/m^3) were added to the batch. Two different points were visually evaluated the just-suspended state defined by modified Zwietering criteria [2], meaning that no particle was stationary for more than 3 s and the fully homogenised state. The necessary power input was calculated from data published by Krupica and Jirout [3]. When the device in question achieved just-suspended state but was unable to fully homogenize the batch due to it being constrained by suction into agitator area/power input limit, the height of the particle cloud was measured instead. When the device was incapable of reaching just-suspended state for given concentration it was discarded and was not used in the following round.

Results and Discussion

The results of the experiments are presented in Table 2 and Figure 1. During the experiments, we were unable to achieve fully homogenised state, as at high rotation speeds a large axial vortex was forming disallowing for a further increase in rotation speed. Additionally, the agitators rarely achieved lifting particles significantly higher than above the middle of the sphere. This may be caused by the curvature of the vessel and by the dominant tangential flow, as the rising particles do not have enough energy in the axial direction and are being redirected from the upward trajectory back down or towards the middle of the vessel. It is apparent from the data that the RT6 is the most optimal agitator type, as it is suited for operation at the widest range of particle concentrations. However, at low particle concentrations other agitators were also capable of suspending particles with a similar result.

Table 1. Geometrical parameters of the used agitators

Agitator Abbreviation	Diameter (mm)	Number of blades	Expected flow pattern	Geometry
PB6	100	6	Axial	Standard pitched blade agitator
RT6	100	6	Radial	Standard Rushton turbine
FB3	100	3	Axial - hydrofoil	[4]
FB6	100	6	Axial - hydrofoil	[4]
A310	100	3	Axial - hydrofoil	Lightnin A310

https://doi.org/10.52202/074122-0050

Table 2. *The results of particle suspension*

Agitator	C_V (% vol.)	n (1/min)	ε (W/m3)	h_p (mm)
PB6	2.44	264	37	110
A310	2.44	468	50	110
RT6	2.44	207	23	115
FB3	2.44	354	42	110
FB6	2.44	300	32	120
PB6	4.77	270	38	110
RT6	4.77	219	27	80
FB6	4.77	309	36	90
RT6	6.98	219	26	100
RT6	9.1	231	26	100
RT6	13	231	28	60 ~180 (wave)

(a) **(b)**

Figure 1. Graphical representation of the results

It is reasonable to assume that the addition of baffles would increase the performance of the various agitators by disrupting the dominant tangential flow and allow the fully homogenised state to occur as was shown by Tacă and Păunescu [5].

The performance of the various jets was hampered by the difference in densities between the particles and the liquid. However, when tested with less dense particles, their performance was comparable to that of agitators. Given by their lower power consumption, they are more suitable for batch homogenisation of such suspensions.

Conclusions

Particle suspension using various agitators was tested on a spherical vessel. Due to the dominance of tangential flow and creation of large axial vortex fully homogenised state was unachievable. However, to just suspend the particles, the Rushton turbine performed the best, as it was working over the largest range of concentrations with lowest required power input. It is reasonable to assume that the addition of baffles would greatly benefit the system as it would allow in increase in rotation speeds and disrupt the hampering tangential flow.

For suspensions of low difference in densities between particles and fluid, such as activated sludge, jet mixing seems like the better option due to the lower power consumption of the method.

Notation

c_v	Volume concentration of particles (% vol.)
h_p	Height of particle cloud (mm)
n	Agitator rotation speed (RPM)
ε	Power density (W/m^3)

Agitators' nomenclature

A310 Ligtnin® A310 agitator
FB3 Pitched 3-blade impeller, with diagonally folded blades, [4]
FB6 Pitched 6-blade impeller, with diagonally folded blades, [4]
PB6 Standard pitched 6-blade impeller, blade angle: 45°
RT6 Standard Rushton turbine, 6 blades

Acknowledgments

This work was supported by the Student Grant Competition of CTU as part of grant no. SGS22/103/OHK2/2T/12.

References

[1] SINGH, Buta, Zoltán SZAMOSI a Zoltán SIMÉNFALVI. State of the art on mixing in an anaerobic digester: A review [online]. *Renewable Energy*. 2019, **141**, 922-936. Available at: DOI: 10.1016/j.renene.2019.04.072.

[2] ZWIETERING, Th.N. Suspending of solid particles in liquid by agitators [online]. *Chemical Engineering Science*. 1958, **8**(3-4), 244-253. Available at: DOI: 10.1016/0009-2509(58)85031-9.

[3] KRUPICA, Adam a Tomáš JIROUT. Blending in a spherical tank. *Chemical Engineering Research and Design*. 2023 (in press).

[4] JIROUT, Tomáš a František RIEGER. Impeller design for mixing of suspensions [online]. *Chemical Engineering*

Research and Design. 2011, **89**(7), 1144-1151. Available at: DOI: 10.1016/j.cherd.2010.12.005.
[5] TACĂ, Constantin Dan a Mihaela PĂUNESCU. Suspension of solid particles in spherical stirred vessels [online]. *Chemical Engineering Science*. 2000, **55**(15), 2989-2993. Available at: DOI: 10.1016/S0009-2509(99)00436-4.

https://doi.org/10.52202/074122-0050

Paddle Agitator for Mixing of Viscoplastic Fluids and Pastes

Poster.5

***Tomáš Jirout*[1]**, *Adam Krupica*[1]. *Czech Technical University in Prague, Faculty of Mechanical Engineering, Department of Process Engineering, Prague, Technická 4, Prague, Czech Republic, (1) tomas.jirout@fs.cvut.cz, (2) adam.krupica@fs.cvut.cz.*

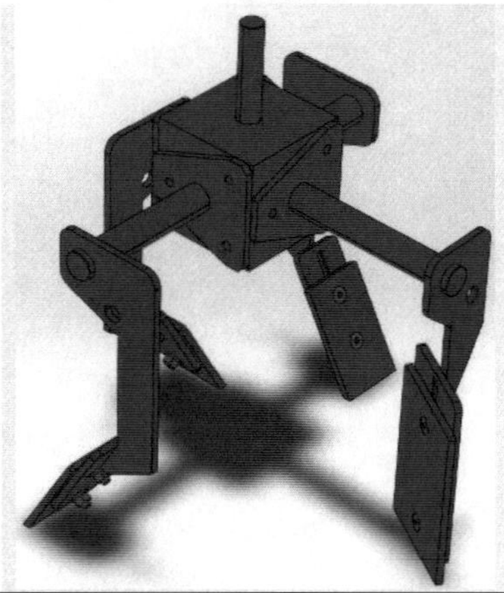

A new paddle mixer suitable for homogenisation of viscoplastic fluids and paste such as modern concrete mixture was developed to improve homogeneity of the product, which is vital for its mechanical properties.
While mixing concentrated suspensions characterised by visco-plastic or pseudoplastic behaviour, there is a risk of non-existent flow in areas distant from the rotating mixing element caused by the low level of shear stress and associated high apparent viscosity. The paddle mixer design is aimed at achieving circulation throughout the whole volume and thus mitigating the problem of caverns and segregation caused by the effect of gravity. The performance of the paddle mixer was evaluated using homogenisation experiments at input power measurements at varying rotation speeds. The paddle mixer was tested and its geometry modified to achieve homogeneity throughout the whole volume.

Introduction

High Performance Concrete is perspective material used in construction. Its mechanical properties (both before and after curing) highly depend on the distribution of reinforcing fibres. However, such suspensions often display a non-Newtonian behaviour which leads to unequal distribution of the said particles, resulting in poor quality. Mixing devices often used in construction provide unsatisfactory homogenisation of such mixtures and it is therefore goal of this work to design and optimize a paddle mixer suitable for homogenisation of such suspensions.

The main design criterion was to facilitate mixing throughout the whole volume of the batch, as due to the viscoplastic nature of the suspension only a small cavern surrounding the impeller is well mixed [1, 2]. This however can be overcome by suitable geometrical configuration of the various paddles leading to mixing throughout the whole batch. Such a result however must be confirmed by an experimental test.

Material and Methods

A paddle mixer consisting of four paddles at various distances and angles from the main axis was made. The configuration of the paddle mixer can be seen in Graphical Illustration. The configuration and geometry were design as to ensure that in one rotation whole volume of fluid is influenced by its motion and that the motion also facilitates radial transport of mass.

The paddle mixer was tested in transparent cylindrical tank with 400 mm diameter and height of 35 mm made from acrylic. A glycerol model fluid of similar properties based on the result of Dostal et al. [3] was used due to its transparent nature, high viscosity, and predictable Newtonian behaviour. To illustrate the flow inside of the vessel Figure 1. is

RPM. The measurement was carried out for all

presented. To measure the homoge-nization time at various rotation speeds, starch indicator method was used as described in Handbook of industrial mixing [4]. This method was chosen due to the low amount of liquid necessary to facilitate the change in colour and its high contrast nature. The experiment was repeated three times for every rotation speed and an average homogenisation time was used in the calculation. The injection of the discolouration solution was performed next to the vessel wall around 15 mm beneath the fluid surface. Whole process was recorded on camera and the homogenisation time was set as difference between moment of injection and moment the fluid became colourless throughout the whole volume. To correct for the change in viscosity due to the addition of water after each round of experiments a sample was taken for viscosity measurement. The lowest speed was set as the lowest when mixing throughout the whole volume was achieved while the highest was set just before aeration behind the paddles occurred.

For the same rotation speeds, power input was also measured as its value is essential for the assessment of the agitator. The amount of power put into the system by the agitator was measured using a torque sensor (precision ± 0.5 Nm) and a tachometer attached to the shaft of the agitator. The data were gathered using PLC and interpreted using an Excel script. The RPM value was displayed on the control panel, but the torque value had to be converted from the electric potential difference measured on the sensor. The RPM value oscillated about 2 RPM around the set value. The measurement of torque value consisted of 400 values taken over the span of 10 s. To determine the drag from other mechanical parts, a torque value for the free-spinning agitator was also measured at each

agitator types.

https://doi.org/10.52202/074122-0051

Figure 1. Illustration of flow inside the batch

Results and Discussion

At the lowest speed, which was set at the limit when mixing in the whole volume was achieved, only slow recirculating flow throughout the volume was observed, which is generally associated with a creeping flow. At the other limit just before aeration occurred, vortex cascade was forming in the wake of the paddle.

The homogenisation results are presented in Figure 2. Even though the batch generally displayed creeping flow behaviour, it is apparen from the gathered data that the mixing was performed under transition condition, as the dimensionless time nt is not constant. A probable cause is while most of the volume is mixed through slow recirculating flow, the region in wake behind the paddles experiences turbulence.

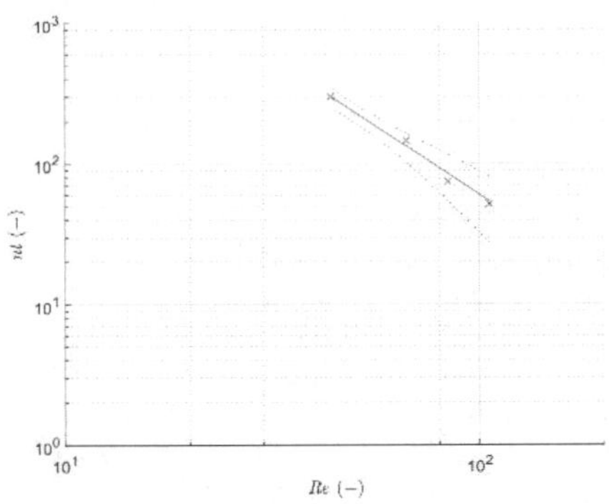

Figure 2. The result of homogenisation experiments

This hypothesis is supported by the observation that the highest amount of mixing occurred in the closest regions surrounding the paddles, an expected result based on the reviewed literature. However, at lowest rotation speeds, the region surrounding the axis of the vessel was very poorly mixed due to an oversight in paddle configuration. This however was overcome by slight modification to one of the paddles after which the paddle mixer was capable of homogenising the whole volume.

Conclusions

A new design of paddle mixer suitable for mixing of suspensions and pastes with non-Newtonian behaviour was produced and its capabilities tested. It was shown that with slight modification to the paddle configuration the mixer was capable of homogenising the whole volume of the batch filled with model fluid at warrying rotation speeds. For this modified configuration homogenisation and power input characteristics were measured and assessed. Based on these results and the observation of the mixing process it was concluded that two main mechanics are at play, slow recirculating flow achieving even distribution of the mixed fluid throughout the volume and more rapid turbulent mixing in the wake of the paddles. The combination of the two resulting in the complete homogenisation of the whole batch.

Acknowledgments

This research has been supported by the Grant Agency of the Czech Republic project No. 21-24070S "Model of fibre segregation in dependence on rheological properties of fresh HPC".

References

[1] CARPENTER, K. J., P. C. LINES, R. J. ALDINGTON, A. KERON a W. A.J. HINDSON. Two examples of designing full scale reactors for multi-stage synthesis involving non-Newtonian mixtures. *AIChE Symposium Series,* 1994, s. 101-105.

[2] JAWORSKI, Z., A. W. PACEK a A. W. NIENOW. On flow close to cavern boundaries in yield stress fluids [online]. *Chemical Engineering Science.* 1994, **49**(19), 3321-3324. Available from: DOI: 10.1016/0009-2509(94)E0075-2.

[3] DOSTÁL, Martin, Jiří MORAVEC, Tomáš JIROUT, Milan RYDVAL, Karel HURTIG a Dita JIROUTOVÁ. *Rheological behaviour of fresh concrete mixtures.* Seville, 26. April 2022. Annual European Rheology Conference 2022.

[4] PERRY, Robert H., Don W. GREEN a James O. MALONEY, ed. *Perry's chemical engineers' handbook.* 7th ed. / prepared by a staff of specialists under the editorial direction of late editor, Robert H. Perry / editor, Don W. Green / associate editor, James O. Maloney. New York, London : McGraw-Hill, 1997. 0-07-049841-5.

https://doi.org/10.52202/074122-0051

Production of loosely packed aggregates using the power number as an index of the salt coagulation process

Poster.6

Rikuto Iwamoto[1], Yoshiyuki Komoda[1], Naoto Ohmura[1]. (1) Department of Chemical Science and Engineering Graduate School of Engineering, Kobe University 1-1 Rokkodaicho, Nada-ku, Kobe, Hyogo 657-8501, Japan, 227t405t@stu.kobe-u.ac.jp.

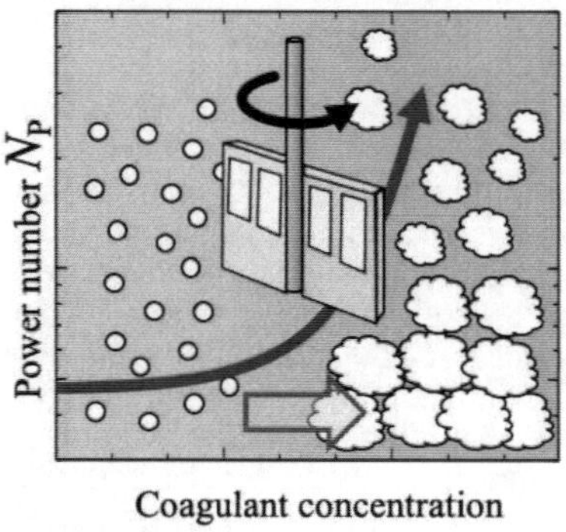

The present study deals with the salt coagulation process of latex particles, aiming to produce loose aggregates. The variation of mixing torque was measured to investigate the effects of the agitation condition and coagulant feeding rate on the coagulation process. The local maximum of the power number indicates that dense aggregates were destructed in the narrow clearance between the impeller tip and vessel wall. The following rapid increase in the power number corresponds to the production of loose aggregates. A higher impeller rotational speed contributes to uniform distribution of the coagulation process but encourages slow aggregation before producing loose aggregates. It is found that a slower feeding rate of the coagulant is more useful for obtaining coarse and loose aggregates.

Introduction

Uniform dispersion of latex particles in a matrix resin can improve the physical properties of the resin. For easier handling, latex particles are converted to aggregates by adding salt, but the resultant densely packed aggregates show poor dispersibility. Therefore, the modified salt coagulation process was proposed to produce loosely packed aggregates, including resin droplets, to improve the dispersibility. In addition, we reported that a moderate fluid flow induced by a rotationally reciprocating impeller is preferable to obtain loose aggregates because of uniform mixing without breaking aggregates [1]. For the particle aggregation process using a steady rotational impeller, it is reported that a larger impeller rotational speed results in small and dense packed aggregates. In contrast, non-uniform aggregates are produced at a low rotational speed due to poor mixing performance [2].

Since a large impeller operated at a low rotational speed is expected to produce loose aggregates, in the present study, we adopt a MAXBLEND impeller to the modified salt coagulation process and examine the effect of the rotational speed. Since the uniformity of the coagulant is another crucial factor in the coagulation process, we change the coagulant feeding rate. When latex particles form aggregates, the latex dispersion becomes more viscous, which increases mixing torque. Therefore, we measure the power number variation to estimate the change in the flow state in the vessel. Finally, we propose a guideline for producing loose aggregates with excellent dispersibility.

Material and Methods

183 mL of 30 wt% latex dispersion (particle diameter 140 nm) and 40 mL of 32 wt% epoxy dispersion (droplet diameter 26 nm) are mixed in a glass vessel (i.d. 0.074 m). The latex particles used in this study have a core-shell structure; the surface of rubber-like core particles is covered with grafted polymers. A MAXBLEND® impeller (diameter d = 51.8 mm) is placed at the center of the vessel and rotated at a specified rotational speed. The off-bottom and side clearances are 4 mm for each. Subsequently, 27 mL of coagulant (1.75 M sodium sulfate solution) is continuously injected into the mixture at a fixed feeding rate while agitating the mixture. The final volume of the mixture is 250 mL, corresponding to the liquid depth of 0.062 m. The final coagulant concentration is 0.212 M. Eqs. (1) and (2) define the Reynolds number and power number for the impeller using a rotational speed n [s^{-1}] and torque T [Nm]

$$Re = \frac{\rho n d^2}{\mu} \tag{1}$$

$$N_P = \frac{2\pi n T}{\rho n^3 d^5} \tag{2}$$

where μ and ρ are the viscosity and density of the mixture of latex-epoxy dispersions (μ = 6.5 mPa•s and ρ = 980 kg/m^3), respectively. The torque exerted on the mixing shaft is measured using a mixing torque meter (ST-4000, Satake Multimix). The power diagram (Fig. 1) of the MAXBLEND impeller created using Newtonian fluid indicates that $N_P > 3$ corresponds to a laminar flow regime with $N_P \propto Re^{-1}$ and $N_P = 1$ in a turbulent flow regime [3].

In the current salt coagulation process, the coagulant concentration increases with time. In the former stage, with a low concentration, aggregation

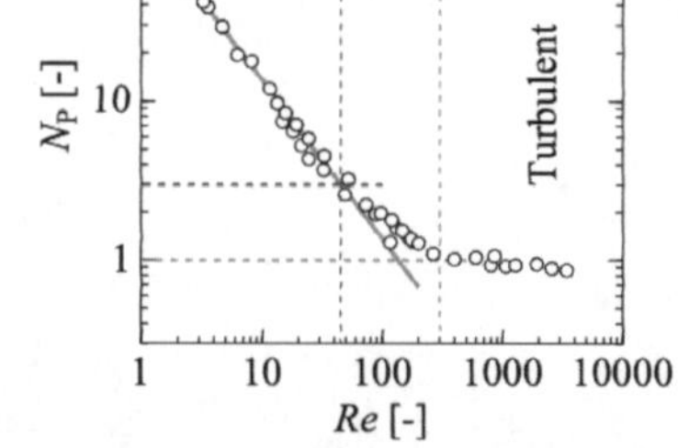

Figure 1. Power curve of MAXBLEND® impeller

https://doi.org/10.52202/074122-0052

proceeds slowly (slow aggregation), producing densely packed aggregates. However, when the coagulant concentration exceeds a critical value, the aggregation rate becomes larger (rapid aggregation), resulting in loosely packed aggregates. Since loose aggregates show good dispersibility [4, 5], suppressing dense aggregates is a key to establishing a desirable coagulation process. Firstly, the coagulation rate was evaluated from the growing experiment of aggregates using DLS and plotted against the coagulant concentration, as shown in Fig.2. It can be evident that the coagulation rate exhibited a drastic increase above a concentration larger than 0.104 M. [6]

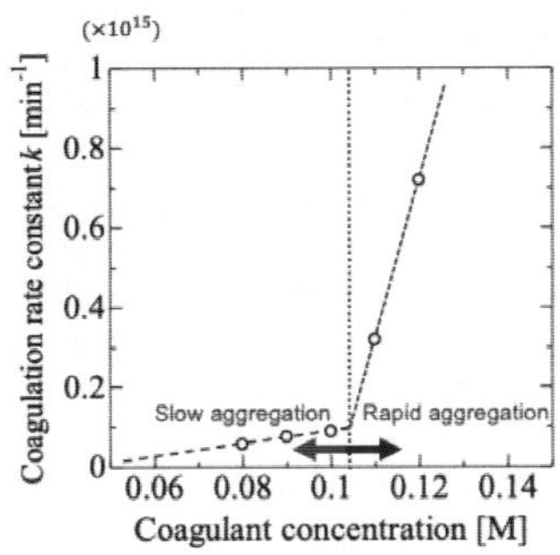

Figure 2. Transition from slow aggregation to rapid aggregation

After all the coagulant is added, aggregates are collected, and the equivalent diameter and circularity are determined by image analysis of microscopic photographs. An aggregate having a long circumferential length exhibits a small circularity, which means a loosely packed state of latex particles. Next, aggregates are dried, crushed into granules, and dispersed in epoxy resin. Finally, the latex/resin mixture is applied over a glass plate, and the turbidity and granule residues are visually examined.

Results and Discussion

Fig.3 shows the effect of the impeller rotational speed on the variation of power number during the coagulation process at a coagulant feeding rate of 5.4 mL/min. The critical coagulant concentration was denoted by a dashed line, whereas a purple line expresses the power number for the transition to the laminar flow regime. The initial power number, approximately 1, indicates that turbulent flow is generated in the vessel before adding the coagulant. As the amount of added coagulant increases, the aggregation of latex particles proceeds, and the viscosity of the dispersion increases. As a result, the fluid flow becomes moderate, and the power number increases under a constant rotational speed. At 450 rpm, the flow state kept turbulent long, and the coagulant was considered to be distributed uniformly. With the decrease in the rotational speed, the power number always took larger values. Still, a significant drop in the power number at the end of the slow aggregation period disappeared. From the shape change of aggregates, it is revealed that the power number drop corresponds to the destruction of dense aggregates into smaller clusters, and the reaggregation of the clusters induces the following second rapid increase in the power number. Therefore, at 450rpm, the second increase in the power number emerged below the critical coagulant concentration, suggesting that clusters of formerly produced dense aggregate again produced dense aggregates. On the other hand, at 150rpm, no obvious drop in the power number means insufficient destruction of dense aggregates. The results of the morphological evaluation of more than 300 aggregates produced at different rotational speeds are summarized in Table 1. At 150 rpm, aggregates became large due to the gentle flow, but a large circularity is caused by insufficient destruction of dense aggregates. At 250rpm, the formation and destruction of aggregates proceed uniformly, and smaller but loose aggregates were produced by the rapid aggregation. However, at 450 rpm, a larger circularity is caused by the slow aggregation of the clusters of dense aggregates formed in the former stage. Therefore. We concluded that 250rpm is the best rotational speed among those investigated.

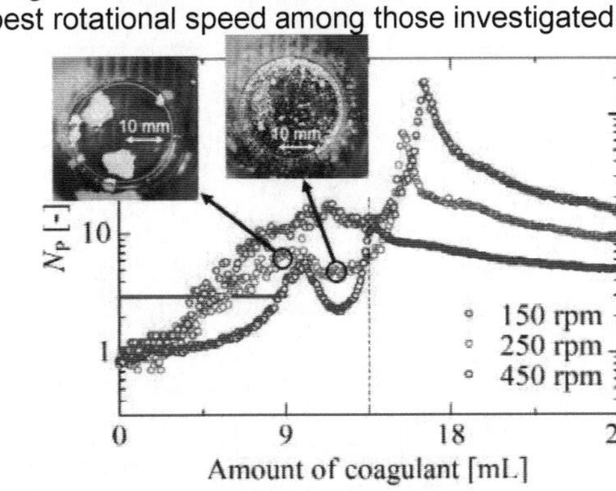

Figure 3. Effect of impeller rotational speed on the variation of power number (feeding rate: 5.4 ml/min) and the variation of aggregates shape

Table 1. Size and circularity of aggregates produced at different rotational speeds

	Rotational speed [rpm]		
	150	**250**	**450**
D_{50}	1129	946	925
ψ	0.58	0.38	0.56

D_{50}: Median diameter [μm], ψ: Circularity [-]

A higher rotational speed is helpful to homogenize the system but also induces shear destruction of aggregates. Therefore, to improve the coagulant mixing under a moderate fluid flow, a slower coagulant feeding rate is a promising operation condition of the coagulation process. Fig. 4 shows the effect of the coagulant feeding rate on the power number variation at a rotational speed of 250 rpm. As predicted from Fig.3, a more significant drop in the power number was observed at the slowest feeding rate, similar to the higher rotational speed. Therefore, it is supported that the uniform distribution of coagulant is related to this behavior of the power number. In contrast, a period showing a stable power number was observed at the highest feeding rate condition, presumably due to the nonuniform distribution of the coagulant. The results of the

https://doi.org/10.52202/074122-0052

morphological evaluation of aggregates are shown in Table 2. As the coagulant feeding rate decreased, the circularity exhibited a notable decrease. Additionally, significantly large aggregates were obtained only at the lowest feeding rate. Under this condition, the coagulant is uniformly distributed, and it takes longer to complete the coagulation process. Therefore, the rapid coagulation of the clusters of densely packed aggregates proceeded sufficiently, producing loose aggregates.

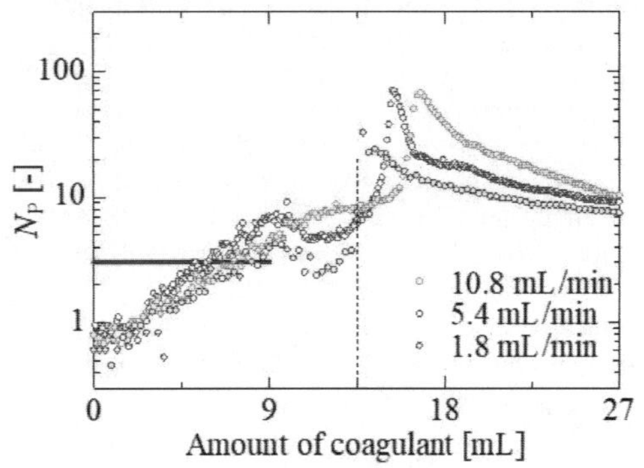

Figure 4. Effect of the coagulant feeding rate on the variation of power number (n = 250 rpm)

Table 2. Size and circularity of aggregates produced at different coagulant feeding rates

	Feeding rate [mL/min]		
	10.8	**5.4**	**1.8**
D_{50}	948	946	1318
ψ	0.60	0.38	0.26

Fig.5 shows the pictures of the coating layers of the latex/resin mixtures using aggregates produced at different coagulant feeding rates. For the aggregates produced at feeding rates of 5.4 and 10.8 mL/min, vertical traces indicate the existence of large residual granules larger than the coating gap (100 μm). In contrast, no vertical streak was observed at a feeding rate of 1.8 mL/min, and a more white and cloudy coating layer means that most of the latex particles are uniformly dispersed. Therefore, a sufficiently slow coagulation feeding rate is required to obtain loosely packed aggregates exhibiting better dispersibility while considering the fluid flow and shear destruction caused by the impeller.

Feeding rate [mL/min]		
10.8	**5.4**	**1.8**

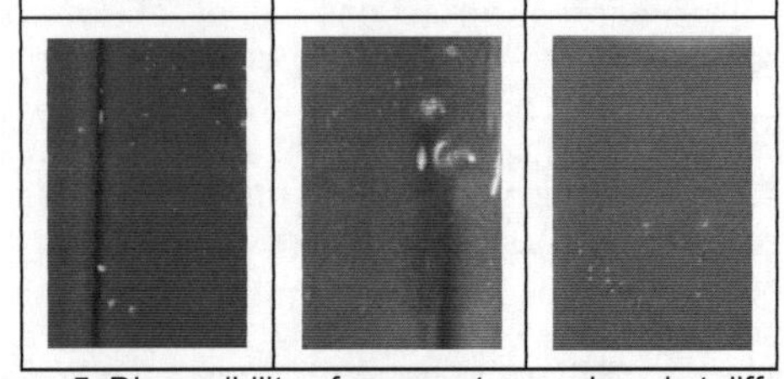

Figure 5. Dispersibility of aggregates produced at different coagulant feeding rates

Conclusions

In this study, we investigated the process for producing loose aggregates with good dispersibility. The effects of the agitation condition and coagulant feeding rate on the fluid flow and aggregate shape were investigated. The power number increases monotonously in the former stage of the coagulation process but suddenly drops before showing a rapid increase depending on the operating condition. It is clarified that the power number drop corresponds to the destruction of dense aggregates into clusters and that the reaggregation of clusters causes the following rapid increase. No obvious power number drop was observed due to nonuniform coagulant distribution at a slower rotational speed or a larger coagulant feeding rate. On the other hand, a higher rotational speed induces a rapid increase in the power number even under the slow coagulation regime, which eventually produces dense aggregates. Therefore, it is concluded that a sufficiently slow feeding rate to keep the coagulant uniform under a moderate fluid flow is required to produce desirable loosely packed aggregates.

Acknowledgments

The authors would like to express our sincere appreciation to Sumitomo Heavy Industries Process Equipment Corporation for providing MAXBLEND® impeller.

References

[1] Takahashi, T., Komoda, Y., Ohmura, N., “Coagulation of latex particles in the aqueous dispersion of epoxy resin” 18th APCChE *Congress*, Sapporo(Japan), PB371, (2019)

[2] Bubakova, P., Pivokonsky, M., Filip, P., "Effect of shear rate on aggregate size and structure in the process of aggregation and at steady state." *Powder Technol.*, 235 540-549,(2013)

[3] Fradette, L., Thome, G., Tanguy, A., Takenaka, K., "Power and mixing time study involving a MAXBLEND® impeller with viscous Newtonian and non-newtonian fluids" *Chem. Eng. Res. Des.,* 85(A11), 1514-1523,(2007)

[4] Ferretti, R., Zhang, J., Buffle, J., "Flocculation of hematite with polyacrylic acid: fractal structures in the reaction-and diffusion-limited aggregation regimes." *J. Colloid Interface Sci.*, 208(2), 509-517, (1998)

[5] Fukasawa, T., Adachi Y., "Direct observation on the Brownian coagulation of PSL particles through optical microscope in the regime near critical coagulation concentration (CCC)." *J. Colloid interface Sci.*, 344(2), 343-347, (2010)

[6] Holthoff, H., Stefan, U., Borkovec, M., Schurtenberger, P., Sticher, H., "Coagulation rate measurements of colloidal particles by simultaneous static and dynamic light scattering." *Langmuir*, 12(23), 5541-5549, (1996).

Impact of Stirred Tank Design and Operating Conditions on Sparger-Generated Foaming

Poster.7

A I Dosumu[1], *E Rundquist*[2], *M Poux*[1], *J Aubin*[1]. *(1) Laboratoire de Génie Chimique, Université de Toulouse, CNRS, INPT, UPS, Toulouse, France, abdulwahidishola.dosumu@toulouse-inp.fr; joelle.aubin@ensiacet.fr; martine.poux@ensiacet.fr; (2) Process Studies Group, Technology and Engineering, Syngenta, Jealott's Hill International Research Centre, RG42 6EY Bracknell, Berkshire, UK, elin.rundquist@syngenta.com*

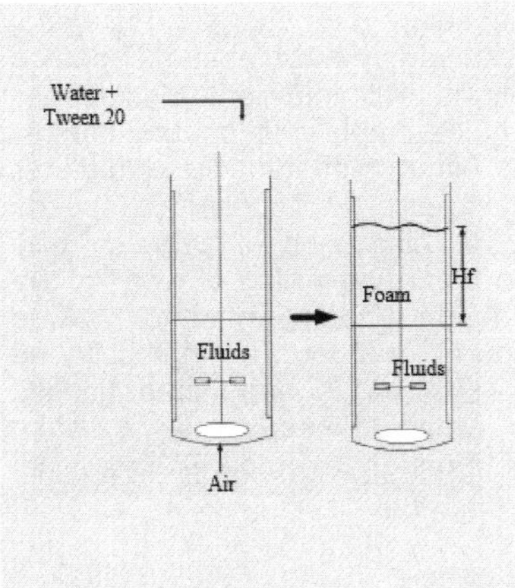

In this study, the impact of stirred tank design and operating conditions on sparger-generated foaming was investigated in a 0.19 m ID tank equipped with a Mixel TT axial flow impeller. The amount of foam generated when the impeller is running in down-pumping mode (MTTD) is compared with that obtained under the up-pumping mode (MTTU) at the same operating conditions. Maximum foam height was established at 10 minutes. The results showed that foam height increases with both increasing impeller speed and gas flow rate. In addition, less foam is produced with the MTTU compared with MTTD. Specifically, at a gas flow rate of 10 L/min and tip speed of 2.49 m/s, the MTTD produces two times more foam than the MTTU. This shows that MTTU has promise for wider industrial applications in foam control.

Keywords: Stirred tank; impeller speed; gas flowrate; foam height

Introduction

Foam formation is a common phenomenon often encountered in many chemical industries. For some processes, for example in the pulp and paper or food industries, the resulting foam is a desirable property [1]. However, foam formation can pose a serious problem in many industries, especially where adequate aeration and agitation of liquid is required. For example, in gas-liquid contacting processes, when gas is sparged in the reactor, unwanted foaming can occur. This leads to loss of capacity of the reactor, thus reducing the batch size and overall productivity [2] and sometimes, it can induce a loss of product. The amount of precautions used in foam control is therefore a critical issue for engineers.

Basically, there are two main methods used to control or reduce foam formation in a reactor, namely chemical and mechanical methods. While mechanical methods involve the use of non-chemical solutions such as changes in working conditions or equipment, chemical methods involve adding antifoaming agents or deformers to the mixture [3]. However, the introduction of antifoaming agents can have some adverse effects on production. For example, they can affect mass transfer rate, toxicity, and reaction rate [4]. They can also adversely affect the separation and purification of the final product, making the chemical methods for foam control not so attractive [5]. The use of mechanical foam control methods has been proposed as an alternative way to manage foam formation in reactors. Most of these studies have focused on mechanical foam breakers [2-8], rather than controlling the formation of foam.

The use of agitators to control foaming began over a decade ago. Notable studies involving the use of various impellers in foam control include Hoeks et al. [9], Boon et al. [10]. These studies showed that the hydrodynamics and the gas-liquid dispersion controlled by the type, size, rotational speed and location of foam control devices affect foaming. However, there is lack of understanding of how the stirred tank design and operating conditions affect the mechanisms of foaming. This present work aims to investigate the effect of mixing tank design and operating conditions on sparger-generated foaming. This is driven by the need to develop enhanced mechanical foam control technique.

Material and Methods

The experiment was carried out in a dished-bottomed cylindrical Perspex tank with internal diameter, T = 0.19 m and height, L = 0.56 m. it was equipped with four transparent, equally spaced baffles of T/10 width. A ring sparger with a diameter T/3 was mounted 0.03 m from the bottom of the tank. The tank was stirred with a 3-blade Mixel TT hydrofoil impeller with a diameter D = T/2 operating in up-pumping and down-pumping mode. The impeller off-bottom clearance, C, was fixed at 0.33T. A variable speed, 900 W Lightnin motor (Model: L5U10F), with a maximum rotational speed of 550 rpm was used to drive the impeller shaft, while gas flow rate was measured using Sho-Ratë rotameter.

The tank was filled with tap water at 22°C to a height H = T (corresponding to 5 L) and 0.05 g/L of polyoxyethylenesorbitan monolaurate (Tween 20) was used as the surfactant. Compressed air was used as the gas phase.

The height of foam formed in the tank was measured by the method of Yuan et al. [11]. For each experimental run, the gas was introduced into the liquid solution at two different flowrates: 6 L/min and 10L/min. For different impeller rotational speeds (ranging from 0 rpm to 500 rpm), the foaming process was monitored and measured with a stop watch while the foam height was measured with a measuring tape (at four positions, mid-way between each baffles) and the average value was taken.

https://doi.org/10.52202/074122-0053

Reproductivity of the experimental data was checked by performing the experiments in duplicate and the average value was recorded. It should be noted that foam formed after each impeller speed was allowed to completely disappear before subsequent experiments were performed. The tank solution was changed after completing a set of experiments for each gas flowrate.

.

Results and Discussion

Determination of the time for stable foam height

Before the principal investigations, the kinetics of the foam formation were studied in order to determine the time at which maximum foam height was achieved at different flow conditions. The results are presented in Figure 1 as a function of gas flowrates and impeller pumping mode.

As expected, foam height increases with time before reaching a maximum value for the range of operating conditions tested. However, for the Mixel TT down-pumping mode (MTTD), it was observed that maximum foam height was achieved at approximately 10 minutes before gradually decreasing. This is because the air content in the foam increases with time before reaching a maximum level. Further increase in time leads to bursting of the foam bubbles due to coalescence, resulting in a decrease in the observed foam height.

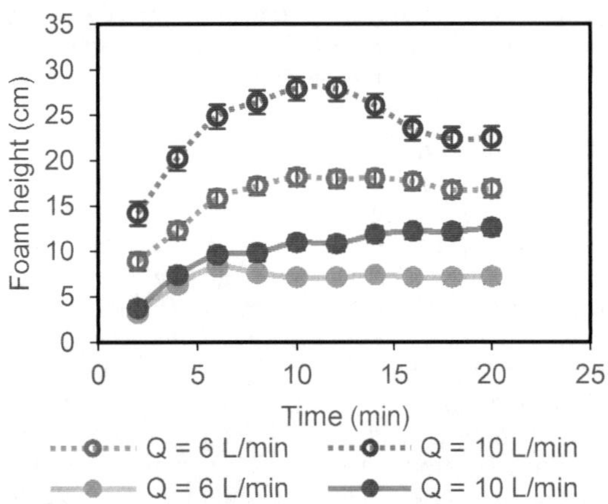

Figure 1. Foam height against time at different gas flow rates and impeller tip speed of 2.49 m/s for MTTD (hollow symbol) and MTTU (filled symbol).

In view of these findings from the preliminary analysis, it is safe to assume that further increases in time beyond 10 minutes will not result in a significant increase in the foam height. Therefore, for the baseline study, foam height was measured 10 minutes after mixing commenced to ensure a stable height was definitely reached.

Effect of operating conditions on foam formation

The effect of impeller tip speed and gas flowrate was investigated to explain their contribution on the foam formed in the stirred tank, and the results are presented in Figure 2. From the figure, it can be seen that foam height increases with an increase in impeller tip speed. This is because at higher impeller speed, there is higher turbulence energy dissipation, which favours bubble breakup and increases the mixing of the dispersed gas and continuous liquid phase. More importantly, the rate of increase of foam height is not uniform at all impeller tip speeds. At lower tip speeds (particularly below 1.5 m/s), little or no increase in foam height was observed. This trend, which was also observed by Aubin et al. [12], can be ascribed to the inability of the impeller to disperse the gas in the liquid at these impeller tip speeds. In most cases, the flow is either in the flooding regime or loading regime. However, at impeller tip speeds above 1.5 m/s, a dispersion regime or fully dispersion regime occurs, allowing the gas to be properly dispersed in the liquid. Furthermore, it was observed from Figure 2 that for a given impeller tip speed, foam height increased with increasing gas flowrate. This is expected because as the gas flowrate increases, more gas interacts with the liquid inside the tank, resulting in an increase in foam height [12].

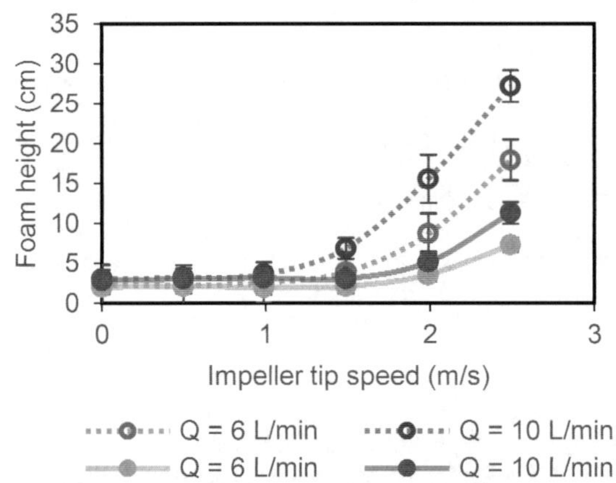

Figure 2. Foam height against impeller tip speed at different gas flow rates for MTTD (hollow symbol) and MTTU (filled symbol).

Comparing the operation mode of Mixel TT impeller at the same operating condition, it was observed that the up-pumping mode (MTTU) gives a lower foam height compared with the down-pumping mode (MTTD). This may be explained by the fact that the turbulence energy dissipation on the fluid is higher with MTTD [12]. This causes smaller and more bubbles, which are more likely to create more foam. The liquid flow patterns generated by up-pumping axial impellers are also significantly different from those of down-pumping impellers. This also certainly plays an important role in the gas dispersion and circulation process, the flow regimes developed and consequently the amount of foam formed. This complex interaction will need to be developed in further studies.

Conclusions

https://doi.org/10.52202/074122-0053

In this study, the impact of stirred tank design and operating conditions on sparger-generated foaming were investigated in a 0.19 m ID tank equipped with Mixel TT axial flow impeller (operating in down-pumping and up-pumping mode). Analysis of the results showed that foam height increases with time until a certain maximum height where coalescence begins. The results also showed that foam height increased with increasing impeller tip speed and gas flowrate, with lower foam height obtained by the up-pumping impeller compared with the down-pumping impeller.

The quantity of foam and the type of foam formed appears to be dependent on a complex combination of gas flow rate, impeller speed, turbulence, gas-liquid dispersion regimes and physical properties of the fluid and understanding the mechanisms still requires significant research. Immediate further work will be dedicated to exploring the impact of geometrical parameters of the impeller (e.g. off-bottom clearance) and the type impeller (radial flow versus axial flow) and to try to link gas-liquid flow regime and turbulence energy dissipation to the amount and type of foam generated.

Acknowledgments
The authors wish to appreciate the financial support of The Nigerian Petroleum Technology Development Fund (PTDF) and Syngenta Agrochemical Company.

References
[1] J. Kouko, B. Prakash, V. Luukkainen, A. Jäsberg, A.I. Koponen, *Chemical Engineering Research and Design*, 167 (2021) 15.
[2] S. Takesono, M. Onodera, A. Ito, M. Yoshida, K. Yamagiwa, A. Ohkawa, *Chemical Technology and Biotechnology,* 36 (2001) 355.
[3] S. Takesono, M. Onodera, M. Yoshida, K. Yamagiwa, A. Ohkawa, *Chemical Engineering Journal of Japan,* 37 (2004) 1488.
[4] A. Ohkawa, K. Sugiyama, N. Sakai, H. Imai, K. Endoh, *Canadian Journal of Chemical Engineering*, 62 (1984) 507.
[5] S. Takesono, M. Onodera, A. Ito, M. Yoshida, K. Yamagiwa, A. Ohkawa, *Chemical Technology and Biotechnology,* 78 (2002) 48.
[6] M. Goldberg, E. Rubin, *Industrial Engineering Chemical Process Design and Development,* 6 (1967) 195.
[7] S. Gutwald, A. Mersmann, *Chemical Engineering and Technology* 20 (1997) 76.
[8] N.S. Deshpande, M. Barigou, *Journal of Chemical Technology and Biotechnology,* 74 (1999) 979
[9] F.W.J.M.M. Hoeks, L.A. Boon, F. Studer, M.O. Wolff, F.V. Schot, P. Vrabel, G.J. Lans, W. Bujalski, A. Manelius, G. Blomsten, S. Hjorth, G. Prada, K. Luyben, A.W. Nienow, *Industrial Microbiology and Biotechnology,* 30 (2003) 118.
[10] L.A. Boon, F.W.J.M.M. Hoeks, G.J. Lans, W. Bujalski, A. M.O. Wolff, A.W. Nienow, *Biochemistry and Engineering Journal,* 10 (2002) 183.
[11] J. Yuan, Y. Li, X. Chen, Y. Zhou, *Advanced Materials Research,* 1104 (2015) 33.
[12] J. Aubin, D. Daniel A. Deu, C. Raffel, W. Shaikh, E. Rundquist, P.M. Piccione, *16th European Conference on Mixing, Toulouse,* France, 2018.

 https://doi.org/10.52202/074122-0053

Gas-to-Liquid Heat Transfer in a Vessel Agitated by a Rushton Turbine

Poster.8

R. Šulc, J. Sieratovský. *Czech Technical University in Prague, Faculty of Mechanical Engineering, Department of Process Engineering, Technicka 4, Prague, Czech Republic, Radek.Sulc@fs.cvut.cz.*

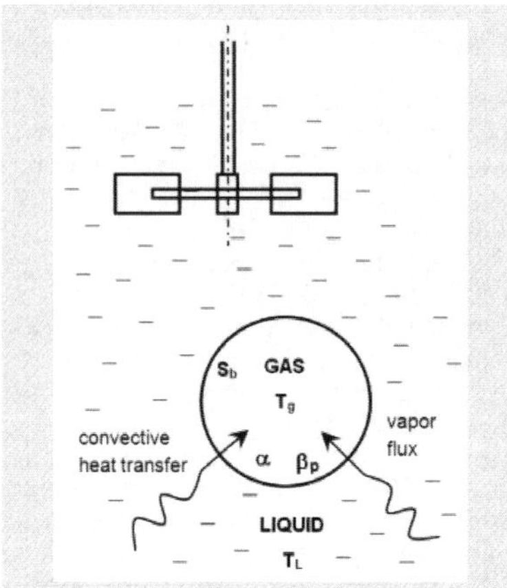

Multiphase contactors, such as bubble columns or agitated gas-liquid reactors, are often used in operations accompanied by heat transfer between the phases. Due to difficultness of gas-to-liquid heat transfer measurement, the correlations for gas-to-liquid heat transfer are not published.

The applicability of the non-steady method for the determination of gas-to-liquid heat transfer developed by Žižka et al. [1] originally for a bubbled column was successfully tested for the determination of gas-to-liquid heat transfer in agitated gas-liquid reactors.

The study was performed on the vessel 0.15 m in diameter agitated by a Rushton turbine. The experiment was carried out for impeller rotational speed N = 360 rev min^{-1} and superficial gas velocity = 0.78 cm s^{-1}.

Introduction

Multiphase contactors, such as bubble columns or agitated gas-liquid reactors, are often used in operations accompanied by heat transfer between the phases. Mostly wall-to-mixture and heat-exchange element-to-mixture heat transfer coefficients have been intensively studied and published. Due to difficultness of gas-to-liquid heat transfer measurement, the correlations for gas-to-liquid heat transfer are not published. Žizka et al. [1] developed a non-steady state method based on measurements of the gas- and liquid- temperature in the time for an evaluation of the heat fluxes and heat transfer coefficients between gas and liquid in a bubbled column.

This contribution aims to test the applicability of the method developed by Žižka et al. [1] originally for a bubbled column for the determination of gas-to-liquid heat transfer in agitated gas-liquid reactors. This study was performed on the vessel 0.15 m in diameter agitated by a Rushton turbine. The experiment was carried out for impeller rotational speed N = 360 rev min^{-1} and superficial gas velocity = 0.78 cm s^{-1}.

Theoretical background

Assuming ideal mixing of the liquid batch and neglecting heat losses to surroundings the enthalpy balance can be written in the case of a liquid being cooled by a gas, i.e. $T_L > T_{gi}$, as follows:

$$m_L \cdot c_{pL} \cdot \frac{dT_L}{dt} = -\dot{m}_{vap} \cdot \Delta h^{vap} - \alpha \cdot S_b \cdot (T_L - T_g) \quad (1)$$

where dT_L/dt is a derivative of liquid temperature T_L with respect to time t (K s^{-1}), m_L is the weight of the liquid batch (kg), c_{pL} is the specific heat capacity of the liquid (J kg^{-1} K^{-1}), α is the gas-to-liquid heat transfer coefficient (W m^{-2} K^{-1}), S_b is the gas bubble interfacial area (m^2), T_L is the liquid temperature (K), T_g is the gas temperature (K), $\dot{m}_{vap}$ is the mass flow of the evaporated liquid (kg s^{-1}) and Δh^{vap} is the heat of vaporization (J kg^{-1}). Žižka et al. [1] reported that directly solving the differential equation (1) was not satisfactory and that an integral formula should therefore be used. They expressed time dependences of $\dot{m}_{vap}$ and T_g indirectly as a 2nd order polynomial functions of instantaneous liquid temperature T_L (t):

$$T_g = a_2 \cdot T_L^2 + a_1 \cdot T_L + a_0 \quad (2)$$

$$\dot{m}_{vap} \cdot \Delta h^{vap} = b_2 \cdot T_L^2 + b_1 \cdot T_L + b_0 \quad (3)$$

These assumptions enable them to integrate Eq. (1) and to obtain for $\Delta > 0$ a solution as follows:

$$\left[\frac{2}{\sqrt{\Delta}} arctg\left(\frac{2 \cdot C_2 \cdot T_L + C_1}{\sqrt{\Delta}} \right) \right]_{TL0}^{TL} = \frac{t}{m_L \cdot c_{pL}} \quad (4)$$

where

$$C_2 = \alpha \cdot S_b \cdot a_2 - b_2 \quad (5a)$$

$$C_1 = \alpha \cdot S_b \cdot a_1 - \alpha \cdot S_b - b_1 \quad (5b)$$

$$C_0 = \alpha \cdot S_b \cdot a_0 - b_0 \quad (5c)$$

$$\Delta = 4 \cdot C_2 \cdot C_0 - C_1^2 \quad (5d)$$

where T_{L0} is the initial temperature of the liquid batch (K) and T_L is the temperature of the liquid batch (K) at time t (s). The changing gas temperature profile along the height of the liquid was respected by a log-mean temperature difference between a liquid batch and a passing gas. The mass flow of the evaporated liquid (kg s^{-1}) was estimated using the mass transfer equation:

$$\dot{m}_{vap} / M_{H2O} = \beta_p \cdot S_b \cdot \Delta p_{\ln - H2O} \quad (6)$$

where β_p is the pressure-based mass transfer coefficient (kmol s^{-1} m^{-2} kPa^{-1}), Δp_{ln-H2O} is the log-mean partial pressure difference between a liquid batch and a passing gas (kPa), and M_{H2O} is the molecular weight of water (kg kmol^{-1}).

The volumetric gas-to-liquid heat and mass transfer coefficients $\alpha \cdot a$ (W m^{-3} K^{-1}) and $\beta_p \cdot a$ (W m^{-3} K^{-1}) are calculated as follows:

$$\alpha \cdot a = \alpha \cdot S_b / V_{L0} \quad (7)$$

$$\beta_p \cdot a = \beta_p \cdot S_b / V_{L0} \quad (8)$$

where V_{L0} is the equipment/bed volume that corresponds to the volume of a non-gassed liquid batch (m^3).

Material and Methods

The experiments were carried out in a laboratory bioreactor of 150 mm in inner diameter agitated by a Rushton turbine. The geometrical parameters of the experimental apparatus are summarized in Table 1. The tap water was used as a model liquid and the air was fed into the reactor via a single nozzle. The scheme of the experimental apparatus is shown in Figure 1. Experimental parameters are summarized in Table 2.

Table 1. Geometrical parameters

Parameter	Symbol	Value
Tank diameter (internal)	T_i (mm)	150
Impeller diameter	D (mm)	50
Impeller bottom clearance	C (mm)	50
Initial liquid level	H_{L0} (mm)	225
Baffle width	W_b (mm)	15

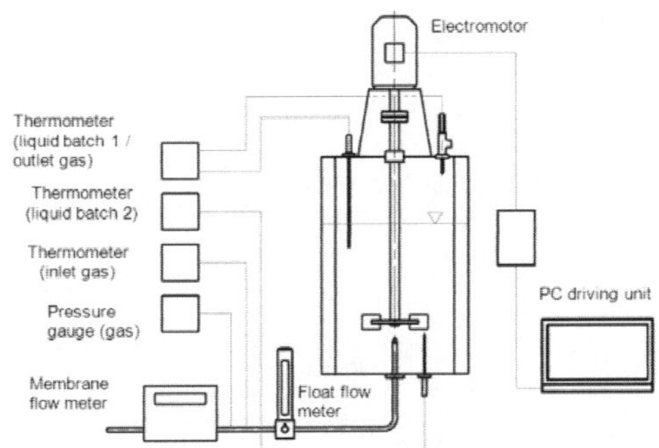

Figure 1. Scheme of experimental apparatus.

Table 2. Experimental parameters

Parameter	Symbol	Value
Impeller rotational speed	N (min^{-1})	360
Gas superficial velocity	u_{g0} (m s^{-1})	0.0078
Inlet gas temperature	T_{gin} (°C)	21.6
Initial liquid temperature	T_{L0} (°C)	43.15
Final liquid temperature	T_{Lend} (°C)	37.6
Liquid cooling time	t_{exp} (min)	30

Results and Discussion

The time courses of gas temperature measured at the inlet ($T_{g\ in}$) and the outlet ($T_{g\ out}$) and liquid temperature measured below the impeller ($T_{L\ down}$) and in the upper part of the liquid batch ($T_{L\ up}$) are presented in Figure 2. Practically the same liquid temperatures measured in the upper and lower part of the liquid batch confirmed the assumption of ideal liquid mixing. The temperature differences were found to be only from 0.12 to 0.52 % of the average value. The average of liquid temperatures was used for calculation. The other two assumptions given by Equations (2) and (3) were also fulfilled as demonstrated in Figures 3 and 4, respectively. The R^2 values were found to be equal to 0.94 and 0.987 for mean gas temperature and evaporative heat flux, respectively. Finally, the Δ value was found to be positive for a given run. Thermo-physical properties of water were calculated using X-steam software [2].

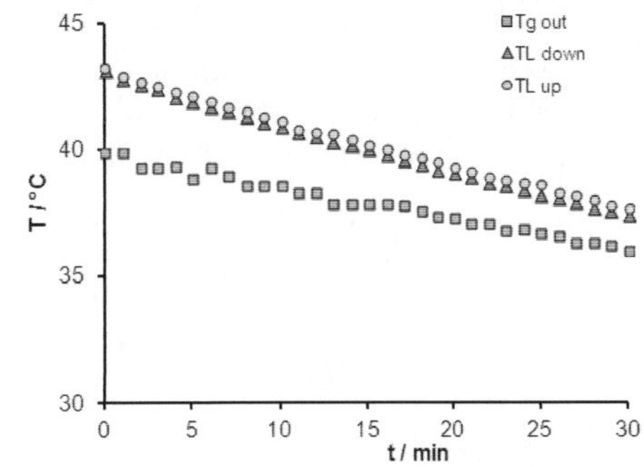

Figure 2. Time course of gas and liquid temperature.

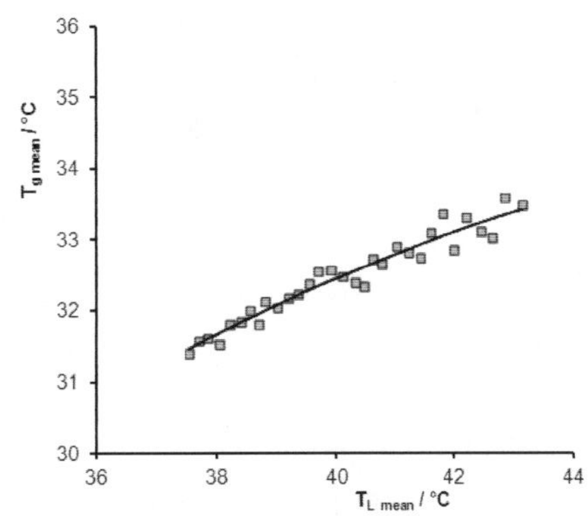

Figure 3. Dependence of gas temperature $T_{g\ mean}$ on liquid temperature.

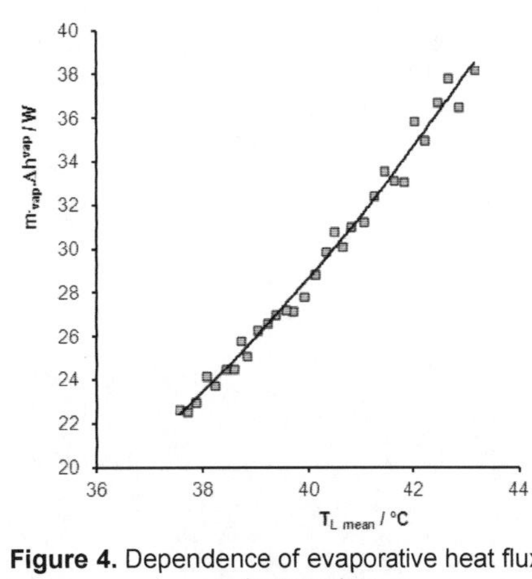

Figure 4. Dependence of evaporative heat flux on liquid temperature.

Using the iterative calculation procedure, the following transfer coefficients were evaluated:

$\beta_p \cdot S_b = 2.272 \cdot 10^{-7}$ kmol kPa^{-1} s^{-1},

https://doi.org/10.52202/074122-0054

$\alpha \cdot S_b = 2.734$ W K^{-1},

and finally volumetric coefficients:

$\beta_p \cdot a = 0.574$ mol kPa^{-1} s^{-1} m^{-3},

$\alpha \cdot a = 691$ W K^{-1} m^{-3}.

Conclusions

The applicability of the non-steady method for the determination of gas-to-liquid heat transfer developed by Žižka et al. [1] originally for a bubbled column was successfully tested for the determination of gas-to-liquid heat transfer in agitated gas-liquid reactors.

The study was performed on the vessel 0.15 m in diameter agitated by a Rushton turbine. The experiment was carried out for impeller rotational speed N = 360 rev min^{-1} and superficial gas velocity = 0.78 cm s^{-1}.

Acknowledgments

This work was supported by the Ministry of Education, Youth and Sports of the Czech Republic under OP RDE grant number CZ.02.1.01/0.0/0.0/16_019/0000753 "Research centre for low-carbon energy technologies".

References

[1] M. Žižka, R. Šulc, P. Ditl, *Chemical Engineering Transactions*, 57 (2017) 1261.

[2] M. Holmgren, *X-steam tables according to IAPWS IF-97; v 2.6.*

Pickering Emulsions with a Nanoclay Using High Intensity Mixing

Poster.9

A. Utomo[1,2], G. A. Padron[1], N. J. Alderman[3], ***N. G. Özcan-Taşkın[4]****. (1) Framatome Limited Cranfield, Bedfordshire MK43 0AJ, UK, (2) Now Independent, (3) Now retired, (4) Now at Loughborough University School of AeroAuto, Chemical and Materials Eng. Department of Chemical Engineering Loughborough LE11 3RQ UK, N.Ozcan-Taskin@lboro.ac.uk*

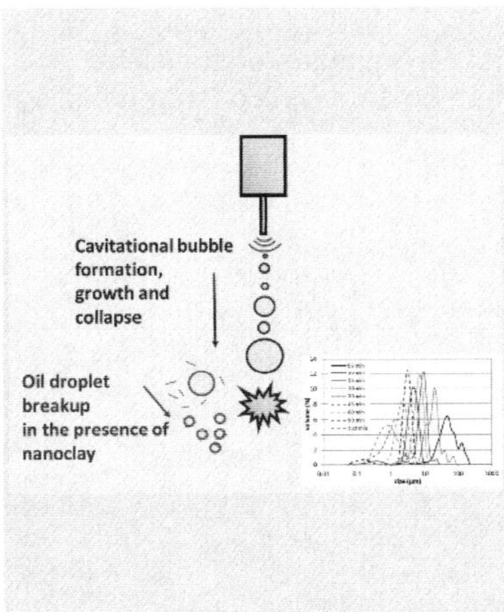

The study explored the potential of using a nanoclay, Laponite RD, as a stabilising agent in comparison to a surfactant, Tergitol TMN6, at dispersed phase volume fraction of 10 to 50%. An energy intensive device, an ultrasonicator was employed covering a specific power input range of ~ 36- 52 W/kg. In both cases, final dispersions obtained could be considered stable over the period of 3 months ageing tests as comparable DSDs were obtained although there were some differences in the rheology at the highest dispersed phase concentration. Nanoclay stabilised emulsions exhibited non-Newtonian behaviour which may be a desirable feature of the final product- certainly one that requires attention in design and scale up. Overall, drop sizes were larger with Pickering emulsions which depended on the dispersed phase concentration whereas with emulsions stabilised with a surfactant drop sizes as low as 240 nm were obtained regardless of the dispersed phase volume fraction. The study also provided insight in terms of the kinetics of breakup in Pickering emulsions.

Introduction

Numerous products in daily use, varying from pharmaceuticals, food to cosmetics and household products, are in the form of an emulsion or nanoemulsion, i.e. with drop sizes in the range of a few to a few hundred nanometers. Nanoemulsions have been of particular interest due to their properties such as transparency and flow behaviour and are produced using an energy intensive process. Conventional stabilisers, surfactants, are required in large quantities to stabilise the interface and due to the concerns relating to their impact on health and the environment, Pickering emulsions, with colloidal particles anchored at the interface have been considered as a good alternative. Pickering emulsions, first reported in the early 1900's [1,2], rely on the principle that solid particles adsorbed at the interface between the dispersed and continuous phases create a kinetic barrier to coalescence owing to the high energy required for the desorption of these particles from the interface. In addition, recent advances in the synthesis of particles with specific properties have paved the path to employ particles which not only stabilise the dispersion but also introduce desired functionalities such as oxidation resistance, UV protection, thereby enhancing final product performance [3]. Pickering emulsions have therefore received increasing attention in the past decade with clear indications of advantages in numerous products varying from active ingredient encapsulation in the pharmaceutical industry, antimicrobial products to food packaging, 3D printing inks and enhanced oil recovery [3,4].

Most previous studies focussed on the specifics of formulations. One of the motivations of this study was to monitor the evolution of the dispersion process. Among inorganic particles used for the purpose, silica nanoparticles have been widely used alone or as a blend with surfactants [3] and [4]. There is relatively less work performed with other solids, such as nanoclays, whilst the potential for applications in different fields is highly promising [5]. This study has been carried out with the objectives of establishing the performance of an energy intensive device, an ultrasonic processor, in generating a fine dispersion and the potential of using a nanoclay as a stabiliser at different dispersed phase volume fractions. The comparative study also included the use of a non-ionic surfactant, importance of dispersion protocol and ageing tests to determine long-term stability of the dispersions.

Materials and Methods

Dispersions of 10 cSt. silicone oil (Xiameter PMX 200 Silicone Fluid 10 cSt equivalent to Dow Corning 200 Fluid 10 cSt) in distilled water were prepared at concentrations of 10, 20 and 50 vol.%. Two stabilisers used were Tergitol TMN6 (Sigma-Aldrich) or Laponite RD (Rockwood, now Altana BYK).
The concentration of Tergitol was varied between 10 and 50 times CMC, depending on the dispersed phase con. Tergitol TMN6 is a branched secondary alcohol ethoxylate with a molecular weight of 543 g/mol. This non-ionic surfactant is soluble in water and has been used widely in paper, textile, paint, coating and agrochemicals industries. The hydrophilic- lipophilic balance (HLB) value is 13.1 The critical micelle concentration (CMC) in silicone oil/water systems is 2.1×10^{-3} mol/L.

Laponite RD is a synthetic smectite clay of high purity, disc shaped with a diameter of around 25 nm and thickness of approximately 0.95 nm and forms clear dispersions in water. A previous study [6] made use of Laponite RD dispersions prepared using an ultrasonicator and the emulsification was with a different device, a batch rotor-stator, operated for

https://doi.org/10.52202/074122-0055

only 2 minutes. Whilst the protocol was different to ours, the overall recommendation relating to formulation was taken into account and as the emulsions were found to be unstable at low Laponite RD concentrations of less than 0.5 wt.% and high concentrations of greater than 4.5 wt.%, a concentration 0.9 wt.% was used. Salts increase the negative charge density of particles and hence the inter-particle repulsion [6], NaCl was added in these dispersions at a concentration of 0.01 M.

Different dispersion protocols tested at 10 vol%, i.e. by making additions either simultaneously or sequentially, did not have a marked effect on DSD; related results will be presented at the conference.

Dispersions were prepared using an ultrasonic processor, Hielscher UP200S, which has a frequency of 24 kHz and a maximum amplitude of 125 µm. The S14D titanium sonotrode (of 14 mm diameter), operated at an amplitude of 100% was immersed in a jacketed cylindrical vessel containing a mixture of about 2500 ml. A pitched blade turbine (PBT), mounted off-centre was operated at 250 rpm to ensure homogeneity within the bulk.

The maximum power input of the ultrasonic disperser in water, determined using the calorimetry method was 90 W (~ 36 W/kg) and the power input of the PBT was only 0.15 W. At higher viscosities of 10 and 100 cP, the power input by the ultrasonic disperser was 124 and 130 W, respectively. The emulsion temperature was maintained at 20 °C by using a chiller.

Drop size distribution (DSD) was determined at ambient temperature using a Beckman Coulter LS230 particle size analyser. Specific details of the set up and measurement protocol will be presented at the conference.

Dispersion rheology was studied using an Anton-Paar Rheolab QC rheometer (Anton Paar, Hertford, UK) with the coaxial cylinder geometry.

Results and Discussion

Dispersion rheology

Dispersions with Tergitol were Newtonian. Shear stress values determined in 10 and 20% dispersions were too low to be measured accurately and hence they can be considered to be of the same viscosity

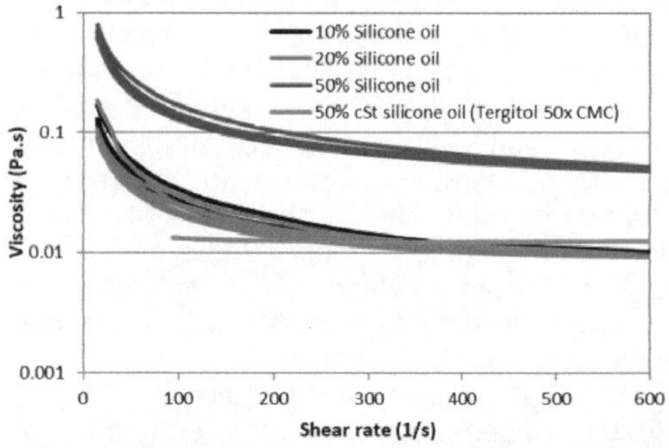

Figure 1. Rheology of dispersions stabilised with Laponite in comparison to that 50 vol.% stabilised with Tergitol.

as the continuous phase. 50% dispersion had a viscosity of 12.5 cP. Dispersions stabilised using Laponite on the other hand had a yield stress value and could be described with the Herschel Bulkley model. Figure 1 shows a comparison; further details on rheology will be presented at the conference.

Evolution of drop breakup

The DSD of the dispersions stabilised with Laponite RD appeared to move to the left gradually as shown in the example for 20 vol% in Figure 2. Those stabilised with Tergitol showed a rapid breakup in the first few minutes of sonication and multiple peaks could be observed prior to reaching an equilibrium size.

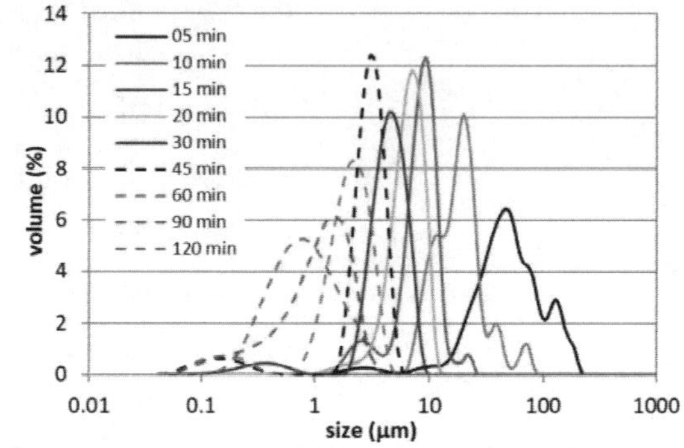

Figure 2. Evolution of DSD with 20 vol% dispersion stabilised with Laponite

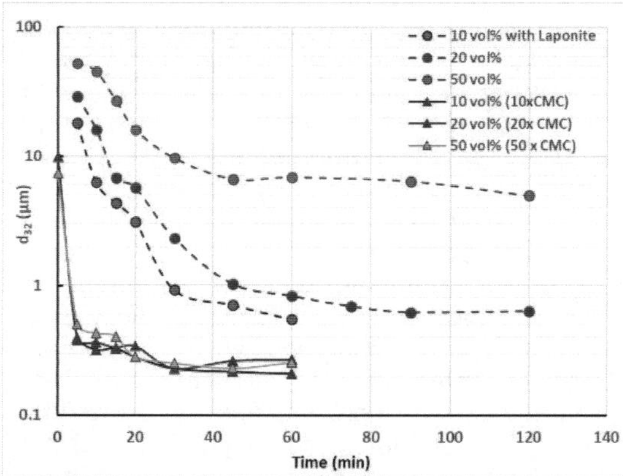

Figure 3. Evolution of d_{32} at different volume fractions for dispersions stabilised with Tergitol and Laponite

An important difference relates to the effect of the dispersed phase volume fraction: with surfactant stabilised dispersions Sauter mean diameters (d_{32}) were around 240 nm regardless of the dispersed phase concentration (Figure 3) whereas for nanoclay based emulsions an effect was noted which was particularly noticeable with 50 vol.% dispersion. This is consistent with previous findings with other formulations stabilised with a nanoclay [6]. The resulting mean sizes at 10 to 50 vol.% were about 0.55, 0.62 and 6.36 µm respectively and hence much larger compared to those stabilised with a surfactant.

Mean drop sizes determined prior to approaching an equilibrium value, normalised with initial mean values are presented in Figure 4. Further details on dispersion kinetics will be shown at the conference.

https://doi.org/10.52202/074122-0055

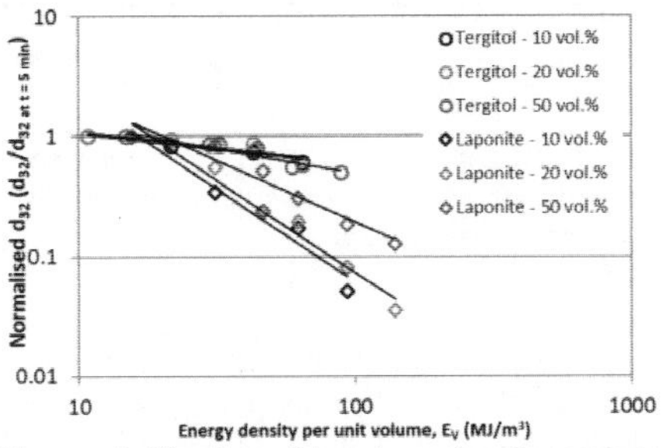

Figure 4. Kinetics of break up for dispersions shown in terms of normalised d_{32}

Dispersion stability

Ageing tests performed over 3 months have shown that the dispersions can be considered stable in this period. There were some differences in the rheology of dispersions in particular at 50 vol % (Figures 5a and b) but the DSDs were comparable (Figures 5c and d).

Conclusions

The comparative study has shown both types of emulsions, one stabilised with a surfactant, another with a nanoclay have long term stability, demonstrating the potential for such Pickering emulsions. Nanoclay stabilised emulsions exhibited non-Newtonian behaviour which may be a desirable feature of the final product- certainly one that requires attention in design and scale up. Overall, drop sizes were larger with Pickering emulsions which depended on the dispersed phase concentration whereas with emulsions stabilised using a surfactant drop sizes as low as 240 nm were obtained regardless of the dispersed phase volume fraction. The study also provided insight in terms of the kinetics of breakup in Pickering emulsions.

Acknowledgements

The authors gratefully acknowledge the support and discussions with the members of the industrial DOMINO consortium.

References

[1] W. Ramsden (1904) *Separation of solids in the surface-layers of solutions and 'suspensions'*, Proc. Roy. Soc. Lond. 72,156–164, https://doi.org/10.1098/rspl.1903.0034

[2] S.U. Pickering (1907) *Emulsions.* Journal of Chemical Society, 91, 2001-2021. https://doi.org/10.1039/CT9079102001

[3] H. Jiang, Y. Sheng, T. Ngai (2020) *Pickering emulsions: Versatility of colloid particles and recent applications.* Current Opinion in Colloid & Interface Science 2020, 49:1–15. Themed issue on Emulsions and Microemulsions. Edited by C. Rodriguez-Abreu and K. Aramaki https://doi.org/10.1016/j.cocis.2020.04.010

[4] G. Gauthier, I. Capron (2021) *Pickering nanoemulsions: An overview of manufacturing processes, formulations and applications,* JCIS https://doi.org/10.1016/j.jciso.2021.100036

[5] T. Lu, H. Gou, H. Rao, G. Zhao (2021) *Recent Progress in nanoclay-based Pickering emulsion and applications* J of Env. Chem Eng., 9, 5, https://doi.org/10.1016/j.jece.2021.105941

[6] N. P. Ashby, B. P. Bicks (2000) *Pickering emulsions stabilised by Laponite clay particles* Phys. Chem. Chem. Phys. https://doi.or/10.1039/b007098j

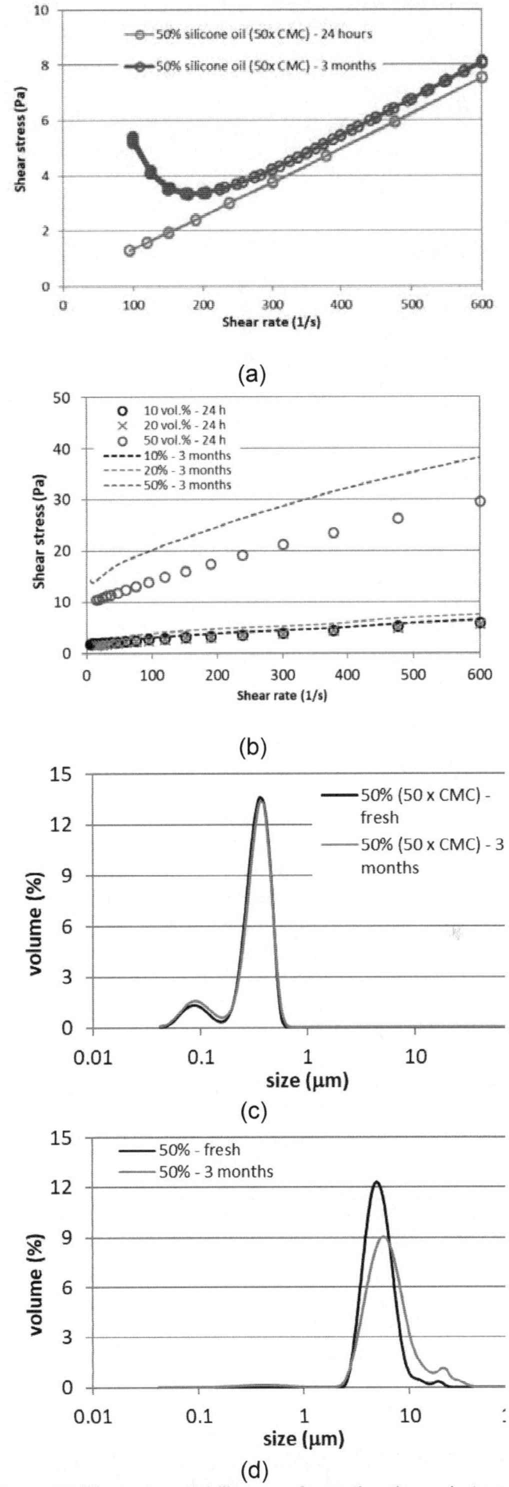

Figure 5. Dispersion stability over 3 months shown in terms of rheology stabilised with (a) Tergitol and (b) Laponite, and DSD using (c) Tergitol and (d) Laponite

https://doi.org/10.52202/074122-0055

Production of Pickering emulsions using a Vibromixer

Poster.10

S. Almasi[1,2], *B. Ghobadian*[2], *M. Poux*[1], *J. Aubin*[1]. *(1) Laboratoire de Génie Chimique, Université de Toulouse, CNRS, INPT, France; sara.almasi@toulouse-inp.fr; joelle.aubin@toulouse-inp.fr ; martine.poux@toulouse-inp.fr; (2) Department of Mechanical and Biosystems Engineering, Tarbiat Modares University, Tehran, Iran ; sara.almasi@modares.ac.ir; ghobadib@modares.ac.ir.*

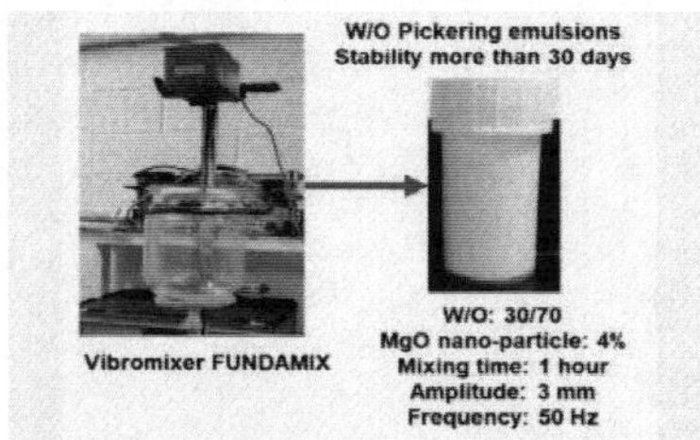

A vibromixer (Fundamix) has been used to produce Pickering emulsions in a 1L vessel. The effect of the mixing time and the concentration of MgO nanoparticles on the stability of the emulsion were studied. The obtained results showed that an increase in MgO nanoparticle concentration (1%wt. to 4%wt.) and mixing time (10 min to 60 min) increased the stability of the emulsions. Interesting results are that for a mixing time of 1 hr and MgO concentration of 4%wt., the Pickering emulsion is stable for at least 6 weeks.

Introduction

Vibromixers are interesting alternative mixing devices to conventional rotating impellers for a number of applications – both single and multiphase – in different industries, e.g. chemical and petrochemical, food and pharmaceutical [1, 2]. Vibromixers have shown to provide effective mixing for liquid reactions and shear-sensitive systems, as well as liquid-liquid dispersion and solid-liquid suspensions [3]. They have shown to ensure equivalent mixing intensity to stirred tanks, but with significantly lower shear rates and up to 40% less energy consumption [4].

In this work, the performance of the Fundamix vibromixer is investigated for a multiphase application: the production of Pickering emulsions. Nowadays, Pickering emulsions (oil-water dispersions stabilized by nanoparticles) have found a special place in industry due to their higher stability than emulsions stabilized by traditional surfactants.

The adsorption of solid particles at the oil-water interface is strong and irreversible. This enables creation of a barrier around the droplets, giving them a strong resistance to coalescence. In this research, the use of a vibromixer to produce a water-in-oil Pickering emulsion stabilized by MgO nanoparticles is explored. The impact of mixing time and nanoparticle concentration on the stability of the emulsion are investigated for different operating conditions: MgO nanoparticle concentration from 1%wt. to 4%wt. and mixing times from 10 min to 60 min.

Material and Methods

1. The FUNDAMIX vibromixer – principle

The principle of the FUNDAMIX vibromixer is based on a vibrating plate immersed in a fluid. The vibration is created and transmitted through a shaft to the horizontal plate (diameter 45 mm) with 20 conical holes that have a base diameter of 10 mm and top diameter of 5 mm, as shown in Fig. 1. As the plate oscillates in the vessel, it pumps the fluid and creates flow. The plate can be installed in the reactor in a type-A or type-B configuration, resulting in an upwards or downwards flow, respectively (Fig. 2). The type-A configuration (Fig. 2a) is recommended for immiscible liquid-liquid mixing, whilst the most common application for the type-B configuration (Fig. 2b) is the suspension of solids. For this purpose, cones are pointing downwards and the liquid is forced towards the vessel bottom, raising the particles. In this work, due to the presence of MgO nanoparticles, the type-B configuration is used to create the Pickering emulsion.

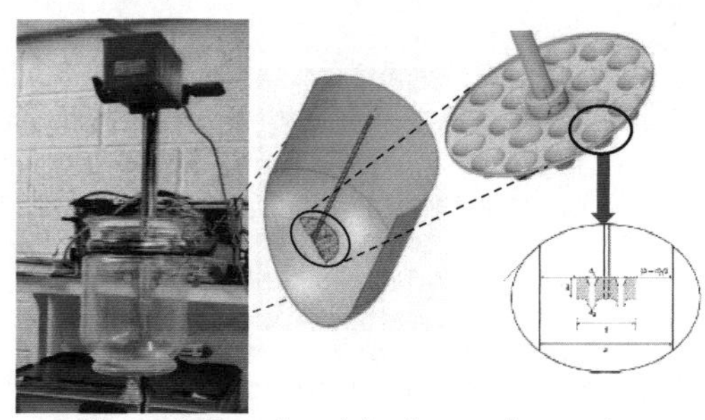

Figure 1. Experimental set-up and procedure.

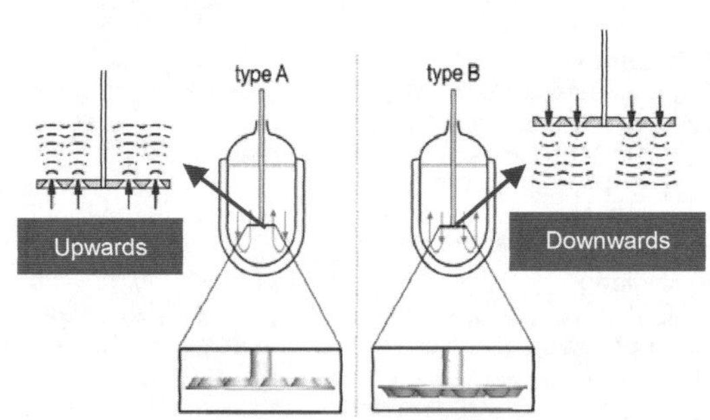

(a) (b)

Figure 2. Plates installed in vessel in type A and type B configuration.

2. Formulation of Pickering Emulsions

Commercial sunflower oil and distilled water were used with a composition of 70 % v/v and 30 % v/v, respectively. All experiments were performed at ambient temperature of 20°C in a 1L curved-bottom cylindrical vessel (diameter 100 mm) and a vibromixer. The frequency of

https://doi.org/10.52202/074122-0056

vibrations was fixed at f = 50 Hz and the vibration amplitude at 3 mm, which was chosen to provide a good suspension of the nanoparticles. For lower values of the amplitude, particles were not entirely suspended and remained at the bottom of the vessel.

The stabiliser used was MgO nanoparticles (Sigma-Aldrich) with particle size of 19 nm and density of 3.58 (g/cm^3). MgO nanoparticles were firstly progressively added to the sunflower oil and this pre-mixture was subjected to mixing for 3 minutes by a magnetic stirrer. Water was then gradually added to the previously prepared suspension and mixed a few minutes. The final three-phase mixture is then poured into the vessel and homogenized by means of the vibromixer.

The Pickering emulsions were prepared for various MgO concentrations and for different mixing times as shown in Table 1.

Table 1. Different proportions of formulations.

Variable	Unit	Value
Water/Sunflower oil	v/v	30/70
Time	Min	10-60
Concentration of MgO	%	1-4

3. Characterization of the Pickering emulsion

3.1. Drop test

The drop test was used to verify the emulsion type produced (O/W or W/O). For this a Pickering emulsion drop is added to a dish of oil and to a dish of water. If the Pickering emulsion drop quickly disperses in the dish of water but remains as drop in a dish of oil, the emulsion is of the oil-in-water type. If the reverse happens, it confirms the water-in-oil type.

3.2. Visual inspection

The emulsions were visually examined after putting them in transparent vials with volume of 40 ml and stored at constant temperature of 20°C just after mixing was stopped; photos were taken for different storage times (every day). If the system is not stable, the ingredients (sunflower, water and MgO nanoparticles) separate according to their different density. The sunflower oil starts to float on top of Pickering emulsion because it is lighter than water and MgO nanoparticles, and the MgO nanoparticles start to settle at the bottom. It is very easy to observe the phase separation and to determine the interface because sunflower oil appears as a very clear liquid at the top of Pickering emulsion.

Results and Discussion

1. Drop test

As can be seen in Fig 3(a), the Pickering emulsion droplet completely dispersed in oil while it did not disperse in water (Fig 3(b)). Therefore, the type of Pickering emulsion is water-in-oil.

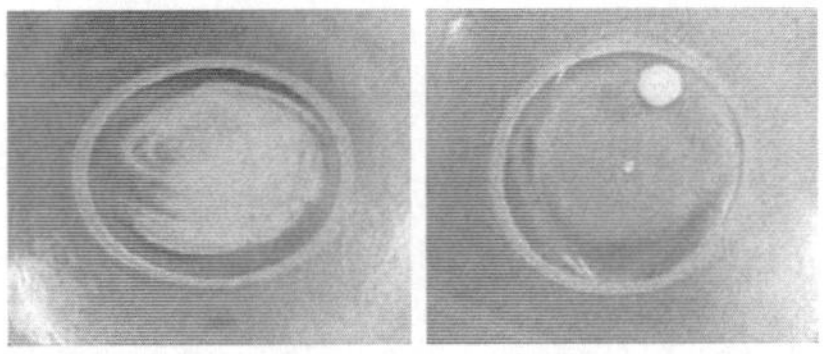

(a) **(b)**

Figure 3. Drop test shows the type of emulsion: drop of Pickering emulsion added to (a) pure sunflower oil and (b) water.

2. Visual inspection

2.1. Effect of MgO concentration

For a fixed mixing time of 1hr, an increase in MgO nanoparticle concentration (from 1%wt. to 4%wt.) increased the stability of the Pickering emulsion significantly. The Pickering emulsion is not very stable when the MgO nanoparticle concentration is low (1%wt.) and duration of stability duration is 4hr. When the MgO concentration was increased, the duration of stability also increased and was still stable, with no separation, after at least 6 weeks for a concentration of 4%wt. (Table 2). This enhanced stability is attributed to the adsorption of MgO nanoparticles at the W/O interface that then prevents droplet coalescence and stabilizes the emulsion [5]. Images of the Pickering emulsion using a MgO nanoparticle concentration of 4%wt. with total stability after one month can be seen in Fig. 4.

Table 2. Duration of stability of Pickering emulsion in MgO concentration (1%wt to 4 %wt) and fixed mixing time of 1 hr.

MgO concentration (%wt)	Duration of stability
1	4 hr
2	1 day
3	3 days
4	> 6 weeks

Day 1 **Day 30**

Figure 4. Photographs of emulsions with 4% wt MgO nanoparticle and 1 hour mixing time on days 1 and 30.

2.2. Effect of time

For a constant concentration of MgO nanoparticles (4% wt.), the stability of the emulsions varied with different mixing times, ranging from 10 min to 60 min. Fig. 5 shows images of the emulsions, which were produced with different mixing times, after 7 days storage. The stability of the emulsions increases when the mixing time increases. This is in agreement with

https://doi.org/10.52202/074122-0056

[6]; indeed, the MgO nanoparticles act as an emulsifier, which becomes more effective with increased mixing time and increases stability of Pickering emulsion. It is worth mentioning that the oil and water phases of the Pickering emulsions started to separate after 3 days storage for a mixing time of 10 min, after 4 days at mixing time of 20 min, after 5 days for a mixing time of 30 min, after 6 days for a mixing time of 40 min, and after 7 days at mixing time of 50 min. As can be seen in Fig. 5, with increasing mixing time, the volume of separated oil phase at the top decreased.

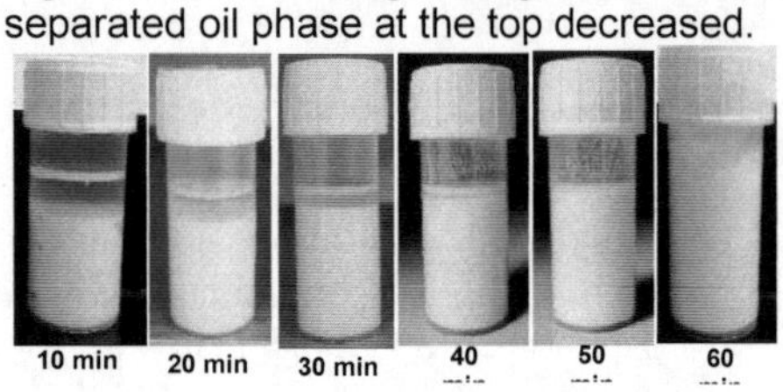

Figure 5. Evolution of the stability of emulsions with 4% wt MgO as a function of the mixing time after 7 days.

Also, for images in Fig. 5, the volume of the clear oil layer and the percentage of the stable emulsion with respect to the total sample volume after one week was determined (Table 3). With increasing mixing time, the volume of separated oil decreased and the volume of stable Pickering emulsion increased. For a mixing time of 60 min, no phase separation is observed. Therefore, under the conditions studied, 60 min mixing time may be considered as the minimal mixing time after which that the emulsions are stable.

Table 3. Stability of Pickering emulsion in fixed MgO concentration (4 %wt) and different mixing time.

Mixing time (min)	volume of oil layer in sample (cm^3)	Amount of stable emulsion in sample (%)
10	3.8	76
20	2.67	83
30	1.51	90
40	0.94	94
50	0.62	96
60	0	100

Future prospects

To complete this study, experiments under different operating including water/oil ratios (10/90, 20/80, 30/70, 40/60, and 50/50), concentration of MgO nanoparticles (1%, 2%, 3%, and 4%), diameter of plate (45 mm, 54 mm, and 65 mm) and mixing time (10 min, 20 min, 30 min, 40 min, 50 min, and 60 min) are yet to be carried out. Also, the Pickering emulsion has to be quantitively characterized via droplet size and droplet morphology measurements. In addition, these results of the Pickering emulsion obtained with the vibromixer will be compared with those prepared in a conventional stirred tank.

Conclusions

This work is the first systematic study of Pickering emulsions preparation by the FUNDAMIX vibromixer. Pickering emulsions stabilized with MgO nanoparticles were successfully produced. The influence of mixing time and the concentration of MgO nanoparticles during the preparation of the Pickering emulsion was investigated. The stability of Pickering emulsion increased with increasing mixing time and MgO concentration. For a mixing time of 1hr and MgO concentration of 4% wt, a Pickering emulsion, which has shown to be stable for at least 6 weeks, has been produced. The findings of this study are very promising and demonstrate the possibility to use the FUNDAMIX vibromixer to make Pickering emulsions as an alternative to a stirred tank.

References

[1] R. Wójtowicz, *Choice of an optimal agitated vessel for the drawdown of floating solids, Industrial & Engineering Chemistry Research 53(36) (2014) 13989-14001.*

[2] R. Wójtowicz, *Flow pattern and power consumption in a vibromixer, Chemical Engineering Science 172 (2017) 622-635.*

[3] P.M. Orlewski, Y. Wang, M.S. Hosseinalipour, D. Kryscio, M. Iggland, M. Mazzotti, *Characterization of a vibromixer: Experimental and modelling study of mixing in a batch reactor, Chemical Engineering Research and Design* 137 (2018) 534-543.

[4] I. de Arcos González-Turmo, *FUNDAMIX® VibromixerCharacterization*, 2014.

[5] X. Li, J. Li, Y. Kuang, S. Guo, L. Mo, Y. Ni, *Stabilization of Pickering emulsions with cellulose nanofibers derived from oil palm fruit bunch, Cellulose* 27(2) (2020) 839-851.

[6] G. Chen, D. Tao, *An experimental study of stability of oil–water emulsion, Fuel processing technology* 86(5) (2005) 499-508.

Investigation of Micromixing of Liquids of Different Viscosity in a Twisted Bend Reactor

Poster.11

J. Kopytowski, *A. Rożeń. Warsaw University of Technology, Faculty of Chemical and Process Engineering, Warszawa, Poland, Janusz.Kopytowski.dokt@pw.edu.pl*

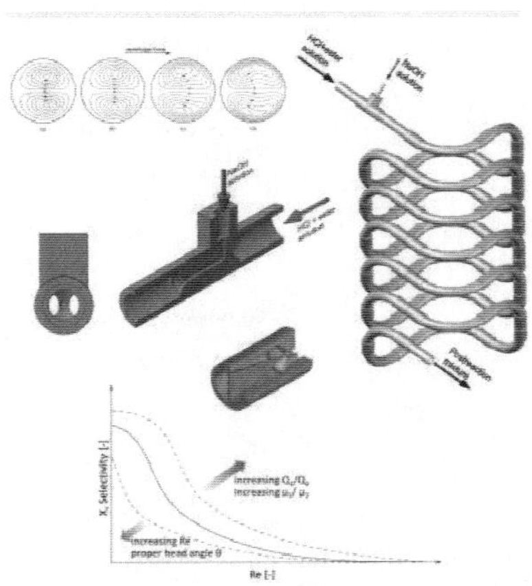

The effect of basic process conditions, viscosity ratio and initial distribution of liquid solutions on mixing on the molecular scale was studied experimentally in twisted bends reactor equipped with a modified L/H Komax distribution head. A reactive tracer method was used to study micromixing in the steady-state, isothermal, laminar and chaotic flow. The fastest mixing was usually observed when the central striation of a limiting reactant, formed in the distribution head, entered the reactor perpendicularly to the symmetry plane of the first pair of Dean vortices and was more viscous than the main stream of other reactants. The worst mixing conditions were obtained for wide range of Reynolds numbers when the central striation of the limiting reactant entered the reactor rotated by 90° to its optimal position and was less viscous than the main liquid.

Introduction

Tube mixers of complex 3D curvilinear geometry provide an alternative for typical continuous flow mixers such as: static mixers, injectors and impinging jet mixers in case of laminar mixing of highly viscous Newtonian and non-Newtonian liquids. They are characterized by head loss compared to that for an empty tube, low shear rates, and higher ratio of heat and mass transfer surface to the active mixer volume comparing to other continuous flow mixers operating in similar process conditions.

A characteristic feature of fluid flow in a curved conduit is the presence of a secondary flow in the channel cross-section, known as Dean vortices or roll-cells [1]. A pair of the counter rotating vortices is formed already at low Reynolds numbers, and it transports fluid elements from the central region of the curved tube to the outer tube wall as shown in the Graphical Illustration. Dean vortices are generated due to an imbalance between inertia and viscous forces. The Dean number expresses the combined effect of inertia and viscous forces:

$$De = Re\sqrt{\delta} \qquad (1)$$

where Re is the Reynolds number and δ is the ratio of pipe radius (d/2) and pipe curvature radius (R_c) Mixing on the molecular level in laminar flow is controlled by the rate of two main processes: streching of the contact surface of mixed fluids and molecular diffusion [2]. The local rate of generation of the intermaterial surface area depends on the rate of deformation of fluid elements and orientation of this surface with respect to the main directions of deformation of the local flow. The most energetically effective and the fastest extension of the intermaterial surface occurs in flow stagnation regions, where trajectories of fluid elements, initially close to each other, move away at an exponential rate. The presence of the stagnation regions is a clear signature of chaotic advection in fluid flow [3]. The rate of molecular diffusion, which is responsible for reducing local variances in concentration of mixture components, depends on the local segregation scale, i.e., on the size of unmixed fluid regions such as striations, filaments or drops.

Chaotic advection is created in the twisted bends mixer in steady laminar flow due to the periodic reorientation of Dean vortices. This process is known to accelerate generation of the intermaterial surface area, reduction of the local segregation scales and enhancement of molecular transport of reactants.

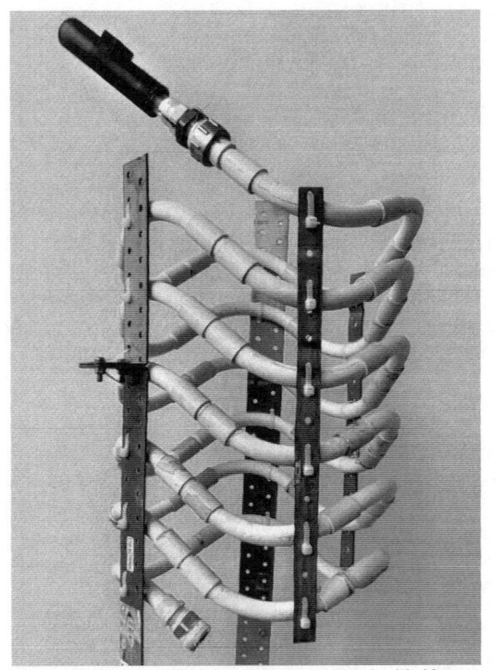

Figure 1. Twisted bend reactor equipped with Komax L/H distribution head.

https://doi.org/10.52202/074122-0057

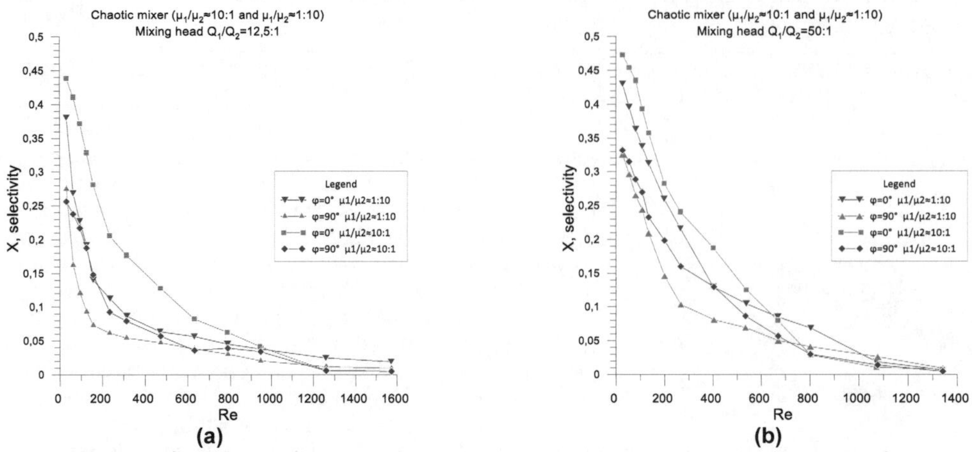

Figure 2. Selectivities of the parallel reactions: (a) volume ratio 12,5:1, (b) volume ratio 50:1

Material and Methods

The main stream pumped in the Komax distribution head was an aqueous solution of hydrochloric acid and ethyl chloroacetate with the initial concentration of the reagents $c_{HCl,0}=c_{ester,0}=0{,}01$ kmol/m^3. The side stream was aqueous sodium hydroxide solution with the initial concentration $c_{NaOH,0}=0{,}125$ kmol/m^3 or 0,5 kmol/m^3. Reactants were fed into the system at a constant rate and in equimolar amounts so that the volumetric flow ratio of the main stream Q_1 to the volumetric flow of the side stream Q_2 was equal to 12.5 and 50, respectively. The viscosity of the reagent solutions was increased to $\mu=10{,}6\cdot$mPa·s by a 12.1% addition of polyethylene polypropylene glycol (Breox 75 W 18000). This addition also increased the density of the solutions up to $\rho=1016{,}9$ kg/m^3.

During mixing, the acid and ester competed with each other in reaction with the base:

$$NaOH + HCl \xrightarrow{k_1} NaCl + H_2O \tag{2}$$

$$NaOH + CH_2ClCOOC_2H_5 \xrightarrow{k_2} CH_2ClCOONa + C_2H_5OH \tag{3}$$

The neutralization reaction is instantaneous ($k_1 \approx 10^8$ m^3/(kmol·s)) and completely controlled by mixing of the reactants. The second reaction is fast enough ($k_2=34{,}5$ m^3/(kmol·s)) to be partially controlled by mixing in the laminar flow regime [4]. Final selectivity X of the parallel reaction system (2) and (3)

$$X = \frac{Q_1 c_{ester,0} - (Q_1+Q_2) c_{ester}}{Q_2 c_{NaOH,0} - (Q_1+Q_2) c_{NaOH}} \tag{4}$$

depends directly on the course of mixing of reactants at the molecular level (micromixing) and indirectly on the course of convective mixing (macromixing). The length of the reactor was chosen so that regardless of the magnitude of the Reynolds number

$$Re = \frac{4(Q_1+Q_2)\rho}{\pi d \mu} \tag{5}$$

and the expenditure ratio Q_1/Q_2 the base conversion was complete ($c_{NaOH}=0$). The concentration of the ester (c_{ester}) in the post-reaction mixture was determined on the basis of HPLC analysis (Agilent 1260 Infinity chromatograph). The experiments were carried out at the room temperature 295±1 K.

Results and Discussion

Micromixing in laminar flow is enhanced by increasing the rate of deformation of fluid elements, which is responsible for creating a new contact surfaces between mixed fluids and reducing local segregation scales in the mixture. The deformation of fluid elements accelerates molecular diffusion, which tends to reduce local variations in components concentration. A simple way to accelerate the deformation of the fluid elements in a continuous flow tubular reactor is to increase the overall throughput of the system. The results of experiments with competitive-parallel reactions carried out in twisted bends reactor are shown in Figs. 2. It can be seen, indeed, that increasing the total flow rate, and thus the Reynolds number, reduces the final selectivity of the tested reactions. The lower selectivity of the parallel reactions occurring between initially unmixed reactants, the faster micromixing in the continuous flow reactor. The course of the test reactions is very sensitive to changes in flow conditions in the twisted bend reactor for Re < 500 (De < 140). However, if Re > 1000 (De > 283), a further increase in the Reynolds number hardly affects the final selectivity because the mixing in the twisted bends reactor becomes so fast that the second slower test reaction has little chance of proceeding (X<0,01). Under these conditions, the the total consumption the limiting reagent is mainly attributed to the acid-base neutralization reaction, and very little to the alkaline hydrolysis of ethyl chloroacetate.

The modified Komax L/H distributor head forms the side stream of the base solution in the shape of the central layer of base solution surrounded on both sides by an acid and ester solution. The lower selectivities, indicating better mixing, are observed when the inclination angle of the central

https://doi.org/10.52202/074122-0057

striation of the base solution towards the tube curvature plane is ϕ = 90°. Then the base striation is bisected and rapidly dispersed in the cross-section of the first bend of the tube by a pair of Dean vortices. If the inclination angle is ϕ = 0°, the contact surfaces of the reactant solutions are parallel to the symmetry plane separating the Dean vortices and the base striation falls between vortices. Then the fluid elements of the base solution are slowly deformed near the reactor wall and transported by the outer edges of the Dean vortices. As a result, the contact surfaces with the ester and acid solution is created slowly, thus increasing the chance for side reaction between base and ester to occur. It is characteristic that the lower Reynolds number, the stronger the influence of the inclination angle on the final selectivity, which proves the increasing importance of the initial orientation of the contact surface in flow field for the course of laminar micromixing. On the other hand, when Re > 1000, the influence of the tilt angle ϕ on the final selectivity becomes practically undetectable in a twisted bend reactor. The experimentally determined relationship between the initial orientation of the intermaterial interface relative to the plane of curvature of the first bend of the pipe and the quality of mixing is consistent with the theoretical findings of [5]. Mixing simulations performed for a helical tubular mixer showed that homogenization of a neutral tracer substance across the channel cross-section is achieved faster when the interface between the tracer solution and the clear liquid is initially perpendicular to the plane of symmetry between the Dean vortices.

Comparison of results obtained for two viscosity ratios 10:1 and 1:10 of the reactant solution and presented in Fig. 2b shows that the effect of the viscosity ratio on the final selectivity is relatively small when the volume ratio equals 50:1. These results are similar to findings obtained for Couette flow reactor [6]. The small effect of the viscosity difference on the final selectivity is most likely due to the antagonistic effect of solution viscosity on the diffusion rate of the reactants in the main stream and the deformation rate of the side stream. Nevertheless, it is possible to determine to some extent which mixing process has a greater impact on micromixing under specific experimental conditions. The solution of acid and ester without the viscosity increasing additive is characterized by higher diffusion coefficients of reactants than the solution with increased viscosity. This should help in reduction of the final selectivity. However, it is more difficult to mix the highly viscous side stream by the Dean vortices, which in turn could increase the final selectivity for the viscosity ratio (1:10). On the other hand, for reversed viscosity ratio (10:1), the diffusion coefficients of the reactants are decreased in the main stream, which negatively affects the rate of micromixing. However the low viscous side stream should be deformed more easily in local flow, which should result in faster micromixing and reduce the final selectivity. The results presented in Fig. 2 show the lower final selectivities for the 1:10 ratio, which can be interpreted as a stronger effect of diffusion mechanism on the final selectivity than the deformation of fluid elements of the side stream. The positive influence of the viscosity difference of the mixed solutions is more visible for the volume ratio 12.5:1 of the reactant solution, especially when the tilt angle is ϕ = 0°. Then the final selectivity values for Re>100 and μ_1/μ_2=1:10 are close to the final selectivities for the ϕ = 90° and μ_1/μ_2=10:1. This seems to confirm the greater influence of the diffusion mechanism on the final selectivity for medium Reynolds numbers than the deformation mechanism when mixing liquids differing in viscosity by one order of magnitude.

Conclusions

Based on the presented results obtained for the tube reactor of curvilinear geometry with the periodically reoriented Dean flow, it can be concluded that:

- the key parameter influencing the final selectivity is the Reynolds number, with the increase of Re, the selectivity of the parallel competitive reactions decreases;
- the final selectivity is also influenced by the volume ratio of the reactant solutions, along with the reduction of the volume ratio. reactants selectivity decreases;
- the tilt angle of the modified Komax distribution head influences final selectivity especially for lower Reynolds number, setting angle to ϕ=90° decrease the final selectivity;
- reversing the viscosity ratio the reactant solutions, has finite influence on selectivity when the high volumetric ratio of the solutions is applied, but reduction of the volumetric ratio increase influence of the viscosity ratio, what translates to decreasing final selectivity even for the tilt angle ϕ=0°.

References

[1] W.R. Dean, Note on the motion of fluid in a curved pipe. Philos. Mag. J. Sci. 1927, 4. pp. 208-223

[2] J. Bałdyga, ,A. Rożeń, F. Mostert, A model of laminar micromixing with application to parallel chemical reactions. Chem. Eng. J. 1998, 69, pp. 7-22

[3] J.M. Ottino, The Kinematics of Mixing: Stretching, Chaos, and Transport. Cambridge University Press, 1989

[4] A. Rożeń, J. Kopytowski, Experimental study of micromixing in curved tube reactors by the reactive tracer method, Chemical Engineering Research and Design, 160, 2020, pp. 335-350

[5] M. Mansour, Z. Liu, G. Janiga, K.D.P. Nigam, K. Sundmacher, D. Thevenin, K. Zahringer, Numerical study of liquid–liquid mixing in helical pipes. Chem. Eng. Sci. 2017, 172, pp. 250-261

[6] A. Rożeń, Investigation of laminar micromixing of liquids differing in viscosity, Chemical and Process Engineering, 2005, 26, pp. 649-670

 https://doi.org/10.52202/074122-0057

Effect of the Number of Blade with Horizontal Load and Torque on The Disk Turbine Impeller

Poster.12

S.Lee[1], *Y.Bando*[1], *K.Takenaka*[1]. *(1) Sumitomo Heavy Industries Process Equipment, 1501 Imazaike Saijo, Ehime, Japan, email address: seungjoo.lee@shi-g.com*

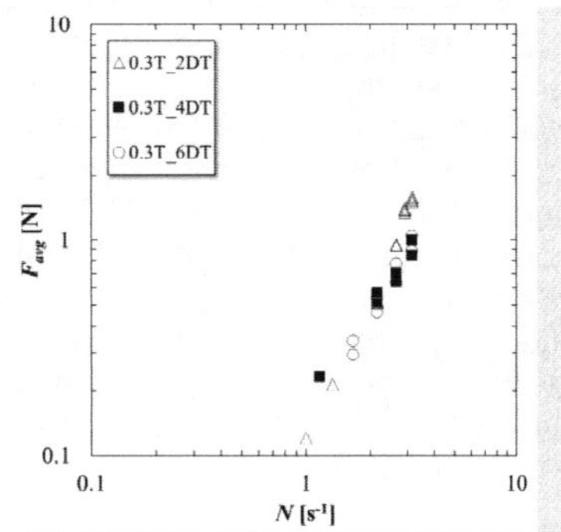

The purpose of this study was to consider whether it is possible to calculate to with impeller blades, one of the shapes of the impeller, could affect the cost of the agitator from a point of view to design shaft diameter. Strain of shaft is used to measure the horizontal loads. Torque was measured with mechanical torque meter. As a result, the number of blades does not affect to horizontal load, but torque was increased with an increasing of the blade numbers. From this result when designing agitator, using only Np regardless of excluding the impeller shape is not appropriate to calculate horizontal load. Therefore, to calculate horizontal load based on impeller shape, other influencing factors should be identified.

Introduction

Mixing is an important unit operation at many kinds of processes in chemical, biochemical dealing with functions such as mass transfer, homogenization, dispersion and so on. In industry on agitator, it values not only the performance of the agitator but also cost. To the cost of agitator, shaft diameter is the key issue, because the shaft diameter affects the specification of shaft seal and other parts.

Shaft design must accommodate allowable stress limits and avoiding natural frequency. Natural frequency is determined by vessel length and material. Therefore, it is the stress limits that decides the shaft diameter in substance.

There are two loads to calculate strength: torque and horizontal load. Torque is torsion load caused by the motor drive turning the impellers. Horizontal load is hydraulic force, which causes bending moment to shaft. Torque, M, can be determined with following equation (1).

$$Torque\ M = \frac{P}{2\pi N} \qquad (1)$$

However, about horizontal loads, it is difficult to define as an equation, because it occurs random. [1]

The purpose of this study is to see how much the number of blades one of the impeller shapes effects to torque and horizontal load.

Experimental

This study investigated in a transparent cylindrical vessel with an ellipsoidal bottom. The vessel diameter was T=0.6m. It was equipped with four equally spaced stainless steel plate baffles. The width of the baffles was 0.08T, and the clearance between the vessel wall and the baffles was 0.02T. Water was used as the mixing liquid, and the liquid was fill with the height equal to T, so the liquid volume was 0.198m^3.

All the impellers were fixed in 2 stages setting to a shaft, with a clearance of T/4 from the bottom and T/2 from the second stage. Figure 1 and Table 1 show the specifications of the impellers used.

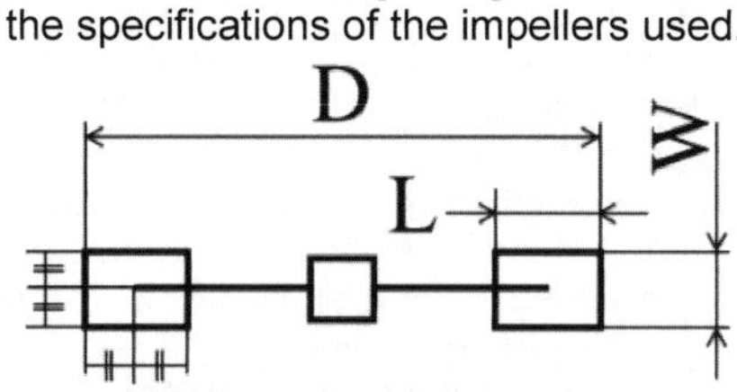

Figure 1. Impeller geometry

To measure horizontal load, *F*, the shaft can suppose as a cantilevered model. Strain is measured at one point on the shaft, and the force acting on the strain can be assumed to be a horizontal load.

To measure composite strain of the shaft, a pair of strain gages were affixed at a distance 0.850m from the end of the shaft, which had a diameter of 0.015m. The torque, M, was measured using a mechanical torque meter (Ono Sokki Co., Ltd). The strain gages detected the shaft strain at two directions of rotation X-direction and Y-direction shown as Figure 2.

The measurement was taken every 180 seconds, repeated three times at each rotation speed. Measured rotation speed was from 1s^{-1} to 3.2s^{-1}.

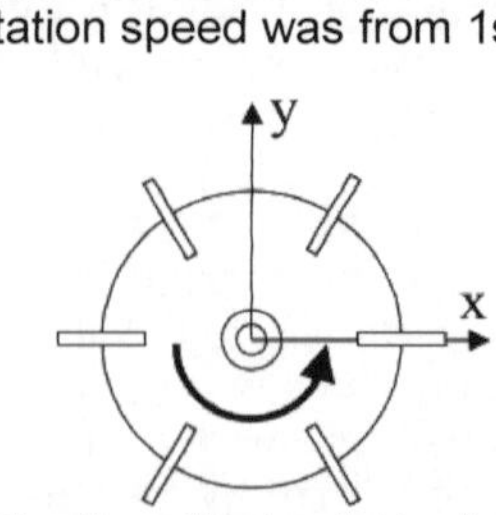

Figure 2. Direction of horizontal load measurement

Impeller Type	Description	D	W	L	Blade numbers	Blade angles
0.3T-2DT	Disc turbine(DT)	0.3T	0.2D	0.25D	2	90
0.3T-4DT	Disc turbine(DT)	0.3T	0.2D	0.25D	4	90
0.3T-6DT	Disc turbine(DT)	0.3T	0.2D	0.25D	6	90

https://doi.org/10.52202/074122-0058

Results and Discussion

To calculate the horizontal load, the composite strain, ε , was calculated by x-direction strain, ε_x , and y-direction strain, ε_y , using equation (2)

$$\varepsilon = \sqrt{\varepsilon_x^2 + \varepsilon_y^2} \quad (2)$$

With composite strain, horizontal load can convert as a bending moment. BM, using equation (3) and equation (4). In the equation (4), l is a distance of the strain gages affixed at from the end of the shaft.

$$BM = EZ\frac{\varepsilon}{2} \quad (3)$$

$$F = \frac{BM}{l} \quad (4)$$

Figure 3. shows the relationship with rotation speed and the horizontal load. Horizontal load increased depend on rotation speed increase entire impeller number.

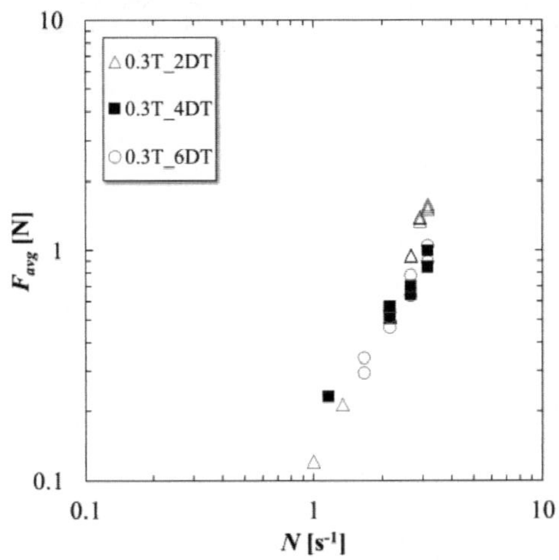

Figure 3. Experimental result of horizontal load

Figure 4. shows the relationship between rotation speed and average of the measured torque, M_{avg}. The average torque increased depending on the rotation speed increase and confirming equation (1). Additionally, the rate of torque increases varied; as impeller blade number increased, the rate of torque increases also increased.

As the Figure 3. and Figure 4. shown, both horizontal load and torque can represent as the relation with rotation speed. Horizontal load can present about rotation speed as below equation (5). The value of coefficient α depends on impeller type.

$$F = \alpha \times N^{\gamma} \quad (5)$$

From the experimental results, the characteristic of horizontal load according to the rotation speed could be created as equation (6). The value of coefficient, α, was depends on impeller type.

$$F = \alpha \times N^{1.814} \quad (6)$$

Torque can present as equation (7) about rotation speed using equation (1). The relation between torque and rotation speed can be represented by equation (8) same as calculation. The value of coefficient β depends on impeller type.

$$P = N_p \rho N^3 D^5 \quad (7)$$

$$M = \beta \times N^2 \quad (8)$$

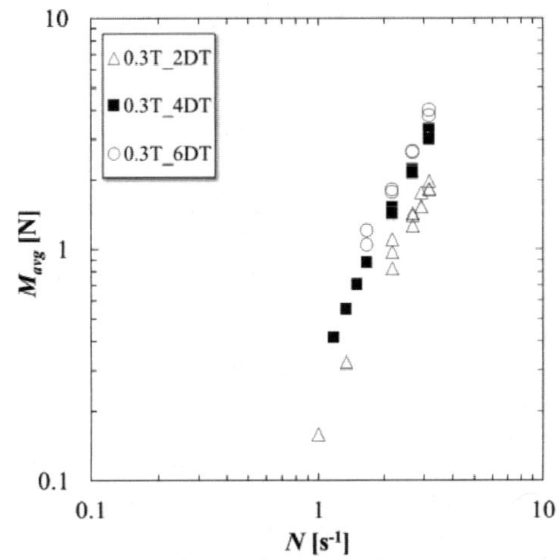

Figure 4. Experimental result of torque

(1) Effect of blade numbers on horizontal load

Figure 5. shows that horizontal load was constant regardless of the blade numbers. About this coefficient of horizontal load *α*, is independent to the impeller blade number. However, the torque is affected by the impeller blade number, and as the impeller blades number increases, the power number, *Np* also increases. This result means that *Np* is not a factor to influence the horizontal load.

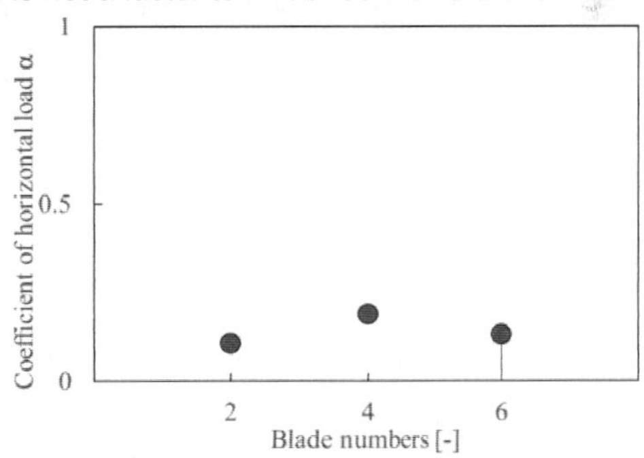

Figure 5. Relation between coefficient of horizontal load and blade numbers

(2) Effect of blade numbers on torque

Figure 6. shows that experimental value and the calculation value using Nagata equation, torque increased with the increase of the impeller blade numbers. Calculation value is as same as the experimental value. From this result, Np is the factor to influence torque, and this means torque can estimate with the shape of the impeller to design agitator.

https://doi.org/10.52202/074122-0058

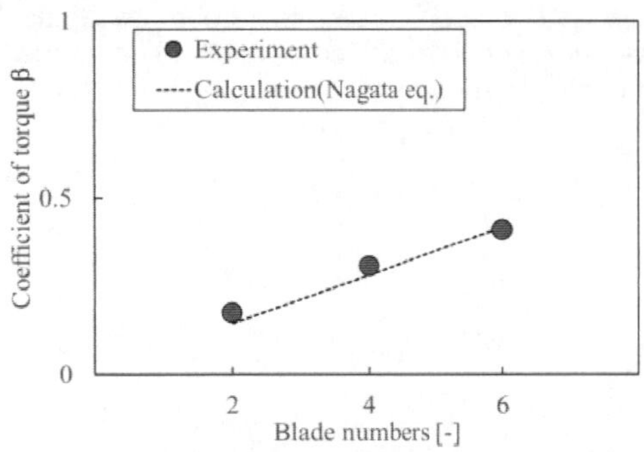

Figure 6. Relation between coefficient of torque and blade numbers

Conclusions

The purpose of this study was to consider whether it is possible to calculate to with impeller blades, one of the shapes of the impeller, could affect the cost of the agitator from a point of view to design shaft diameter. As a result, the number of blades does not affect to horizontal load, but torque was increased with an increasing of the blade numbers. From this result when designing agitator, using only Np regardless of excluding the impeller shape is not appropriate to calculate horizontal load. Therefore, to calculate horizontal load based on impeller shape, other influencing factors should be identified

NOMENCLATURE

BM	bending moment	[N m]
D	diameter of impeller	[m]
F	Horizontal load	[N m]
L	Blade length	[m]
N	Rotation number	[s^{-1}]
Np	Power number	[-]
M	Torque	[N m]
P	Power consumption	[W]
T	Diameter of tank	[m]
W	Blade width	[m]
α	Coefficient in equation (5)	
β	Coefficient in equation (8)	
γ	Coefficient in equation (5)	
ε	Strain of shaft	[m]

References

[1] H.Tanabe, R.Misumi, M.Kaminoyama and K.Nishi, "Investigation of torque and horizontal load on a paddle impeller in eccentric mixing". *J. Chem. Eng. Jpn*, **51**, 197 (2018)

[2] V.Hudcova, V.Machon and A.W.Nienow, "Gas-Liquid Dispersion with Dual Rushton Turbine Impellers", *Biotechnology and Bioengineering*, **34**, 617 (1989

Modelling of Phase Inversion through Coalescence Cascades by Separation Induced Coalescence

Poster.13

Arunachalam S[1], Sanjeev Kumar[2]. *(1) Indian Institute of Science, Bengaluru, India. sarunachalam@iisc.ac.in*
(2) Indian Institute of Science, Bengaluru, India. sanjeev@iisc.ac.in

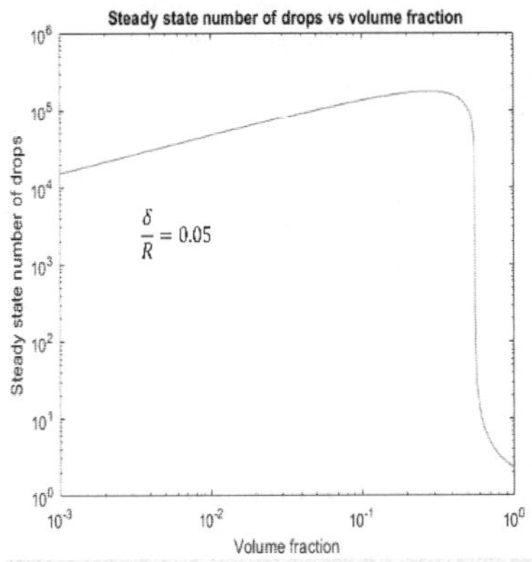

A model for predicting phase inversion in agitated liquid-liquid dispersions through population balance framework has been developed. The model captures an inversion like instability by the addition of coalescence cascade through separation induced coalescence. It is further able to capture the invariance of inversion fraction with volume, breakup/collision frequencies which is consistent with experimental findings on the asymptotic value of inversion point. The inversion fraction is found to depend on the critical distance parameter for coalescence, which removes the limitation of invariant inversion fraction with the idea of coalescence cascades.

Introduction

Agitated liquid-liquid dispersions are widely encountered in various applications like mixer settlers, suspension polymerization, liquid-liquid extractions etc., These are dispersions of two immiscible liquids where one liquid is suspended in the other in the form of drops. The drop phase is called the dispersed phase and the other phase is known as the continuous phase. These systems are stable till an external source of energy input, like stirring, is provided. Phase inversion is the catastrophic phenomena in which the dispersed phase and the continuous phase gets interchanged. This can happen due to a variety of reasons like addition of dispersed phase, change in stirrer speed etc., The dispersed phase volume fraction at which phase inversion occurs is known as the inversion fraction. The inversion fraction is a function of stirrer speed till a certain stirrer speed beyond which it becomes invariant with respect to it. This inversion fraction is known as asymptotic value of inversion fraction. This asymptotic inversion fraction was found experimentally to be independent of stirrer and tank geometry, flow field but is dependent on some physical property of the liquid-liquid system itself [1]. With many more such observations, understanding the mechanism of phase inversion is still a challenge [2]. The most widely accepted mechanism is the drop breakup vs. drop coalescence dynamics. Away from inversion volume fraction, both drop breakup and drop coalescence happen in the turbulent dispersions and reach a steady state drop size distribution. At the inversion fraction, it is believed that the drop coalescence overwhelms the drop breakup process and phases invert. Various methods have been proposed in the literature to model phase inversion [3,4]. Most of these methods fail in capturing the inversion point for different systems while providing consistent explanation for the same. The dynamics of the liquid-liquid dispersions can be captured by population balance equations (PBE) framework. Works in the literature using PBE resulted in the drop size seems to either reach a steady state or no steady state at all volume fractions, thus unable to predict an inversion-like transition. These models had considered drop breakup and drop coalescence to be independent processes. This consideration was changed by the idea of coalescence cascade and a model proposed with this linked breakup and coalescence processes, was able to capture an inversion like transition. It could show that the inversion was not dependant on the breakup/coalescence frequencies but rather only on the proximity of the drops which triggers coalescence cascades leading to inversion. The drawback of the model was that it predicted phase inversion to happen at a critical volume fraction of 0.1 irrespective of the system considered [5]. The model was not able to capture the idea that the inversion fraction is also a property of the liquid-liquid system.

The main mode of coalescence considered in these frameworks were the collision mode of coalescence. Works in microfluidics have shown that drops which were almost in collision contact, stabilised by surfactants, could undergo coalescence when they were pulled away from each other [6]. This is referred to as 'separation induced coalescence' mode. It has further been shown that this coalescence mode can trigger avalanche of coalescences in the system and has been compared to inversion like behaviour [7]. Here, we have incorporated the idea of separation induced coalescence in the model through the mechanism of coalescence cascades.

Material and Methods

The Model

Apart from drop breakup and collision coalescence, drops can also undergo coalescence through the idea of coalescence cascades. The term Coalescence Cascade' refers to a breakup/collision event leading to an extra coalescence event along with cascades essentially provide the dependence

https://doi.org/10.52202/074122-0059

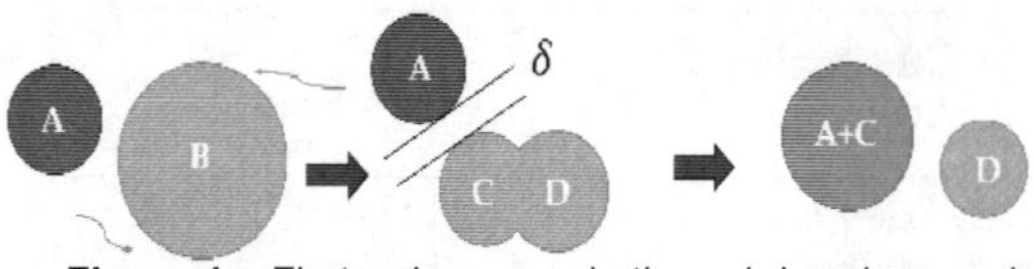

Figure 1a: First order cascade through breakup mode

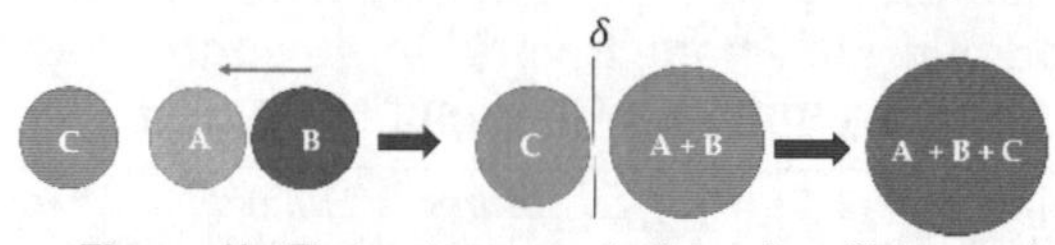

Figure 1b: First order cascade through collision mode

between the breakup, collision and coalescence in the system. A first order coalescence cascade through breakup/collision coalescence would indicate one coalescence event has also accompanied a breakup/collision coalescence process. The order of cascade represents the number of coalescence events which are triggered due to a breakup or collision event.

The cascades are visualized to occur through the separation induced mode of coalescence. Figure 1 shows a schematic of a cascade event. In figure 1a, the drop B is undergoing breakup process to form drops C and D. If there is a drop A in proximity of drop C within a distance δ, then, drops A and C can coalesce through separation - induced mode to form a new drop A+C. Effectively, the breakup process which is supposed to increase the number of drops can rather decrease it with increasing order of cascades. Same could occur through collision coalescence as well. Figure 1b shows drops A and B undergoing collision coalescence. If there is a drop C in the proximity of the drops within the distance δ, then, it can undergo separation induced coalescence leading to formation of drop A+B+C instead of just A+B. The above cases are examples of first order coalescence cascade induced by either a breakup process or a collision coalescence process. In our model, we have accounted for the possibility of infinite orders of cascades happening induced by either of the events in a mean field way. A PBE is written considering drop breakup, drop collision coalescence and coalescence cascades triggered by them respectively.

$$
\begin{aligned}
\frac{\partial n(v,t)}{\partial t} = \quad & -\Gamma(v)n(v,t) + \left(1-p(v)\right)B^{0+}(v) \\
& + \left(1-p(v)\right)B^{1+}(v) - B^{1-}(v) + \;.. \\
& + \left(1-p(v)\right)B^{i+}(v) - B^{i-}(v)\;... \\
& - \int_0^\infty w(v,v')n(v,t)n(v',t)dv' \\
& + \left(1-p(v)\right)C^{0+}(v) \\
& + \left(1-p(v)\right)C^{1+}(v) - C^{1-}(v) + .. \\
& + \left(1-p(v)\right)C^{i+}(v) - C^{i-}(v)\;...
\end{aligned}
$$

Here, *n(v,t)* is the number of drops of size '*v*' per unit system volume per unit drop volume. *B* and *C* refer to the breakup and collision processes leading to cascades respectively. Superscript ***i+/-*** refers to i^{th} order of cascade leading to increase/decrease in number of drops of a particular size '*v*' respectively. The terms represent the increase/decrease in number of drops of size '*v*' due to breakup/collision events and the respective orders of cascades induced by them. The cumulative probability of encountering another drop within a distance δ is given by *p(v)*. Since it is not a function of '*v*' and it turns out to be a function of δ, it is referred to as $p(\delta)$. This probability is derived in a mean field way by assuming uniformly distributed drops as done in [8]. The final form of probability is given by:

$$p(\delta) = \frac{12\delta\emptyset}{R}\left(1-\frac{1}{VM_0}\right)$$

where $\emptyset$ is the volume fraction, *V* is the volume of the system, *R* is the mean drop radius and M_0 is the number of drops per unit system volume.

It is assumed that coalescence will occur if the nearest drop is at a distance of δ or less.

Results and Discussion

The moment equation of the PBE was solved for simple breakup and coalescence kernels to get steady state number of drops ($\Gamma(v) = kv \text{ and } w(v,v') = b$) and an analytical solution was obtained. The predictions of the model are shown in figures and the value of parameters in table 1.

Table 1. Parameters in the model[a]

Parameters	Symbol	Value
Volume	V	2500
Breakup frequency	k	0.0525
Distance ratio	δ/R	7%

[a] for the figures, these values are used unless specified otherwise.

The graphical illustration in the abstract shows the prediction of the model as we increase the volume fraction. We can see that the steady state number of drops increases with volume fraction steadily till a critical volume fraction. Until this point, the breakup and coalescence processes reach a steady state. At this critical volume fraction ($\emptyset$ =0.56), it can be seen that the number of drops drop drastically from 10^5 drops to almost 10 drops. Such a large decrease in number of drops leads to a large increase in the mean size of drops which is observed near to inversion. This is indicative of an inversion like instability. The model also predicts that the change in volume of the system does not change the inversion fraction as shown in figure 2.

Figure 3 shows the variation of the predictions with breakup frequency k. On increasing the breakup frequency by two orders of magnitude, the number of drops before inversion point has increased as the breakup process becomes more dominating relatively. But the critical inversion point remains unchanged. The curves essentially start overlapping from the critical inversion point. The model predicts that the critical volume fraction is not affected by breakup/collision frequency. Figure 4 shows that on

https://doi.org/10.52202/074122-0059

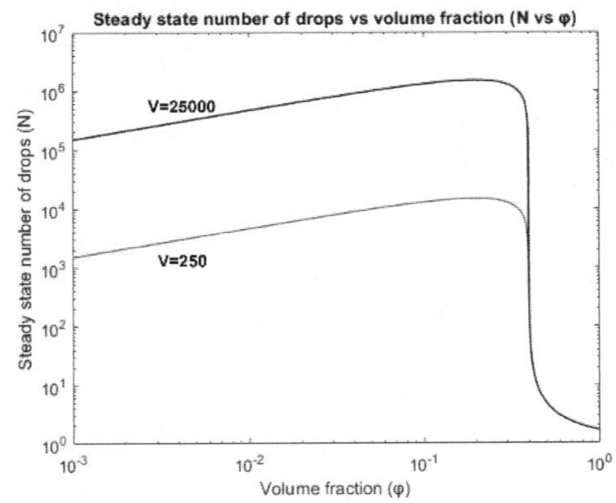

Figure 2. Inversion points for varying volumes.

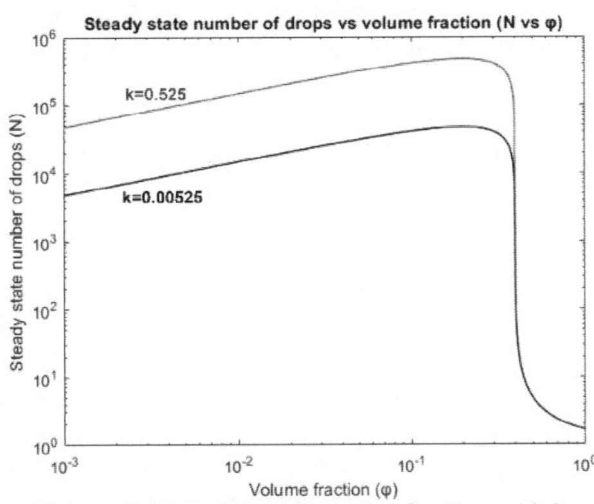

Figure 3. Variation of inversion fraction with breakup frequency

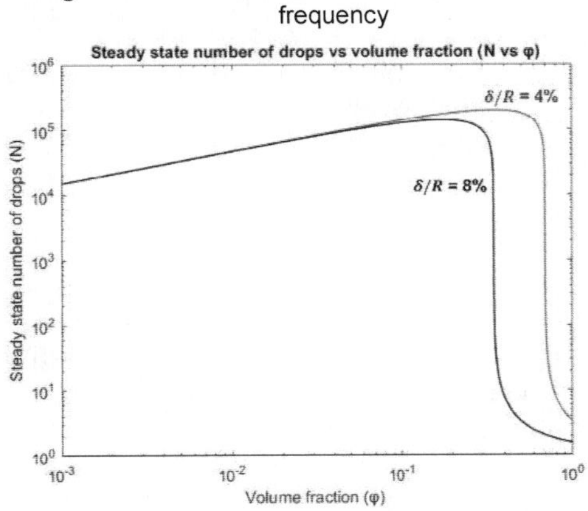

Figure 4. Variation of inversion fraction with crit. distance δ

increasing the critical distance to radius ratio δ/R, before the critical point, there is no change between the cases. But the critical inversion fraction shifts to lower values. So, the model predicts variant inversion points for systems with different δ/R ratios.

At low volume fractions, the drop breakup and the coalescence process compete to increase/decrease the number of drops and they lead to a steady state number of drops. As the volume fraction increases, the cascades become more prominent and at the critical volume fraction, the cascades dominate the processes such that the net coalescence process overwhelms the breakup process leading to large amount of coalescence and rapid decrease in the number of drops in the system leading to inversion like instability. On changing the volume of the system, the number of drops scales with it but this does not affect the dynamics of the system and the inversion fraction does not change. Interestingly, the model predicts invariant inversion fraction with change in breakup/collision frequency. This implies that it is not the individual breakup or collision processes, but it is the cascade which triggers an inversion like instability in the system. A hundred-fold increase in breakup frequency '*k*' does not change the inversion fraction, indicating that even if number of drops increases drastically, these drops need to be in proximity for inversion to occur, failing which inversion does not happen. Similar effect is observed with changing the collision frequency as well. Finally, the inversion fraction predicted seems inversely related to δ/R. On increasing the value of δ/R from 4% to 8%, the inversion fraction decreases from 0.69 to 0.35. This decrease can be related to the propagation of cascade as follows: as we increase the critical distance δ/R, drops which are relatively farther apart can now coalesce, effectively increasing the cascades at a lower volume fraction. Hence, more coalescence happens, leading to inversion happening at relatively lower volume fractions. It can be inferred from the model that it is not only the proximity of drops which triggers cascades leading to inversion but the characteristic of the liquid-liquid dispersion itself (through δ/R) which decide the inversion point together. These observations are consistent with the experimental observations with the asymptotic inversion fraction mentioned earlier.

Conclusions

The developed model predicts phase inversion point. It further captures the invariance of asymptotic inversion point with changing volume, breakup/collision frequencies. It is also able to predict varying inversion fractions for different systems by change in corresponding δ/R. The inversion point predicted is a function of both the proximity of drops and a characteristic of the liquid-liquid system itself. It has been shown that by incorporation of physics like separation-induced coalescence, the limitation of invariant inversion point can be removed, and phase inversion can indeed be predicted for different systems using the idea of coalescence cascades.

Acknowledgments

The authors acknowledge PMRF for the financial support provided.

References

[1] B. Deshpande, K and Kumar, S.,*Chemical Engineering Science*, 58(2003) 3829.
[2] Yeo L Y, Matar O K, de Ortiz ESP, Hewitt GF, *Multiphase science and technology*, 12(2000) 66.
[3] Arashmid M, Jeffreys G V,*AIChE Journal*,26:1 (1980)51.
[4] Vigil R D, Ziff R M, Journal of Colloid and Interface Science, 133:1(1989) 257.
[5] *Ramana Reddy VRC,.PhD thesis, IISc Bangalore (2008)*
[6] Bremond N, Thiam A R, Bibette J., *Physical Review Letters*, 100(2008) 024501.
[7] Bremond N, Domejean H., Bibette J, *Physical Review Letters*, 106(2011) 214502.
[8] Bansal P P, Ardell A J, *Metallography*, 5:2(1972) 97.

https://doi.org/10.52202/074122-0059

An alternative approach to develop analytical solutions for breakup process using eigenvectors

Poster.14

Anjali Kumari[1], *Sanjeev Kumar*[2]. *(1) Indian Institute of Science, CV Raman Road, Bengaluru, India, anjalikumari@iisc.ac.in. (2) Indian Institute of Science, CV Raman Road, Bengaluru, India, sanjeev@iisc.ac.in.*

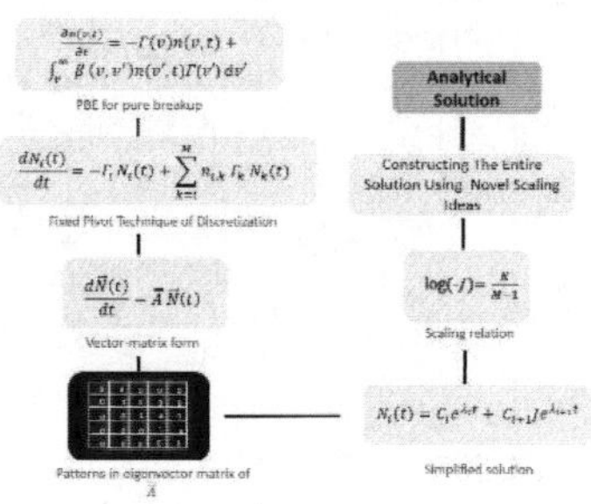

Proposed Solution Framework

Population balance equations (PBEs) are a set of mathematical equations used to describe the evolution of a population of particles or droplets in a mixing process. In the present work, a new approach is proposed for finding the analytical solution of population balances for pure breakup problems. The PBEs is discretized using fixed pivot technique [1], reducing it to a set of coupled ODEs which is then cast in vector-matrix form. This approach relies on finding patterns in the eigenvectors of this matrix when the degree of discretization is changed. These patterns are mathematically formulated as 'scaling relations' which can map coarse grid solution to fine grid solution, and in the limit of infinite discretization, lead to analytical solutions. Analytical solutions for an entire class of breakup problems are derived. Also, for highly complex daughter size distributions given by Hsia & Tavlarides [2] with power law breakup rate, the first-ever series solution is proposed.

Introduction

Breakup of drops/ particles is one of the most basic processes in most of the mixing applications such as, grinding of solids, rolling mills, polymer blends etc. This continuous breakup leads to a dynamic change in the population of entities in the system. A mathematical framework to capture this evolution is given by population balance equations (PBEs). The population balance equation for a pure breakup process is given by

$$\frac{\partial n(v,t)}{\partial t} = -\Gamma(v)n(v,t) + \int_{v'=v}^{\infty} \Gamma(v')n(v',t)\beta(v,v')dv' \quad \dots(1)$$

A numerical solution of this equation by discretization for the conservation of mass and the exact preservation of increase in number of particles accompanying breakup is available from Kumar and Ramkrishna [1] for a general initial distribution, breakup rate $\Gamma(v)$, and daughter distribution function $\beta(v,v')$. The method combined with geometric grids allows a large size to be covered with varying degrees of fineness. The process of discretization leads to M coupled linear ordinary differential equations. Each equation represents time variation of particle population in a discrete size range, represented by a fixed pivot, as

$$\frac{dN_i(t)}{dt} = -\Gamma_i N_i(t) + \sum_{k=i}^{M} n_{i,k}\Gamma_k N_k(t) \qquad i = 1 \; to \; M \quad (2)$$

The coupled M equations can be compactly represented in the vector-matrix notation as

$$\frac{d\vec{N}(t)}{dt} = \bar{\bar{A}}\,\vec{N}(t) \qquad (3)$$

In recent years, there has been a growing interest in developing analytical solutions for pure breakup processes (Equation (1)), with only a few solutions available till now [5,6,7]. The current approach can be extended to various complex breakup kernels that have not been attempted using the existing techniques.

The current work focusses on unravelling the unusual structure of eigenvectors of $\bar{\bar{A}}$ in infinite dimensional space, which very elegantly facilitates analytical solutions to be extracted for a general class of problems. The patterns in the eigenvector structure helps to construct solution in fine discretization using coarse discretization solution. In the limit of very high discretization, fixed pivot technique is known to approach the exact solution [4]. Using the scaled values of eigenvectors and eigenvalues, in the limit of $M \to \infty$, the exact solution is constructed.

Method

The matrix $\bar{\bar{A}}$ is upper diagonal with distinct eigen values λ_i and the corresponding eigen vectors $\vec{e_i}$. Solution of equation (3) is given by (4) where λ_i is the ith eigenvalue and $\vec{e_i}$ is the corresponding eigenvector of matrix $\bar{\bar{A}}$.

$$N_i(t) = \sum_{j=1}^{M} C_j \vec{e_j}\, e^{\lambda_j t} \qquad (4)$$

Here C_j depends on the initial condition. The eigenvector matrix V, given by (5), has ith column as the eigenvector corresponding to λ_i.

$$\bar{\bar{V}} = [\vec{e_1}\,\vec{e_2}\,\vec{e_3} \dots \dots \vec{e_M}] \qquad (5)$$

This matrix is expected to have all non-zero upper diagonal elements, reflecting that the particle population in the smaller size bins is impacted by the breakup of particles in all the large size bins. Equation (4) is thus expected to have a large number of terms on the R.H.S as the fineness of discretization increases. For a fixed size range, if we keep on increasing the number of discrete points (pivots), M, the eigenvector matrix keeps changing. This is

https://doi.org/10.52202/074122-0060

equivalent to making the discretization fine. According to (4), the solution in depends on C_j, $\vec{e_i}$ & λ_i, and since eigenvalues are simply the diagonal elements of matrix, only the C_j and $\vec{e_i}$ are to be predicted with changing M. It is to be noted that C_j are also deduced from a combination of initial condition and eigenvector matrix. Therefore, the prediction of eigenvector matrix is the key to this approach. The structure of eigenvector matrix is studied for the hidden patterns which can be used to construct the analytical solutions.

Results and Discussion

Case 1:

$$\Gamma(v) = v^2 \text{ and } \beta(v, v') = \frac{2}{v'}$$

Figure 1. shows the eigenvector matrix structure. Each eigenvector lies on the 2-dimensional planes of the M-dimensional hyperspace. Here M can be extremely large. This simplifies the solution for $N_i(t)$ to a two-term solution given by equation (6).

$$N_i(t) = C_i e^{\lambda_i t} + C_{i+1} J e^{\lambda_{i+1} t} \qquad (6)$$

This is quite interesting as the rate of breakup of larger size particles, beyond (i+1) does not appear directly through exponential terms. The matrix structure is preserved on changing M. Only the value of J on the super diagonal changes. Therefore, finding how J scales with M can be used to generate the solution on a fine grid, as fine as desired, without even obtaining eigenvectors for large matrices.

1	J	0	0	.	.	0
0	1	J	0	.	.	0
0	0	1	J	0	.	0
.	.	0	.	.	.	.
.	.	.	0	.	J	.
.	.	.	.	0	1	J
0	0	0	.	.	0	1

Figure 1. Eigenvector matrix $\bar{\bar{V}}$ showing a unique pattern

In the present case, we find that a plot of $\log(-J)$ vs. M-1 for several small values of M satisfies the following scaling relationship where K is a constant.

$$\log(-J) = \frac{K}{M-1} \qquad (7)$$

Figure 2a shows a comparison of the size distributions obtained for M=25 using the above approach (for K values obtained for M in single digits) and for the analytical solution provided by Ziff and McGrady [3].

Case 2:

$\Gamma(v) = v^a$ and $\beta(v, v') = \frac{Cv^{b-1}}{v'^b}$ for $a - b = 1$ and $b > 0$

The eigenvector matrix structure is same as shown in Figure 1. The scaling relation is also same as given by (7).

Figures 2 (b) show a comparison of results obtained using the scaling approach outlined above and the integration of coupled ODEs for b=0.5. The figures show that the results are in very good agreement.

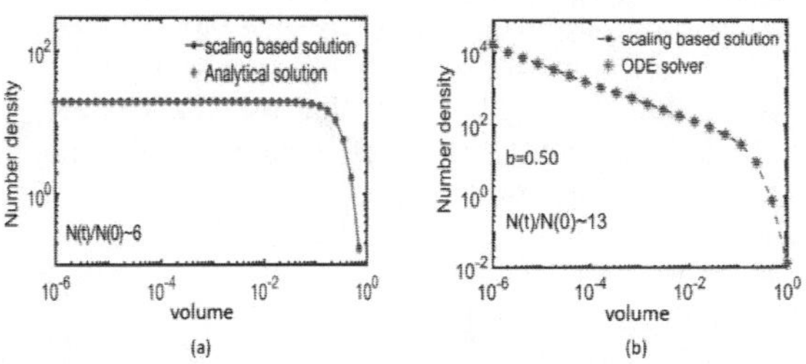

Figure 2. (a) shows a comparison with analytical solution by Ziff and McGrady (1985) [1]. (b) shows a comparison of the scaling-based solution and ODE solver solution for size distribution, for b=0.5.

Analytical Solution The internally consistent discretization schemes, such as the fixed pivot technique, approach the exact solution in the limit of very high discretization [4]. Using the scaled values of eigenvectors and eigenvalues, in the limit of $M \to \infty$ we can construct the exact solution.

$$n(v, t) = \lim_{M\to\infty} \frac{N_i(t)}{v_{i+1} - v_i} \qquad (8)$$

Solving equation (8) for monodispersed initial condition we find

$$n(v,t) = \begin{cases} 2tv_{max}v^{b-1}\left(\frac{b+1}{2}\right)e^{-tv^{b+1}} & v < v_{max} \\ \delta(v - v_{max})e^{-tv_{max}^{b+1}} & v = v_{max} \\ 0 & v > v_{max} \end{cases}$$

..(9)

Case 3:

$\Gamma(v) = v^{-1}$ and $\beta(v, v') = \frac{4.8}{v'}\exp\left(-4.5\frac{(2v-v')^2}{v'^2}\right)$

The $\beta(v, v')$ is the truncated normal distribution given by Hsia and Tavlarides [2].

The eigenvector structure is shown in Figure 3. The number of scaling relations to predict such a structure will increase in the order of M. Thus, mapping solution in infinite dimensional hyperspace will need infinite scaling relations.

A relation that can determine all eigenvectors components is given by

$$\bar{\bar{V}}(1, x-1) = \tilde{a}(M) - \tilde{b}(M)\, e^{\tilde{c}(M)x} \qquad (10)$$

Since this generates 1st row of $\bar{\bar{V}}$, it is sufficient for generating entire $\bar{\bar{V}}$.

1	a	b	c	.	.	.	z
0	1	a	b	c	.	.	.
0	0	1	a	b	.	.	.
0	0	0	.	.	.	.	.
0	0	0	0	.	.	.	c
0	0	0	0	0	.	a	b
0	0	0	0	0	0	1	a
0	0	0	0	0	0	0	1

Figure 3. Eigenvector matrix $\bar{\bar{V}}$ showing all super diagonals with non-zero values

https://doi.org/10.52202/074122-0060

A series solution is developed for monodispersed I.C using the same approach for this very complex breakup function. The equation was scaled with v_{max}. The solution given by (11) has only three parameters, $\tilde{a}(M), \tilde{b}(M)$ and $\tilde{c}(M)$, which scale with M rather than M parameters as expected.

$$N_i(t) = C_i\, e^{\lambda_i t} + \tilde{a}(M)\left(C_{i+1}\, e^{\lambda_{i+1} t} + C_{i+2}\, e^{\lambda_{i+2} t} + C_{i+3}\, e^{\lambda_{i+3} t} + \ldots . C_M\, e^{\lambda_M t}\right) - \tilde{b}(M)\left(C_{i+1}\, e^{\lambda_{i+1} t+(M-1)\tilde{c}(M)} + C_{i+2}\, e^{\lambda_{i+2} t+(M-2)\tilde{c}(M)} + \ldots . C_M\, e^{\lambda_M t+i\tilde{c}(M)}\right) \quad (11)$$

$$N_M(t) = C_M\, e^{\lambda_M t} \quad (12)$$

Here,

$$C_{M-(i-1)} = -\sum_{j=1}^{i-1}\left(\tilde{a}(M) - \tilde{b}(M)\, e^{\tilde{c}(M)(M-j+1)}\right) C_{(i-j+1)} \quad ..(13)$$

Equation (8) gives number density function.

Figure 4 shows the series solution results compared with the solution produced by solving the coupled ODEs using an ODE solver in MATLAB.

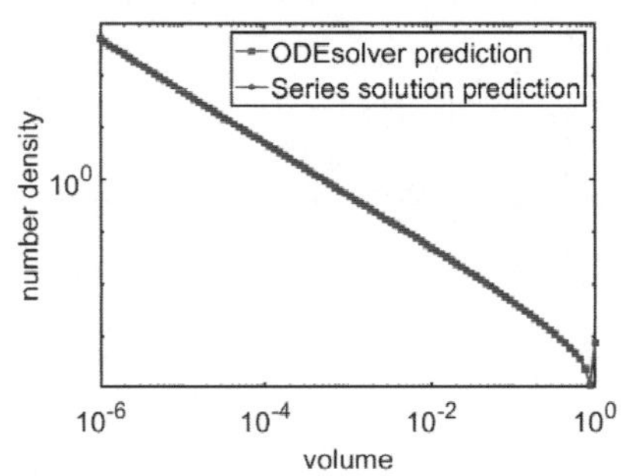

Figure 4. Comparison of number density given by series solution with solution produced by ODE solver

Case 4:

$\Gamma(v) = v^2$ and $\beta(v, v') = \frac{4.8}{v'} \exp\left(-4.5 \frac{(2v-v')^2}{v'^2}\right)$

This is one of the most widely used breakup kernel and too complex to solve for even semi-analytical solution by existing mathematical techniques.

The eigenvector matrix structure looks as given by Figure 3. But its variation can't be captured by a simple mathematical relation like above. This bottleneck can be solved by the patterns present in its inverse matrix. The inverse matrix structure is same as $\bar{\bar{V}}$ but shows more elegant patterns. The process of constructing $\bar{\bar{V}}$ using $\bar{\bar{V}}^{-1}$ is done by developing a new recurrence relation given by

$$\bar{\bar{V}}(1, i) = \sum_{j=1}^{i-1} \bar{\bar{V}}(1, j)\, \bar{\bar{V}}^{-1}(1, i-j+1) \qquad i > 1 \quad (14)$$

Using this recurrence relation, we could recover $\bar{\bar{V}}$ using patterns in $\bar{\bar{V}}^{-1}$.

Equation (8) gives number density function.

The solution using this recurrence along with scaling technique is compared against the solution produced by ODE solver in MATLAB in Figure 5 showing a good agreement.

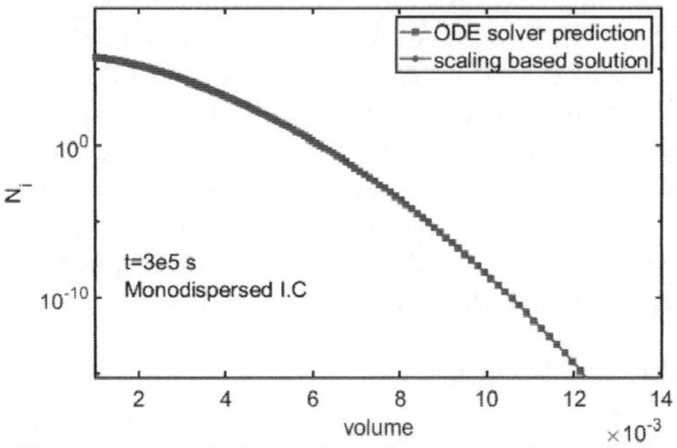

Figure 5. Comparison of numbers given by series solution with solution produced by ODE solver

Conclusions

The current approach of finding patterns in eigenvectors and scaling them with discretization provides a systematic methodology for solving the size distribution in fragmentation processes with monodispersed initial conditions. The monodispersed I.C solution can be easily superimposed to find solutions for any initial distribution as its pure fragmentation process. The method was shown to reproduce previously known solutions and was used to produce new results for the power-law kernel in fragmentation for an entire class. Also, for the first time, an explicit approximate series solution was constructed for daughter size distribution given by Hsia and Tavlarides [2] which is too complex for existing mathematical methods. The method offers the possibility of constructing the solution of Hsia and Tavlarides daughter size distribution with power law breakup rates. A compact mathematical series solution for this case is under progress. The number density plots using scaling-based technique shows very good agreement with ODE solver solution.

References

[1] *Kumar, S., Ramkrishna, D., 1996. On the solution of Population Balance Equations by Discretization - I. A Fixed Pivot Technique. Chem. Eng. Sci. 51, 1311–1332.*

[2] *Hsia, M.A. and Tavlarides, L.L., 1983. Simulation analysis of drop breakage, coalescence and micromixing in liquid-liquid stirred tanks. The Chemical Engineering Journal, 26(3), pp.189-199.*

[3] *Ziff, R. M. and Mcgrady, E. D., 1985, The kinetics of cluster fragmentation and depolymerization. J. Phys. A: Math. Gen. 18, 3027-3037.*

[4] *Arnold, D.N., 2015. Stability, consistency, and convergence of numerical discretizations. Encyclopedia of Applied and Computational Mathematics, pp.1358-1364.*

[5] *Simha, R., 1956. On the degradation of branched chain molecules. The Journal of Chemical Physics, 24(4), pp.796-802.*

[6] *Singh, P., and Hassan, M.K., 1996. Kinetics of multidimensional fragmentation. Physical Review E, 53(4), p.3134.*

[7] *Ziff, R.M. and McGrady, E.D., 1986. Kinetics of polymer degradation. Macromolecules, 19(10), pp.2513-2519.*

Experimental Study on Heat Transfer Performance of Oscillatory Baffled Reactor

Poster.15

T. Horie[1], *R. Morishita*[1], *R. Murotani*[1], *E. Okita*[1], *M. Yasuda, N*[1]. *Ohmura*[2]. *(1) Osaka Metropolitan University, 1-1 Gakuen-cho, Naka-ku, Sakai-shi, Osaka, Japan, horie@omu.ac.jp (2) Kobe University, 1-1 Rokkodai-cho, Nada-ku, Kobe, Hyogo, Japan, ohmura@kobe-u.ac.jp*

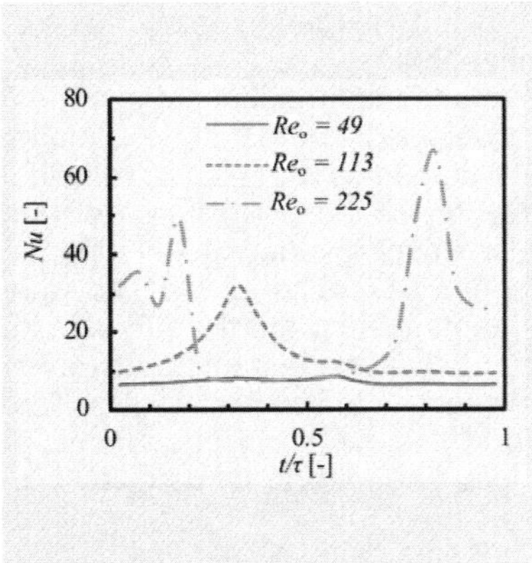

To evaluate the heat transfer performance of the continuous Oscillatory Baffled Reactor, thermocouples were installed at several locations in the cross-section of the reactor. Based on the temperature measurements at the center between the baffles, the time averaged Nusselt number at that location was calculated. The average Nusselt number increased slightly for oscillatory Reynolds numbers, Re_o, below 80, but increased significantly from 80 to 160, and became constant above 160. The local Nusselt number, *Nu*, was calculated within one period of the oscillatory flow, and as shown in the figure, instantaneous and rapid increases could be observed. PIV showed that at Re_o = 113, the donut-ring vortex generated by the oscillatory flow propagated and the *Nu* increased as it passed through the temperature measurement position. At Re_o = 225, two peaks of the *Nu* were obtained due to the passage of the vortices generated in the forward and reverse flow.

Introduction

In the chemical industry, converting a batch process into a continuous process has many advantages, including increased efficiency through automation of operations, reduced labour costs, and fewer human errors. One of the ideal means of continuous reaction processes is the tubular reactor. Oscillatory Baffled Reactor (OBR) was proposed to further improve mixing and temperature control [1]. Multiple baffles are placed at equal intervals in the reactor tube, which further imparts an oscillatory flow to the mainstream. It is known that unsteady vortices are generated, giving an efficient mixing zone, and improving heat transfer performance due to the secondary flow near the wall surface of each baffle section. Although recent studies have been conducted to investigate the periodic variation of the Nusselt number, *Nu*, in response to vortex generation in the OBR by numerical simulation [2], experimental investigations have been limited to the heat transfer performance of the entire reactor.

In this study, the temperature of the fluid inside the OBR was measured at multiple points to evaluate the time variation of local heat transfer performance in various flow conditions with varying oscillatory Reynolds number, Re_o.

Material and Methods

1. Heat transfer experiments

A schematic diagram of the experimental setup is shown in Figure 1. A reactor tube with a total length of 500 mm was constructed by connecting cylindrical acrylic tubes with a length of 100 mm and an inner diameter of D_{in} = 14 mm, an outer diameter of D_{out} = 18 mm. Orifice baffles with a thickness of 2 mm and an aperture ratio of 0.25 were placed at intervals of 20 mm. 6 thermocouples were placed in each of the 4 cross sections of the connecting tubes at 100 mm intervals. The position of the 6 thermocouples is shown in Figure 2. The 4 sets of the thermocouples were placed in the middle position between the baffles. The entire reactor tube was submerged in a water bath kept at 0°C, and water kept at 55°C in a thermostatic bath flowed inside the tube at a net flow rate of 0.17 mL·s^{-1}. The oscillatory Reynolds number ($Re_o = \rho(2\pi f x_o)D_{in}/\mu$), a dimensionless number representing the intensity of the oscillatory flow, was varied at $0 \le Re_o \le 300$ where ρ [kg·m^{-3}]: fluid density, f [Hz]: oscillation frequency, x_o [m]: oscillation amplitude, μ [Pa·s]: fluid viscosity.

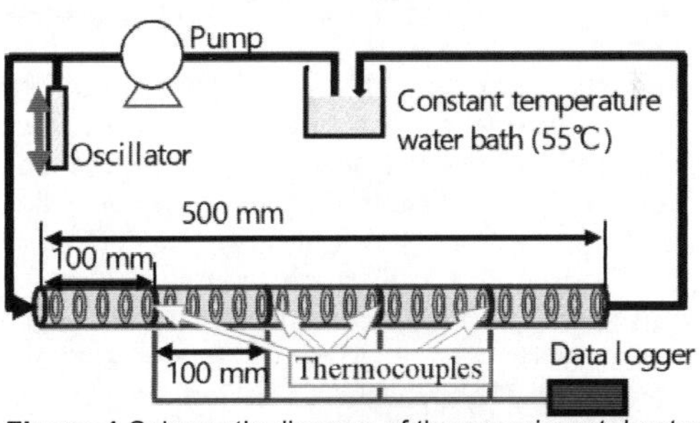

Figure 1 Schematic diagram of the experimental setup.

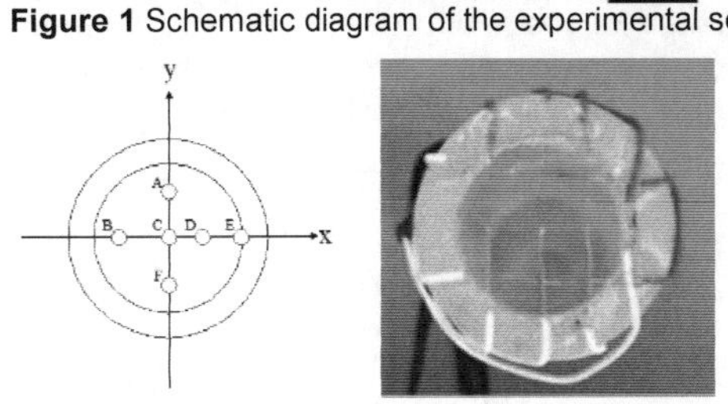

Figure 2 Positions of the 6 thermocouples at cross sectional view of the reactor tube.

The heat transfer performance was evaluated by determining the local Nusselt number from the temperature measurements. Q [W], the amount of heat lost from the fluid in one tube section where the interval distance is l [m] (l = 100 mm), is obtained as follows

$$Q = mC_p\Delta T_b = q\pi D_{in} l \quad (1)$$

where m [kg·s^{-1}] is the mass flow rate, C_p [J·(kg·K)$^{-1}$]

https://doi.org/10.52202/074122-0061

is the specific heat, ΔT_b [K] is the bulk temperature difference in the tube sections, q [W·m^{-2}] is the heat flux. The velocity-weighted average temperature was used for the bulk temperature, T_b. The heat transfer coefficient h [W·m^{-2}·K^{-1}] can be obtained from Eq. (2)

$$h = \frac{q}{T_w - T_b} \quad (2)$$

where T_w [K] is the wall surface temperature. The heat transfer performance under each oscillatory condition was evaluated by calculating local Nu.

$$Nu = \frac{hD_{in}}{k} = \frac{mC_p}{\pi lk}\frac{\Delta T_b}{T_w - T_b} \quad (3)$$

where k [W·m^{-1}·K^{-1}] is the thermal conductivity.

Flow visualization

Flow visualization experiments were performed using a baffled reactor tube with a refraction-suppressed correction vessel around it. Water was supplied with a 0.2 wt% concentration of tracer particles, i.e., Diaion particles. Tube geometry and flow conditions were the same as in the heat transfer experiments. Laser sheet light was irradiated, and images were taken using a high-speed camera at 500 fps. 2-D fluid measurement software (KATOKOKEN CO., LTD) was used to obtain flow velocity vector distributions at the centre of the baffle section and near the wall surface.

Results and Discussion

1. Heat transfer experiments

The average Nusselt number, $\overline{Nu}$, was also calculated from Eq. (3) using the time-averaged temperatures in one oscillatory flow cycle. The average Nu plotted against each Re_o is shown in Figure 3. The dashed line in Figure 3 shows Nu is 3.66 for laminar flow in a smooth circular tube under constant wall temperature. The average Nu without oscillation and with net flow only is $Nu \cong 5.3$ ($Re_o = 0$). The presence of the baffle alone is shown to improve heat transfer performance. As for the oscillatory condition, when $0 \leq Re_o \leq 80$, the average Nu increased only slightly with increasing Re_o, while for $80 \leq Re_o \leq 160$, the average Nu increased much more with increasing Re_o. As the Re_o was further increased ($160 \leq$ Reo), the average Nu became constant.

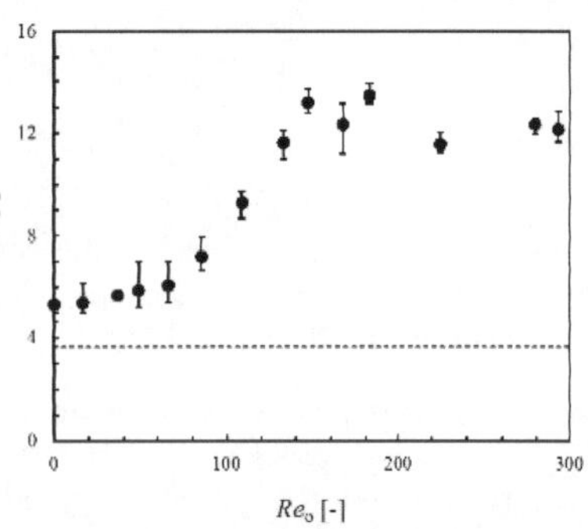

Figure 3 The averaged Nusselt number obtained by the time-averaged bulk and wall temperature against the oscillatory Reynolds number.

The temporal variation of Nu during one oscillatory flow cycle at the centre between the baffles is shown in the "Graphical Illustration". The elapsed time t [s] was normalized by the oscillation period τ [s]. For $Re_o < 100$, there was almost no time variation of Nu, but for $Re_o > 100$, there was an instantaneous and rapid increase in Nu with increasing Re_o. Furthermore, the Nu was unimodal at $Re_o < 200$ but became bimodal when $Re_o = 225$.

2. Flow visualization

Figure 4 shows the longitudinal streamlines between baffles obtained by Particle Image Velocimetry (PIV). When Re_o was 117, a symmetric doughnut-ring vortex generated downstream of the baffle was observed traveling downstream (to the right) with a period of 0.375. When the period was 0.750, the vortex flow moved toward the left, but did not reach the centre between the baffles. At $Re_o = 229$, the vortex flow reached the center of the baffles at both 0.375 and 0.750. At $Re_o = 277$, the symmetry of the vortex flow was broken, and random vortices were observed.

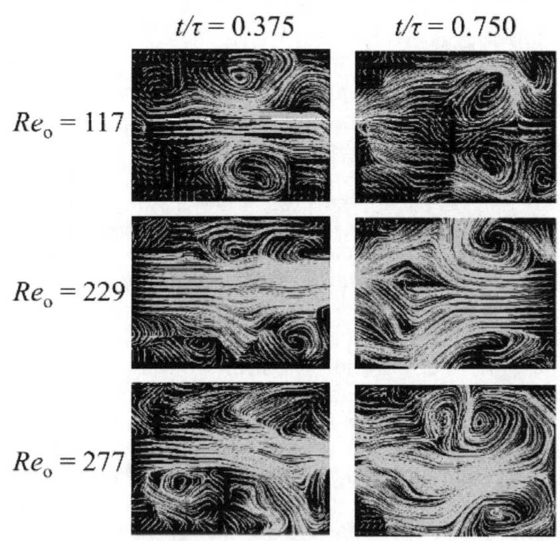

Figure 4 Longitudinal streamlines (visualized vortex flow) between baffles at various Re_o and at oscillatory flow periods of 0.375 and 0.750.

Figure 5 shows the time variation of the radial velocity during one oscillatory flow cycle: for $Re_o > 100$, the velocity was instantaneously increased around $t/\tau = 0.25$. It indicates that the vortex generated near the upstream baffle during forward flow reached the center of the baffle section (It is also the location of the section where the thermocouples were installed.). Therefore, the local Nu increased during the passage of the vortex flow. Furthermore, for $Re_o > 190$, the vortex generated near the downstream baffle also reached the center of the baffle section during the backflow due to the oscillatory flow, resulting in a bimodal Nu profile.

https://doi.org/10.52202/074122-0061

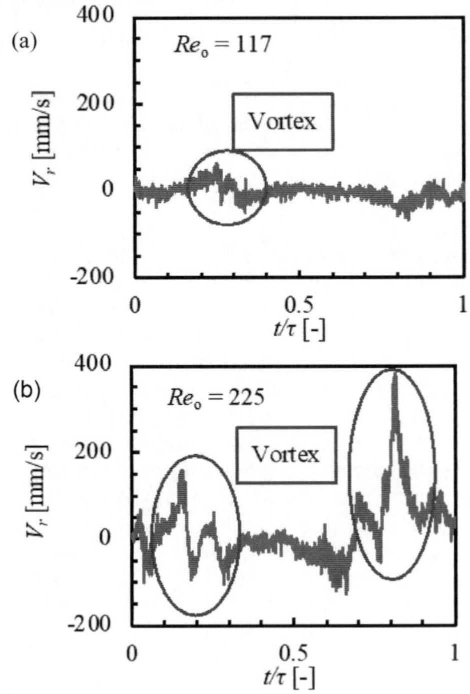

Figure 5 Radial component of the velocity vector in the 2.7 mm square area near the wall of the middle of the baffle section, at Re_o (a) 117 and (b) 225.

Conclusions

For $Re_o > 100$, a rapid increase in radial velocity at the vicinity of the wall due to vortex generation could be observed resulting in an instantaneous and rapid increase in *Nu*. The increase in Re_o generates an unsteady vortex, which promotes radial mixing. Accordingly, the heat transfer performance was also found to improve instantaneously.

Acknowledgments
This research was supported by JSPS KAKENHI Grant No. JP18H03853 and JP 19KK0127.

References

[1] X. Ni, H. Jian. A.W. Fitch, Chemical Engineering Science, 57 (2002) 2849-2862
[2] A. W. Fitch H. Jian, X. Ni, Chemical Engineering Journal, 112 (2005) 197-210

https://doi.org/10.52202/074122-0061

Power, mixing and flow dynamics of the novel single-use Allegro™ stirred tank reactor

Poster.16

***J. Delbridge*[1]**, *T.A. Barrett*[2], *A. Ducci*[3], *M. Micheletti*[1]. *(1) Dept. of Biochemical Engineering, University College London, Gower Street, London WC1E 6BT, UK, j.delbridge.17@ucl.ac.uk. (2) Pall Corporation, Southampton Road, Portsmouth PO6 4BQ, UK, timothy_barrett@europe.pall.com. (3) Dept. of Mechanical Engineering, University College London, Gower Street, London WC1E 6BT, UK, a.ducci@ucl.ac.uk.*

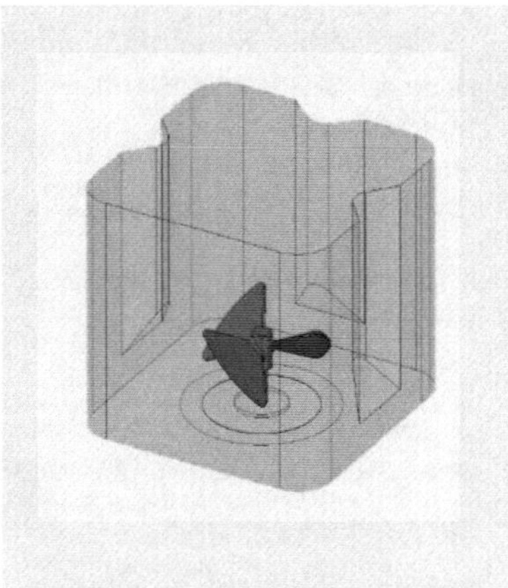

Novel single-use bioreactor (SUB) designs can prove highly beneficial in biotherapeutics production, although unlike traditional STRs they lack extensive characterisation which can inform operating conditions and scale-up. A scale-down 1L mimic (abbreviated A-STR) of Pall Corporation's 50-2000 L Allegro STR SUB range was produced for characterisation of impeller power input, mixing dynamics and flow structure. In addition, an equivalent volume 'standard' cylindrical STR was characterised for comparison. The corners of the A-STR were found to increase impeller power number by acting as an additional form of baffling. Asymmetric baffle placement resulted in a flow structure differing from the norm for the Elephant Ear (EE) impeller, with the wedge shaped baffles redirecting flow upwards. The EE impeller performed best in the A-STR in down-pumping mode, with significantly faster mixing times compared to the standard configuration.

Introduction

Biopharmaceuticals currently dominate the global pharmaceutical market, with therapeutics in the form of monoclonal antibodies (mAbs), vaccines, and cell and gene therapies to name a few [1, 2]. Screening for highly productive protein producing cell lines and optimising cell culture conditions in initial small-scale studies is vital in developing new therapeutics. Despite these efforts, scale-up to industrial scale stirred tank reactors (STRs) is often associated with a decrease in cell productivity. Economic feasibility of large scale production is improved by implementation of single-use bioreactors (SUBs), which have reduced downtime, contamination risks, and lower costs compared to traditional stainless steel STRs. Pall Corporation's Allegro™ STR range of 50-2000 L is one such single-use bioreactor range. The Allegro STR was notably used in the development of the ChAdOx1 nCoV-19 vaccine, achieving low costs and high productivity with scale-up.

Like many single-use bioreactors, the Allegro STR range has a novel design differing from the standard cylindrical dished head tank. A single-use biocontainer of square cross-section improves practicality for installation into the surrounding support structure, which forms rounded corners and three asymmetrically positioned wedge-shaped baffles leaving an open front viewing face. While the traditional cylindrical baffles stirred tank has been used for decades of operation, characterisation of novel geometries to determine the implications on mixing and flow structure is key to understand the microenvironment experienced by circulating cells.

By developing a 1 L scale-down mimic of the Allegro STR range, this work aims to characterise the implications of this novel geometry on impeller power consumption, mixing, and flow dynamics. Moreover, a cylindrical tank with standard baffling stirred by the same elephant ear impeller is characterised for comparison.

Material and Methods

A transparent 1 L prototype (Figure 1), hereafter referred to as the A-STR, of the Allegro STR 50-2000 L range was fabricated at University College London using vendor provided dimensions of the Allegro STR 50. Characteristic ratios (D/T=0.5, C/T=0.25, W/D=0.5) were kept constant with scale-down and in the equivalent volume cylindrical STR. The maximum working volume, where $H_L/T=1.1$ is used as a reference condition. The cylindrical vessel used for comparison as a 'standard' stirred tank (referred to as S-STR) had four baffles of width Bw/T=0.1 and 2 mm thickness. A hemispherical dished bottom was used in the cylindrical tank given its prevalence over the flat bottom configuration in commercial bioreactors [3].

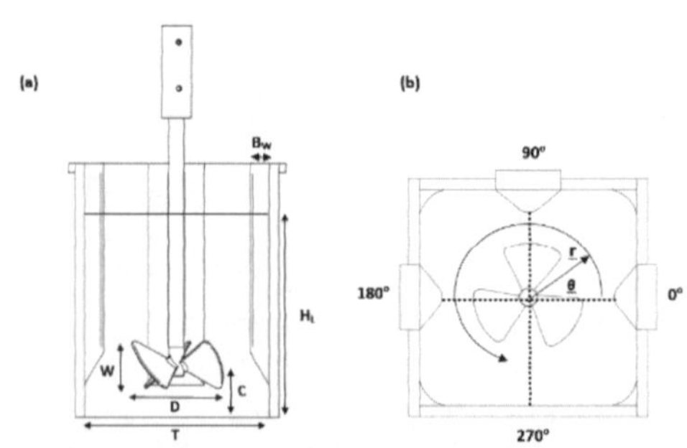

Figure 1. Schematic of scale-down A-STR (a) side view (b) top view with dashed lines indicating PIV measurement planes (θ=0°, 90°, 180°, 270°).

The 3-bladed 45° pitched elephant-ear impeller was 3D printed to a diameter of 50 mm with a blade thickness of 2 mm. Anti-clockwise impeller rotation corresponds to up-pumping (UP) mode while down-pumping (DP) is clockwise.

To acquire impeller power number data, the STRs were fitted to a frictionless air-bearing system. Pressurised air supplied at 0.25 bar eliminate friction from measurements and allowed the tanks to freely rotate. A digital force gauge (Omega Engineering, UK) was used to measure force readings from the laminar (Re ~ 1) to turbulent (Re~2.4x10^4) regimes. Reynolds number of the working fluid was varied by adjusting impeller agitation rate and rheological properties of the working fluid by using mixtures of Milli-Q water and glycerol.

Flow characterisation was carried out using Particle Image Velocimetry (PIV) measurements. A pulsed Nd:Yag laser with wavelength λ=532 nm and power output 2x50 mJ was used with a cylindrical lense to generate a vertical laser sheet of approximately 1 mm thick. The tank was filled with Milli-Q water and seeded with Rhodamine-coated polymethyl methacrylate particles of 1-20 µm diameter. The 8x8 pixel interrogation area had a final resolution of 4x10^{-4} m. Ensemble-averaged values were determined from 3000 image pairs, with a timestep of 700 ms between pairs. For flow characterisation the A-STR and S-STR were operated at equivalent power inputs in the typical range for mammalian cell culture (~45 W/m^3). This corresponds to agitation at 220-240 RPM (Re ~1.1x10^4).

The Dual Indicator System for Mixing Time (DISMT) was used for mixing characterisation. A fast acid-base reaction in the presence of methyl red and thymol blue indicators results in a colour change from red to yellow. Recorded images of the colour change are processed using an in-house developed MATLAB code to determine the standard deviation in green pixel intensity from which mixing time is estimated.

Results and Discussion

Impeller power number was investigated across a range of Re=1.0-2.5x10^4. A summary of impeller power number in the turbulent regime is presented in Figure 2.

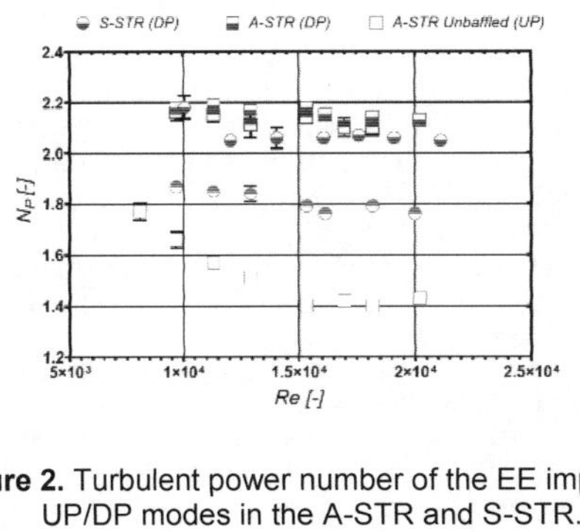

Figure 2. Turbulent power number of the EE impeller in UP/DP modes in the A-STR and S-STR.

The A-STR in UP and DP modes had a turbulent power number of Np = 2.11 ± 0.01 and 2.17 ± 0.02 respectively. These were in agreement with the values of 1.93 ± 0.4 and 2.19 ± 0.2 determined from manufacturer provided torque readings on the Allegro STR 200 L bioreactor, indicating good scalability of the 1 L prototype. Operation without baffles (UP) resulted in a 34% decrease in power number to N_P=1.4. The plateau of the power number, which would typically continually decrease in unbaffled tanks [4], [5], is indicative of the natural baffling provided by the square cross-section. Comparison of N_P of the A-STR with that of the S-STR shows a significantly higher power number due to the presence of baffles as well as the tank corners, causing additional disturbance of flow. N_P in the S-STR (UP) was determined to be 1.81 ± 0.04, 14% lower than that in the A-STR (UP). In DP mode, the power number of the S-STR is 2.09 ± 0.04, more similar to that of the A-STR.

Ensemble-averaged velocity magnitudes and vector fields of the A-STR and S-STR were compared to determine the influence of vessel geometry. Asymmetric baffle placement in the A-STR results in a varied flow structure across the measured planes. The plane at θ=270°, perpendicular to the unbaffled front face of the vessel contains a strong circulation loop in the lower half of the vessel, and a weak anticlockwise circulation loop above z/H=0.5. This is similar to the typical velocity field of the EE impeller seen in the S-STR. Counter clockwise rotation of the impeller through the baffled planes at θ=0° to θ=90° and subsequently θ=180° results in impeller discharge orientation to the upward direction. Baffled planes at θ=90° and θ=180° exhibit a single large circulation loop rather than a primary and secondary recirculation loop.

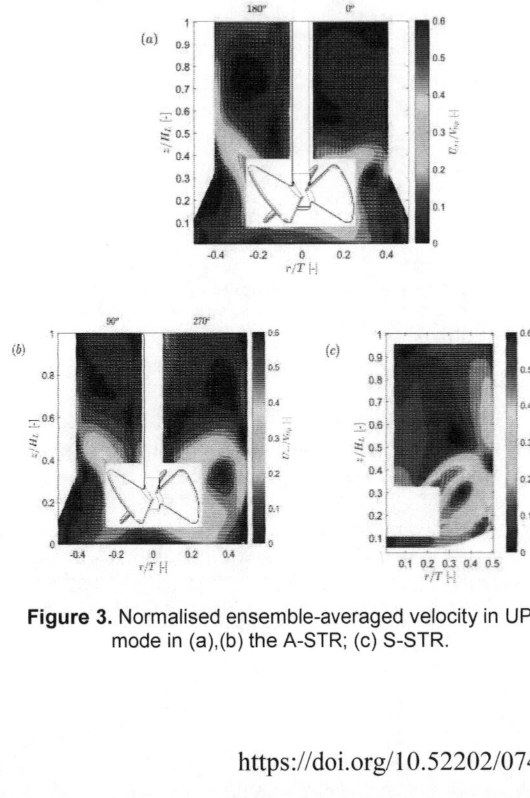

Figure 3. Normalised ensemble-averaged velocity in UP mode in (a),(b) the A-STR; (c) S-STR.

https://doi.org/10.52202/074122-0062

Flow structure asymmetry was less prominent in DP mode in the A-STR geometry. All measured planes indicate one large anticlockwise circulation loop generated by the EE impeller (Figure 4a). High velocities near the liquid surface of the A-STR indicate good macromixing, unlike the S-STR (DP) which exhibits low velocity magnitudes above z/H_L=0.7 (Figure 4b).

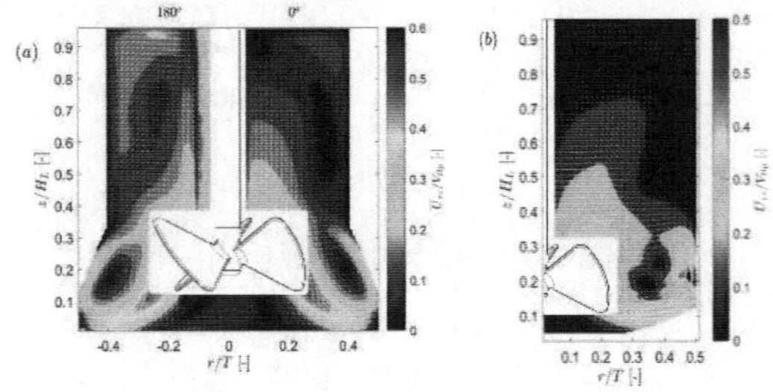

Figure 4. Normalised ensemble-averaged velocity in DP mode in (a) the A-STR; (b) S-STR.

Analysis of kinetic energy distribution (data not shown) indicated that the A-STR (DP) had a more homogenous distribution compared to the other configurations, where kinetic energy was more localised to the impeller discharge region. The implications of this are evident in comparison of mixing efficiency (Figure 5). Mixing performance across the investigated configurations are compared on the basis of non-dimensional mixing time, Nt_m, representing the number of impeller revolutions to achieve 95% homogeneity.

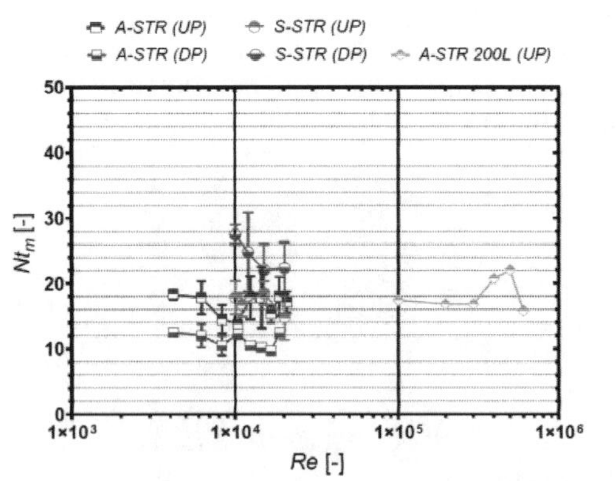

Figure 5. Non-dimensional mixing number in the A-STR in UP (⬒) and DP (⬓) modes compared to that of the S-STR in UP (◓) and DP (◒) modes. Data on the commercial 200 L STR (UP ⬘) indicates scalability.

Mixing time data from past work on the Allegro STR 200 L [6] was used to determine the non-dimensional mixing time in UP mode. The mixing number of 18.2 ± 2.5 revolutions was in agreement with the value of 16.7 ± 1.7 determined for the 1 L A-STR in this work, indicating good scalability. DP mixing time data at 200 L scale were not available for comparison. While mixing performance of the A-STR (UP) and S-STR (UP) were roughly equivalent, the S-STR (DP) performed relatively poorly with the longest mixing times. This is likely due to the poor circulation in the upper tank region as exhibited in Figure 3b. The A-STR (DP) outperformed all other configurations, with rapid mixing time, most likely due to the homogenous distribution of kinetic energy.

Conclusions

Comparison of power and mixing number data with that of the industry scale Allegro STR 200 illustrate that the scale-down 1 L A-STR is an appropriate mimic of industry scale conditions. Power number of the EE impeller was higher in the A-STR relative to the S-STR due to the presence of corners acting as an additional form of baffling. The novel geometry of the A-STR results in an asymmetric flow structure differing to that in the standard configuration in UP mode. In DP mode, velocity magnitudes in the A-STR were higher near the tank surface. Kinetic energy was found to be more homogenously distributed in the A-STR (DP), which also had the fastest mixing time when comparing configurations. Operation of the A-STR in DP mode is recommended for suspended mammalian cell culture, where rapid mixing and homogenous kinetic energy distribution will reduce exposure of cells to spatial heterogeneities and improve product quality and yields.

Acknowledgments

The authors would like to acknowledge the funding and support of the University College London – Pall Corporation Centre of Excellence and the UKRI Engineering and Physical Sciences Research Council (EPSRC) CDT Bioprocess Engineering Leadership Grant EP/S021868/1. Special thanks to the UCL mixing group.

References

[1] R. M. Lu *et al.*, "Development of therapeutic antibodies for the treatment of diseases," *J. Biomed. Sci.*, vol. 27, no. 1, pp. 1–30, 2020.

[2] H. Narayanan, M. Sponchioni, and M. Morbidelli, "Integration and digitalization in the manufacturing of therapeutic proteins," *Chem. Eng. Sci.*, vol. 248, no. 117159, pp. 1–24, 2022.

[3] C. Schirmer, R. W. Maschke, R. Pörtner, and D. Eibl, "An overview of drive systems and sealing types in stirred bioreactors used in biotechnological processes," *Appl. Microbiol. Biotechnol.*, vol. 105, no. 6, pp. 2225–2242, 2021.

[4] S. C. Kaiser, S. Werner, V. Jossen, M. Kraume, and D. Eibl, "Development of a method for reliable power input measurements in conventional and single-use stirred bioreactors at laboratory scale," *Eng. Life Sci.*, vol. 17, no. 5, pp. 500–511, 2017.

[5] W. Bujalski, A. W. Nienow, S. Chatwin, and M. Cooke, "The dependency on scale of power numbers of Rushton disc turbines," *Chem. Eng. Sci.*, vol. 42, no. 2, pp. 317–326, 1987.

[6] A. W. Nienow, B. Isailovic, and T. A. Barrett, "Design and Performance of Single-Use, Stirred-Tank Bioreactors," *Bioprocess Int.*, vol. 14, no. 10, pp. 12–21, 2016.

https://doi.org/10.52202/074122-0062

Case Study: How to realize the full potential of your heterogeneous catalytic hydrogenation

Poster.17

***K. Epp*[1]**, *K. Moebus*[2], *(1) EKATO Ruehr- und Mischtechnik, Hohe-Flum-Strasse 37, Schopfheim, Germany, konstantin.epp@ekato.com. (2) Evonik Operation GmbH, Rodenbacher Chaussee 4, 63457 Hanau-Wolfgang, Germany, konrad.moebus@evonik.com*

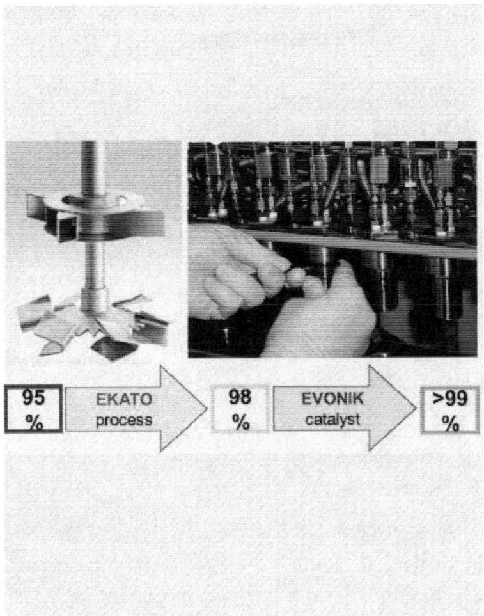

In this work a novel heterogenous catalytic hydrogenation process developed by a customer in lab scale should be optimised to achieve a conversion of >95% and scale up to commercial scale. In the pilot phase an optimised gas-liquid mixing stirrer including a self-aspirating turbine achieved excellent results (98% conversion) and the reaction time was reduced by half. Next to this mechano-physical mass transfer limitations associated with power/volume ratio and stirrer geometry; chemical limitations rooted from the catalyst were observed. In collaboration with the catalyst manufacturer Evonik a detailed catalyst screening study elaborated the most promising catalyst candidate, resulting in a record conversion of 99%. Since not only a catalytic high activity is important but also the lifetime over several catalytic cycles a subsequential study tackling the reusability of the catalyst indicated a good lifetime demonstrating also lowered operating costs. These combined approaches considering the mechano-physical but also the chemical limitations allowed a strong improvement of the catalytic conversion up to 99%.

Introduction

A multi-phase gas-liquid-solid heterogeneously catalysed reaction is intrinsically complex, since the reaction rate is influenced by many variables such as mechanical (impeller type, power input), physical (temperature, pressure, viscosity), as well as chemical (catalyst type, solvent, educt concentration) variables [1]. To optimise a reaction, it is necessary to look at the whole reaction system and to consider all mechanical, physical and chemical parameters and their interdependencies. Generally, the main reaction physical and chemical parameters are screened in a feasibility study in lab scale.

In this work the objective was to optimise an existing hydrogenation process which was developed in lab scale with respect to conversion (>95%) and its scale up to commercial scale. By achieving this goal, the subsequent purification of the resulting hydrogenated product would be facilitated, saving time and costs.

One major task is to detect possible mass transfer limitations and to use tools to decrease or ideally, completely remove these mass transfer limitations.

Material and Methods

In this work a home-built 5-liter stainless steel autoclave (EKATO ELA-5) including impeller, 4 baffels, hydrogen dip-pipe, various temperature and pressure indicators was used to perform the hydrogenation test work. Heating and cooling of the autoclave was done via the autoclave double-jacket using an external heating/cooling system and a standard heat transfer fluid. Importantly, the autoclaved was equipped with a mass flow controller (MFC) to measure the hydrogen flow during the reaction. The educt was an organic, bio-derived material. Hydrogen was used a reaction gas. As solid catalysts, a variety of Pd-based catalysts supported on various supports, such an activated carbon and alumina were used.

Results and Discussion

In the first hydrogenation the parameters of the existing process were reproduced 1:1 and the hydrogen flow or hydrogen consumption was measured during the reaction (Figure 1).

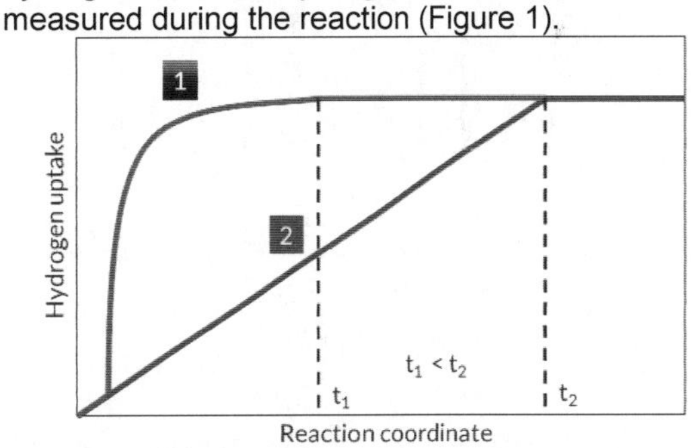

Figure 1. Blue line (1): Visualisation of the cumulative hydrogen uptake in an ideal, unlimited reaction. Red line (2): Visualisation of the cumulative hydrogen uptake in the exp. hydrogenation reaction.

In contrast to the ideal hydrogen uptake behaviour in a theoretically unlimited reaction, a linear hydrogen uptake was observed (red line), indicating a mass transfer limitation. Assuming a gas-liquid mass transfer limitation, the impeller speed was increased, resulting in an accelerated hydrogen uptake rate. Due to the increased power input and higher recirculation of secondary hydrogen gas, the reaction time was halved.

https://doi.org/10.52202/074122-0063

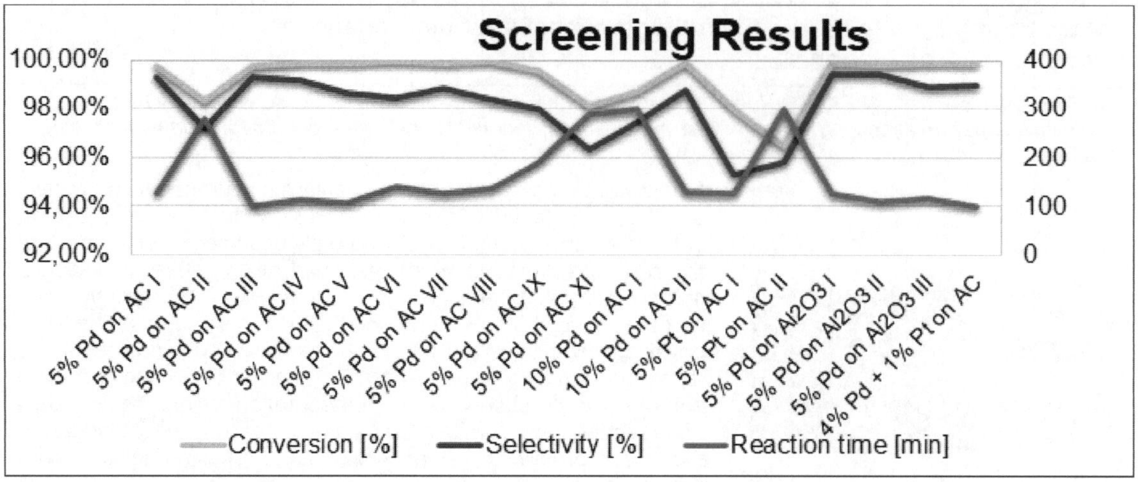

Figure 2. Catalyst screening study using a variety of different noble metals supported on activated carbon (AC) or alumina performed by Evonik Operations GmbH. Secondary y-axes display the reaction time in minutes.

In the catalyst screening study performed at Evonik facilities, a broad variety of catalysts were tested using a high-throughput screening system performing 8 parallel reactions under the same conditions. The two most promising catalyst candidates were selected, and their reusability was further evaluated in a subsequent reusability study.

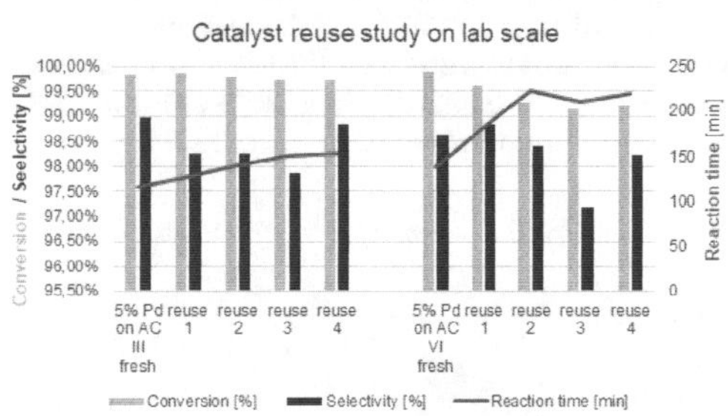

Figure 3. Results of the reusability of the two most promising catalysts.

Catalyst "AC III" could be successfully reused four times without any drop in the catalyst conversion and almost constant reaction time. In contrast, catalyst candidate "AC VI" revealed a negative trend of the conversion with increasing reuses and also a significant increase in reaction time.

Conclusions

In this study the conversion of a given catalytic process was optimised in pilot trials using more suitable impellers and higher power input. The basis for this optimisation was the determination of the linear hydrogen uptake rate which indicated a possible gas-liquid mixing limitation. Moreover, a comprehensive catalyst screening study revealed even more appropriate catalyst candidates to further push the conversion to >99%. Two of the most promising candidates were further tested in four consecutive reuses, highlighting that only one of the candidates showed constant conversion and its potential to be used in a cost-effective industrial process.

Acknowledgments

We kindly thank Dr. Konrad Möbus from Evonik Operations GmbH for his valuable contributions in this study.

References

[1] EKATO the book, Schopfheim, Germany 2012.

Solid-Liquid Suspension in Continuous Oscillatory Baffled Reactors – Experimental Study

Poster.18

Margarida S.C.A. Brito[1,2], *António M. Ferreira*[1,2], *Fernando A. Rocha*[1,2], *Ricardo J. Santos*[2,3], *Joelle Aubin*[5]
(1) Laboratory of Process Engineering, Environment, Biotechnology and Energy (LEPABE), Faculty of Engineering, University of Porto, Rua Dr. Roberto Frias, 4200-465 Porto, Portugal, mbrito@fe.up.pt. (2) ALiCE –Associate Laboratory in Chemical Engineering, Faculty of Engineering, University of Porto, Rua Dr. Roberto Frias, 4200-465 Porto, Portugal. (3) Laboratory of Separation and Reaction Engineering –Laboratory of Catalysis and Materials (LSRE-LCM), Faculty of Engineering, University of Porto, Rua Dr. Roberto Frias, 4200-465 Porto, Portugal. (5) Laboratoire de Génie Chimique, Université de Toulouse, CNRS, INPT, UPS, Toulouse, France

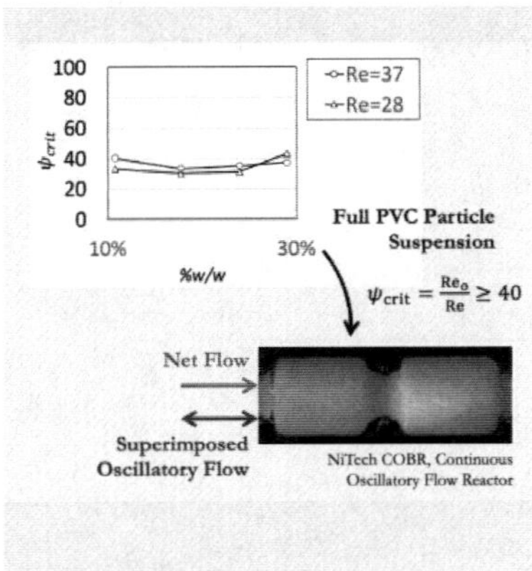

This work addresses an experimental study on the suspension of particles in a continuous oscillatory flow NiTech reactor. The main objective is the understanding of the flow interaction of time dependent flow and solid particle suspensions. A system of PVC particles and water was studied since it mimics the flow condition in crystallization reactors, which is one of the main application of this work. Four mass fractions of particles between 10% and 30% (w/w) were studied because it also corresponds to the typical operating conditions in pharmaceutical crystallization. This work studies the impact of mass fraction of solids, the oscillatory conditions and the net flow Reynolds number on the fluid-particle suspension. Results show that the ratio between net velocity and oscillatory velocity must be at least above 40 to guarantee the suspension of particles, which is in agreement with previous fundamental works on crystallization.

Introduction

Conventionally, solid-liquid mixing operations are performed in batch stirred tanks and a vast amount of knowledge on the design and performance of stirred tanks has been generated over the last 70 years. Process Intensification (PI) technologies as a means for more sustainable and green chemical processes has been the focus for solid suspensions over recent years. Continuous oscillatory flow reactors (COFRs), including both continuous oscillatory baffled reactors (COBRs) and pulsed helix reactors, have demonstrated successful operation for solid suspension because the superimposed periodic oscillatory flow upon a net flow enhances mixing and prevent clogging, solid deposition or sedimentation at reactor walls [1]. COFRs have been used for various applications including dispersion and slurry dissolution, crystallization and precipitation, solid catalysed reactions, suspension polymerization and slurry processing [2].

The oscillatory flow is characterized by a frequency and amplitude, that interacts with the geometrical structure of the tube, causing effective transverse flow and enhanced transport and mixing. The net flow controls the residence time inside the reactor. Since the oscillatory flow typically dominates the net flow rate, mixing and residence time are uncoupled, and COFRs have been shown to provide good radial mixing, limited axial dispersion and long residence times in compact geometries, i.e. short reactor lengths [3-5].

COBRs are characterized by different dimensionless numbers, such as the net flow Reynolds number (Re), oscillatory flow Reynolds number ($\mathrm{Re_o}$), Strouhal number (St) and velocity ratio (ψ), which are defined by $\mathrm{Re} = \frac{\rho v_{\mathrm{inj}} D}{\mu}$, $\mathrm{Re_o} = \frac{2\pi \rho x_0 f D}{\mu}$, $\mathrm{St} = \frac{D}{4x_0}$ and $\psi = \frac{\mathrm{Re_o}}{\mathrm{Re}}$, where ρ is the density of the fluid, μ is the viscosity of the fluid, D is the reactor's diameter, v_{inj} is the net velocity, x_0 is the oscillation amplitude and f is the frequency of the oscillation [6].

In $\mathrm{Re_o}$, the characteristic velocity is the maximum oscillatory velocity. $\mathrm{Re_o}$ describes the intensity of mixing in COBR. For $\mathrm{Re_o} < 250$, the flow is essentially 2D and axi-symmetric with low mixing intensity; for $\mathrm{Re_o} > 250$, the flow becomes 3D and mixing is more intense; when $\mathrm{Re_o} > 2000$ the flow is turbulent [7].

The Strouhal number measures the effective eddy propagation with relation to the tube diameter. The most common range of St number is $0.15 < \mathrm{St} < 4$ [8].

The recommended velocity ratio, ψ, is greater than 1 to ensure that the oscillatory flow dominates the superimposed net flow. This ratio has been used largely to describe the plug flow behaviour in the COBR [7]. Previous studies recommend to work at conditions when the radial flow dominates and limits axial dispersion that corresponds to $2 < \psi < 4$ [7]. However, these recommendations are not always used in practice, particularly in solid-liquid suspensions where COBRs have been operated with velocity ratios near the upper limit of the recommended range, or even at significantly higher values, to ensure the solid suspension and uniform particle size distribution [9-13]. The challenge of solid-liquid suspensions in COFRs involves overcoming sedimentation, which depends on the interaction of oscillatory conditions (amplitude frequency), solid-phase characteristics (size, shape, density, concentration), and liquid properties [1].

This work focuses on the suspension of particles

https://doi.org/10.52202/074122-0064

in a COBR, which aims at understanding the complex interaction of time dependent flow and solid particle suspensions. Experiments show the impact of volume fraction, frequency, amplitude and net flow rate on the fluid-particle interaction.

Material and Methods

The geometry studied in this work was the NiTech COBR, which is a single orifice baffled reactor with smooth constrictions. The reactor has a diameter of 15 mm with 7.5 mm diameter orifices and the distance between orifices is 16.9 mm.

The solid-liquid suspension consists of a system of water and PVC particles. The experimental conditions were chosen to simulate the typical conditions in pharmaceutical crystallization: diameter of the particles between 100 μm and 200 μm, density of the fluid close to 1 g/cm3 and concentration of solids between 10% (w/w) and 30% (w/w).

The mass fractions of PVC particles in the liquid studied in this work were 11% (w/w), 18% (w/w), 24% (w/w) and 30% (w/w). For each mass fraction, frequency was studied at 1.75 Hz, 1.5 Hz, 1.25 Hz and 1Hz and at amplitude range between 2.1 mm and 27.7 mm in order to visualize the solid suspension at different $\mathrm{Re_o}$ conditions. For all conditions studied in this work, the oscillatory Reynolds number is $1500 < \mathrm{Re_o} < 2000$, which means that the flow is 3D but mixing is not under fully turbulent flow conditions. The net flow Reynolds number was operated at two conditions $\mathrm{Re} = 37$ and $\mathrm{Re} = 28$.

Images of the suspension were captured using a Thorlabs camera.

Results and Discussion

Figure 1 shows the 11% (w/w) suspension at $\mathrm{Re} = 37$, $f = 1.75$ Hz for $\psi = 53$, $\psi = 42$ and $\psi = 35$. At $\psi = 53$ (Figure 1a), the particles are fully suspended. At $\psi = 42$ (Figure 1b), it is possible to visualize the formation of recirculating flow at the top of the reactor, and at $\psi = 35$ (Figure 1c), the sedimentation of particles can be observed.

The critical velocity ratio (ψ_{crit}) and the amplitude ($x_{0,crit}$) for the transition between the total settling and suspension were identified by the visual inspection.

Figures 2 and 3 show the experimental results for x_0 and ψ_{crit}, respectively. Figure 2 shows that x_0 for suspension decreases with the increase of frequency. Studies on particle suspension in a batch OBR also showed the critical amplitude decreases with frequency [14-17].

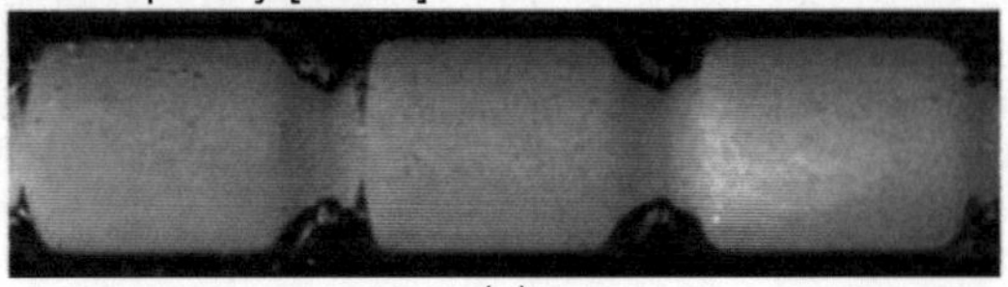

(a)

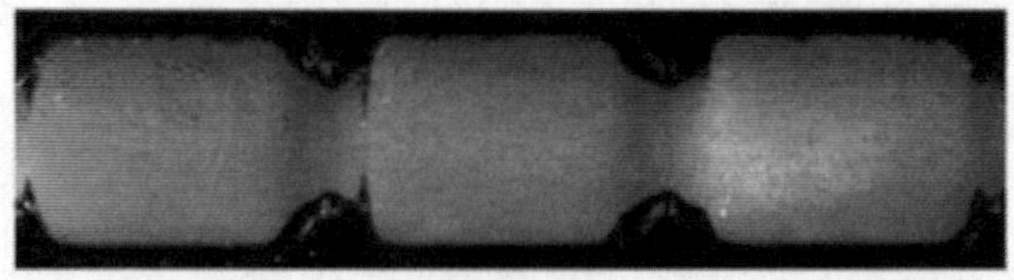

(b)

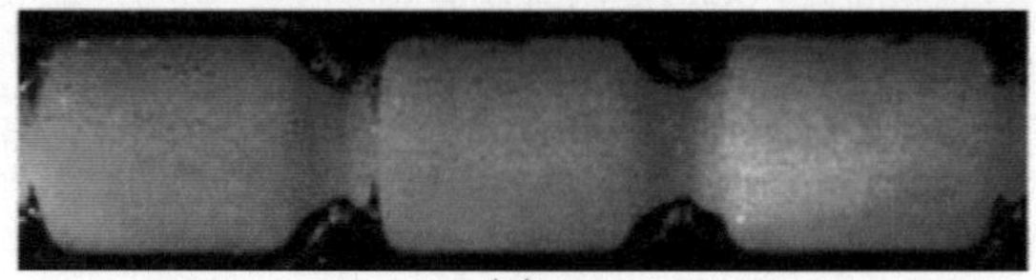

(c)

Figure 1. Images of 11% (w/w) suspension at Re=37 and at $f = 1.75$ Hz for different velocity ratios (a) $\psi = 53$; (b) $\psi = 42$; (c) $\psi = 35$.

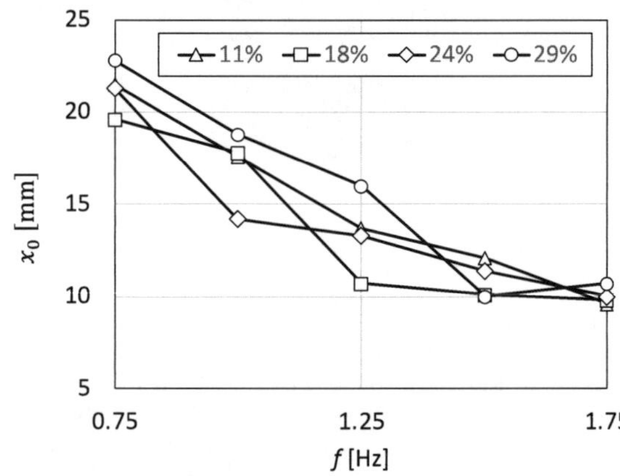

Figure 2. Critical amplitude of oscillation for total suspension of solids as a function of frequency of oscillation in a NiTech reactor at Re=37 with 11% (w/w), 18% (w/w), 24% (w/w) and 29% (w/w).

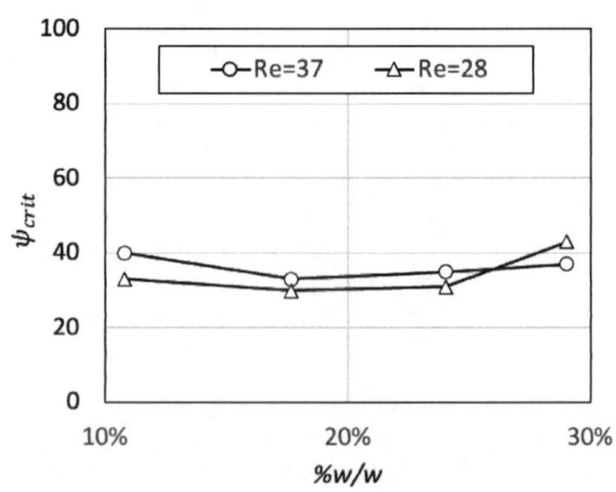

Figure 3. Critical velocity ratio for total suspension of solids as a function of mass fraction of particles 11% (w/w), 18% (w/w), 24% (w/w) and 29% (w/w) at Re=37 and Re=28.

Figure 3 shows the results of the critical velocity ratio (ψ_{crit}) as a function of mass fraction of solid particles. For the studied range of mass fraction (10-30% w/w) and both net flow Reynolds number, the velocity ratio is the same to keep the particles suspended. Results show that $\psi \geq 40$ is the range to enable the full suspension of particles. These results are in agreement with previous works that reported the full particle suspension in COBRs is achieved from significantly high velocity ratios, which contradicts the general recommendation of $2 < \psi < 4$ that ensure plug flow conditions.

Conclusions

This work addresses the experimental work on solid-liquid suspensions in a NiTech continuous oscillatory baffled reactor. Two net flow Reynolds numbers ($\mathrm{Re} = 37$ and $\mathrm{Re} = 28$) and oscillatory Reynolds numbers in the range $1500 < \mathrm{Re}_o < 2000$ were studied The oscillation frequency was varied from 0.75 to 1.75 Hz and amplitude range was set between 2.1 mm and 27.7 mm. Results show that the oscillatory velocity has to be at least 40 times higher than the net flow velocity to guarantee the full suspension of particles. The critical amplitude to avoid the settling of particles decreases with the oscillation frequency.

Acknowledgments

This work was financially supported by: LA/P/0045/2020 (ALiCE), UIDB/50020/2020 e UIDP/50020/2020 (LSRE-LCM), UIDB/00511/2020 and UIDP/00511/2020 (LEPABE) by national funds through FCT/MCTES (PIDDAC), by NORTE-01-0145-FEDER-000077

References

[1] Slavnić, D., B. Bugarski, and N. Nikačević. *Chemical Engineering and Processing - Process Intensification*, 135 (2019),

[2] Avila, M., B. Kawas, D.F. Fletcher, M. Poux, C. Xuereb, and J. Aubin. *Chemical Engineering and Processing - Process Intensification*, (2021),

[3] Mazubert, A., D.F. Fletcher, M. Poux, and J. Aubin. *Chemical Engineering and Processing: Process Intensification*, 108 (2016),

[4] McDonough, J., A. Phan, and A. Harvey. *Chemical Engineering Journal*, 265 (2015),

[5] Ni, X., H. Jian, and A.W. Fitch. *Chemical Engineering Science*, 57 (2002), 14.

[6] Avila, M., D.F. Fletcher, M. Poux, C. Xuereb, and J. Aubin. *Chemical Engineering Science*, 219 (2020),

[7] Stonestreet, P. and P.M.J. Van Der Veeken. *Chemical Engineering Research and Design*, 77 (1999), 8.

[8] Abbott, M.S.R., A.P. Harvey, G.V. Perez, and M.K. Theodorou. *Interface Focus*, 3 (2013), 1.

[9] Briggs, N.E.B., U. Schacht, V. Raval, T. McGlone, J. Sefcik, and A.J. Florence. *Organic Process Research & Development*, 19 (2015), 12.

[10] Kacker, R., S.I. Regensburg, and H.J.M. Kramer. *Chemical Engineering Journal*, 317 (2017),

[11] Peña, R., J.A. Oliva, C.L. Burcham, D.J. Jarmer, and Z.K. Nagy. *Crystal Growth & Design*, 17 (2017), 9.

[12] Siddique, H., C.J. Brown, I. Houson, and A.J. Florence. *Organic Process Research & Development*, 19 (2015), 12.

[13] Zhao, L., V. Raval, N.E.B. Briggs, R.M. Bhardwaj, T. McGlone, I.D.H. Oswald, and A.J. Florence. *CrystEngComm*, 16 (2014), 26.

[14] Cruz, P., C. Silva, F. Rocha, and A. Ferreira. *Industrial & Engineering Chemistry Research*, 60 (2021),

[15] Ejim, L.N., S. Yerdelen, T. McGlone, I. Onyemelukwe, B. Johnston, A.J. Florence, and N.M. Reis. *Chemical Engineering Journal*, 308 (2017),

[16] Reis, N., A.P. Harvey, M.R. Mackley, A.A. Vicente, and J.A. Teixeira. *Chemical Engineering Research and Design*, 83 (2005), 4.

[17] Takriff, M.S., R. W., D. W., F. Z., and A. Z. *Journal of Industrial Technology*, (2000), 9.

https://doi.org/10.52202/074122-0064

Characterization of Gas-Liquid Mass Transfer in a 15L Stirred Tank: Identifying K_L and a.

Poster.19

***Johan L.N.D Carfort*[1]**, *Tiago Pinto*[,2], *Ulrich Krühne*[1]. *(1) Technical University of Denmark, Department of Chemical and Biochemical Engineering, PROSYS, Søltofts Plads building 228A, Kongens Lyngby, Denmark, jlne@kt.dtu.dk, ulkr@kt.dtu.dk. (2) R&D Department, UNIBIO A/S, Langebjerg 1, Roskilde, Denmark, tp@unibiogroup.com.*

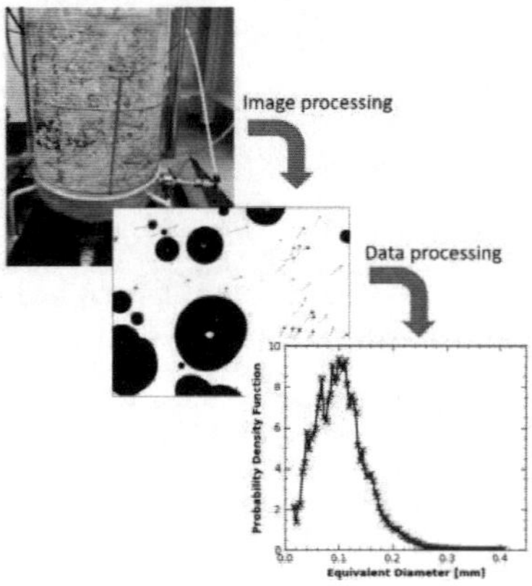

Gas-liquid mass transfer is an important transport phenomenon in a wide range of bioprocesses such as aerobic fermentations, to provide the necessary oxygen to the growth of microorganisms. In this study, the mass transfer in a small scale stirred tank is measured at different operating conditions and the individual contributions from the interfacial area and mass transfer coefficients are quantified, using modern measurement techniques. The results show that the higher $k_L a$ values reached with higher stirring speed and aeration rate are caused by a larger interfacial area, while the mass transfer coefficient k_L is surprisingly negatively correlated with the two operating conditions. Furthermore, the higher interfacial area reached at higher aeration rates is explained by a higher gas holdup, as the aeration rate does not have an impact on the bubble size in the investigated conditions.

Introduction

Stirred tanks are widely used for biocatalytic and fermentative productions to mix and aerate a reacting liquid phase. Despite the suboptimal mass transfer reached by such design, the stirred tank is considered as a standard, and is the most widely used reactor type in the biomanufacturing industry [1]. Mass transfer is a critical parameter in aerobic and gas fermentations and has been extensively studied and described in the literature [2, 3, 4]. However, how the operation and design parameters affect the two components of the mass transfer coefficient (k_L and a) has not been descried, due to the lack of measurement method for identifying the interfacial area and the mass transfer coefficient separately. With recent advances in fluidic measurement, dispersed multiphase systems can be characterized using optic based measurement techniques. The objective of this work is to measure the two components of the volumetric mass transfer coefficient, namely the interfacial area and the overall liquid-phase mass transfer coefficient and investigate how the stirring speed and the aeration rate affects the two terms.

Material and Methods

-Experimental Setup-

The experiments are performed in a 15 L stirred tank with the dimensions described in table 1. The gas is added to the tank with a linear sparger located at the bottom of the tank, through 10 equally spaced 1[mm] holes. Here, both air and pure nitrogen can be sparged. The rotor used for stirring the impeller has a maximum rotational speed of 1000 [rpm]. The air flowrate is adjusted using a Mass Flow Controller, with a maximum output of 5 [NL/min]. For this study, the liquid is tap water (conductivity ≈ 40 [mS/cm]). Measurements are performed at different aeration rates (4 and 5 [NL/min]) and different agitation speeds (from 300 to 500 [rpm]).

Table 1. Tank Dimensions

Parameter	Value
Tank height	500 [mm]
Tank diameter	240 [mm]
Tank Volume	24 [L]
Working volume	15 [L]
Impeller diameter	80 [mm]
Impeller height	20 [mm]
# of blades	4
Impeller type	Paddle
Baffle height	252 [mm]
Baffle width	15 [mm]
# of baffles	4
Distance between baffles and tank	5-10 [mm]
Bottom type	Rounded

-$k_L a$ measurement-

The volumetric mass transfer coefficient $nitrogen\,and$ easured using the "gassing-out" method, which consists of sparging the liquid with pure nitrogen, and change to air. The dissolved oxygen concentration is reduced by sparging pure nitrogen in the tank, then air is sparged, where the concentration rises with a logarithmic profile towards the saturation value [5, 6].

According to the general mass balance for the dissolved oxygen in the tank (eq.1) the $K_L a$ can be determined from the slope of equation 2.

$$\frac{dC}{dt} = k_L a(C^* - C) \quad (1)$$

$$\ln\left(\frac{C^* - C_0}{C^* - C}\right) = -k_L a(t - t_0) \quad (2)$$

The dissolved oxygen is measured using a fibre-

https://doi.org/10.52202/074122-0065

optic DO sensor (*PYROSCIENCE, Germany),* with a response dynamic around 3 [s]. The probe is fixed on one of the baffles in the tank (Mid-height).

-Bubble size measurement-

The bubble size distribution is measured using a shadowgraphic endoscope (DANTEC DYNAMICS, Denmark). Here, an LED is synchronised with a high-speed camera to create short light pulse during the camera exposure time. The LED light is shining through the measurement volume into the camera lens. Any bubbles present in the measurement volume will scatter the LED light and appear dark (figure 1). For each measurement, 300 frames are captures at 800 [fps]. For each set of images, identical bubbles are identified to only account for each bubble once. This also allows for tracking individual bubbles and determine their velocities.

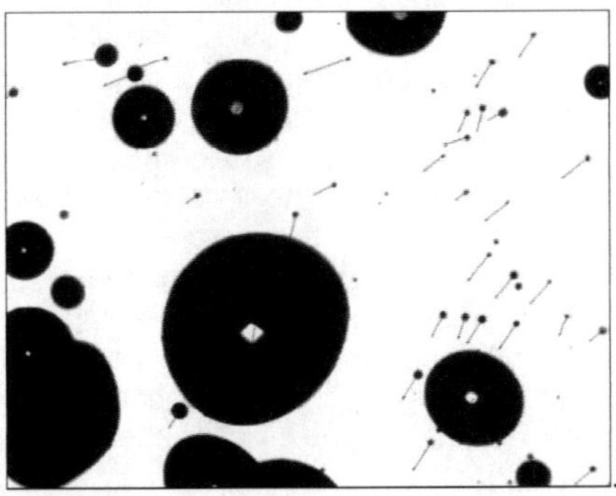

Figure 1. Example of image from the shadowgraphic endoscope, black spots are the shadows from the bubbles, red contour lines, blue velocity vector.

-Gas holdup measurement-

The gas holdup in the tank is determined based on the total volume change with and without gas based on the liquid height.

$$\epsilon = \frac{\Delta V}{V_{final}} \tag{3}$$

Where ϵ is the gas holdup, ΔV the difference in volume with and without the sparging, and V_{final} the total volume with sparging. The volume is estimated from the liquid level in the tank.

-Calculations-

From the bubble images, the diameter of each bubble is determined and used to calculate the Sauter mean diameter according to equation 4.

$$d_{23} = \frac{\sum_i^N d_i^3}{\sum_i^N d_i^2} \tag{4}$$

Where $d_i$is the diameter of bubble i, and N the total number of bubbles. From the Sauter mean diameter, the volumetric interfacial area a $[m^2/m^3]$ can be determined using equation 5.

$$a = \frac{6\epsilon}{(1-\epsilon)d_{23}} \tag{5}$$

Finally, the liquid-phase mass transfer coefficient K_L can be determined from K_La and a using equation 6.

$$k_L = \frac{k_La}{a} \tag{6}$$

Results and Discussion

The k_La is determined for all the test conditions and reported in figure 2. It can be observed that the k_La increases with the aeration rate and the stirring speed.

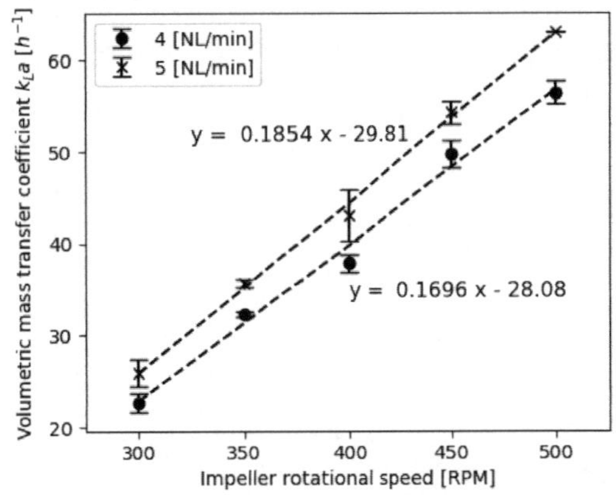

Figure 2. k_La measurement at different aeration rates and agitation speeds, with linear regression, n=2, error bar= 1*σ.

Assuming a linear correlation between the K_La and both the stirring speed and the aeration rate, the K_La can be expressed as a linear combination both parameters ($R^2 = 0.98$):

$$k_La = 5F_{gas} + 0.176N_i - 51.6 \tag{7}$$

F_{gas} being the aeration rate in [NL/min] and $N_i$the stirring speed in [rpm].

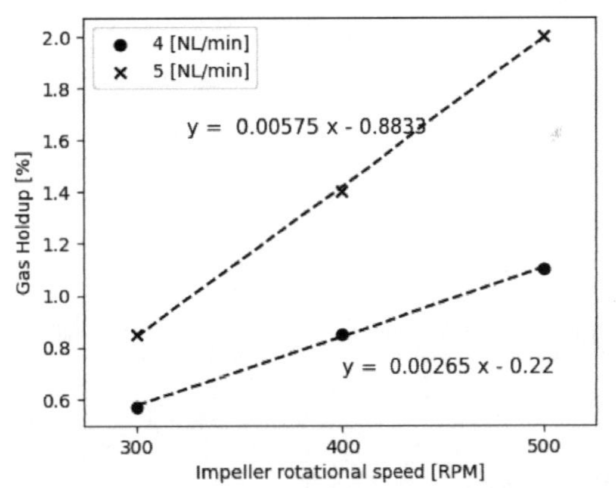

Figure 3. Gas Holdup at different aeration rates and agitation speeds, with linear regression.

The measured gas holdups are reported in figure 3. There is a clear linear correlation between the stirring speed and the gas holdup, however, the aeration rate seems to have a nonlinear impact on the gas holdup. Due to the low accuracy (eye reading of the level), the measurements could not be distinguished between the different replicates. At all investigated operating conditions, 8 sets of images are captured with the high-speed camera and the Sauter mean diameter is calculated according to equation 4. The results are reported in figure 4. The bubble diameter decreases with higher stirring speed, while the aeration rate does not seem to have a significant impact on the resulting bubble Sauter diameter.

https://doi.org/10.52202/074122-0065

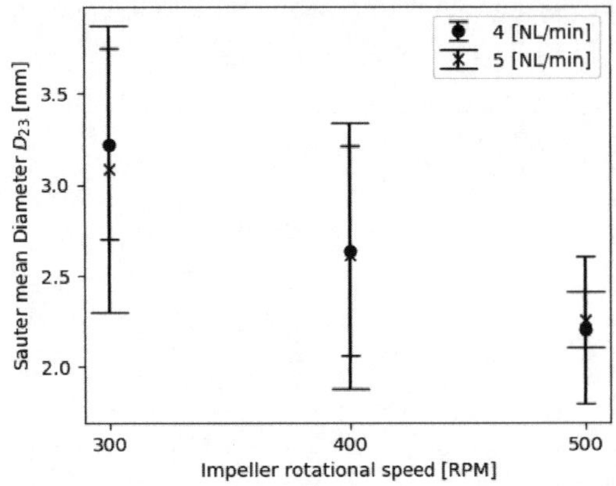

Figure 4. Sauter mean diameter as function of stirring speed and aeration rate, n=8, error bar=1*σ.

The large standard deviation in the measurement reflects the high variability between replicates. This can be reduced by capturing more frames per measurement, at multiple positions in the tank, to make sure enough bubbles have been captured to have a representative sample the bubbles.

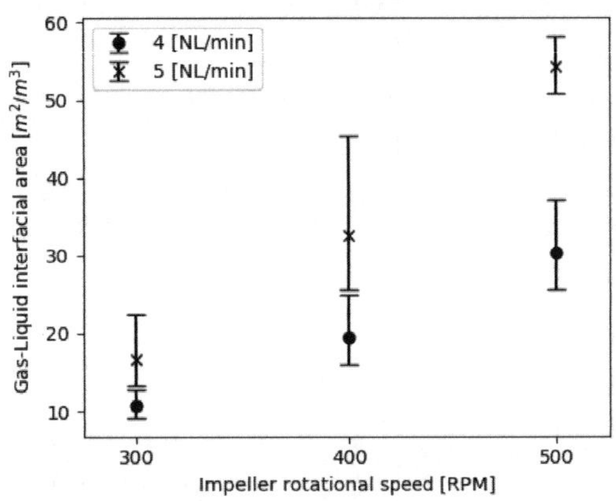

Figure 5. Gas-liquid interfacial area as function of stirring speed and aeration rate, error bar: calculated σ.

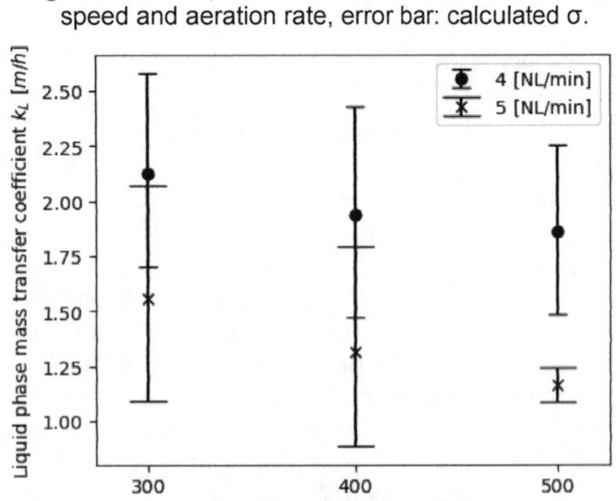

Figure 6. Overall liquid-phase mass transfer coefficient as function of stirring speed and aeration rate, error bar: calculated σ.

The measured gas holdup and bubble Sauter mean diameter are used to calculate the interfacial area using equation 5 and reported in figure 5. Finally, the mass transfer coefficient is calculated using equation 6 and reported in figure 6. Higher stirring speed seem to have a slight negative impact on k_L, while higher aeration rates have a clear negative impact. According to the Higbie equation for turbulent Newtonian fluids [7], K_L is proportional to the fourth root of the turbulent energy dissipation rate. This suggests that a higher aeration rate leads to lower energy dissipation. This can potentially be explained by the higher gas fraction around the impeller leading to a less efficient momentum transfer from the impeller into the liquid, leading to less turbulence in the liquid phase.

Conclusions

Mass transfer in the studied stirred tank was characterized at different stirring speeds and aeration rates. For each condition, the k_La was determined, as well as its individual components, k_L and a. The main findings are listed below:

- k_La is linearly correlated with stirring speed and aeration rate in the investigated ranges.
- Higher stirring speed is the main driver for bubble breakup and the aeration rate has no significant impact on the bubble size.
- The higher interfacial area reached at higher aeration rates is caused by higher gas holdup.
- The higher interfacial area reached at higher stirring speeds is caused a combination of higher gas holdup and smaller bubbles.
- The liquid phase mass transfer coefficient is mostly dependent on the aeration rate.
- The higher k_La values reached at higher aeration rates and higher stirring speeds is explained by the higher interfacial area.

Further work will focus on a wider range of operational conditions, and design features (dimensions/impeller type), as well as improving the bubble size measurement method to decrease the error. Also, more accurate measurements of the gas holdup are necessary to improve the estimation of the interfacial area, as the gas holdup is a large contributor to the interfacial area.

References.

[1] Yatmaz, E., & Turhan, I. (2019). *Liquid State Bioreactor*.
[2] Orgill, J. J., Atiyeh, H. K., Devarapalli, M., Phillips, J. R., Lewis, R. S., & Huhnke, R. L. (2013). A comparison of mass transfer coefficients between trickle-bed, Hollow fiber membrane and stirred tank reactors. *Bioresource Technology*, *133*.
[3] Garcia-Ochoa, F., & Gomez, E. (2004). Theoretical prediction of gas-liquid mass transfer coefficient, specific area and hold-up in sparged stirred tanks. *Chemical Engineering Science*, *59*(12).
[4] Yawalkar, A. A., Heesink, A. B. M., Versteeg, G. F., & Pangarkar, V. G. (2002). Gas-liquid mass transfer coefficient in stirred tank reactors. *Canadian Journal of Chemical Engineering*, *80*(5).
[5] Garcia-Ochoa, F., & Gomez, E. (2009). Bioreactor scale-up and oxygen transfer rate in microbial processes: An overview. In *Biotechnology Advances* (Vol. 27, Issue 2).
[6] Titchener-Hooker, N. (1995). Bioprocess Engineering Principles (by Pauline M. Doran). *TRENDS IN BIOTECHNOLOGY*, *13*(1).
[7] Prasher, B. D., & Wills, G. B. (1973). Mass Transfer in an Agitated Vessel. *Industrial and Engineering Chemistry Process Design and Development*, *12*(3).

Mathematical model of two-phase L-L Taylor flow hydrodynamics for non-Newtonian fluids in microchannels

Poster.20

Rufat Sh. Abiev[1,2]*. (1) St. Petersburg State Institute of Technology (Technical University), Moskovskiy pr. 26, St. Petersburg, Russia. (2) Ioffe Physical Technical Institute, Russian Academy of Sciences, Politekhnicheskaya ul. 26, St. Petersburg, Russia, abiev.rufat@gmail.com*

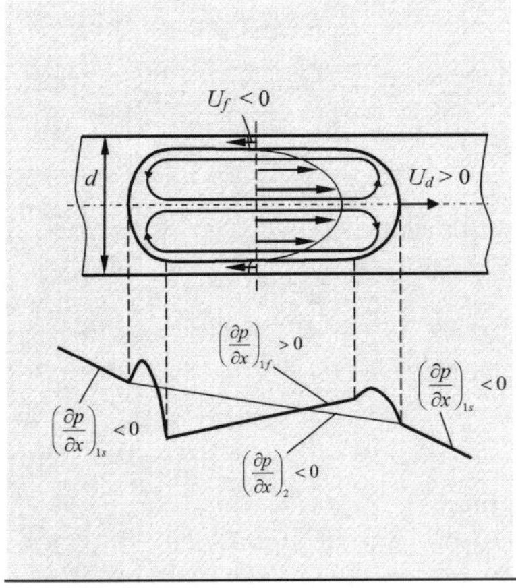

A mathematical model of Taylor flow hydrodynamics for combinations of non-Newtonian and Newtonian fluids in microchannels: (i) non-Newtonian/non-Newtonian, (ii) Newtonian/non-Newtonian, (iii) non-Newtonian/Newtonian, (iv) Newtonian/Newtonian, is proposed. The output parameters are: velocity profiles and mean velocities in liquid film, plug of dispersed phase, and pressure gradients in the liquid film, in the liquid slug and in the plug. These three pressure gradients have different values, which allowed to explain previously observed phenomena and the results of CFD modeling. The results of calculations for the case (iv) were compared with the experimental data [1]. Calculations were made for the other cases (i)-(iii); due to the lack of experimental and CFD data for non-Newtonian fluids these results should be verified by future studies. It was demonstrated that the positive value of pressure gradient in the film in the majority of studied cases correlates well with the negative film velocity.

Introduction

Hundreds of papers published within the past few decades were devoted to hydrodynamics as well as heat and mass transfer of two-phase (L-G and L-L) flows in micro- and milli-channels, mainly for Newtonian media. Despite that, the number of solutions having high molecular structure and, therefore, non-Newtonian rheology that is required in microreactors, is growing. The objectives of this work is to build up a mathematical model for a two-phase flow hydrodynamics in the case of non-Newtonian liquids moving in the micro- (milli-) channels in the form of a segmented (Taylor) flow. This will allow to find all necessary parameters of three-layer model of Taylor circulations and, finally, to predict theoretically mass transfer characteristics of two-phase flow in microchannels.

Material and Methods

For the continuous phase, tangential stresses τ are linked with shear rate $\dot{\gamma}$ by the equation (1):

$$\tau_{xr} = \tau_1 = K_1 \dot{\gamma}^n , \quad (1)$$

or for the axisymmetric flow:

$$\tau_1 = K_1 \left(\frac{\partial u_x}{\partial r} \right)^n . \quad (2)$$

The analytical solution will be found for axisymmetric flow, assumed that the bubble (or droplet) of the dispersed phase is long enough (at least $L_d/d_d > 2$), so that the end effects could be disregarded.

The Navier-Stokes equations (in an axisymmetric steady-state form) along with continuity equation for two-phase flow were used to solve the problem. More details, including boundary conditions, could be found in [2, 3].

Results and Discussion

Velocity profile in the liquid slug as well as velocity profile for the non-Newtonian droplet in the non-Newtonian continuous phase (film around it) were found theoretically.

Four combinations of non-Newtonian fluids that were considered and used in calculations:

(a) – dispersed phase – shear thinning fluid in shear thinning liquid (continuous phase) (Fig. 1 a);

(b) – dispersed phase – shear thickening fluid in shear thinning liquid (continuous phase) (Fig. 1 b);

(c) – dispersed phase – shear thinning fluid in shear thickening liquid (continuous phase);

(d) – dispersed phase – shear thickening fluid in shear thickening liquid (continuous phase).

For the case shown in Fig. 11a the velocity profiles of both liquids ($u_{1s}(r)$ and $u_2(r)$) are flattened, for the cases with shear thickening fluid velocity profiles are elongated (e.g. Fig. 12b for both $u_{1s}(r)$ and $u_2(r)$).

Radius $R_{1.s}$ characterising the borderline between the circulation area within Taylor vortex and the by-pass flow in the surrounding film, could be found

$$R_{1.s} = R \left[\frac{(3n+1) - (n+1)\eta}{2n} \right]^{\frac{n}{n+1}} . \quad (3)$$

The radius $R_{0.s}$ corresponding to the center of circulations of Taylor vortex

$$R_{0.s} = R \left(1 - \frac{n+1}{3n+1} \frac{U_d}{U_s} \right)^{\frac{n}{n+1}} \quad (4)$$

https://doi.org/10.52202/074122-0066

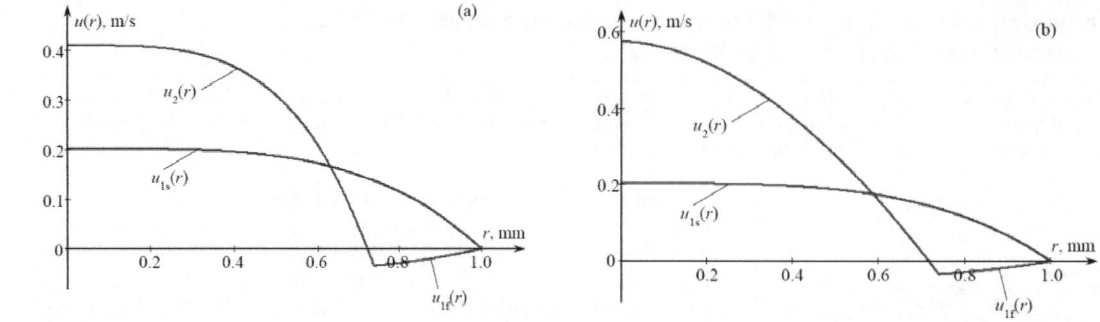

Figure 1. Calculated velocity profiles for liquid slug $u_{1s}(r)$, liquid film $u_{1f}(r)$, and plug (droplet) $u_2(r)$ (a) for non-Newtonian fluids (Taylor flow) in microchannels: d = 2 mm, j_c = 0.029 m/s, j_d = 0.103 m/s. (a) – shear thinning fluid in shear thinning liquid, continuous phase: K_1 = 0.85, n = 0.36, ρ_1 = 1144 kg/m^3; dispersed phase: K_2 = 0.85, m = 0.36, ρ_2 = 1144 kg/m^3; σ = 0.0307, ε_L = 0.734; (b) – shear thickening fluid in shear thinning liquid, continuous phase: K_1 = 0.85, n = 0.36, ρ_1 = 1000 kg/m^3; dispersed phase K_2 = 0.0085, m = 1.2, ρ_2 = 1144 kg/m^3; σ = 0.0307, ε_L = 0.734.

Similarly the radius $R_{0.d}$ was defined from the condition of zero relative velocity in the plug of the dispersed phase as follows:

$$R_{0.d} = R_d\left(\frac{2m}{3m+1}\right)^{\frac{m}{m+1}}, \qquad (5)$$

whereas the radius $R_{1.d}$ was found to be equal to the droplet radius: $R_{1.d} = R_d$.

Mean velocity of circulations in the liquid slug of the continuous phase was found as:

- in the direct flow:

$$W_{c.s.for} = U_s\left[1-\frac{n+1}{3n+1}\eta\right]. \qquad (6)$$

- in the backflow:

$$W_{c.s.back} = \left[\left(\frac{3n+1}{2n}\right)^{\frac{2n}{n+1}}-1\right]^{-1} W_{c.s.for}. \qquad (7)$$

Conclusions

A mathematical model of two-phase Taylor flow hydrodynamics for four combinations of non-Newtonian and Newtonian fluids in microchannels (dispersed phase/continuous phase): (i) non-Newtonian/non-Newtonian, (ii) Newtonian/non-Newtonian, (iii) non-Newtonian/Newtonian, (iv) Newtonian/Newtonian is proposed on the basis of general conservation (momentum and continuity) equations along with correlations for film thickness.

Input parameters are: diameter of microchannel, angle between directions of gravity acceleration and the velocity of two-phase flow, densities and rheological properties of liquids, superficial velocities of phases, relative length of the plugs (or their mean velocity). Output parameters are: U_f, U_d, U_s, $(\partial p/\partial x)_{1f}$, $(\partial p/\partial x)_{1s}$, $(\partial p/\partial x)_2$, $\tau_{1s}(r)$, $\tau_{1f}(r)$, $\tau_2(r)$.

The difference between the three values of pressure gradients: in the liquid slugs $(\partial p/\partial x)_{1s}$, in the liquid film $(\partial p/\partial x)_{1f}$, and in the plug of continuous phase $(\partial p/\partial x)_2$, was explained theoretically. This result confirms previous discussions and assumptions presented in [4].

The model proposed in this work allows to predict a set of the most important hydrodynamic characteristics of two-phase Taylor flow for Newtonian and non-Newtonian liquids with rheological properties described by eqs. (1) and (2).

The following parameters could now be calculated theoretically for two-phase Taylor flow of non-Newtonian fluids (both for liquid slug and liquid plug): (i) the center of Taylor vortices, (ii) their boundaries, (iii) mean velocities and frequency of circulation, (iv) mean velocities in by-pass and transit films, and (v) transition from circulation to by-pass flow conditions. The coincidence with the Newtonian fluids as a limiting case was evaluated. The obtained results represent crucial input data for the three-layer mass transfer model, the frequency of circulation-based mass transfer model, and are useful for the understanding of the influence of various parameters on heat and mass transfer intensity in Taylor flows.

The vortex center in non-Newtonian plugs is almost independent of the rheological properties (in the frames of the shear-thinning or shear-thickening fluids). The film around the liquid droplet (plug) was found to be the area of transit flow not involved in Taylor circulations.

The frequency of circulations in the droplet (plug) of the dispersed phase was defined.

Acknowledgments

The author would like to thank the Russian Science Foundation (project 20-63-47016) for financial support.

References

[1] H. Liu, C.O. Vandu, R. Krishna, *Ind. Eng. Chem. Res.* 44 (2005) 4884-4897.

[2] R.Sh. Abiev, *Chem. Eng. Sci.*, 247 (2022) 116930.

[3] R.Sh. Abiev, *Chem. Eng. Sci.*, 250 (2022) 117380.

[4] R.Sh. Abiev, *Chem. Eng. Sci.*, 174C (2017) 403.

Effect of Mixing in Highly Viscous Media on Volumetric Mass Transfer Coefficient

Poster.21

***Emilie O. Willer*[1]**, *Johan le Nepvou de Carfort*[1], *Ulrich Krühne*[1], *John M. Woodley*[1], *(1) Department of Chemical and Biochemical Engineering, Technical University of Denmark, 2800 Kgs. Lyngby, Denmark, eoje@kt.dtu.dk*

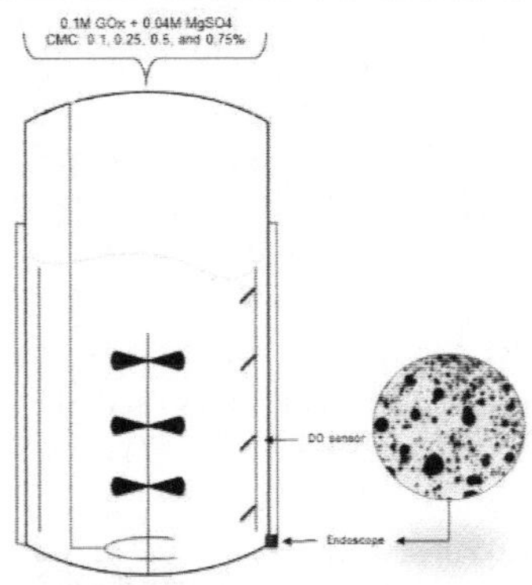

Insight into viscosity impact on mixing and oxygen distribution

Multiphase (gas-liquid) systems operated in stirred tank reactors are widespread in the bio-industry, creating considerable interest in understanding the complexity of mass transfer. One of the greatest challenges is understanding the influence of mixing on mass transfer in highly viscous media.
The main novelty here is to identify the influence of changing apparent viscosity on multiphase systems, mainly by focusing on the oxygen distribution and the homogeneity of the liquid at pilot scale (200 L). The main results conclude that the mixing in highly viscous media significantly impacts the transfer of oxygen. However, the distribution of oxygen observed by the reactor wall is still maintained homogenous. This work could potentially lead to enhanced oxygen mass transfer by improving the reactor design and models for fermentation upscaling.

Introduction

Products produced by microbial growth via aerobic fermentation are widespread in the bio-industry [1].

Many fermentation processes have been operated aerobically successfully, supplying oxygen by sparging of air and dispersing the bubbles by agitation [2].

However, in previous work, it has been found that the viscosity of the fermentation media increases through the process due to microbial growth, creating non-ideal mixing and oxygen distribution challenges, which in turn impacts oxygen mass transfer [3,4].

Many fermentation broths are non-Newtonian pseudoplastic fluids, showing shear thinning behaviour. Most commonly, the fermentation broth starts with Newtonian properties but changes into non-Newtonian pseudoplastic behavior along the process. This is also the case with yeast fermentation, in which broth viscosity depends on increasing cell concentration described by the Vand equation [1,5].

The work described here was conducted in a stirred tank reactor at pilot scale, focusing on pseudoplastic fluids, determined by Ostwald-de Waele power law, Equation 1 [6].

$$\mu_{app} = K \cdot \left(\frac{dv}{dx}\right)^{n-1}$$

Equation 1: Ostwald-de Waele power law. K flow consistency index and n flow behaviour index.

During fermentation upscaling, a widespread problem is the process impact on the volumetric mass transfer coefficient, K_La, among other the changing viscosity, making it a crucial parameter to investigate.

This work aims to investigate the mixing properties of a highly viscous liquid and its impact on oxygen mass transfer in a 200L working volume bioreactor. However, the main objective is to investigate the mixing properties and the ability to distribute oxygen into the highly viscous liquid.

Materials and Methods

The following experiments were all performed in a 200L pilot-scale bioreactor equipped with a ring sparger, four baffles, and a three-blade segment Rushton 6-blade disk impeller.

Solutions of 0.1M Glucose oxidase (GOx), 0.04M magnesium sulfate, and various carboxymethyl cellulose (CMC) concentrations mimicking a viscous media were analyzed.

The bioreactor was supplied with 1 VVM (volume per volume per minute) air, operated at 25°C, and under two agitation speeds, 250 and 400 RPM, corresponding to a power input of 1 and 4 W/L, respectively.

Determination of liquid hydrodynamics

The liquid hydrodynamics of the four CMC solutions (0.1, 0.25, 0.5, and 0.75%) were analyzed. Determination by a rheometer was performed to obtain the apparent viscosity using Equation 1, which further has been used to characterize the liquid flow by Reynolds number.

Determination of volumetric mass transfer coefficient

To establish the liquid's apparent viscosity influence on oxygen mass transfer, one of the four baffles was equipped with four oxygen probes to obtain the oxygen concentration at various heights in the bioreactor. The gassing-out dynamic method was used to determine the volumetric mass transfer coefficient, Equation 2 [7].

$$ln\left(\frac{C_s}{C_s - C_l}\right) = -K_l a * t$$

Equation 2: The gassing-out dynamic method. C_s fully

https://doi.org/10.52202/074122-0067

saturated oxygen concentration, C_l liquid oxygen concentration at a given time, t.

Results and Discussion

From the determination of the liquid hydrodynamics, the Reynolds number displayed in Table 1 shows that the fluid viscosity greatly influences the flow regime, and for highly viscous liquids, the flow behavior changes. As the agitation speed increases, so does power number and also the corresponding Reynolds number.

Table 1. Reynolds number for all solutions at two agitation speeds.

Solution	**Reynolds number at:**	
	1 W/L Power	**4 W/L Power**
Water	166300	266000
0.1% CMC	34646	53200
0.25% CMC	16798	27708
0.5% CMC	**5961**	11271
0.75% CMC	**2618**	**5587**

[a] Marked with bold out of turbulent regime

However, to better understand the impact of the flow on oxygen transfer, the oxygen concentration was investigated at four heights in the bioreactor to determine the homogeneity of the oxygen concentration.

The determined K_La is significantly affected by viscosity creating critical low values for oxygen transfer. However, by increasing the agitation speed a change in the apparent viscosity is observed due to shear thinning properties, thereby enhancing K_La.

Variable agitation speed is often a common option in the industry; therefore, the further aim of the work is to establish a better understanding of the apparent viscosity influence on the oxygen distribution under influence of changing agitation speeds.

Conclusions

The results lead to the conclusion that the apparent viscosity significantly impaired K_La.

However, by increasing the agitation speed, an improvement can be obtained, due to a change in fluid properties.

Future work must be conducted to establish the homogeneity of the two-phase solutions under the influence of the increased viscosity.

Future work will focus on investigating the gas bubble size distribution to better understand the influence of viscous media and mixing.

Acknowledgments

The authors would like to thank the Technical University of Denmark for access to pilot-plant facilities and financial support for the project.

References

[1] M. C. Sadino-Riquelme, J. Rivas, D. Jeison, A. Donoso-Bravo, R. E. Hayes. (2022) Investigating a Stirred Bioreactor: Impact of Evolving Fermentation Broth Pseudoplastic Rheology on Mixing Mechanisms. Fermentation, 8 (3), 102-131

[2] Principles of Fermentation Technology, P. F. Stanbury, A. Whitaker, S. H. Hall, (3rd ed), ChemE, 2017

[3] K. Schuegerl. (1981) Oxygen transfer into highly viscous media. Advances in biochemical engineering, 19, 71-174

[4] C. Bach, J. Yang, H. Larsson, S. M. Stocks, K.V. Gernaey, M. O. Albaek, U. Krühne, (2017), Evaluation of mixing and mass transfer in a stirred pilot scale bioreactor utilizing CFD, Chemical Engineering Principles, 171 19-26

[5] Bioprocess engineering principles, P. M. Doran (2nd ed), Elsevier, 2013

[6] W. D. Deckwer, K. Nguyen-Tien, A. Schumpe, Y. Serpemen (1982). Oxygen mass transfer into aerated CMC solutions in a bubble column, Biotechnology and Bioengineering 24 (2) 461-481

[7] M. O. Cerri, M. N. Esperanca, A. C. Badino, M. P. Ribeiro, (2017). Automated algorithm to determine kLa considering system delay: Automated kLa estimation, *Chemical Technology and Biotechnology*, 92 (7) 1630-1637.

Effect of pH gradients on the hydrolysis of ethyl acetate by *Candida antarctica* lipase B (CALB)

Poster.22

Caroline Hamelmann[1], *Juliet J. Victoria*[1], *John M. Woodley*[1], *(1) Department of Chemical and Biochemical Engineering, Technical University of Denmark, 2800 Kgs Lyngby, Denmark, carham@kt.dtu.dk*

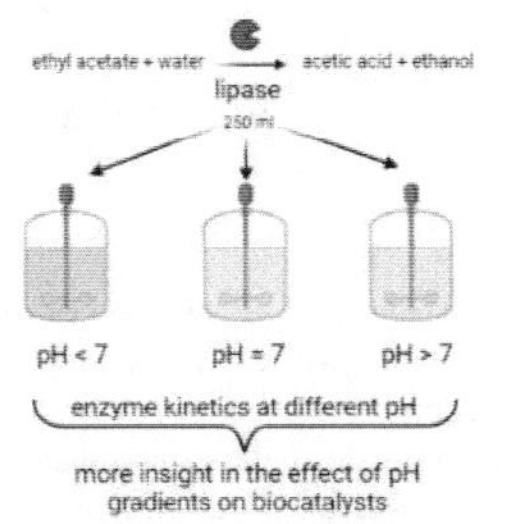

Biocatalysts have a great potential to replace ordinary catalysts as the environmentally friendly alternative. However, as this is still an emerging technology, more research is required in order to successfully use biocatalytic reactions in large scale production processes. More precisely, insight is needed in the effect of gradients in large scale reactors on the biocatalyst, gaining this insight is the purpose of this work.
The main novelty is to identify the specific enzyme kinetics at the different pH levels in small scale experiments (250 ml). The results will aid in understanding the impact of pH gradients on biocatalytic reactions better. In the future, this work could lead to the definition of a CFD simulation which could also include the effect of pH gradients on the reaction rate.

Introduction

Biocatalytic reactions are reactions catalysed by an enzyme, which, similar to chemical catalysts, are not used up or altered during the process. However, biocatalysts have the great advantage of requiring only mild reaction conditions, being highly specific in their catalytic function and having a fast reaction rate [1]. Due to those abilities, the industrial interest in such reactions is rising and the need is created to increase the production scale [2].

However, an increase in scale usually results in the occurrence of gradients throughout the reactor as non-ideal mixing is occurring [3,4]. Those gradients will expose the biocatalyst to extreme conditions most likely have a negative effect on the biocatalyst itself and the reaction it is catalysing [5]. Many studies have been conducted on the effect of such gradients on fermentation processes [3,4]. Yet, this topic is unexplored for biocatalysts, even though the effect of gradient formation is more extreme as the reaction rates are faster compared to fermentations [2,5].

In order to find a solution to minimize these effects, they need to be better understood, which is what this work aims to do.

More precisely, small scale experiments are being executed to understand the impact of gradients on the biocatalytic reaction of interest and to determine its kinetics within the gradient range.

The kinetics of interest are the Michaelis-Menten kinetics, a model which can be used to describe the reaction rate (v) by being equal to the maximum reaction rate (V_{max}), multiplied with the quotient of the limiting substrate concentration ($[S]$) and the sum of the Michaelis constant (K_M) and limiting substrate concentration ($[S]$), see Equation 1 [6].

$$v = V_{max} \frac{[S]}{K_M * [S]}$$

Equation 1: Michaelis-Menten kinetics, with reaction rate (v), maximum reaction rate (V_{max}), Michaelis constant (K_M) and substrate concentration ($[S]$)[6].

In the rate law, Equation 1, V_{max} is a function of pH and thereby the kinetics are sensitive to pH gradients in a reactor.

In this work a model reaction will be used to understand these effects further. The model is the hydrolysis of ethyl acetate ($C_4H_8O_2$) to ethanol (C_2H_6O) and acetic acid (CH_3COOH) by the enzyme *Candida antarctica* lipase B (CALB), see Scheme 1. It has been chosen as a model reaction [7].

$$C_4H_8O_2 + H_2O \rightarrow CH_3COOH + C_2H_6O$$

Scheme 1: Hydrolysis of ethyl acetate to acetic acid and ethanol by the lipase CalB [7].

As can be seen in Scheme 1, the reaction produces an acid which will accumulate in the reactor and cause the pH to drop and most likely denature the enzyme, but at least alter the kinetics. Therefore, a base is added to the tank to buffer the effect of the produced acid.

The base will enter the reactor through a dosing point and will then be spread in the reactor through the mixing by the stirrer. At a large scale with insufficient mixing this will lead to base gradients throughout the reactor [3].

Material and Methods

The experiments described in this section are all executed with a liquid lipase (CalB).

Analysis of enzyme activity, substrate and product concentration

In order to analyse the reaction rate (activity) of the lipase (CalB) an amount of p-nitrophenyl hydrolysed from p-nitrophenyl butyrate by the enzyme is measured by UV-spectroscopy. This is a standard assay.

In the real reaction, the substrate (ethyl acetate) and the products (ethanol and acetic acid) are analysed using HPLC.

Determination of experimental conditions

Prior to starting the key experiments, the optimal enzyme and substrate concentration as well as temperature were determined.

This is done by testing a range of different

https://doi.org/10.52202/074122-0068

substrate (ethyl acetate) concentrations of around 8% and their solubility in water, as the aim is to make the system a single phase [8].

Additionally, as the enzymes optimum temperature is at 55 ℃, but ethanol, as one of the products is highly volatile the right temperature is determined in order for the enzyme to work ideally while ensuring the ethanol does not evaporate from the reaction mixture. This is achieved by testing different temperatures in the range of 40 – 60 ℃ [9].

Finally, the right enzyme concentration is identified, so that the reaction is completed within a reasonable time without requiring excessive amounts of enzyme.

Testing the effect of pH on the biocatalytic reaction

As the overall aim of the project is to better understand the effect of different pH concentrations on the biocatalytic reaction, the reaction is introduced to different pH values. As the optimum pH for CalB is at around 7, the tested pH ranges from 5 to 9 [10].

Determination of reaction kinetics

In order to better understand the effect of different pH values on the biocatalytic reaction the relevant kinetic parameters are identified. This is done by running the reaction at different pH values and different enzyme and substrate concentrations respectively. This allows for the reaction rate to be determined at different concentrations which is subsequently used to calculate the V_{max} and K_m values for the rate law.

Results and Discussion

Initial experiments have shown that a change in pH affects the enzyme kinetics. More precisely, the parameters V_{max} and K_m from the Michaelis-Menten kinetics vary with a change in pH. In the specific case of ethyl acetate hydrolysis, the reaction rate decreases when the pH deviates from the lipase optimum pH of 7.

As these tested pH ranges will also be found in large-scale reactors in terms of gradients due to non-ideal mixing, the results indicate that gradient formation is one of the key problems when scaling up biocatalytic reactions.

The impact of gradient formation in bioreactors on fermentation process has been frequently reported in literature [3,4]. However, this is not the case for biocatalytic reactions, even though the effect of gradient formation is assumed to be more extreme as reaction rates are faster than in fermentations [2,5]. Due to this, further investigation into this topic is highly relevant.

Conclusion

The results of the experiments described in this work show that different pH values have an effect on the enzyme kinetics of the hydrolysis reaction described here. Due to gradient formation in large-scale reactors, the results lead to the conclusion that these gradients have an impact on the enzyme kinetics of such a hydrolysis reaction.

Future work will focus on using computational fluid dynamics (CFD) simulations, to further understand the formation of such pH gradients better. The enzymatic kinetics at different pH values, which have been found in this work, could be implemented in the CFD simulation to give a better idea on the effect of gradients formation throughout a bioreactor with a biocatalytic reaction.

Additionally, such a simulation could allow experimentation with the positioning of the dosing point in the reactor.

Acknowledgments

I would like to express my gratitude to Novozymes, who generously donated the CalB enzyme for the experiments. Furthermore, I would like to thank the Technical University of Denmark for financial support. This project has received funding from the European Research Council (ERC) under the European Union's Horizon 2020 research and innovation programme. (Grant agreement No. 101021024)

References

[1] Woodley, J. M. (2020). New frontiers in biocatalysis for sustainable synthesis. Current Opinion in Green and Sustainable Chemistry, 21, 22–26. https://doi.org/10.1016/j.cogsc.2019.08.006

[2] Wu, S., Snajdrova, R., Moore, J. C., Baldenius, K., & Bornscheuer, U. T. (2020). Biocatalysis: Enzymatic Synthesis for Industrial Applications. Angewandte Chemie International Edition. https://doi.org/10.1002/anie.202006648

[3] Nadal-Rey, G., McClure, D. D., Kavanagh, J. M., Cornelissen, S., Fletcher, D. F., & Gernaey, K. V. (2021). Understanding gradients in industrial bioreactors. Biotechnology Advances, 46, 107660. https://doi.org/10.1016/j.biotechadv.2020.107660

[4] Crater, J. S., & Lievense, J. C. (2018). Scale-up of industrial microbial processes. FEMS Microbiology Letters, 365(13). https://doi.org/10.1093/femsle/fny138

[5] Erdem, E., & Woodley, J. M. (2022). Industrially useful enzymology: Translating biocatalysis from laboratory to process. Chem Catalysis, 2(10), 2499–2505. https://doi.org/10.1016/j.checat.2022.09.037

[6] Johnson, K. A., & Goody, R. S. (2011). The Original Michaelis Constant: Translation of the 1913 Michaelis–Menten Paper. Biochemistry, 50(39), 8264–8269. https://doi.org/10.1021/bi201284u

[7] Chiplunkar, P. P., Zhao, X., Tomke, P. D., Noro, J., Xu, B., Wang, Q., Silva, C., Pratap, A. P., & Cavaco-Paulo, A. (2018). Ultrasound-assisted lipase catalyzed hydrolysis of aspirin methyl ester. Ultrasonics Sonochemistry, 40, 587–593. https://doi.org/10.1016/j.ultsonch.2017.08.004

[8] Altshuller, A. P., & Everson, H. E. (1953). The solubility of ethyl acetate in water. Journal of the American Chemical Society, 75(7), 1727–1727. https://doi.org/10.1021/ja01103a501

[9] Zhang, X.-guang, Xue, Y., Lu, Z.-ping, Xu, H.-jin, & Hu, Y. (2022). Chemical modification for improving catalytic performance of lipase b from Candida Antarctica with hydrophobic proline Ionic liquid. Bioprocess and Biosystems Engineering, 45(4), 749–759. https://doi.org/10.1007/s00449-022-02696-x

[10] Monteiro, R. R. C., Neto, D. M. A., Fechine, P. B. A., Lopes, A. A. S., Gonçalves, L. R. B., dos Santos, J. C. S., de Souza, M. C. M., & Fernandez-Lafuente, R. (2019). Ethyl Butyrate Synthesis Catalyzed by Lipases A and B from Candida antarctica Immobilized onto Magnetic Nanoparticles. Improvement of Biocatalysts' Performance under Ultrasonic Irradiation. International Journal of Molecular Sciences, 20(22), 5807. https://doi.org/10.3390/ijms20225807

Transition flow regimes in Baffled Reactors

Poster.23

Sofia RSP Brandão[1,2], *Margarida SCA Brito*[1,3], *António Ferreira*[1,3], *Ricardo J Santos**[1,2] *(1) Faculty of Engineering of University of Porto, Rua Dr. Roberto Frias, 4400-465, Porto, Portugal, *rsantos@fe.up.pt. (2) Laboratory of Separation and Reaction Engineering – Laboratory of Catalysis and Materials, University of Porto, Rua Dr. Roberto Frias, 4400-465, Porto, Portugal. (3) LEPABE - Laboratory for Process Engineering, Environment, Biotechnology and Energy, Rua Dr. Roberto Frias, 4400-465, Porto, Portugal.*

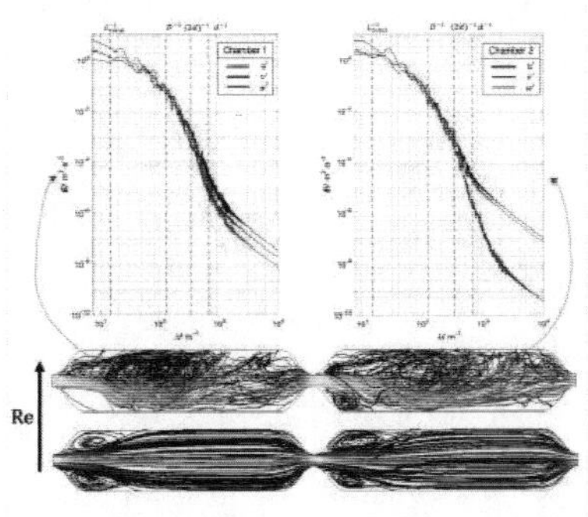

This work focuses on the identification of different laminar flow regimes (steady to chaotic) in Baffled Reactors operating without any external stimuli. It seeks full comprehension of the mixing mechanisms and dynamics in these reactors for subsequent optimization of the active mixing parameters, when applying the external oscillatory stimulus. Computational Fluid Dynamics (CFD) simulations were performed to understand the flow behaviour under different operational Reynolds (Re) numbers. Different 3D geometries were studied, a planar and a cylindrical form containing smooth periodic constrictions (SPC). For the planar geometry, the influence of the reactor's depth on the flow dynamics was also assessed. The dominant flow scales at different regimes were identified through spectral analysis.

Introduction

Mixing is a critical step in many industries, strongly impacting product quality, processes reproducibility and scale-up [1]. Alternative mixing devices as the Oscillatory Flow Reactors (OFRs) have been developed to overcome these drawbacks.

Oscillatory Baffled Reactor (OBRs) are a type of OFR composed of a tube with equally spaced baffles, to which an oscillatory flow is imposed. The flow is defined by a net Reynolds number (Re) and an oscillatory Re.

Application of these devices has been reported on batch processes such as: bioreactions, polymerizations, photocatalysis, flocculation, gas-liquid contacting, phase-transfer catalysis, hydrate formation, and mitigation of wax deposition [2]. OBRs have also been used for ozonation water treatments [3], [4].

The latest studies focus on continuous operation, being one of the most relevant applications the continuous crystallization of active pharmaceutical ingredients [5], [6]. Nevertheless, the control of mixing scales in OBRs is not completely mastered, which, for instance, compromises the nucleation stage during crystallization, leading to lower selectivity and product quality [7].

Characterization of flow regimes in several chemical reactors, such as CIJs and NETmix, has been widely studied at LSRE-LCM since it enables control of mixing scales and improve mixing efficiency [8]. The key factor to onset efficient mixing in these meso sized mixers is the flow transition from a segregated regime, characterized by a steady flow where almost no mixing occurs, to a chaotic state where vortices shedding occurs. The chaotic flow promotes convective mixing while keeping the Re within the laminar flow range. When setting an external stimulus is crucial to understand the natural frequencies of the flow at different regimes, to improve the control of mixing scales [8], [9].

This work aims to provide a complete description of flow regimes in a Baffled Reactor (BR), i.e. an OBRs without external stimuli. The goal is to identify the natural frequencies of oscillation to enhance the frequencies applied in the oscillatory stimulus and to study the main mechanisms for triggering different flow regimes. As future work, a comparison between the Strouhal number (St) that induces flow resonance and the characteristic St determined from the natural frequency in non-modulated self-sustainable chaotic flow regimes will be assessed. Computational Fluid Dynamics (CFD) simulations were conducted using two distinct 3D geometries: a planar and a cylindrical patented configuration with Smooth Periodic Constrictions (SPC) [10]. Various Reynolds numbers were tested, and the influence of the reactor's depth was evaluated for the planar geometry.

Material and Methods

The flow fields were obtained through laminar transient 3D CFD simulations using the commercial package Ansys Fluent 2021 R2. Figure 1 shows the cylindrical and planar BR and the respective dimensions. The 3D CFD model corresponds to a simplified geometry, where Periodic Boundary Conditions (PBCs) were imposed at inlet and outlet. No-slip boundary condition was imposed on the walls. For the planar geometry, the depth values (ω) tested are 15 mm, 8 mm and 5 mm. The following continuity and Navier-Stokes equations are solved for an incompressible and isothermal flow,

$$\nabla \cdot \vec{v} = 0 \text{ and } \rho\left(\frac{\partial \vec{v}}{\partial t} + \vec{v} \cdot \nabla\vec{v}\right) = -\nabla p + \mu \nabla^2 \vec{v}$$

where $\vec{v}$ is the velocity vector and p is the static pressure. The fluid's density and viscosity are $\rho = 998.2\ \mathrm{kg/m^3}$ and $\mu = 1.003 \times 10^{-3}\ \mathrm{Pa \cdot s}$.

https://doi.org/10.52202/074122-0069

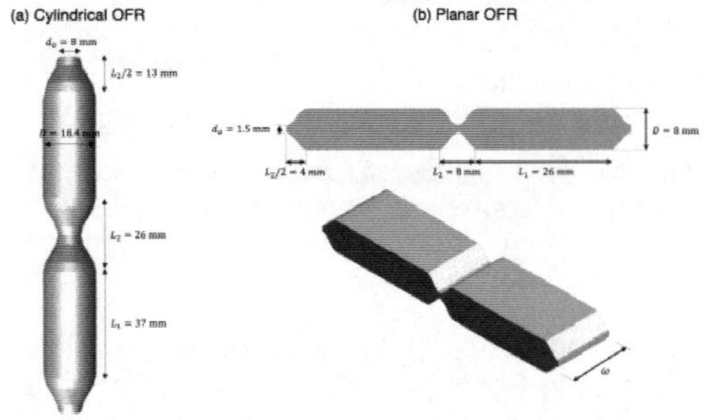

Figure 1. (a) Cylindrical and (b) planar BR geometries, and respective dimensions.

The convergence criterium was defined as the residuals below 10^{-5} in each time step. The Courant number was set to 0.5 for all simulations.

Results and Discussion

The Reynolds numbers were computed considering the hydraulic diameter of the constriction, $Re = \rho d_h v/\mu$ where $d_h = 4A_{cs}/P_{cs}$, and the values tested range from 50 to 1000.

3D CFD Simulation – Cylindrical BR

Figure 2 shows the pathlines obtained from the 3D CFD simulations at different Re numbers for the cylindrical geometry.

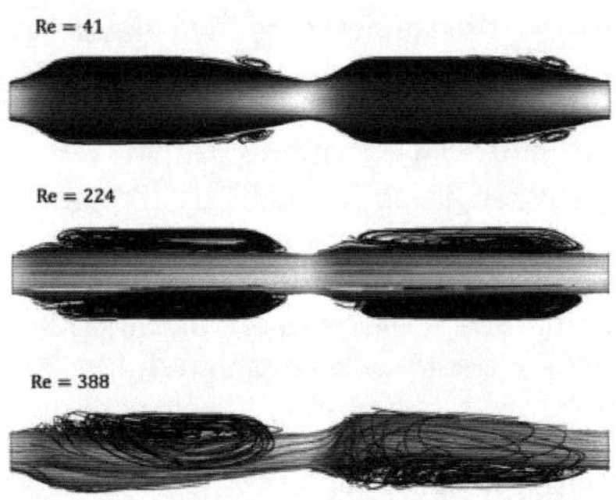

Figure 2. Streamlines for cylindrical BR at different Re.

At $Re = 41$, a segregated flow regime is identified since no dynamic flow structures are formed. At $Re = 224$, two symmetrical vortices are formed in each chamber. However, the flow symmetry does not enable mixing, and a steady flow with recirculation zones is observed. At $Re = 388$, the symmetry is broken. The consecutive expansions and contractions generate a vortex within each chamber that matches the length of the respective chamber. This vortex has a 3D shape that engulfs the jet being issued from the constriction. At this Re number, the flow transitions from a steady to a chaotic state, marked by convective mixing.

The turbulence spectra of velocity fluctuations at the midpoint of each chamber for a cylindrical geometry at $Re = 437$ are depicted in Figure 3.

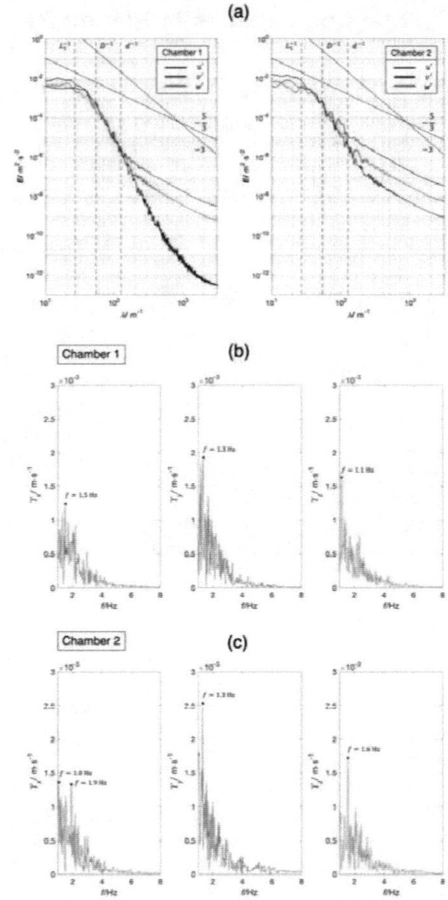

Figure 3. (a) Power spectrum, (b) frequency spectrum first chamber, and (c) frequency spectrum second chamber.

It is possible to see an inversion of the energy cascade marked by a change in the slope in Figure 3a. This is the energy injection scale that corresponds to approximately d^{-1}, as expected. The turbulence is fed by the small scale, the inlet jet, that feeds the larger vortices. This constitutes an inversion of the 3D turbulent energy cascade where the large scales inject energy into the smaller ones [11].

The flow natural frequencies are in the range between 1 and 2 Hz, as seen in Figure 3b, that correspond to the formation of the vortices with characteristic lengthscale L_1^{-1}. This highlights the significance of this lengthscale as the primary mixing mechanism.

The Strouhal number $St = fD_h/v$, varies between 1.8 and 3.4, which is within the most common range used in the literature for cylindrical BRs, $0.15 < St < 4$ [7].

3D CFD Simulation – Planar BR

Figure 4 shows the results for the flow fields at different Re numbers for the planar geometry with a depth of $\omega = 15$ mm.

https://doi.org/10.52202/074122-0069

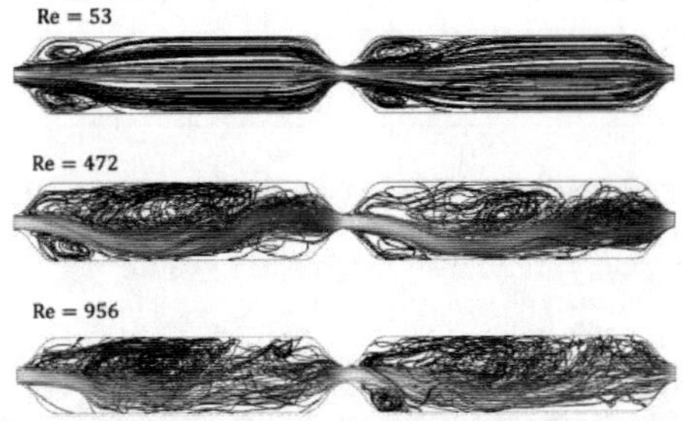

Figure 4. Streamlines for planar BR at different Re.

Although the transition for a chaotic flow regime in planar BR differs significantly from the cylindrical BR, the flow fields obtained exhibit similarities.

Figure 4 shows that the transition occurs between $Re = 472$ and 956. At $Re = 472$, it is possible to identify a jet passing through the reactor and the formation of vortices in each chamber. At $Re = 956$, the jet loses definition and a complex flow pattern characterized by chaotic motion is generated, promoting mixing by flow engulfment in the chambers.

Figure 5 shows the turbulence spectra for the planar BR at $Re = 956$. An inversion of the energy cascade at a scale between D^{-1} and $(2d)^{-1}$ is observed. The flow natural frequencies are approximately 2 - 4 Hz (as seen in Figure 5b), and the z-component of the velocity presents a more chaotic behaviour than in the cylindrical case. The St for the planar BR varies between 0.29 and 0.40. The wave number associated to this St is bellow D^{-1}, which is the scale of the larger vortices fed by the jets and that contain most of the turbulent energy.

As previously mentioned, different values of the planar BR depth were tested. Results show that the wall effect hinders the onset of vortex shedding causing the transition regimes to occur at higher Reynolds numbers.

Conclusions

Transition flow regimes were identified for a cylindrical and a planar BR with SPC operating without external oscillation. The flow transitions from steady to chaotic between $300 < Re < 400$ for the cylindrical BR and between $500 < Re < 1000$ for the planar OFR. At the transition regime, the Strouhal number of the unforced flow rounds $1.8 < St < 3.4$ for the cylindrical and $0.29 < St < 0.40$ for the planar OFR. It was also concluded that deeper planar geometries favor the flow dynamics. The inversion of the turbulent energy cascade that was reported in CIJs was also clearly observed in BRs.

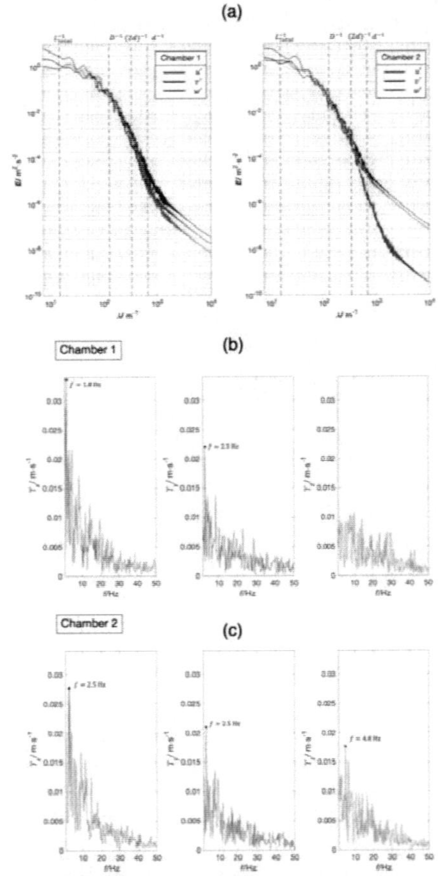

Figure 5. (a) Power spectrum, (b) frequency spectrum first chamber, and (c) frequency spectrum second chamber.

References

[1] A. Ferreira, J. A. Teixeira, F. Rocha, *Chemical Engineering Journal*, 262 (2015) 499.
[2] T. McGlone, N. E. B. Briggs, C. A. Clark, C. J. Brown, J. Sefcik, A. J. Florence, *Org Process Res Dev*, 19 (2015) 1186.
[3] A. Ferreira, P. O. Adesite, J. A. Teixeira, F. Rocha, *Chem Eng Sci*, 170 (2017) 400.
[4] C. A. L. Graça, R. B. Lima, M. F. R. Pereira, A. M. T. Silva, A. Ferreira, *Chemical Engineering Journal*, 389 (2020)124.
[5] P. Cruz, F. Rocha, A. Ferreira, *CrystEngComm*, 18 (2016) 9113.
[6] P. Cruz, C. Alvarez, F. Rocha, A. Ferreira, *Chemical Engineering Research and Design*, 175 (2021) 115.
[7] M. Avila, B. Kawas, D. F. Fletcher, M. Poux, C. Xuereb, J. Aubin, *Chemical Engineering and Processing - Process Intensification* (2021)
[8] E. Erkoç, C. P. Fonte, M. M. Dias, J. C. B. Lopes, R. J. Santos, *Chemical Engineering Research and Design*, 106 (2016) 74.
[9] F. Okkels, P. Tabeling, *Phys Rev Lett*, 92 (2004).
[10] F. Rocha, A. Ferreira, J. A. Teixeira, A. Vicente (2015).
[11] N. D. Gonçalves, H. M. Salvador, C. P. Fonte, M. M. Dias, J. C. B. Lopes R. J. Santos, AIChE Journal, 63 (2017) 2335

https://doi.org/10.52202/074122-0069

Intelligent terminal of mixing reactor with real-time monitoring and feedback control functions

Poster.24

***Songsong Wang*[1]**, *Peiqiao Liu*[1], *Yundong Wang*[2], *Zuohua Liu*[1,*].
(1) School of Chemistry and Chemical Engineering, Chongqing University, 55 University City South Road, Shapingba District, Chongqing, China, liuzuohua@cqu.edu.cn.
(2) Department of Chemical Engineering, Tsinghua University, Beijing, China, wangyd@tsinghua.edu.cn.

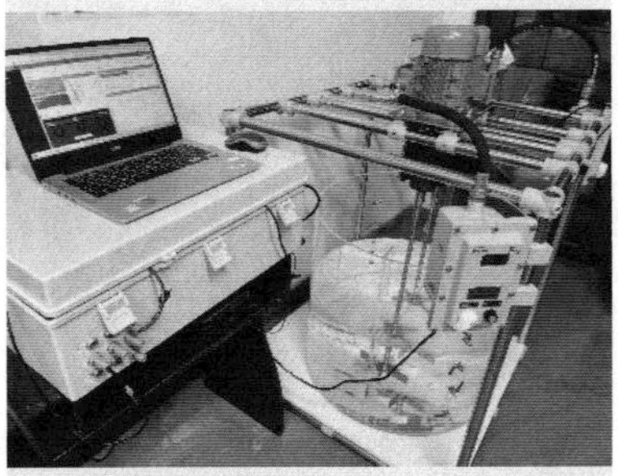

Combined with the chaotic mixing characteristics, macro mixing behavior, structural parameters, operational parameters of the stirred reactor, and big data analysis was carried out to reveal the correlation between the chaotic characteristic parameters, mixing effect and safety control, and establish the operation platform of intelligent regulation and feedback of the stirred reactor. It mainly includes three functional modules: dynamic risk control, online monitoring and early warning and real-time feedback control. The results showed that the application of intelligent terminal in stirred reactor can realize the online monitoring and dynamic monitoring and early warning of stirring reactor speed, fluid chaos characteristic parameters, temperature, pressure, pH and other parameters, and realize the visualization and feedback control of historical data, which are of great significance for promoting energy conservation and emission reduction, developing intelligent chemical equipment and safety.

Introduction

Traditional chemical, metallurgical and other process industries are the basis of manufacturing industry, and their economic efficiency affects the speed of social development to a certain extent. As one of the pillar industries of national economic development, the chemical industry is in urgent need of industrial technology renewal. Stirred reactor fluid mixing is the basic chemical operation unit, and its future trend is to develop high-end and intelligent stirred reactors with high efficiency and energy saving [1-4]. Chaotic mixing of fluids in stirred reactors is a typical nonlinear process and far from equilibrium, which necessarily contains a large number of nonlinear kinetic mechanisms, including the formation, transport and evolution of multi-scale flow fields [5,6], as well as vortex aggregation and rupture processes characterized by spatio-temporal chaos and cross-scale correlation coupling, resulting in a lack of synergy between the transfer mixing law of the reaction system and the reaction process [7,8].

However, because the fluid mixing performance is closely related to the structural and operational parameters of the stirred reactor, as well as the complex correlation and highly sensitive dependence of the evolutionary behavior of the flow field, traditional stirred reactors often suffer from high energy consumption, low efficiency, and difficult regulation. Therefore, carrying out theoretical innovation, technology development and application of high-efficiency, energy-saving and intelligent stirred reactors based on industrial Internet will help promote energy-saving and emission reduction, intelligent manufacturing, and facilitate the transformation of related industries to energy and resource-saving, environment-friendly and intelligent production modes.

There is a lack of research and development on intelligent online acquisition of chaotic characteristic parameters and remote intelligent feedback regulation system for stirred reactors. Therefore, the acquisition of chaotic parameters of mixing system and the establishment of online chaotic intelligent regulation operation mode and feedback mechanism are the urgent scientific problems to be solved in the development of large-scale, high-end and intelligent stirring reactors. it is of great academic significance and potential economic value to comprehensively improve the intelligent diagnosis, control and matching capability of chemical process equipment, more accurately predict the future state changes of equipment, and collect, analyze and process the operating mode matching and health mode trajectory of stirred reactors over time by combining with big data-Internet.

Material and Methods

In a stirred reaction system, the flow field transport process has the mutual transfer and coupling of material flow, energy flow, information flow and chaotic flow, which can be described by the Largest Lyapunov exponent (LLE), chaotic attractor and other chaotic characteristic parameters. It is known that when the Lyapunov exponent > 0 in the stirring system, it indicates that the system is in a chaotic state, and the larger the value, the greater the degree of chaos. Therefore, the data of time-varying signals and fault information sources in the mixing process of stirrer fluid are collected by using big data platform, and the original basic data collected are intelligently analyzed and processed in real time to form a Big Data-Internet-based technology system for collecting and analyzing time-varying signals of stirrer fluid. The intelligent feedback and synchronization control system based on the fluid

https://doi.org/10.52202/074122-0070

chaotic characteristics of the stirred reactor was established, and the intelligent control terminal was developed by combining the intelligent controller of the Internet of Things, online remote control and monitoring, computer language and Internet signal transmission.

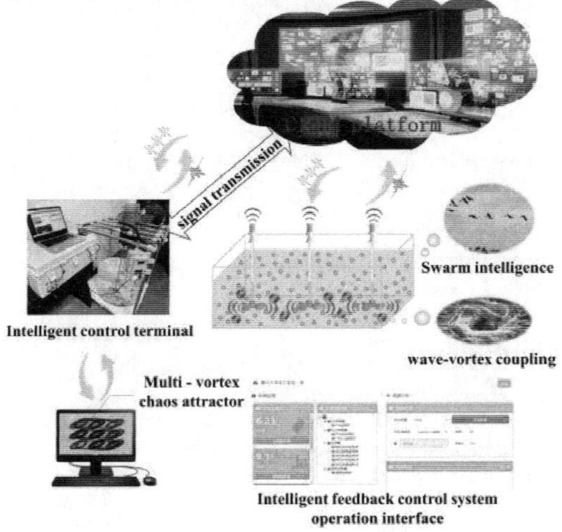

Figure 1. Intelligent multi-axis stirring and group intelligent feedback control system.

Results and Discussion

As shown in Figure 2, the system is divided into three levels: data collection, information collection, and dispatching command center. The platform adopts the structure design of C/S+B/S. With the help of various testing equipment facilities (such as torque sensors, temperature transmitters, pressure transmitters, etc.) and many technologies such as GIS technology, big data computing, network communication technology, etc., it can quickly discover the abnormal situation of the mixing equipment, and can quickly visualize and locate the location of the alarm equipment and send SMS notifications to promptly remind the relevant responsible persons of the equipment to deal with.

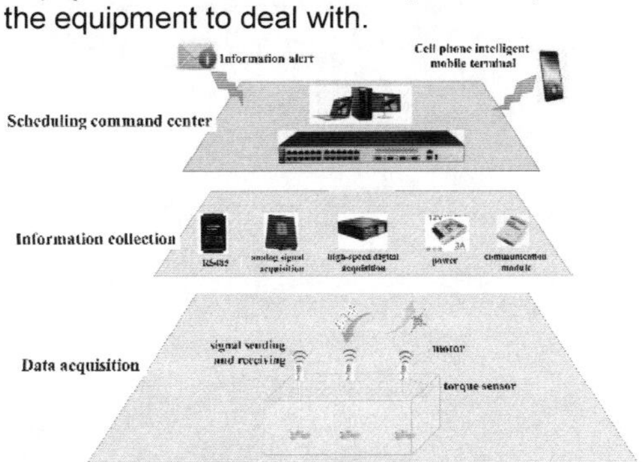

Figure 2. System Levels

During the operation of the platform, we need to collect in real time the common data that affect the operation status of the mixing equipment: mixing speed, temperature, pressure, viscosity of the mixing medium, etc. At the same time, certain key parameters (torque pulsation signal, wall pressure pulsation signal, etc.) must be collected at high frequency and data processing. In addition, the system needs to be able to distinguish and alert the user of the real-time operating status of the equipment (normal, alarm, fault, stop).

Taking all aspects into consideration, the system is divided into four parts in terms of real-time monitoring.

(1) Public environment parameters

The common environmental parameters such as temperature and humidity of the environment in which the mixing equipment is located are displayed on the interface of the system in real time, and the change curve of the relevant parameters in a certain cycle is synchronized, so that the user can understand the current environmental status of the operation of the equipment in a multi-dimensional and intuitive way.

(2) Key parameters of equipment

The key operating parameters of the mixing equipment (such as torque pulsation signal, wall pressure pulsation signal, etc.) and common data are extracted and displayed on the interface in real time after data processing and calculation, and other functions are similar to the common monitoring.

(3) Monitoring

The system can provide the specific equipment operating conditions of all mixing equipment under each site.

(4) Equipment operation status monitoring and control

The system uses four different states to distinguish the current operating status of the equipment. In the default normal operation state, the equipment status point is green; in the alarm state, the equipment status point is pink, and an alarm bubble will be generated to display the alarm content and alarm time to remind the user of the alarm behavior of the current equipment, so that the first time to alarm and rushed to the equipment site for timely handling; in the fault state, the equipment status point showing the corresponding fault information is red, and the fault event is also displayed; in the equipment out of service state, the equipment status point is blue. In the fault state, the status point of the device showing the corresponding fault information is red and the fault event is displayed at the same time; in the equipment out of use state, the device status point is blue, indicating that the equipment has stopped running. Users can easily locate the current fault information according to the different status prompts, so as to remotely control the equipment to continue operation or disable operation. The functions of the real-time monitoring and control module are shown in Figure 3.

https://doi.org/10.52202/074122-0070

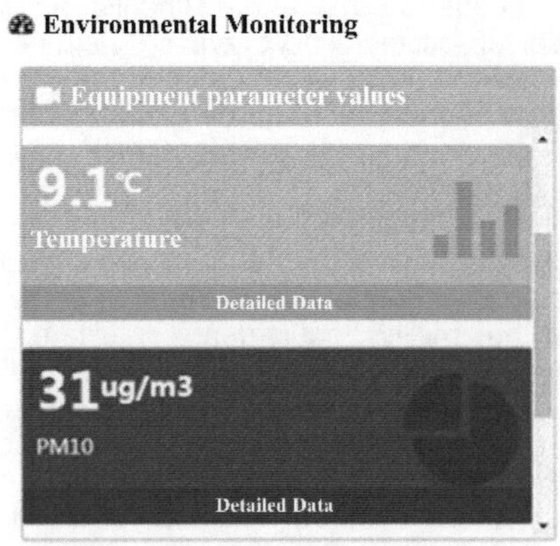

Figure 3. Platform environment monitoring page.

Conclusions

(1) Employing industrial Internet of Things technology, we developed a rigid-flexible stirred reactor fluid time-varying signal acquisition and analysis terminal that integrates online data acquisition, automatic analysis and calculation, remote monitoring and control.

(2) Based on the Internet signal transmission and computer code to automatically analyze and process the original data, it realized the remote acquisition and feedback regulation of chaotic characteristic parameters, vibration signal, stirring speed, temperature, pH and other data in the rigid-flexible stirred reactor.

(3) This terminal forms the operation mode of rigid-flexible stirring reactor with real-time view, real-time alarm, remote monitoring and regulation, and predicts the change trend of the future state of the equipment, and can be used wirelessly for dynamic monitoring of pilot line (temperature, torque, environmental detection, etc.). This saves comprehensive wiring and decoration costs and avoids line operation and maintenance costs due to line damage at a later stage.

Acknowledgments

This work was supported by the National Natural Science Foundation of China (22078030, 52021004), Fundamental Research Funds for the Central Universities (2022CDJQY-005), National Natural Science Foundation of Chongqing(2022NSCQ-LZX0271), Key Technologies Research and Development Program (2019YFC1905802, 2022YFC3901204), Key project of independent research project of State Key Laboratory of coal mine disaster dynamics and control (2011DA105287-zd201902).

References

[1] FENG Zhongxu, ZHAO Lijun. *Intelligent stirring equipment. Journal of Chang'an University (Natural Science Edition)*, 2004, (06): 77-79.

[2] ALJAAFREH A. *Agitation and mixing processes automation using current sensing and reinforcement learning. Journal of Food Engineering*, 2017, 203:53-57.

[3] SUN Hongwei, ZHANG Guojun. *Chemical engineering: from basic research to industrial application. Science in China: Chemist*ry, 44 (2014) 1375-1384.

[4] WANG Hui. *Smart Industry*. Beijing: CITIC Press, 2018.

[5] *Reactor*. FENG Lianfang, WANG Jiajun. *Beijing: Chemical Industry Press*, 2010.

[6] KELLEY D H, OUELLETTE N T. *Separating stretching from folding in fluid mixing. Nature Physics*, 7 (2011) 477-480.

[7] TONG Binggang, YIN Xieyuan, ZHU Keqin. *Vortex motion theory*. Anhui: *China University of Science and Technology Press*. 2012.

[8] *Fluid chaotic mixing and stirring process strengthening method*. LIU Zuohua, WANG Yundong, TAO Changyuan. *Chongqing*: *Chongqing University Press*. 2016.

Mass transfer limitations in photocatalytic mesostructured reactors for green ammonia production

Poster.25

M. Teresa Oliveira[1,2], *Isabel S.O. Barbosa*[1,2], *Margarida S.C.A. Brito*[1,2], *Cláudia G. Silva*[1,2], *Ricardo J. Santos*[1,2]
(1) Laboratory of Separation and Reaction Engineering-Laboratory of Catalysis and Materials (LSRE-LCM), Rua Dr. Roberto Frias, 4200-465 Porto, Portugal; (2) ALiCE Associate Laboratory in Chemical Engineering, Faculty of Engineering, University of Porto, Rua Dr. Roberto Frias, 4200-465 Porto, Portugal.

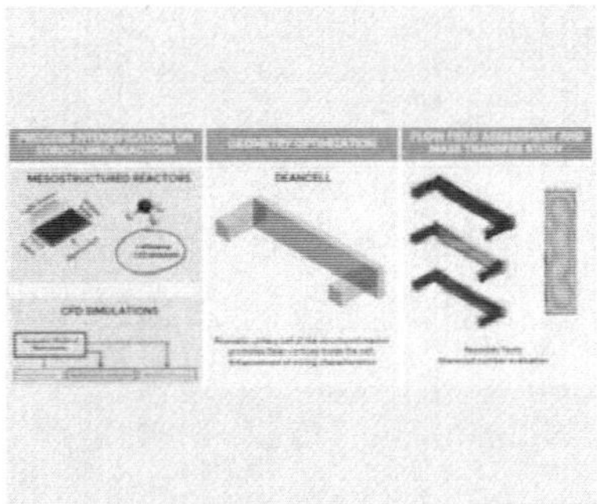

An elementary configuration of a unit-cell of a mesostructured photocatalytic reactor for green-ammonia production was designed and optimized through computational fluid dynamic (CFD) simulations. A parametric study assessed the best geometry of the cell based on vorticity and velocity profiles, which ensured higher mixing rates inside the reactor. Mass transfer studies were carried out to evaluate mass transfer limits in the cell that favor ammonia diffusion and maximize the efficiency of the photocatalytic reaction. The designed cell seems to be effective as an elementary cell of a mesostructured photocatalytic reactor, showing that ammonia diffusion through the reactor is only slightly dependent on the location of the immobilized catalytic material.

Introduction

Micro and mesostructured reactors have been under popular research in chemical engineering due to their potential to enhance mixing, reaction rates, and easy scalability. This type of reactor shows high surface-to-volume ratio, allowing for safer operation and more precise control of reaction variables [1].

Aside from efficient catalysts and adequate light sources, photocatalytic reactions, in particular, are much conditioned by the photoreactor's fluid dynamics and the dimensions and geometry. Small characteristic lengths, which are typical of mesostructured reactors, enable more uniform light and flow distribution (usually in the laminar regime), mitigating the impact of possible stagnant or non-irradiated zones on the reaction efficiency. Light attenuation effect when scaling-up the photoreactor must also be kept in consideration [2]. The scale-up from laboratory scale to pilot and industrial devices is usually performed not through a dimension increase but by numbering-up the unit elements of the mesostructured network, which keeps constant the surface/volume ratio and does not affect the high characteristic mass and heat transfer performance of these reactors.

One example of this type of reactor is the NETmix, a static mixer micro/meso-structured reactor, developed by Laranjeira et al. [3] and patented at University of Porto, which has successfully handled heterogeneous photocatalytic reactions. This reactor's unique inner geometry enables high fluid mixing, even under laminar regimes, and contributes to high mass and photon transfer rates.

Besides integrating flow field dynamics, photocatalytic systems must account for mass transport, reaction kinetics and photon flux distribution, increasing these models' complexity. Computational fluid dynamics (CFD) tools arise as an effective numerical approach to the problem that accounts for the interaction of all these phenomena and an interesting design strategy to predict and optimize fluid behavior inside the photoreactor. This work reports a CFD study on the design of a structured and immobilized photoreactor with direct application in green-ammonia synthesis, a field in recent advancement, as the search for a more sustainable and low-carbon alternative to Haber-Bosch industrial-scale process is widened.

Material and Methods

The base case geometry of the DeanCell, the structured reactor's unit-cell, was designed to promote secondary flow in the form of dean vortices and enhance fluid mixing inside the reactor.

This prismatic geometry, schematized in Figure 1, shows the inlet and outlet channels and a main body on which the immobilization of the catalyst will occur. The main inlet channel was fixed, presenting a cross-section area of 1x1 mm^2 (d and w) and 3 mm length (l_{in}) to promote a developed flow at the entrance.

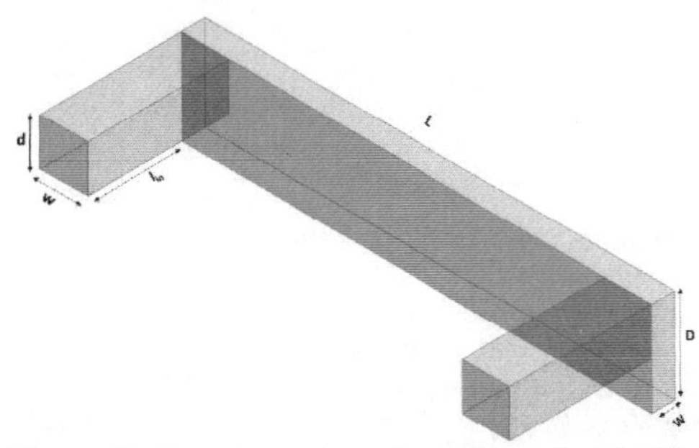

Figure 14. Base case geometry of the unit-cell of the mesostructured reactor.

To assess the geometry and the dimensions of the best unitary element configuration, a parametric study was developed using ANSYS Fluent R2 2022. Simulations were performed in laminar flow regime at a Reynolds number (Re) of 100, using the Coupled scheme method for

https://doi.org/10.52202/074122-0071

pressure-velocity coupling. Water was set as the working fluid, and the following default physical properties were defined: ρ = 998.2 kg·m^{-3}, and μ= 1 x10^{-3} Pa·s.

For all simulations, a mesh validation was carried out assessing the velocity profiles, and the final chosen mesh element size was 0.02 mm.

The dimensions of the main channel (*D*, *W* and *L*) were under analysis, and vorticity maps were traced in order to optimize these parameters.

After that, a Re range between 1 and 1000 was tested to study its influence on the generation of Dean vortices. Mass transfer simulations were also performed to set the mass transfer upper limit for ammonia production. Last, for further analysis of the hydrodynamics, the results of dynamic simulations of tracer injection are reported.

Results and Discussion

Unit-Cell Geometry

A simplification of DeanCell geometry was used for this study. This geometry is similar to the one sketched in Figure 1 but without the outlet square pipe.

Several geometries were tested for Re=100, and the main results are summarized in Figure 2. The dimensions are identified in legend of Figure 2, e.g., D2W2 corresponds to $D = 2$ mm and $W = 2$ mm.

Vorticity profiles along the length of the cell show that the steep expansion from the inlet channel to the chamber causes an increase in flow rotation. This flow pattern enhances mixing in the Dean cell and mass transfer to reactor walls. The x-vorticity magnitude is higher for channel heights equal or above 2mm.

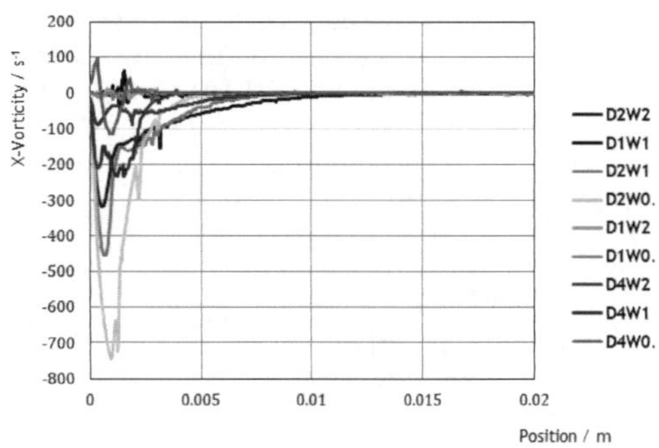

Figure 2. Vorticity profiles in the axis of the Dean cell along flow x-direction.

The case that promotes the highest x-vorticity values was a cell with 10mm of length *L*, *D* of 2mm and *W* of 0.5mm (D1W0.5 in Figure 2), since the vorticity is higher for this case. It was also observed that for a length above 10mm, the vorticity is close to zero, and so this was considered the mixing length. Because of that, the geometries studied in this work have a length of 10 mm.

Flow Field Assessment

After reaching the best length of DeanCell (L=10 mm), Re numbers between 1 and 1000 were evaluated by the streamlines and vorticity.

For 1< Re <125, the vortices generation is not detectable. For Re=250, Figure 3 shows the vortices formed along the main channel, through a surface streamline, obtained from *y* and *z* velocity components in a section of the main chamber. This illustration, done with CFD-Post, enables to clearly draw the flow rotation path.

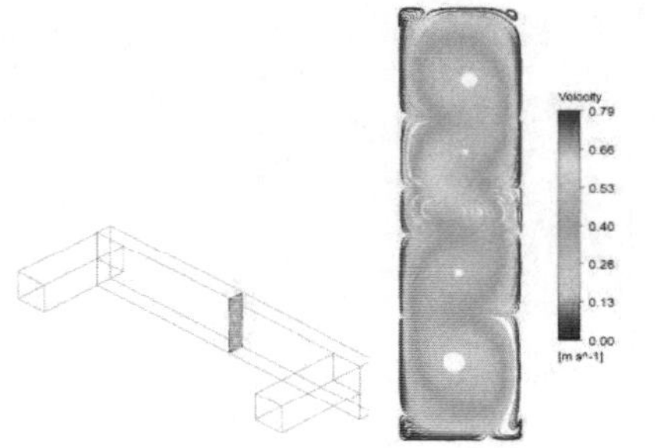

Figure 3. Velocity streamline in a mid-section plane for Re=250.

Mass Transfer

Mass transfer from the catalytic plate to the bulk of the cell is evaluated through steady-state simulations at different Re. To assess mass transfer limits, a case of infinite reaction was modelled. The catalyst is immobilized on the two side walls of the main channel, walls normal to z. The maximum mass fraction of NH_3 in water was set at these walls considering an immediate conversion of N_2 to NH_3, according to

$$N_2 + 3H_2O \rightarrow 2\ NH_3 + \frac{3}{2}O_2 \quad (1)$$

The operating conditions were 1 atm and 20 °C.

Different catalytic immobilization configurations are studied, and the efficiency of the NH_3 diffusion from the catalytic surfaces to the fluid phase is monitored through a contour of the ammonia concentration at the outlet. Figure 4 shows a side-view of the unit-cell, corresponding to the case of immobilization of catalyst in the interior and exterior walls, which are highlighted in red in the figure below.

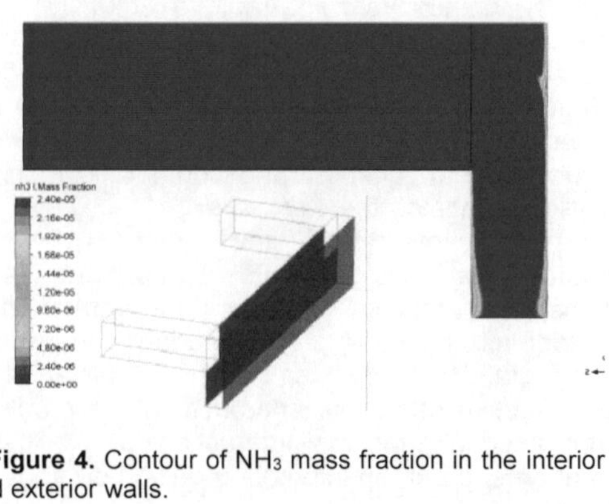

Figure 4. Contour of NH_3 mass fraction in the interior and exterior walls.

https://doi.org/10.52202/074122-0071

Considering a well-mixed solution, ΔC, the molar concentration gradient between the catalyst and the bulk of the reactor is given by the difference between the concentration in the catalytic plate, C_{CP}, and the monitored concentration value at the outlet face, C_{out}. Equation 2 can be written, where $\dot{N}$ is the molar flux, and k is the mass transfer coefficient.

$$\dot{N} = k \cdot \Delta C = k \cdot (C_{CP} - C_{out}) \qquad \textbf{(2)}$$

The extent of mass transfer in the cell was further defined by calculating Sherwood numbers (Sh) through Equation 3, in which l_{cp} is the catalytic plate length, and D is the diffusivity of NH_3 in water, estimated in 2.32×10^{-9} $m^2\cdot s^{-1}$.

$$Sh = \frac{k\, l_{cp}}{D} \qquad \textbf{(3)}$$

Residence Time Distribution

For further analysis of the hydrodynamics of the system, the Residence Time Distribution (RTD) curve, $E(\theta)$, was obtained from the derivative of the $F(\theta)$ curve, shown in Figure 5b and 5a, respectively. It considers the different times that different fluid elements take to travel through the cell, given as function of θ, a dimensionless time with respect to the passage time.

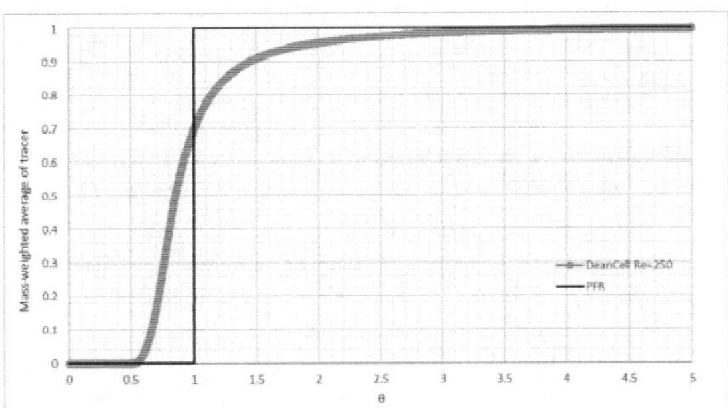

(a)

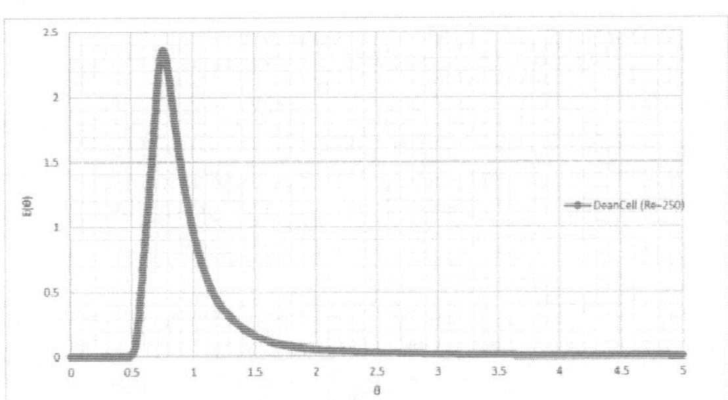

(b)

Figure 5. (b) RTD curve for the DeanCell.

The $E(\theta)$ peak shows that most elements leave the reactor before the reactor's passage time: $\theta = 1$. The main flow path provides a shortcut to the flow that is not too pronounced (approximately 75%), probably due to the higher flow velocity away from the walls. The absence of additional peaks indicates that the Dean vortices are promoting a good dispersion of mass in the cell. Otherwise, fluid elements in different regions would leave the reactor at different times, creating an RTD with several modes.

This narrow RTD curve indicates uniformity of the residence time across all fluid elements, as well as good fluid homogenization inside the reactor.

Conclusions

Based on the simulations performed, the designed cell appears effective for use as an elementary cell of a mesostructured photocatalytic reactor.

Mass transfer studies show ammonia diffusion through the reactor, which, according to NH_3 mass fraction reports at the outlet, is only slightly dependent on the localization of the catalytic material. The maximum value reported for the Sh number was of 950, for Re=350, corresponding to the case of catalyst immobilization in the exterior and top walls of the main channel.

Further studies will assess the mass transfer limiting rates that guarantee the operation of the system under chemical regime.

Acknowledgements

This work was financially supported by LA/P/0045/2020 (ALiCE), UIDB/50020/2020 and UIDP/50020/2020 (LSRE-LCM), and the project SuN2Fuel 2022.04682.PTDC - funded by national funds through FCT/MCTES (PIDDAC). I.S.O. Barbosa acknowledges her FCT grant UI/BD/151092/2021.

References

[1] Barbosa, Isabel S.O., João M. Costa, Yaidelin A. Manrique, Madalena M. Dias, Joaquim L. Faria, Ricardo J. Santos, Cláudia G. Silva, and Margarida S.C.A. Brito. *Chemical Engineering & Technology.* 2023.

[2] Kayahan, Emine, Mathias Jacobs, Leen Braeken, Leen CJ Thomassen, Simon Kuhn, Tom van Gerven, and M Enis Leblebici. *Beilstein Journal of Organic Chemistry.* 2020. 16. 2484–2504.

[3] Laranjeira, Paulo Roberto, António A. Martins, Maria Tereza Nunes, José Manuel Lopes, and Madalena M. Dias. *Aiche Journal.* 2011. 57, no. 4. 1020–32.

https://doi.org/10.52202/074122-0071

Batch (rotor-stator) *vs* continuous (NETmix): mixing devices comparison for Pickering emulsions production

Poster.26

A. Ribeiro[1,2], *Y.A. Manrique*[1,2], *M.F. Barreiro*[3,4], *J.C.B. Lopes*[1,2,5], *M.M. Dias*[1,2]. *(1) LSRE-LCM – Laboratory of Separation and Reaction Engineering – Laboratory of Catalysis and Materials, Faculdade de Engenharia da Universidade do Porto, Rua Dr. Roberto Frias, 4200-465, Porto, Portugal. (2) ALiCE – Associate Laboratory in Chemical Engineering, Faculdade de Engenharia da Universidade do Porto, Rua Dr. Roberto Frias, 4200-465, Porto, Portugal. (3) CIMO – Centro de Investigação de Montanha, Instituto Politécnico de Bragança, Campus de Santa Apolónia, 5300-253, Bragança, Portugal. (4) SusTEc – Associate Laboratory for Sustainability and Technology in Mountain Regions, Instituto Politécnico de Bragança, Campus de Santa Apolónia, 5300-253, Bragança, Portugal. (5) CoLAB Net4CO2 – Network for a Sustainable CO2 Economy, Rua Dr. Júlio de Matos 828-882, 4200-355, Porto, Portugal. asribeiro@fe.up.pt*

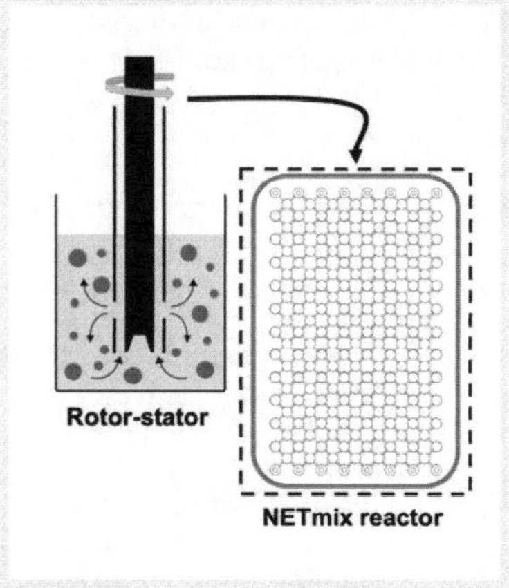

Emulsions are used in various applications. Pickering emulsions, stabilised by solid particles, are emerging as clean alternatives to create sustainable products since it is a surfactant-free product. Continuous production devices are required to achieve the industrial production of these emulsions. In this context, a mesostructured static mixer and reactor, NETmix, was tested and compared in nano-hydroxyapatite (n-HAp) Pickering emulsions production. The results indicated the feasibility of using NETmix to produce Pickering emulsions. Compared with traditional devices, rotor-stator homogenisers, NETmix ensures easy control of the mixing parameters, providing good reproducibility between assays, does not limit the volume, and allows production in continuous mode.

Introduction

Pickering emulsions are stabilised by solid particles, inorganic or organic, constituting an alternative to the traditional stabilisation by emulsifiers. Pickering emulsions, due to their emulsifier-free composition and high physical stability, have received increased interest at both academic and industrial levels, becoming very attractive solutions for various applications, such as cosmetics, pharmaceuticals, and foods [1]. Additionally, the formation of a protective shell around the dispersed phase can be explored in the development of encapsulation systems for drugs and bioactive compounds [2,3].

According to the literature, most of the previous works are focused on producing Pickering emulsions in batch mode using high-speed homogenisers or ultrasonic devices to ensure mechanical dispersion [4,5]. However, considering the high applicability of the Pickering emulsions in several emergent fields, it is important to promote the use of continuous devices enabling the control of the final emulsion characteristics, namely droplet diameter, and process variables like emulsification temperature to ensure a more feasible industrial process. NETmix is a structured mixer and reactor comprising a network of static mixing chambers interconnected by transport channels [6]. The used NETmix is made of stainless steel and comprises 25 rows and 8 columns; each chamber has 3.3 mm of diameter and channels of 0.5 mm depth, summing up a total volume of 1 mL.

In this context, this work shows a new alternative to produce Pickering emulsions in continuous mode using the NETmix technology, discussing the advantages of this process over the previously tested batch mode. The Pickering emulsions were compared in terms of stability and droplet dimensions.

Material and Methods

To produce the Pickering emulsions, sunflower oil (20%, *v/v*) and water containing dispersed n-HAp (5%wt.) were used as oil and aqueous phases, respectively. For comparison purposes, two production processes were used:

i) Batch mode: the mixing system comprised a rotor-stator homogeniser, a thermostatic bath, and a peristaltic pump; the sunflower oil was injected at 43 mL/min into the n-HAp aqueous phase and homogenised at 11000 rpm during 6 min [7]

ii) Continuous mode: both phases were fed into NETmix in a pre-mixed mode, and the droplet size reduction achieved through recirculation. The Pickering emulsion was produced using a Reynolds number (Re) = 400 and 17 recirculation cycles [8]

Results and Discussion

Figure 1 compares the two produced emulsions obtained by rotor-stator and the other with NETmix, presenting the optical images, size distributions, and cryo-SEM images right after production. From the comparison, it is possible to perceive the lower size achieved by the continuous process, which was not possible to reach in batch mode, respectively ~7 µm and ~18 µm (~60% lower in the case of NETmix). This effect was also accompanied by the narrowing of the size distribution. Cryo-SEM images show Pickering droplets with an n-HAp layer around the oil core [7,8]. Thus, the Pickering emulsions had similar

https://doi.org/10.52202/074122-0072

round-shape morphologies in both production modes and proved to be of O/W type. Moreover, they were stable for more than 8 weeks. These results show that Pickering emulsions can be produced using traditional methods or a static mixer, such as NETmix. Identical Pickering emulsions stability was obtained independently of the production method.

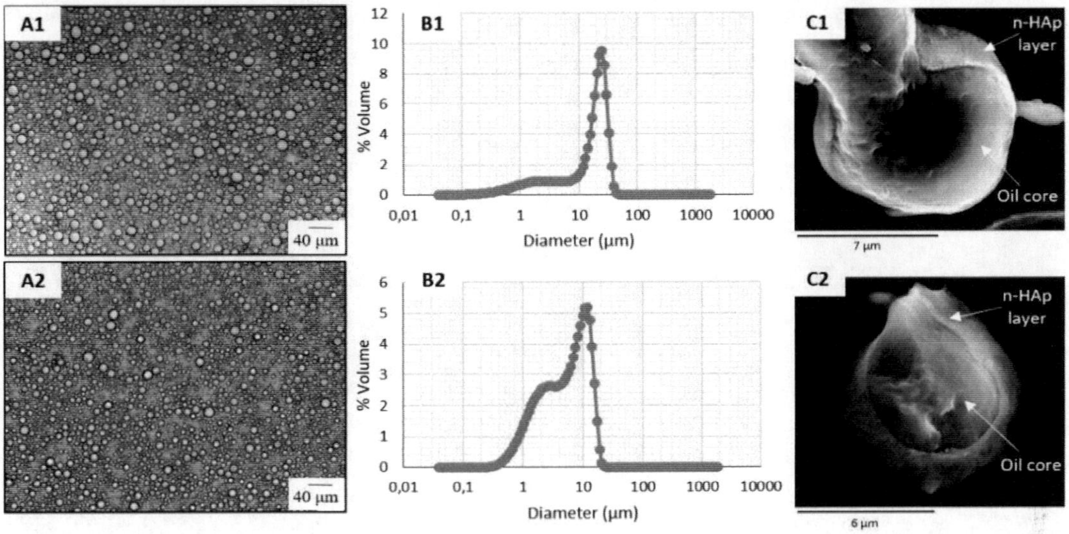

Figure 1. Optical images, size distributions and cryo-SEM images comparison obtained from batch using rotor-stator (A) and continuous using NETmix (B). Pickering emulsions were obtained with 20/80 (sunflower oil/water, *v/v*) and 5 %wt. of n-HAp.

NETmix presents several advantages over using rotor-stator homogenisers in the production of the Pickering emulsions, with the main ones summarised in Table 1.

Table 1. Advantages comparison of NETmix over to rotor-stator in Pickering emulsion production.

	Continuous - NETmix	Batch – Rotor-stator homogenizer
Easy set-up	+/-	+/-
Size droplet control	+	+/-
Risk of disruption	-	+/-
Risk of temperature rise	-	+
Energy consumption	low	high
Space-time yield (kg/m^3/day)	3.0×10^7	2.4×10^5
Industrialization possibilities	+	+/-

+ = positive; - = negative; +/- = intermediate.

Briefly, it can be pointed out the feasibility to use the NETmix technology in Pickering emulsions production, the droplet size control that can be adjusted by the used number of cycles and Re, the capacity to achieve a production scale two orders of magnitude higher than with rotor-stator (compared by space-time yield values), the capacity to avoid the overheating, which is not easy to achieve when using rotor-stator homogenisers, and high-volume-scale capacity [7,8]. These emulsions are currently being considered for developing functionalised food products incorporating bioactive extracts or compounds. Considering this point, the NETmix reactor can be an advantage since it enables the mixing temperature control avoiding thermosensitive active compounds degradation.

Conclusions

This work compares two different types of Pickering emulsions production processes, namely rotor-stator homogeniser in batch mode and NETmix in continuous mode. Pickering emulsions stabilised with n-HAp particles were successfully produced in continuous mode using NETmix with similar characteristics to those produced by a rotor-stator homogeniser. The use of NETmix can ensure an effective size and temperature control, overcoming the main constraints of the rotor-stator homogenisers.

The production of Pickering emulsions in continuous mode can support high-volume scale production, which is compatible with high demand and industrial purposes.

Acknowledgements

This work was financially supported by LA/P/0045/2020 (ALiCE), UIDB/50020/2020, UIDP/50020/2020 (LSRE-LCM), LA/P/0007/2021 (LA SusTEC), UIDB/00690/2020 and UIDP/00690/2020 (CIMO), funded by national funds through FCT/MCTES (PIDDAC).

References

[1] C. Berton-Carabin, K. Schroen, Annu. Rev. Food Sci. Technol., 6 (2015) 263-297.

[2] A. Ribeiro, R. Gonçalves, A. Pinheiro, Y. Manrique, M. Barreiro, J. Lopes, M. Dias, LWT – Food Sci. Technol., 154 (2022) 112706-112714.

[3] F. Ngwabebhoh, S. Erdagi, U. Yildiz, Carbohydr. Polym., 201 (2018) 317-328.

[4] Y. Yang, Z. Fang, X. Chen, W. Zhang, Y. Xie, Y. Chen, W. Yuan, Front. Pharmacol., 8 (2017) 287.

https://doi.org/10.52202/074122-0072

[5] X Li, J. Zhu, Y. Pan, R. Meng, B. Zhang, H. Chen, Food Hydrocoll., 90 (2019) 19-27.
[6] P.E. Laranjeira, A.A. Martins, J.C.B. Lopes, M.M. Dias, AIChE J., 55 (2009) 2226-2243.
[7] A. Ribeiro, Y.A.Manrique, F.Barreiro, J.C.B.Lopes, M.M. Dias, J. Disper. Sci. Technol., 43 (2020), 814-826.
[8] A.Ribeiro, Y.A.Manrique, F.Barreiro, J.C.B.Lopes, M.M. Dias, Colloid. Surf. A, 616 (2021), 126365-126377.

The NETmix technology, a novel approach for continuous production of microcapsules for textile application

Poster.27

Isabel M. Martins[1,2], *Ana Catarina Moreira*[1,2], *Yaidelin A. Manrique*[1,2], *Alírio E. Rodrigues*[1,2], *José C. B. Lopes*[3], *Madalena M. Dias*[1,2]. *(1) Laboratory of Separation and Reaction Engineering – Laboratory of Catalysis and Materials (LSRE-LCM), Faculdade de Engenharia da Universidade do Porto, Rua Dr. Roberto Frias, 4200-465 Porto, Portugal. (2) Associate Laboratory in Chemical Engineering (ALiCE), Faculdade de Engenharia, Universidade do Porto, Rua Dr. Roberto Frias, 4200-465 Porto, Portugal. (3) CoLAB NET4CO2, R. Dr. Júlio de Matos 828 882, 4200-355 Porto, Portugal. isa@fe.up.pt*

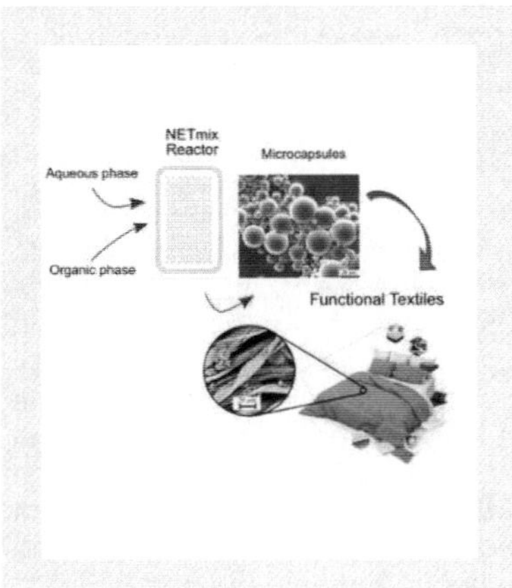

In this work, the production of microparticles (melamine-formaldehyde - MF and cellulose acetate - CA) for textile applications was developed by adapting the NETmix technology for the steps more commonly made by batch processes. MF microcapsules were obtained through a polycondensation process; different temperatures and recirculation times were tested in the curing step. Microcapsules presented diameters around 20 µm and were successfully impregnated on cotton fabrics by padding. CA microspheres were produced through a solvent evaporation method. Emulsions were produced at Reynolds numbers (Re) ranging from 200 to 1200 to evaluate its effect on the resulting microspheres' size, all characterised by a well-defined spherical shape with a rough and porous shell. The NETmix reactor can reduce the cure time relative to traditional batch processes, producing microcapsules with similar characteristics in a continuous process.

Introduction

Microencapsulation processes consist of coating an active ingredient by a membrane of another material, thus protecting the active principle, the functional material, from the surrounding environment. The active ingredient is referred as the core, and the surrounding material forms the shell [1]. The textile industry is an important area of interest for microencapsulation since functionalised textiles with microcapsules can promote comfort and protection. Carrier oils such as triglycerides are non-irritating skin emollients, encapsulated due to their hydrating properties and good skin absorption [2]. Melamine resin was used in this work as the shell material due to its good resistance against water and heat, mechanical robustness, and surface smoothness, an advantage for textile applications [3]. Cellulose acetate is a biodegradable thermoplastic polymer and the most abundant natural polysaccharide [4]. It is widely used in pharmaceutical formulations for sustained-release [5], tastes masking products [6] and in the manufacture of various commercial products such as fabrics, plastics,

Emulsification, with the formation of a stable emulsion, is the crucial step in microencapsulation since the oily droplets' diameters define the microcapsules' diameter [7]. Droplet size control is challenging when the emulsification process is scaled from laboratory to industrial scale, commonly resulting in polydisperse emulsions [7, 8]. Furthermore, in batch processes, care must be taken to obtain a homogeneous and stable emulsion in order to produce microcapsules with similar properties [8]. Various types of stirring systems can be used to get a suitable emulsion, and it must be ensured that conditions such as the power of the stirrer promote good mixing through the whole volume of the vessel [7].

The NETmix static mixer is a patented technology [9] developed at LSRE-LCM/FEUP, and it consists of a network of interconnected chambers and channels where chambers operate as mixing zones and channels behave as plug flow. Above a critical Reynolds number of ca. 150, this system evolves to a self-sustained oscillatory laminar flow regime inducing local strong laminar mixing [10]. Two different set-ups were used in this work. LabNETmix has a capacity of 23 mL and presents cylindrical chambers with a diameter of 6.5 mm and depth of 3 mm, and rectangular cross-section channels with a width of 1 mm and length of 2 mm. With smaller dimensions, MicroNETmix has a total volume of 1 ml and chambers with a diameter of 3 mm and 0.5 mm of depth and channels with 0.5 mm of width and 2 mm of length.

The production of microparticles (MF and CA) for textile applications was developed by adapting the NETmix technology for the steps more commonly made by batch processes.

For the MF microcapsules, carrier oil is used as the core material and melamine resin is used as the shell material. Thus, it was intended to produce a stable oil-in-water emulsion and to cure the shell of microcapsules by polycondensation in a continuous process, using the two NETmix devices described above. The resulting microcapsules were impregnated into cotton fabrics by padding to obtain functionalised textiles. Fig.1 shows a schematic illustration of the emulsification and cure steps in the NETmix reactor.

https://doi.org/10.52202/074122-0073

For the CA microspheres, eugenol (EG), the major component of clove essential oil, was used as core material; the emulsion step was transposed to continuous mode using the LabNETmix technology, and the influence of the Reynolds number was studied. Selected microspheres were impregnated by padding to obtain functionalised textiles. Treated textiles were observed with Scanning Electron Microscopy, and release studies of EG to the air were evaluated.

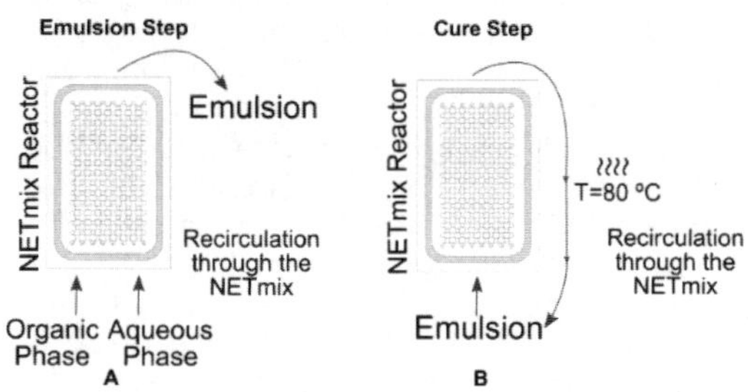

Figure 1. NETmix Reator. A – Emulsion step, B – Cure step (Adapted from Moreira et al. [11]).

Material and Methods

For the production of MF microcapsules, the emulsion step (Fig.1A) was carried out by mixing a melamine-based pre-polymer (aqueous phase) and a carrier oil+Vitamin E (organic phase) at room temperature. Afterwards, the shell cure step was done by recirculating the emulsion at 80 °C for 30 minutes as shown in Fig.1B. In LabNETmix both steps were performed at Re=200 and for MicroNETmix was ensured that both steps were performed at Re above 350. The cured microcapsules were then submitted to the coating and stabilising steps.

The microcapsules were characterised by optical microscopy and scanning electron microscopy (SEM) to evaluate morphology and size. Particle size distribution was also analysed by laser diffraction. After impregnation with microcapsules, textiles were analysed by SEM.

Oil-in-water emulsion of CA microspheres was carried out using the LabNETmix reactor at room temperature [11]. AP (aqueous phase) and DP (dispersed phase) were fed to the NETmix and recirculated. The recirculation period was 1/3 of the emulsification time for the corresponding batch process [12]. Figure 1A shows the NETmix set-up used for the emulsion step scheme. Reynolds number between 200 and 1200 were tested to study the effect on the mean diameter of microspheres. At the end of the recirculation, the emulsion was collected in the product vessel and then undergoes solvent evaporation of the ethyl acetate under lower mechanical stirring for 18 h. During this period, the DP becomes rich in CA and EG due to the removal of the solvent and starts to crystallise, forming polymeric microspheres with an active principle entrapped.

The functionalisation of cotton fabrics was made through padding (Fig. 2), a traditional process of textile impregnation.

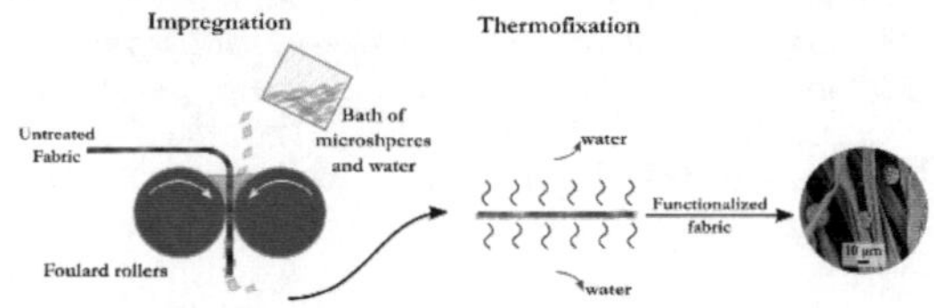

Figure 2. Foulard technique used for the impregnation of microspheres [13].

Results and Discussion

For MF microcapsules produced in LabNetmix, the emulsions were obtained with either a single pass or by recirculating the mixture through the NETmix, whereby it was shown that recirculation promotes a decrease of the diameter of the oily droplets to circa 25 µm. These microcapsules were then cured and formed. After coating and stabilising steps, microcapsules present a mean diameter of 30 µm, remain stable, do not disintegrate, and show a smooth surface with no membrane fragments (Fig.3A). In MicroNETmix, emulsion with recirculation promoted a mean droplet diameter of 9 µm. After the curing step at 80 °C, microcapsules showed a slight increase of diameter, 13 µm, which remains after coating and stabilising steps. In MicroNETmix, Re>350 ensures small diameters avoiding the breakage of microcapsules. Microcapsules are spherical, with a smooth surface, and no rupture is observed (Fig.3C). Microcapsules produced in both reactors show mean diameters within the range is 10-40 µm, the preferred size [14] for textile application (Fig. 3B/D). Noteworthy is that, either in LabNETmix or in MicroNETmix, the curing step is reduced from 150 minutes, typically observed in the batch process, to 30 minutes.

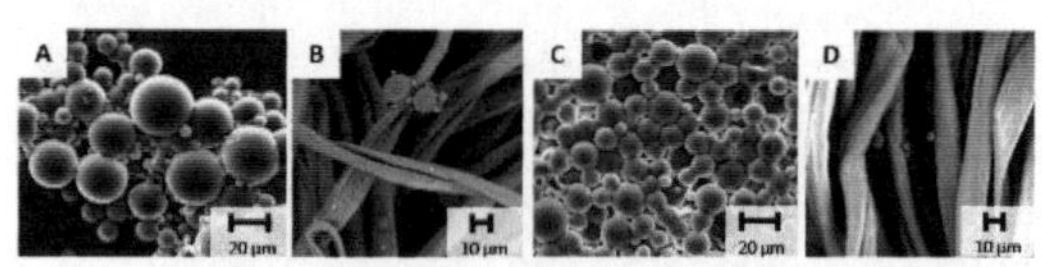

Figure 3. SEM images of melamine-based microcapsules: A and B - produced in LabNETmix and its impregnation in cotton fabrics, respectively. C and D - produced in MicroNETmix and its impregnation in cotton fabrics, respectively. A, C- resolution of 2000x; B, D– resolution of 2500x.

For the CA microspheres, the morphology of the particles is independent of the Reynolds number. Figure 4 shows high-resolution SEM images for larger diameter CA microspheres (CA200 to CA400).

All microspheres present a well-defined spherical shape, with a porous and rough shell contrasting with the smooth surface microspheres produced in batch conditions [12]. This porosity and roughness of the shell are likely due to the low CA concentration used and to the solvent evaporation step. With low CA concentration, less mass of CA is available to cover the DP droplets, thus explaining the porosity. This is augmented during solvent evaporation since the solvent, which is firstly entrapped into the DP matrix, is then evaporated, going through the polymer shell that is being formed [15], destroying it in the process. In some cases, microspheres appear to be broken or hollowed (black holes), mostly for higher Reynold numbers. The continuous mode enables the production of microspheres with a specific range of diameters by changing Reynolds number.

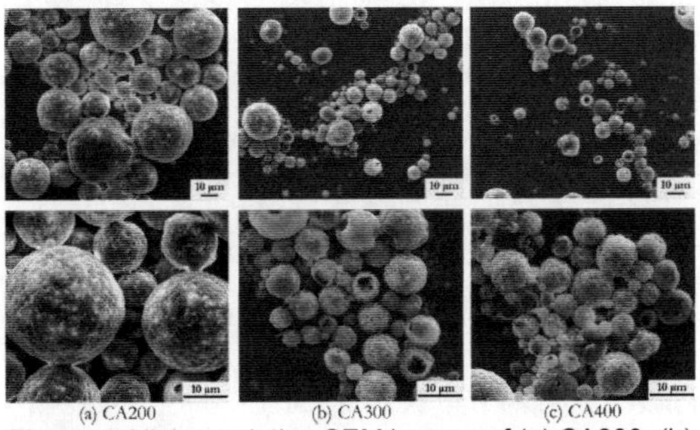

Figure 4. High-resolution SEM images of (a) CA200, (b) CA300 and (c) CA400. Resolutions of 2000× (first row of images) and 5000× (second row of images).

Conclusions

In conclusion, the emulsion and cure steps in the production of microcapsules can be achieved using the NETmix technology. Microcapsules are stable after production and keep the spherical shape and morphology after impregnation on fabrics. The continuous process of melamine-based microcapsules leads to microcapsules with the same characteristics as batch production, with a significative (80%) reduction of curing time when compared with the batch procedure.

CA microspheres containing EG were produced in a continuous mode using NETmix technology. From a known batch formulation based on the solvent evaporation method, the emulsion step was transposed to the NETmix reactor. Emulsions were formed at Reynolds number ranging from 200 to 1200 to study its influence on the mean diameter of microspheres. Microspheres from all trials exhibit well-defined spherical shapes with a porous and rough shell, the latter explained by the evaporation of the entrapped solvent through the shell and the low CA concentration.

With NETmix, it is possible to control the diameter by changing the Reynolds number to obtain microspheres with similar morphology and required size for textile applications.

Acknowledgments
This work was supported by: ProjectTexBioPro, POCI-01-0247-FEDER-003472, fundedby FEDER, through COMPETE 2020, under PT2020. This work was partially supported by: LA/P/0045/2020 (ALiCE), UIDB/50020/2020 and UIDP/50020/2020 (LSRE-LCM) funded by national funds through FCT/MCTES (PIDDAC). Authors thank Centro de Materiais da Universidade do Porto (CEMUP) for the services provided with high resolution SEM.

References
[1] R. Dubey, T.C. Shami, K.U.B Rao, *Defence Science Journal*, 59 (2009) 82-95.
[2] M. Mahjour, B.E. Mauser, Z.A. Rashidbaigi, M.B Fawzi, *International Journal of Pharmaceutics*, 95 (1993) 161-169.
[3] U.S. Elesini, J. Švarc, B. Šumiga, R. Urbas, *Textile Research Journal* 87 (2016) 1-14.
[4] M. Pinto, D. Silva, A. Gomes, V. Leite, A. Moraes, R. Novais, J. Tronto, F. Pinto, *RSC Advances*, 9 (2019). 5620–562.
[5] K.S. Soppimath, A.R. Kulkarni, T.M. Aminabhavi, *Drug Development and Industrial Pharmacy*, 27 (2001). 507–515.
[6] R.C. Rowe, P.J. Sheskey, M.E. Quinn, Handbook of pharmaceutical excipients, 6th Ed. Pharmaceutical Press and American Pharmacists Association, USA, 2009.
[7] S. Benita, Microencapsulation Methods and Industrial Applications, 2nd ed., CRC Press, Boca Raton, FL, USA, 2006.
[8] R. Heusch, Emulsions in Ullmann's Encyclopedia of Industrial Chemistry, Wiley-VCH Verlag GmbH & Co. KGaA, Weinheim, 2005, 1-40.
[9] J.C.B Lopes, P.E. Laranjeira, M.M. Dias, A.A. Martins, Network mixer and related mixing process. WO 2005/077508 A1. 2005.
[10] P.E. Laranjeira, A.A. Martins, J.C.B. Lopes, M.M. Dias, *AIChE*, 55 (2009) 2226-2243.
[11] A.C.G. Moreira, Y.A. Manrique, I.M. Martins, I.P. Fernandes, A.E. Rodrigues, J.C.B. Lopes, M.M. Dias, *Ind Eng Chem Res*, 59 (2020) 18510–18519.
[12] M.G. Simões, P. Coimbra, A.S. Carreira, M.M. Figueiredo, M.H. Gil, P.N. Simões, *Cellulose*, 27 (2020) 4109–4121.
[13] A.C.G. Moreira, Y.A. Manrique, I.M. Martins, I.P. Fernandes, A.E. Rodrigues, J.C.B. Lopes, M.M. Dias. *Cellulose*, 29 (2022) 3595–3612.
[14] S. Bône, C. Vautrin, V. Barbesant, S. Truchon, I. Harrison, C. Geffroy, *CHIMIA - International Journal for Chemistry,* 65 (2011) 177-181.
[15] M. Li, O. Rouaud, D. Poncelet, *International Journal of Pharmaceutics*, 363 (2008) 26–39.

https://doi.org/10.52202/074122-0073

CFD modelling of vertical flow and mixing in Raceway Pond Reactors

Poster.28

Margarida L.R. Peixoto [1,2], *Margarida S.C.A. Brito* [1,2], *Ricardo J. Santos* [1,2], *Vítor J.P. Vilar* [1,2] *(1) LSRE-LCM - Laboratory of Separation and Reaction Engineering – Laboratory of Catalysis and Materials, Faculty of Engineering, University of Porto, Rua Dr. Roberto Frias, 4200-465 Porto, Portugal. (2) ALiCE - Associate Laboratory in Chemical Engineering, Faculty of Engineering, University of Porto, Rua Dr. Roberto Frias, 4200-465 Porto, Portugal.*

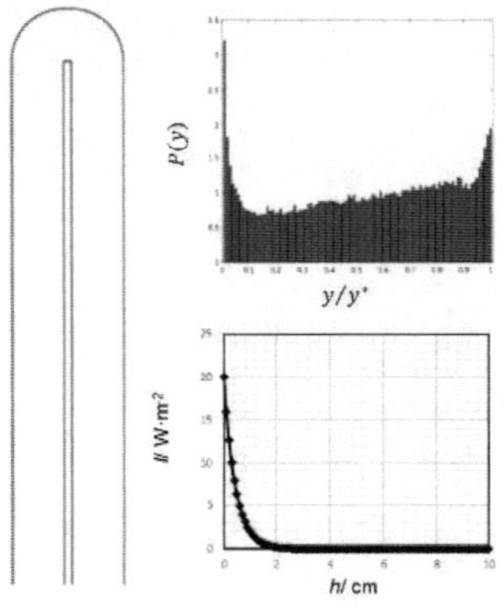

In this work, the vertical flow and the mixing degree in raceway pond reactors (RPRs) were studied through computational fluid dynamics (CFD). The radiation profile over the fluid depth was also analysed for a typical case study. The results show that there is a good distribution of the fluid elements over the RPR vertical direction. Nonetheless, these have some propensity to remain longer on the surface and at the base of the reactor.
Regarding the intensity of radiation, it can be reported that RPRs are promising devices that allow the fluid within the reactor to absorb a substantial amount of energy. Additionally, this quantity can be adapted by changing the residence time in the reactor according to the intended application.

Introduction

Raceway pond reactors (RPRs) are extensive non-concentrating photoreactors, with a very low construction cost, consisting of open channels forming a raceway where the stream is set in motion by a paddlewheel. They have been widely used for large-scale production of microalgae and more recently for disinfection of secondary effluents in WWTPs [1, 2].

The application of this type of reactor is very promising, given its innumerable advantages. For example, their configuration provides a large area of solar exposure, which can be very useful for radiation dependent uses. Besides that, they can be suitable for processes that highly depend on the effects caused by mechanical forces, such as shear stresses. They also have the ability to vary the liquid depth operation according to the radiation availability [1]. Moreover, they generally present a low construction cost and low energy consumption, being able to yield high productivity rates, as they present an easy scale-up [3, 4]. In this sense, the design of these reactors is crucial, since it may interfere with several variables, such as the degree of mixing, and the existence of property gradients, such as temperature, solar penetration, dissolved oxygen concentration, among others. To optimise the design, some aspects like the geometry, the configuration of bends, and the specific design of the paddlewheel have to be taken into account, as they may affect both the mass transfer and the hydrodynamic behaviour of the reactor [2]

In this work, the vertical flow and the radiation exposure of the fluid in the RPR were analysed using Computational Fluid Dynamics (CFD).

Material and Methods

CFD simulations were performed using the ANSYS 2022 R2 software. Figure 1 shows a 2D configuration of the RPR top view. It should be noted that only half of the reactor is represented, as there is a symmetry axis on the *inlet*/*outlet* plan.

Figure 15. Geometry of the overall 2D domain.

The computational domain geometry used in this work consists of a cross sectional of the RPR on the *yz* plan, as shown in Figure 2. A length, L, of 1 m and a height, h, equivalent to the reactor fluid level, of 0.1 m were considered.

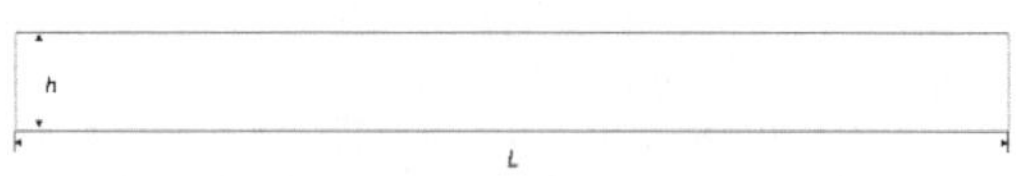

Figure 16. RPR depth plan 2D geometry.

Reynolds Average Navier-Stokes equations were used to simulate the turbulent flow and Standard k-ε model was set for the subgrid modelling. The working fluid was set as water at 20 °C, where ρ = 998.2 kg·m^{-3} and μ = 1.003 mPa·s.

Regarding the boundary conditions, at the bottom wall, a no-slip condition was established. At the top, X and Y shear stresses of zero were specified, as the top is a free surface exposed to the atmosphere. Furthermore, on the left and right-hand sides, translational periodic boundary conditions were implemented. For that purpose, the value of the pressure drop per unit length, $\Delta P/L$, was estimated considering Manning's equation,

$$v = \frac{k_n}{n} R_h^{\frac{2}{3}} S^{\frac{1}{2}} \qquad (1)$$

https://doi.org/10.52202/074122-0074

where v is the cross-sectional fluid velocity; k_n is a constant (1.0 for SI units); n is the Manning coefficient of roughness, ranging from 0.01 (for a clean and smooth surface) to 0.06; R_h is the hydraulic radius, and S represents the slope in m m^{-1} [5]. The $\Delta P/L$ value was determined according to

$$\frac{\Delta P}{L} = S\rho g \qquad (2)$$

where g is the acceleration of gravity, 9.8 m s^{-2}.

To evaluate the vertical mixing in the RPR, a particle tracking simulation was made, so the Discrete Phase Model with discrete random walk was used over the flow simulation with the k-ε model. Accordingly, the trajectories of 29 particles were monitored to obtain a probability density function (PDF), that describes the probability that the y position of a particle, $y(t)$, assumes a value between y and $y + \Delta y$ at any instant of time.

Additionally, the radiation exposure of the fluid throughout the RPR depth was analysed. As so, the Lambert Beer's law was used in the following form:

$$I = I_0 \times 10^{-\frac{A}{l} \times h} \qquad (3)$$

in which I is the intensity of the radiation passing through the fluid in the direction of solar light incidence, A is the absorbance of the sample, I_0 is the intensity of the incident radiation, l is the width of the cuvette, typically 1 cm, and h is the fluid depth at which I is being determined.

Results and Discussion

With the aim of analysing the turbulence distribution in the flow, the turbulent kinetic energy and the turbulent dissipation rate maps are shown in Figures 3 and 4, respectively.

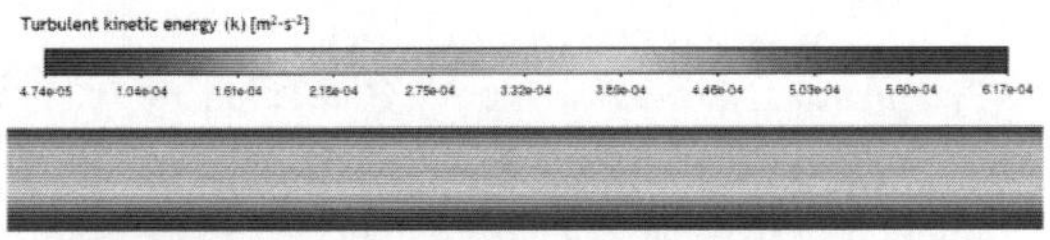

Figure 17. Turbulent kinetic energy contours.

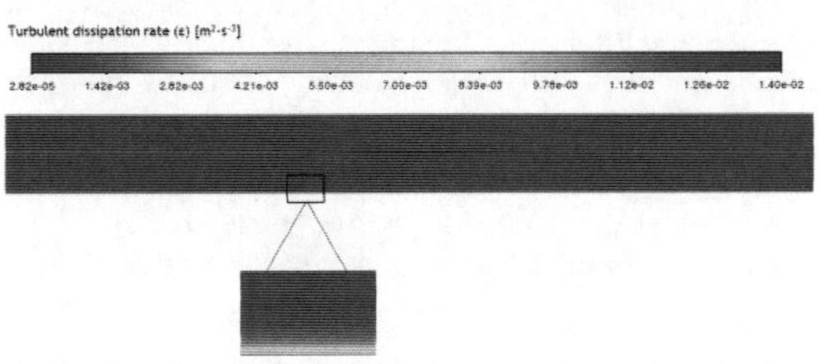

Figure 18. Turbulent dissipation rate contours.

According to Figure 3, one can conclude that the turbulent kinetic energy is mainly concentrated near the bottom wall of the reactor. This energy is transferred from larger eddies to smaller eddies through a direct turbulent energy cascade. Below the Kolmogorov scale viscosity dissipates the eddies to heat. This dissipation is more intense at the bottom wall, as seen in the close-up in Figure 4.

To analyse the degree of vertical mixing, a methodology based on the velocity fluctuating component in the y direction, $\overline{v'}$, was implemented. Its value was determined considering the mathematical definition of turbulent kinetic energy,

$$k = \frac{1}{2}\left(\overline{u'^2} + \overline{v'^2} + \overline{w'^2}\right) \qquad (4)$$

where $\overline{u'}$, $\overline{v'}$, and $\overline{w'}$ are the velocity fluctuating components in the cartesian x, y, and z directions, respectively.

For the turbulent kinetic energy modelled with k-ε, $\overline{u'} = \overline{v'} = \overline{w'}$. So, for this case, $\overline{v'} = 1.84 \times 10^{-2}$ m s^{-1}.

Once the height of the fluid, h, is known, in this case 0.1 m, it becomes possible to determine the time it takes for a fluid element to travel through the vertical section of the reactor, $\tau_{vertical}$, as described in equation 5.

$$\tau_{vertical} = \frac{h}{\overline{v'}} \qquad (5)$$

On the other hand, knowing the volume occupied by the fluid in the reactor, V, and its volumetric flow rate, Q, one can calculate the average passage time in the reactor, $\tau_{circulation}$. For a circulation time much longer than the vertical flow time, $\tau_{circulation} >> \tau_{vertical}$, it can be assumed that the fluid is uniformly distributed in the direction of solar light incidence, i.e., all the fluid elements end up spending about the same time in the illuminated region of the RPR.

For the case under study, the above assumption is valid since the $\tau_{circulation}$, about 144.8 s, is much higher than the $\tau_{vertical}$, approximately 5.4 s. Thus, it is confirmed that the implemented velocity and fluid height allow to obtain a good level of vertical mixing, which is very important to avoid property gradients and ensure a proper exposition of the fluid to solar radiation.

In order to consolidate the results shown above, it was possible to determine the probability of a particle to be in each millimetre of the RPR depth at any instant of time. This probability is given by the area under the PDF curve between the two desired values. The PDF plot is shown in Figure 5. It should be noted that the point $y/y^* = 0$ is equivalent to the top of the RPR, while the point $y/y^* = 1$ corresponds to the bottom.

Observing the graph, one can ascertain that in general there is a good distribution of the particles in the vertical direction of the RPR. Although, as expected, they are more likely to move to the extremes, i.e., to the top and bottom of the RPR.

In the area in between there is a higher kinetic energy and as such a larger mixing degree due to the vortices formed, which causes the particles to acquire higher velocities and spend less time on the same position. Moreover, due to the boundary conditions imposed at the base and top of the RPR, in these regions they have more difficulty to move and, as such, tend to remain there for longer periods. This way, the results deviate from the initial premise that mixing in the vertical direction is perfect. Nonetheless, there is a good degree of

https://doi.org/10.52202/074122-0074

uniformity, so the conclusions earlier drawn are not too far from reality.

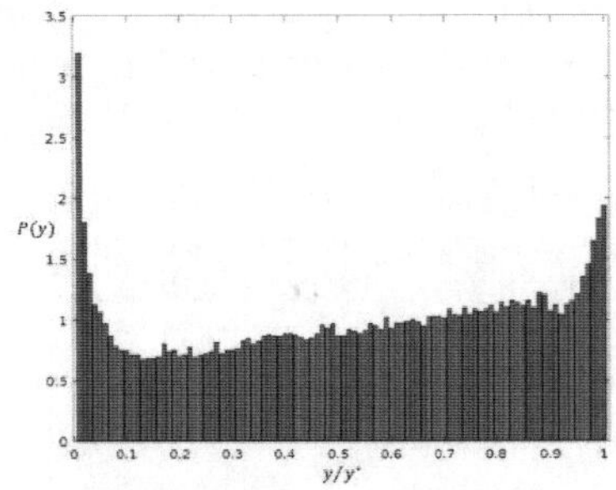

Figure 19. PDF of particles' y position.

Since solar radiation plays a crucial role in RPRs applications, as it may influence the rate of microalgae growth or interfere with the efficiency of secondary effluents disinfection, the importance of its study in this work becomes evident.

For that purpose, firstly, it is necessary to know the absorbance of the fluid under study in the range of the desired radiation. For example, for wastewater treatment applications, UVA radiation (315 - 400 nm) strongly affects the Photo-Fenton process. Thus, assuming a case study where the sample to be treated has an average absorbance of 1 in the UVA spectrum and I_0 corresponds to the medium UVA incident radiation in Portugal, 20 W m^{-2} [6], the radiation profile along the vertical direction of the fluid can be determined by equation (3) and is shown in Figure 6.

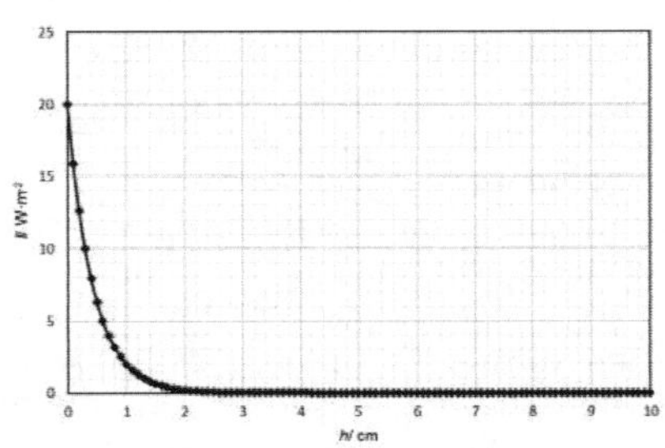

Figure 20. Radiation profile over the fluid depth.

Figure 6 shows that the penetration height of UVA radiation is substantially reduced in the first 2 cm, being practically zero from that point onwards. This emphasises the importance of obtaining a good degree of vertical mixing, as only in this way is possible to guarantee that all the elements present in the RPR receive the same intensity of radiation throughout the day.

Along with the intensity of radiation, the total energy received by the fluid during a given treatment cycle, E_t, may also be determined. For that purpose, the data obtained from the PDF becomes useful. Hence, E_t can be calculated as follows,

$$E_t = \int_0^{t_c} \frac{1}{100} P(y)\, I(y)\, dt \qquad (6)$$

where t_c corresponds to the duration of a treatment cycle, that depends on the application.

Thus, it is possible to adapt t_c in order to obtain a higher or lower energy input depending on the desired treatment level and the process conditions.

Conclusions

The RPR under study has proven to have a good performance in mixing vertically the fluid. Accordingly, the velocity in the *y* direction has shown to be sufficiently higher to promote the mixture of the fluid, avoiding property gradients in the light incidence direction. Similarly, the PDF of the particles' *y* position indicates a good distribution. Nevertheless, the most energetic eddies in the middle section of the reactor depth force the particles to move to the extremes, remaining there longer comparatively with the rest of the domain.

Concerning the radiation analysis, the RPR seems very advantageous for solar applications as it enables the energy input to be easily adjusted by changing the process conditions. Additionally, there is the advantage of the energy source being endless, which may substantially reduce energy consumption in the process.

Acknowledgments

This work was financially supported by LA/P/0045/2020 (ALiCE), UIDB/50020/2020 and UIDP/50020/2020 (LSRE-LCM), funded by national funds through FCT/MCTES (PIDDAC).

References

[1] Gualda-Alonso, E., Soriano-Molina, P., Casas López, J.L., García Sánchez, J.L., Plaza-Bolaños, P., Agüera, A., and Sánchez Pérez, J.A.: 'Large-scale raceway pond reactor for CEC removal from municipal WWTP effluents by solar photo-Fenton', Applied Catalysis B: Environmental, 2022, 319, pp. 121908

[2] Inostroza, C., Solimeno, A., García, J., Fernández-Sevilla, J.M., and Acién, F.G.: 'Improvement of real-scale raceway bioreactors for microalgae production using Computational Fluid Dynamics (CFD)', Algal Research, 2021, 54, pp. 102207

[3] Liffman, K., Paterson, D.A., Liovic, P., and Bandopadhayay, P.: 'Comparing the energy efficiency of different high rate algal raceway pond designs using computational fluid dynamics', Chemical Engineering Research and Design, 2013, 91, (2), pp. 221-226

[4] Mendoza, J.L., Granados, M.R., de Godos, I., Acién, F.G., Molina, E., Banks, C., and Heaven, S.: 'Fluid-dynamic characterization of real-scale raceway reactors for microalgae production', Biomass and Bioenergy, 2013, 54, pp. 267-275

[5] Bonetti, S., Manoli, G., Manes, C., Porporato, A., and Katul, G.G.: 'Manning's formula and Strickler's scaling explained by a co-spectral budget model', Journal of Fluid Mechanics, 2017, 812, pp. 1189-1212

[6] Silva, T.F.C.V., Fonseca, A., Saraiva, I., Boaventura, R.A.R., and Vilar, V.J.P.: 'Scale-up and cost analysis of a photo-Fenton system for sanitary landfill leachate treatment', Chemical Engineering Journal, 2016, 283, pp. 76-88

A Reduced Geometrical Model for the Design of NETmix Reactors

Poster.29

***Isabel S. Fernandes**[1,2], Madalena M. Dias[1,2], José Carlos B. Lopes[1,2,3], Ricardo J. Santos (1) LSRE-LCM - Laboratory of Separation and Reaction Engineering – Laboratory of Catalysis and Materials, Faculty of Engineering, University of Porto, Rua Dr. Roberto Frias, 4200-465 Porto, Portugal, (2) ALiCE – Associate Laboratory in Chemical Engineering, Faculty of Engineering, University of Porto, Rua Dr. Roberto Frias, 4200-465 Porto, Portugal, (3) CoLAB NET4CO2 - Network for a Sustainable CO2 Economy, Rua Júlio de Matos, 828-882, 4200-355 Porto, Portugal*

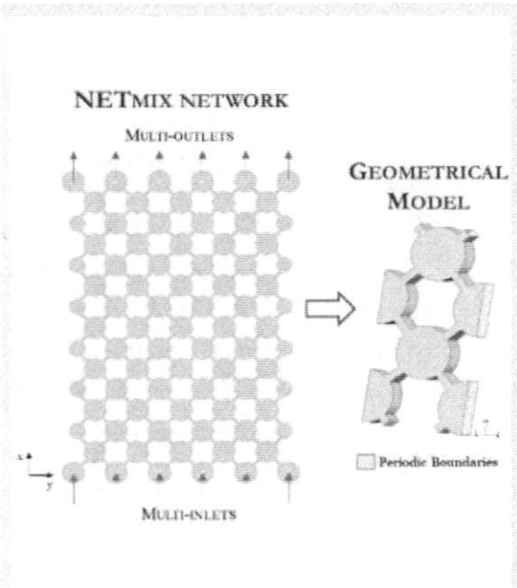

A reduced geometrical model is introduced in this work to study the hydrodynamics in NETmix reactors. The numerical investigation of the proposed geometry is done using Computational Fluid Dynamics. NETmix is a static mixer comprising a network of mixing chambers and channels. This novel geometrical model simulates a representative central section of the NETmix reactor and considers Periodic Boundary Conditions (PBCs) at the sidewalls and between the inlets and outlets and is known as LoopNUB. Different configurations of the LoopNUB model were tested to determine which is less affected by imposing inlet and outlet PBCs. The LoopNUB model composed of four lines proved suitable for simulating flow dynamics and structures in NETmix regions downstream of row 8 of previously validated geometry, NUB, that considers only PBCs in sidewalls. This new model reduces simulation times and can be used to deepen the knowledge regarding heat transfer mechanisms in NETmix reactors.

Introduction

NETmix is a mesostructured static mixer introduced and patented in 2005 at the University of Porto [1] that has been proven to surpass microreactors regarding mixing and heat transfer efficiency [2]. Because of that, industry has been pushing to apply NETmix technology for highly endothermic or exothermic single-phase and multiphase flow processes. Its structure consists of transport channels interconnected by mixing chambers. The network is generated by linking and repeating a unit cell along the x and y directions (see Figure 1a). Each unit cell is composed of a cylindrical chamber and four rectangular cross-section half-channels at 45° from the main flow direction. A mixing chamber contacts fluids from two chambers of the previous row of unit cells and splits the flow into two streams feeding two chambers in the next row. This flow pattern repeats throughout the mixer. The successive splitting and recombination of the fluid increases the contact area and contributes to an intense rate of fluid renewal. However, to attain mixing in these reactors, it is necessary to operate at Reynolds numbers above 150 ($\mathrm{Re} = \frac{\rho v_{\mathrm{in}} d_{\mathrm{h}}}{\mu}$, where ρ and μ are the density and viscosity of the fluid, respectively; v_{in} is the velocity in the inlet channel and d_{h} is the hydraulic diameter of the channels) [3]. This critical Reynolds number marks the transition from a symmetric laminal flow, where no advective mixing is observed, to a chaotic laminar flow regime with strong laminar mixing dynamics.

The flow field in NETmix reactors has been characterised by Computational Fluid Dynamics (CFD) simulations validated from experimental results. Due to the network symmetry and periodicity of the flow, reduced geometrical models have been used to decrease the simulation times. CFD simulations of NETmix reactors have been performed using the NETmix Unit Block (NUB) introduced by Fonte et al. [3]. This geometrical model simulates a representative central section of the NETmix reactor. The lateral regions are considered by setting Periodic Boundary Conditions (PBCs) at the sidewalls of this section (see Figure 1b). From statistical analysis of 2D turbulent flow, Torres [4] concluded that the flow dynamics in NUB geometries composed of one central column, two half columns, and at least nine rows, are not affected by the forcing of the lateral PBCs. However, the simulation of 3D domains using this model remains time-consuming. The possibility of using NUB reduced geometries without compromising the numerical accuracy was evaluated. In addition to the PBCs in the sidewalls, PBCs were defined between the inlets and outlets of the geometry. This new geometrical model is called LoopNUB (see Figure 1c). Physically the LoopNUB model corresponds to the simulation of a NETmix comprised of infinite number of rows and columns. The LoopNUB geometries tested have the same number of columns as the NUB model, and the number of rows was varied to determine which configuration is less affected by imposing inlet and outlet PBCs. The number of rows tested was 2, 4, 6 and 12. The models are identified by the name "NUB" or "LoopNUB" followed by a number that indicates the number of rows of the geometrical model. The characteristic dimensions of the models are shown in Figure 1b. The geometry depth was fixed at 2 mm.

https://doi.org/10.52202/074122-0075

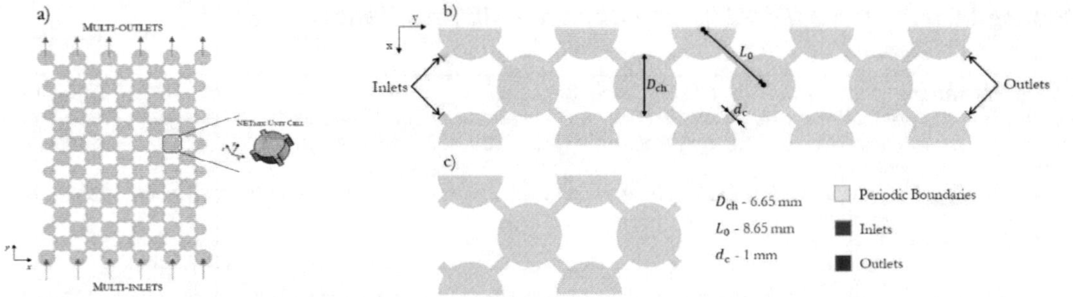

Figure 1. (a) Scheme of a NETmix network consisting of an array of cylindrical chambers interconnected by prismatic channels, (b) NETmix unit block (NUB) composed of nine rows and (c) LoopNUB composed of four rows and surfaces identification.

Material and Methods

A 3D single-phase CFD model using ANSYS/Fluent software was implemented in LoopNUB and a NUB model composed of twelve rows (NUB12) to assess the adequacy of LoopNUB to study the hydrodynamics in NETmix. The working fluid was liquid water at 20 °C and atmospheric pressure.

The flow field was evaluated for a channel's Reynolds number of 300 (above the critical value to onset chaotic flow regime). For the NUB model, a flat velocity profile with $u_{in}=2.16\times10^{-1}$ m·s^{-1} was specified at the inlets. For LoopNUB models, a mass flow rate corresponding to $u_{in} = 2.16\times10^{-1}$ m·s^{-1} was set in *the y-axis* direction PBCs.

Results and Discussion

Flow dynamics characterisation studies were performed to evaluate if the flow is fully developed using the LoopNUB model. Time histories of velocity components at the middle point of each complete mixing chamber were obtained from the CFD simulations and the respective Probability Density Function (PDF), power spectra and turbulence intensity were evaluated.

Regarding the PDFs, it was observed that for all LoopNUB configurations, the velocity fluctuations follow an approximately normal distribution, which indicates that the flow is fully developed. For NUB12, the flow behaviour evolves along the mixing chambers. The NUB12 model only reached a PDF with normal distribution for the mixing chambers at rows 8, 10, and 12. Figure 2 shows the PDF of LoopNUB4 at Ch2 and NUB12 at Ch8. The abbreviation Ch stands for chamber and the number is the row where the mixing chamber is located. A NUB model with at least eight rows is required to describe the chaotic flow dynamics in NETmix. Therefore, the LoopNUB model accurately describes higher order statistical quantities pertaining the chaotic flow in NETmix.

The turbulence intensity of x -velocity component, I_x , was calculated for NUB12, LoopNUB12, and LoopNUB4 at the centre of all complete mixing chambers. The results are shown in Figure 3. It is observed that the turbulence intensity for NUB12 grows with the row number and presents a plateau after row 8. The turbulence intensity stabilises at the same level as the values obtained for the LoopNUB models. Therefore, at least eight complete mixing chambers are necessary without setting the periodic boundary conditions between the inlets and outlets to reach the fully developed chaotic flow regime. Using the LoopNUB models, the turbulence intensity is nearly constant in all mixing chambers.

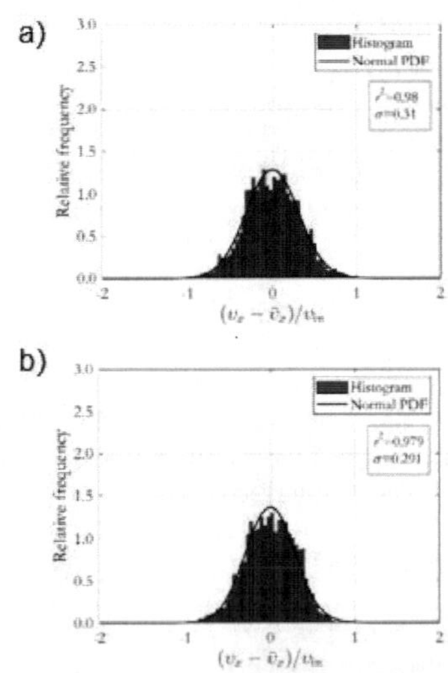

Figure 2. PDF of (a) LoopNUB4 at Ch2 and (b) NUB12 at Ch8.

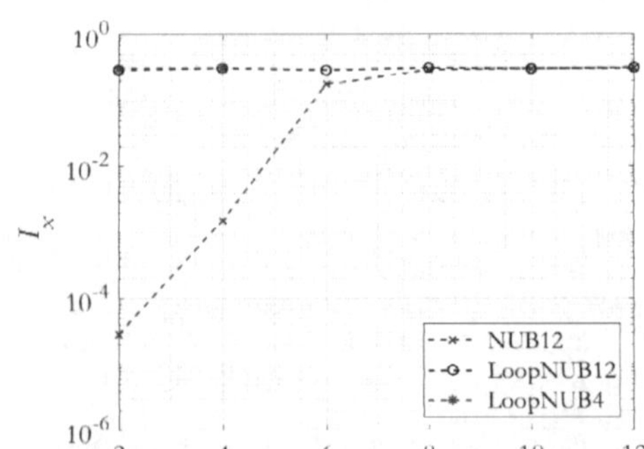

Figure 3. Turbulence intensity of x-velocity component along the chambers of NUB12, LoopNUB12 and LoopNUB4.

The contour maps of the z-component of the normalised vorticity ($\zeta^* = \frac{\vec{\zeta} d_c}{v_{in}}$, where $\vec{\zeta}$ is the vorticity field computed as $\nabla\times\vec{u}$) in the middle plane of geometry depth at the complete mixing chambers for LoopNUB4 and NUB12 are shown in Figure 4. The positive and negative sign of the vorticity is related to the direction of the flow rotation. The maps were obtained for a time step equal to 10 times the passage time. The flow patterns with the LoopNUB model are asymmetrical and dynamic in both mixing chambers. For NUB12, the flow is symmetrical in the first complete mixing chamber (Ch2) and increases the flow dynamics with the row number. The flow in chambers 8, 10, and 12 is fully chaotic with the occurrence of unstable vortices.

The LoopNUB4 reduces the CFD grid domain by a factor of 2.3 in relation to the NUB model, which has been used for single-phase [5] and two-phase [6] studies in NETmix.

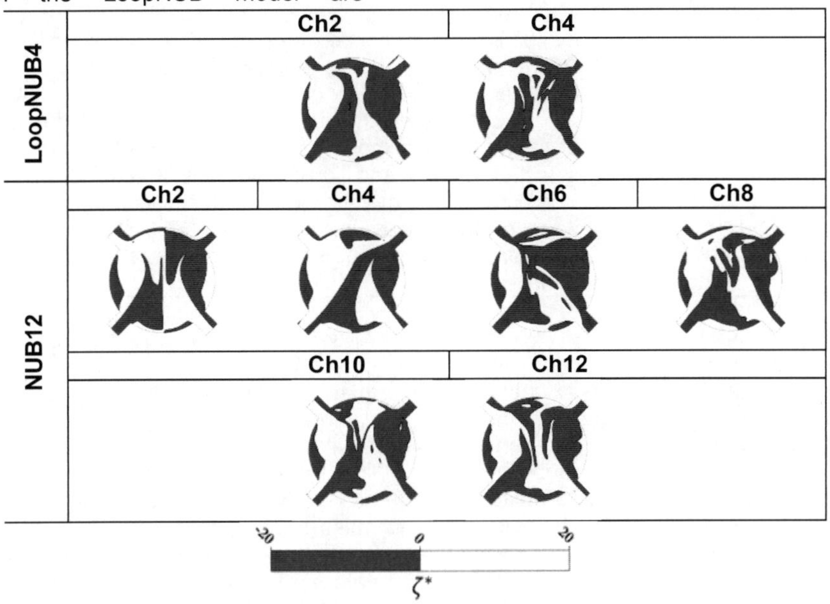

Figure 4. z-component vorticity contour maps in the middle of geometry depth for LoopNUB4 and NUB12 at each complete mixing chamber for a time step of 10 times the passage time.

Conclusions

The velocity dynamics analysis for a Reynolds number above the critical value, where a chaotic flow regime is observed, shows that the flow inside of the LoopNUB4 model is similar to the flow observed in the mixing chambers downstream of row 8 in a previously validated NUB model with 12 rows. The LoopNUB4 model is composed of four rows and three columns, where the middle column has two complete mixing chambers and the side columns two half mixing chambers. If PBCs between inlets and outlets are not considered, a geometry composed of at least eight lines is necessary to simulate the chaotic flow regime in NETmix. LoopNUB4 reduces the physical domain of simulations enabling the use of more refined grids for mass transfer studies in single-phase and multiphase processes or the computation of longer flow periods for dynamics analysis.

Acknowledgments

This work was financially supported by national funds through the FCT/MCTES (PIDDAC), under project PTDC/EAM-AMB/4702/2020 (OZONE4WATER), LA/P/0045/2020 (ALiCE), UIDB/50020/2020 and UIDP/50020/2020 (LSRE-LCM), and by project NORTE-01-0145-FEDER-000074 (S4Hort_Soil&Food), supported by Programa Operacional Regional do Norte (NORTE 2020), through Portugal 2020 and Fundo Europeu de Desenvolvimento Regional (FEDER). I.S. Fernandes acknowledges her FCT scholarship SFRH/BD/143958/2019.

References

[1] J.C.B. Lopes, P.E.L. Laranjeira, M.M.Q. Dias, A.A. Martins, Network Mixer and Related Mixing Process, in: F.d.E.d.U.d. Porto (Ed.) PCT/IB2005/000647. US Patent 8434933 B2, September 2008, granted May 2013. EU Patent 1720643, September 2006, granted October 2008. Portuguese Patent 103072, February 2004, granted December 2000, 2005.

[2] M.F. Costa, C.M. Fonte, M.M. Dias, J.C.B. Lopes, Heat transfer performance of NETmix-A novel micro-meso structured mixer and reactor, AIChE Journal 63(6) (2017) 2496-2508.

[3] C.M. Fonte, M.E. Leblebici, M.M. Dias, J.C.B. Lopes, The NETmix reactor: Pressure drop measurements and 3D CFD modeling, Chemical Engineering Research and Design 91(11) (2013) 2250-2258.

[4] J.P. Torres, Proper Orthogonal Decomposition of Turbulent Flows, Departamento de Engenharia Química, Universidade do Porto, 2017.

[5] J. Matos, R.J. Santos, M.M. Dias, J.C.B. Lopes, Mixing in the NETmix Reactor, Frontiers in Chemical Engineering 3 (2021) 771476.

[6] I.S. Fernandes, M.S.e.C.A. de Brito, Y.A. Manrique, M.M. Dias, J.C.B. Lopes, R.J. Santos, Computational Fluid Dynamics Model of Two-Phase Flow in NETmix Reactors, Chemical Engineering Technology 44(11) (2021) 2002-2010.

https://doi.org/10.52202/074122-0075

Assessment of the Effect of Channel Length on the Gas-Liquid Mass Transfer: Enhancing the NETmix technology for Ozonation

Poster.30

P. H. Marrocos[1,2]*, L. Silva*[1,2]*, M. M. Pituco*[1,2]*, I. S. Fernandes*[1,2]*, R. J. N. Santos*[1,2]*, V. J. P. Vilar*[1,2]*. (1) LSRE-LCM – Laboratory of Separation and Reaction Engineering – Laboratory of Catalysis and Materials, Faculty of Engineering, University of Porto, Rua Dr. Roberto Frias, 4200-465 Porto, Portugal. (2) ALiCE – Associate Laboratory in Chemical Engineering; Faculty of Engineering, University of Porto, Rua Dr. Roberto Frias, 4200-465 Porto, Portugal.*

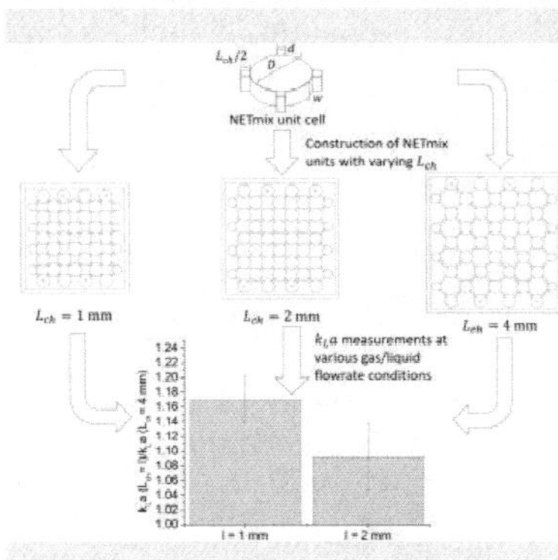

The innovative geometry of the micro/meso-structured static mixer NETmix provides a powerful gas/liquid mixing intensity, enabling its use for ozonation through a sidestream configuration. Thus, this work aims the enhancement of the NETmix for ozonation by evaluating the influence of the mixer's channel length on the gas-liquid mass transfer. This assessment was done by the construction of three similar NETmix units, differing only in their channel length - it was considered 1 mm, 2 mm and 4 mm –, and the measurement of the volumetric mass transfer coefficient for each NETmix at a selection of gas and liquid flowrates. It was shown a significant influence of the channel length on the gas-liquid mass transfer, whereby there was an increase of 17% in the volumetric mass transfer coefficient when the length was reduced from 4 mm to 1 mm.

Introduction

Ozonation is a powerful method for water/wastewater treatment due to the oxidation potential of ozone molecules enabling the degradation of persistent organic molecules and disinfection. Traditionally it is used bubbling devices, such as fine bubble diffusers, to combine two processes in the same contact chamber: transferring the ozone gas into the water/wastewater to be treated, along with the reaction process. Although common, this approach carries several disadvantages, such as a large footprint. An alternative approach considers the use of a sidestream configuration, consisting in the splitting-off a fraction of the main flow, into which the ozone is mixed by efficient equipment such as a venturi or a static mixer, and subsequently, this stream saturated with ozone is injected back into the main flow, consequently reducing the total size of the equipment [1,2].

The micro/meso-structured static mixer, the NETmix, consists of a network of unit cells composed by a cylindrical chamber connected to four prismatic half-channels, each at a 45° of the flow direction. Due to its innovative geometry, the NETmix presents interesting mixing capabilities for one-phase and two-phase flows [3,4], displaying a huge potential for ozone dissolution using a sidestream configuration. Hence, this work aims to investigate the effect of a NETmix geometrical property, the channel length, on the gas-liquid mass transfer, to optimize its application for ozone dissolution.

Material and Methods

In order to evaluate the effect of the NETmix channel length (L_{ch}) on the gas-liquid mass transfer, three NETmix units, made of acrylic, were conceived, each of which was identical in their depth ($w = 2$ mm), chamber diameter ($D = 6.75$ mm) and channel width ($d = 1$ mm), differing only in L_{ch}. Hence, in this work, it was adopted 4 inlets and 9 rows of chambers, in addition to three values of channel length: 1 mm, 2 mm and 4 mm, as it can be seen in Figure 1. As such, the gas-liquid mass transfer for each NETmix unit was assessed by the measurement of the volumetric mass transfer coefficient ($k_L a$) associated to co-current downwards gas-liquid flows at chosen values of liquid and gas flowrates (Q_L and Q_G, respectively). To perform these measurements, it was used a dissolved oxygen desorption method through nitrogen injection technique, as described elsewhere [5]. The experimental apparatus utilized is displayed in Figure 1. Consequently, for each experiment, 5 dm^3 of demineralized water, saturated with oxygen, was placed inside a jacketed tank at 20 °C. Once the desired temperature was attained, the water was pumped into the NETmix through a venturi injector, in which it was simultaneously injected nitrogen gas, thus initiating the desorption of the oxygen dissolved in the water. Thereafter, samples of the gas-liquid mixture were collected at the outlet of the NETmix at fixed time intervals. Subsequently, a dissolved oxygen probe was utilized to measure the dissolved O_2 concentration ($C_{Out}^{O_2}$) associated with each sample, by which the related $k_L a$ to the evaluated condition was computed, as given by Equation 1. This equation was used to calculate the $k_L a$ value for each sample, whereby an average of the computed values was used as the result associated with the gas/liquid flowrate conditions at which the experiment was performed.

$$k_L a = -\frac{1}{\tau_L} \ln\left[\left(1 + \frac{Q_L}{Q_G H}\right)\left(\frac{C_{Out}^{O_2}}{C_{In}^{O_2}}\right) - \frac{Q_L}{Q_G H}\right] \qquad (1)$$

https://doi.org/10.52202/074122-0076

In which $C_{In}^{O_2}$ is the dissolved oxygen concentration at the venturi inlet; H, the Henry coefficient for oxygen based on molar concentrations, measured to be 27 at room temperature [5]; and τ_L, the liquid passage time, given by Equation 2.

$$\tau_L = \frac{1-\alpha_G}{Q_L} V_{NETmix} \left(1 + \frac{Q_L}{Q_G H}\right) \quad (2)$$

Whereby V_{NETmix} is the NETmix volume and α_G is the gas hold-up, which was numerically estimated through CFD simulations.

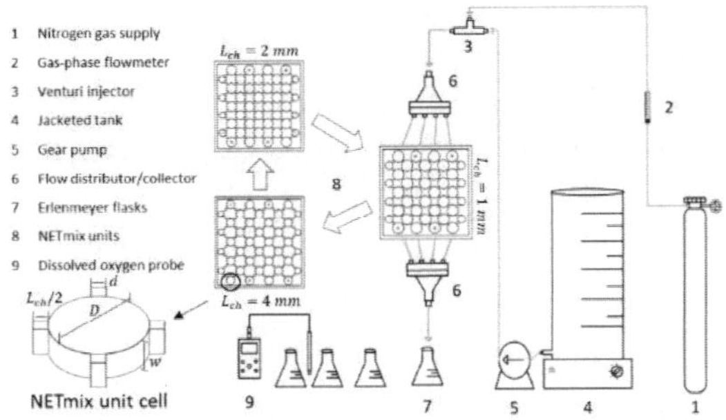

Figure 21. Illustration of the experimental setup used in this work, along with each NETmix unit employed.

Hence, the described procedure was performed for liquid and gas flowrates, both ranging from 30 Ndm^3·h^{-1} to 60 Ndm^3·h^{-1}, for each of the three NETmix units, as indicated in Table 1.

Table 1. Gas and liquid flowrate conditions selected to measure the volumetric mass transfer coefficient for each of the three NETmix units.

Q_L (Ndm^3·h^{-1})	Q_G (Ndm^3·h^{-1})
30	
45	30, 45, 60
60	

Results and Discussion

According to Higbie's penetration theory, the liquid-side mass transfer coefficient k_L is given by Equation 3:

$$k_L = 2\sqrt{\frac{D_g}{\pi\theta}} \quad (3)$$

in which D_g is the diffusion coefficient of the gas species in the liquid and θ is the time during which the body of gas is in contact with the liquid.

Therein, θ is often interpreted as the inverse of a surface renewal rate S, such that $S = 1/\theta$, meaning that a body of gas in contact with a high velocity liquid leads to lower exposure times, and, consequently, larger surface renewal rates: hence, Higbie's theory predicts an enhancement of the mass transfer in these situations due to an increase in k_L. As a consequence of the NETmix micro/meso scale, the mass transfer enhancement phenomenon is expected to occur; however, as the mixer is associated with different length scales between the chambers and the channels, the strength of the phenomenon is also expected to vary between these two elements of the unit cell. To illustrate this statement, it is shown in Figure 2 a photograph of a gas-liquid flow inside the NETmix.

Figure 22. Photograph of a gas-liquid flow inside the NETmix.

It can be seen that the remaining space available for the liquid flow around the bubbles is much larger at the chambers than at the channels. As such, due to the minimal space available for the liquid flow, the liquid velocities around the bubbles at the channels are much larger than at the chambers, leading to small θ and, consequently, large values of surface renewal at the channel. Thus, the surface renewal rate in the NETmix - and, by extension, the gas-liquid mass transfer - is controlled by the dimensions of the channel. As is displayed in Figure 3, the averages and respective standard deviations, of the ratio of the $k_L a$ measurements for channel lengths 1 mm and 2 mm in relation to those for 4 mm, obtained from the experiments varying the injected gas and liquid flowrates, confirm that the variation of L_{ch} leads to a significant effect on the mass transfer. Hence, these results show that reducing the channel length, thus reducing θ and increasing S, leads to a gas-liquid mass transfer enhancement of the NETmix, which is in agreement with Higbie's theory.

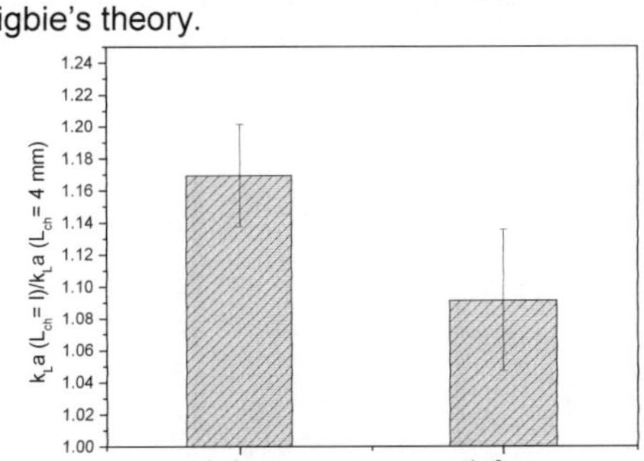

Figure 23. Average and standard deviation results of the ratio of the $k_L a$ measurements, encompassing all flowrate conditions tested, for 1 mm and 2 mm channel length in relation to measurements for 4 mm length.

https://doi.org/10.52202/074122-0076

Conclusions

This study evaluated the effect of the channel length of a NETmix static mixer on the gas-liquid mass transfer through the measurement of the volumetric mass transfer coefficient as a function of three values of channel lengths - 1 mm, 2 mm and 4 mm – for various gas and liquid flowrates. The results confirmed that the channel length of the NETmix has a significant role on the gas-liquid mass transfer, in which the reduction of the length led to an enhancement of the transfer: there was a 17% increase on the volumetric mass transfer coefficient as the length was reduced from 4 mm to 1 mm.

Acknowledgments

This work is financially supported by national funds through the FCT/MCTES (PIDDAC), under the project PTDC/EAM-AMB/4702/2020 (OZONE4WATER), LA/P/0045/2020 (ALiCE), UIDB/50020/2020 and UIDP/50020/2020 (LSRE-LCM). P.H. Marrocos acknowledges FCT - Fundação para a Ciência e Tecnologia - for his PhD scholarship (2022.10437.BD). V. Vilar acknowledges the FCT Individual Call to Scientific Employment Stimulus 2017 (CEECIND/01317/2017).

References

[1] K. L. Rakness, *Ozone in Drinking Water Treatment: Process Design, Operation, and Optimization*, First Ed., no. C. American Water Works Association, 2005.

[2] B. Langlais, D. A. Reckhow, and D. R. Brink, *Ozone in Water Treatment: Application and Engineering*, First Ed. CRC Press, 1991.

[3] M. F. Costa *et al.*, "Carbon Capture and Storage Toward Industrialization: A Novel Continuous Process for the Production of Carbon Dioxide Clathrates," *Energy Technol.*, vol. 10, no. 6, 2022.

[4] I. S. Fernandes, M. S. e. C. A. de Brito, Y. A. Manrique, M. M. Dias, J. C. B. Lopes, and R. J. Santos, "Computational Fluid Dynamics Model of Two-Phase Flow in NETmix Reactors," *Chem. Eng. Technol.*, vol. 44, no. 11, pp. 2002–2010, 2021.

[5] *A. M. Al Taweel, J. Yan, F. Azizi, D. Odedra, and H. G. Gomaa, "Using in-line static mixers to intensify gas-liquid mass transfer processes," Chem. Eng. Sci., vol. 60, no. 22, pp. 6378–6390, 2005.*

Hydrodynamics and mixing in an oscillatory flow reactor for application in the pharmaceutical industry

Poster.31

F. Almeida[1,2]*, F. Rocha*[1,2]*, A. Ferreira*[1,2]*. (1) LEPABE – Laboratory for process Engineering, Environment, Biotechnology, and Energy, Faculty of Engineering, University of Porto, Rua Dr. Roberto Frias, 4200-465 Porto, Portugal, (2) ALiCE – Associate Laboratory in Chemical Engineering, Faculty of Engineering, University of Porto, Rua Dr. Roberto Frias, 4200-465 Porto, Portugal*

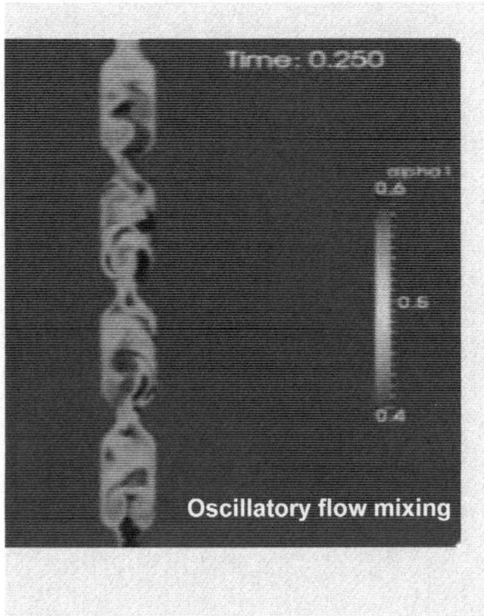

In the pharmaceutical industry is crucial an efficient mixing, as it influences the reaction yield, heat and mass transfer, as well as the final product quality. Despite the numerous developments in this area, some of the intrinsic difficulties of the mixing process have not yet been overcome. In order to optimize the mixing, in the present work, an oscillatory flow reactor provided with 2D-smooth periodic constrictions (OFR-SPC) with two geometries is proposed to study the influence of oscillatory conditions (i.e., oscillation amplitude and frequency) on the residence time distribution (RTD) and axial dispersion. Two cases, without and with oscillation were studied and compared. The results show that either with or without oscillation, the two geometries did not exhibit the presence of dead zones and bypass. Notably, with the liquid oscillation, the mixing intensification was reached even at lower flow rates. The OFR-SPC demonstrated an excellent capacity to provide an efficient mixing with low axial dispersion.

Introduction

The mixing is one of the most relevant operations in the chemical engineering affecting the mass and heat transfer, reaction performance and the product quality. Various problems associated to the bad mixing are not yet solved, which leads to high operational costs. In USA around 600 million dollars were spent due to poor mixing in the pharmaceutical industry [1].

Traditionally, the industry has been using the stirred tank reactor. However, this reactor presents problems associated to bad mixing, product quality, reproducibility of the processes and scale-up. To overcome these limitations, static mixers in continuous mode have been adopted. However, they present small size for providing a mixing intensification and an improvement in heat and mass transfer. Moreover, as the mixing in these units depends on the superficial velocity, to obtain the desired mixture, it is necessary to increase the liquid flow rate or the number of mixers, which can become disadvantageous in some processes [2].

The economic changes seen in the last decade have driven the pharmaceutical industry to adopt efficient processes, enabling lower consumption of reagents, energy, waste and costs [3].

In order to achieve the mixing optimization and to respond to these industry challenges an oscillatory flow reactor provided with smooth periodic constrictions (OFR-SPC) was developed [EP3057694 (B1)]. The OFR-SPC consists of a column provided with smooth periodic constrictions equally spaced. It operates under oscillatory conditions (i.e., different amplitude (x_0) and frequency (f)) originating a unique and efficient mixing, the oscillatory flow mixing (OFM).

The OFM mechanism is illustrated in Figure 24.

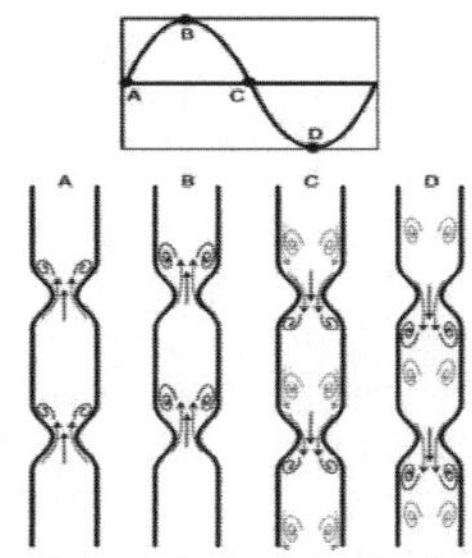

Figure 24. Oscillatory flow mixing in the OFR-SPC [4].

This mixing mechanism is controlled by the x_0 and f imposed on the liquid. This liquid movement originates the formation of eddies in each SPC and their propagation in radial and axial directions. As the liquid passes through the SPC, it is accelerated and creates eddies downstream (A) and (B). In opposite direction, when the liquid is accelerated the eddies move from the wall to the centre of the tube (C). Then, the liquid is accelerated, and the eddies are destroyed (D). Previous studies of the OFM in a tube with periodic baffles demonstrated that applying oscillatory flow coupled with continuous net flow in the baffle tube increases the residence time distribution (RTD) [5]. The OFM in the OFR depends on several key geometrical parameters such as, the baffle spacing, l , orifice diameter, d_0 , reactor diameter, D, open cross-sectional area $a = (d_0/D)^2$. The constriction spacing, l should be in the range of $l = 1.5D$ [6]. Baird and Stonestreet [7] reported a typical value of a is in the range of 0.2 to 0.4.

So far, all the studies published in the literature have been using a single orifice as constriction [5–7]. In the present work an OFR-2D-SPC (EP3439773 B1), with 2 different geometries was used to evaluate the best operational conditions (flow rate, oscillation amplitude and frequency) that ensure the minimum

https://doi.org/10.52202/074122-0077

axial dispersion, a plug flow, and also the existence of dead zones and bypass. Indigo carmine was used as a tracer.

Material and Methods

The experimental setup consists of a vessel, OFR-2D-SPC, a feed tank, a peristaltic pump, syringe, probes, spectrometer, linear motor, light source UV-Vis and a computer, shown in Figure 25. An OFR-2D-SPC constituted by two geometries providing a total length of 2 m and a total volume of 47 ml was used.

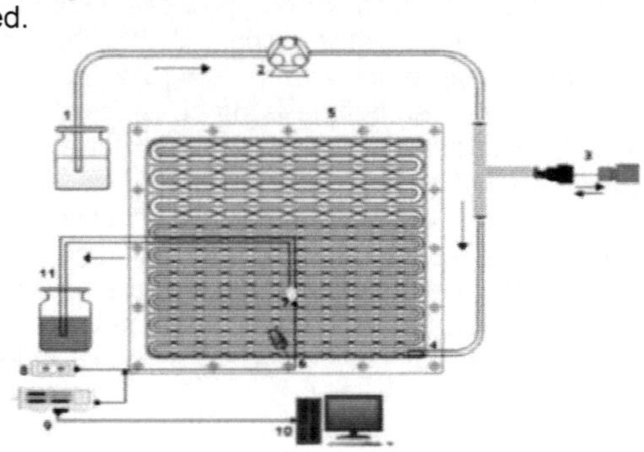

Figure 25. Schematic representation of the experimental setup: 1 – Feed tank; 2 – peristaltic pump; 3 – Linear motor; 4 – Water inlet; 5 – 2D-OFR-SPC; 6 – tracer injection; 7 – Probe; 8 – Light source UV-Vis; 9 – Spectrometer; 10 – Computer; 11 – Vessel.

The liquid oscillation was controlled by a linear motor (LinMot, PS01-23x160) for a range of f from 0 to 7 Hz and x_0 from 0 mm to 6 mm. The continuous net flow was controlled using a peristaltic pump (Longerpump, BT300-2J), which can feed the system with a net flow of 2.5 ml min^{-1} and 59.4 ml min^{-1}. The syringe (Terumo) was used to inject a pulse tracer (0.2 mL) of indigo carmine (Merck), with a concentration of 0.2 g l^{-1}. Also, experimental setup incorporates a probe, where the absorbance is monitored using a spectrometer (Sarspec, spec res+) and a light source UV-Vis (Sarspec). The calibration of the spectrometer was performed according to [3].

After washing the OFR with water to eliminate air bubbles, the operational conditions were settled. Then, at initial time, it was injected rapidly a 0.2 ml of tracer into the reactor in order to achieve a perfect pulse. Hereby, the response was measured using a spectrometer and light source UV-Vis and the data were recorded in a computer. The data recording speed was 3 data points per second, and each data point was an average of 3 scans of the spectrometer. The experimental data were normalized for a pulse entrance according to [8], to obtain ($E(t)$). The $E(t)$ curve was fitted to a plug flow with axial dispersion and tanks-in-series models, as shown in Figure 26. In the plug flow with axial dispersion model, the parameters to be estimated are the Peclet number (Pe), the axial dispersion coefficient (D_c) and the residence time (τ). A $Pe > 100$ is ideal, since small deviations occur to a plug flow behaviour [8]. In the tank-in-series model the number of stirred tanks (N) and the residence time in each stirred tank (τ_i) are estimated.

The mean residence time (τ_m) was also calculated to evaluate the existence of dead zones ($\tau_m < \tau$) or bypass ($\tau_m > \tau$) [8].

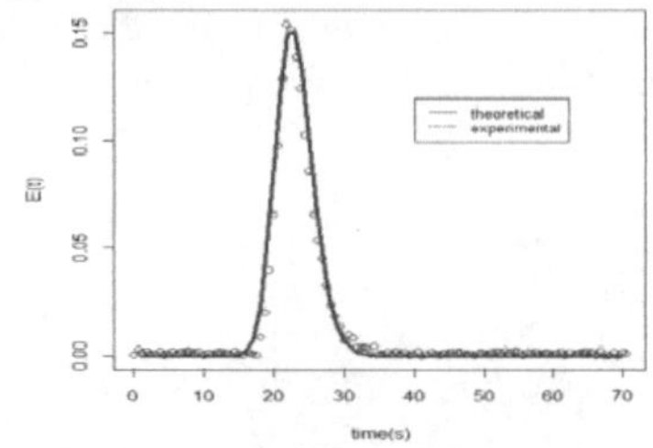

Figure 26. RTD for $x_0 = 2$ mm, $f = 7$ Hz and $Q = 59.4$ ml min^{-1}.

In Figure 27 an illustration of the 2D-SPC with a rectangular cross section is shown

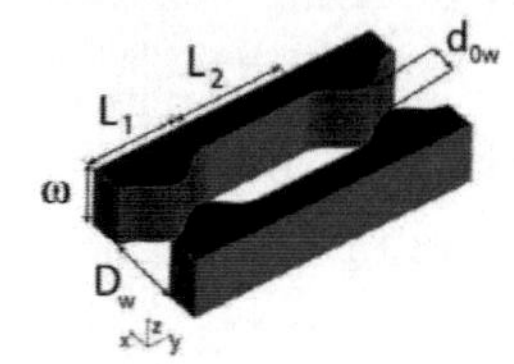

Figure 27. 2D smooth periodic constriction: D_W –tube width; d_{OW} – the shortest tube width in the constriction; ω – thickness; L_1 – constriction length; L_2 –Straight tube length.

The 2D smooth periodic constriction characteristics are presented in Table 2.

Table 2. Parameters of the 2D-SPC

Geometries	D_W (mm)	d_{OW} (mm)	L_1 (mm)	L_2 (mm)	ω (mm)
Geometry 1	4	1.5	6.76	12.4	5
Geometry 2	4	2	6.76	12.4	5

Results and Discussion

The results show that without oscillations the optimum conditions to ensure a plug flow behaviour, mixing intensification and low axial dispersion coefficient are: (i) geometry 1: $Q = 39.4$ ml min^{-1}, $D_c = 2.9 \times 10^{-4}$ m^2 s^{-1}, $N = 58$, $Pe = 115$; (ii) geometry 2: $Q = 19.4$ ml min^{-1}, $D_c = 1.8 \times 10^{-4}$ m^2 s^{-1}, $N = 46$; $Pe = 91$. Comparing both geometries, the geometry 1 achieved a better mixing ($N = 58$), compared with geometry 2 ($N = 46$). This difference is due to the tight constriction in geometry 1, resulting in a high velocity in this zone, thus forming eddies in the lower pressure zones, responsible by the mixing in geometry 1. The optimum conditions assure a plug flow behaviour and a mixing intensification.

When oscillations are imposed on the liquid the optimum conditions for geometry 1 are: (i) $f = 3$ to 7 Hz for $x_0 = 2$ mm; (ii) $f = 3$ to 5 Hz for $x_0 = 4$ mm; (iii) $f = 3$ to 4 Hz for $x_0 = 6$ mm. Relatively to the geometry 2, the optimum conditions are: (i) $f = 1$ to 3 Hz for $x_0 = 2$ mm; (ii) $f = 5$ to 7 Hz for $x_0 = 4$ mm; (iii) $f > 7$ Hz for $x_0 = 6$ mm. Globally, the best operating conditions in this OFR-2D-SPC is $Q = 15.6$ or 19.4 ml min^{-1}, $f = 7$ Hz and $x_0 = 6$ mm. It is noted that geometry 2 presents a plug flow behaviour and an improvement in mixing intensification for lower flow rates. Therefore, the geometry 2 presents a positive effect for lower flow rates, because when a low liquid flow rate passes the constriction causes a lower velocity in axial direction, since d_{0w} is larger than the geometry 1, consequently, the D_c is reduced.

For the two 2D-SPC geometries, when an oscillation frequency is introduced, for the same Q and x_0 the D_c tends to decrease to a minimum value. This decrease occurs because the increase of the frequency increases the radial dispersion rather than axial dispersion, promoting the formation, and the destruction of the eddies, which becomes stronger, making the flow regime more chaotic. The oscillations improved the mixing and remarkably with low energy consumption.

The Figure 28 shows the RTD curves obtained with and without oscillation of the liquid for the geometry 2. From the analysis of Figure 28b, for $Q = 2.5$ and $Q = 5.0$ ml min^{-1}, it is concluded that the reactor behaves as a laminar flow, because the RTD profile is similar to that of a laminar flow reactor [8]. This is characterized by the low eddies formation which are responsible for the low mixture. So, it can be affirmed that the liquid oscillation plays an important role in the mixing intensification. Furthermore, the RTD curve presented in Figure 28b can be adjusted to the model of the two ideal reactors disposed in series (stirred tank and plug flow reactors) or with the stirred tank reactor in the presence of dead zones and bypass. However, in all the conditions tested the $\tau_m \sim \tau$, meaning that the presence of dead zones or bypass was negligible. Therefore, the laminar flow model is the most plausible explanation, since the flow rates and oscillations are low, and the RTD was adjusted through this model as Figure 28b illustrates.

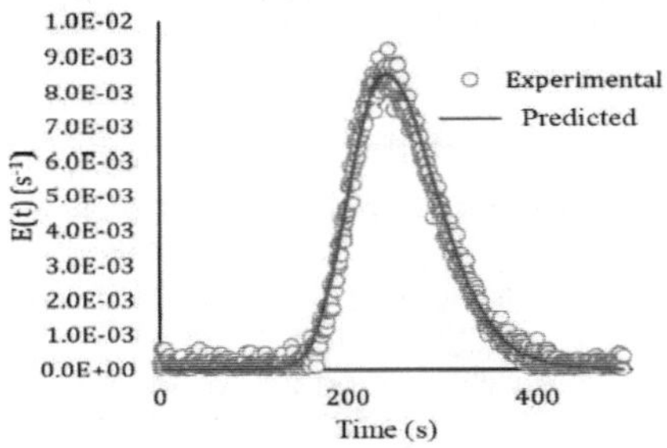

(a)

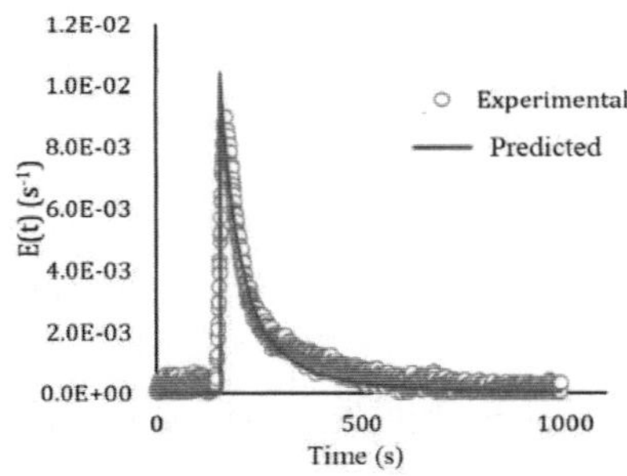

(b)

Figure 28. RTD for $Q = 5.0$ ml min^{-1} and geometry 2: a) with oscillation - $x_0 = 2$ mm $f = 1$ Hz and— - predicted by axial dispersion model [8]; b) without oscillation and—- predicted by laminar flow model [8].

Conclusions

A continuous OFR-2D-SPC is proposed to intensify the mixing, in order to evaluate its potential for application in the pharmaceutical industry.

The performance of the RTD is an important parameter in the characterization of the OFR, since the RTD reflects the behaviour of the mixing that occurs inside the OFR. It was noted a mixing intensification even at low flow rate when oscillation of the liquid was applied. The frequency has a radial dispersion effect, while amplitude has an axial dispersion effect. It was verified that the geometry 2 exhibited a positive effect, since it was achieved better results than geometry 1 at low flow rates with oscillations.

Generally, the optimum conditions that assure a plug flow behaviour, an efficient mixing and a low axial dispersion are: (i) geometry 1: $Q = 5$ ml min^{-1}, $x_0 = 2$ mm, $f = 5$ Hz; (ii) geometry 2: $Q = 15.6$ and 19.4 ml min^{-1}, $x_0 = 6$ mm, $f = 7$ Hz. No dead zones or bypass phenomena were found. Without oscillations and low flow rates a laminar flow was achieved for geometry 2, as result of the large d_{0W}.

The OFR-2D-SPC demonstrated a great potential in increasing mixing intensification (with moderate energy consumption) indicating it as a promising technology for application in the pharmaceutical industry.

Acknowledgments

This work was financially supported by LA/P/0045/2020 (ALiCE), UIDB/00511/2020 and UIDP/00511/2020 (LEPABE), funded by national funds through FCT/MCTES (PIDDAC). F. Almeida would wish to thank to FCT for PhD scholarship 2020.05246.BD.

References

[1] E.. Paul, V.A. Atiemo-Obeng, S.M. Kresta, Handbook of Industrial Mixing, in: J. Willey (Ed.), 1ª edição, New Jersey, 2004: pp. 3–5.

[2] A. Ferreira, F. Rocha, J.A. Teixeira, A. Vicente, WO/2015/056156, 2015.

[3] F. Almeida, F. Rocha, A. Ferreira, U.Porto J. Eng. 4 (2018) 1–15.

[4] A. Ferreira, F. Rocha, A. Mota, J. Teixeira, Nonmechanically Agitated Bioreactors, in: A.P. Christian Larroche, M. Angeles Sanroman, Guocheng Du (Ed.), Curr. Dev. Biotechnol. Bioeng. Bioprocesses, Bioreact. Control., First Edit, Elsevier B.V., Amesterdam, 2017: pp. 217–233.

[5] A.W. Dickens, M.R. Mackley, H.R. Williams, Chem. Eng. Sci. 44 (1989) 1471–1479.

[6] C.R. Brunold, J.C.B. Hunns, M.R. Mackley, J.W. Thompson, Chem. Eng. Sci. 44 (1989) 1227–1244.

[7] M.H. Baird, P. Stonestreet, Chem. Eng. Res. Des. 73 (1995) 503–511.

[8] H.S. Fogler, Distributions of residence times for chemical reactors, Elem. Chem. React. Eng. (2006) 867–944.

https://doi.org/10.52202/074122-0077

CFD studies of mass transfer, flow distribution and flow dynamics in a NETmix technology-based crystalliser for struvite precipitation

Poster.32

Laura J. R. Cullen[1,2,*], *Isabel S. Fernandes*[1,2], *Ricardo J. Santos*[1,2], *Vítor J. P. Vilar*[1,2]. *(1) LSRE-LCM - Laboratory of Separation and Reaction Engineering – Laboratory of Catalysis and Materials, Faculty of Engineering, University of Porto, Rua Dr. Roberto Frias, 4200-465 Porto, Portugal. (2) ALiCE - Associate Laboratory in Chemical Engineering, Faculty of Engineering, University of Porto, Rua Dr. Roberto Frias, 4200-465 Porto, Portugal. (*) up201806139@edu.fe.up.pt.*

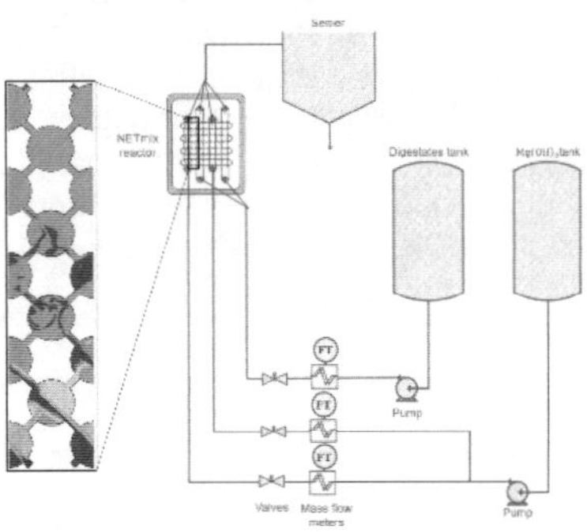

The main objective of this work was to validate the use of the NETmix technology to produce sustainable nitrogen-phosphorus based fertilisers from the liquid fraction of digestates. With this purpose in mind, Computational Fluid Dynamics (CFD) tools were employed to study the flow distribution, flow dynamics and mass transfer in this reactor. A simplified model, the 2D ExtendedNUB, was used to reduce the computational time of dynamic tracer experiments for various asymmetric inlet flow conditions. Several parameters were determined and analysed, namely time-averaged flowrates along the channels, x and y-velocity fluctuations at a midpoint in the upper chamber of the ExtendedNUB model, and time-averaged intensities of segregation along the channels calculated from computed mass fractions of tracer.

Introduction

The use of natural or synthetic fertilisers is usually necessary to ensure proper crop growth [1]. Phosphate rocks are one of the major raw materials for synthetic fertiliser manufacture, and are extracted from geological phosphorus reserves. Recently, there has been a growing concern regarding the non-renewability of these resources and their growing depletion worldwide, owing to an intensification of agricultural practices and poor management of organic P resources. Moreover, the excessive use of these fertilisers is also linked to the eutrophication of water streams, which causes severe harm to aquatic ecosystems [2]. Hence, it is mandatory to recycle and recover this element so that its use becomes more sustainable.

Wastewater is known to have significant amounts of certain nutrients, such as nitrogen (N) and phosphorus (P). These effluents are subject to several treatment processes in wastewater treatment plants before being discharged. P is usually removed from the liquid fraction of a side-stream generated during anaerobic digestion, the digestate. This recovery can be done through a crystallisation process, in which N and P precipitate in the form of an insoluble mineral called struvite with the addition of magnesium (Mg) salts [3]. Other P recovery methods also exist, namely adsorption and electrochemical precipitation, but these are still premature from a scientific point of view. Also, there are already some crystallisation-based technologies, but they are not economically feasible [4].

This work pretends to evaluate the performance of the NETmix reactor for the production of struvite from wastewater effluents through crystallisation with Computational Fluid Dynamics (CFD) tools. NETmix is a patented static mixer consisting of a network of cylindrical mixing chambers interconnected by prismatic channels. It is well known for its great mixing capacities above a critical Reynolds number ($\mathrm{Re}_c = 200$), due to the formation of a self-sustained chaotic laminar flow [5]. This work will focus on the following aspects: flow distribution, mass transfer, and flow dynamics.

Material and Methods

Two-dimensional dynamic tracer experiments were simulated in ANSYS/Fluent 2022 R2. Two working fluids (water and tracer) were defined with the properties of water at 20 °C: density of 998.2 $\mathrm{kg \cdot m^{-3}}$ and viscosity of 1.003 mPa.s. Also, the continuity and the Navier-Stokes equations were solved during the simulations assuming an incompressible and isothermal flow. A reduced geometric model of the NETmix, the 2D ExtendedNUB, was used, in which translational periodic boundary conditions were set at the sidewalls. The geometry domain is depicted in Figure 1, in which surfaces and characteristic dimensions are identified.

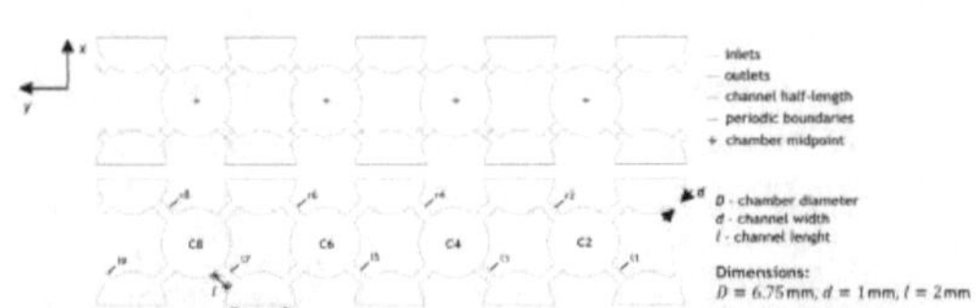

Figure 1. Surfaces of the ExtendedNUB geometry and characteristic dimensions and channels and chambers identification. Letters '*r*' and '*l*' indicate the right and left-hand side channels.

The flow inside the NETmix reactor was simulated for asymmetric flowrates at the two inlets. A laminar regime was set on the right-hand side inlet (tracer input) and a self-sustained fully developed chaotic regime on the left-hand side inlet (water input). This was done by defining a

https://doi.org/10.52202/074122-0078

Reynolds number larger and lower than Re_c in the left inlet (Re_l) and in the right inlet (Re_r), respectively. The objective was to predict the performance of this reactor when mixing a dilute stream with the nutrients to be precipitated (the digestate liquid fraction) and a stream with a high concentration of Mg. The simulations were run for ten passage times, τ.

Results and Discussion

Regarding flow distribution, the time-averaged volumetric flowrate along the left and right-hand side NETmix channels ($q_{l,i}$ and $q_{r,i}$, respectively, in which i is the channel number) was determined and normalised by the average outlet flowrate. Then, the variation of this normalised variable, q^*, was analysed for different sets of inlet flowrates. Regarding mass transfer studies, the mass fraction of tracer, x_t, along each channel was computed over time, which enabled the determination of a time-averaged intensity of segregation, I_s, for the left and right-hand side channels. This statistical measure (Equation 1) was proposed by Danckwerts in 1952 to describe the deviation of the concentration of a portion of fluid from homogeneity. For a completely segregated state, I_s assumes a value of 1 and, for a perfectly mixed state, it assumes a value of 0 [6].

$$I_s = \frac{\sigma_a^2}{\bar{x}_a \cdot \bar{x}_b} = \frac{\sigma_b^2}{\bar{x}_a \cdot \bar{x}_b} = \frac{\sigma_a^2}{\bar{x}_a(1-\bar{x}_a)} \qquad (1)$$

Variables σ_a^2 and σ_b^2 represent the variance of the mass fraction of a species A and B, respectively, in a mixture of both fluids, and $\bar{x}_a$ and $\bar{x}_b$ are the average mass fractions of A and B [6]. In this work, the average mass fraction of tracer, $\bar{x}_t$, that is, the mass fraction of tracer obtained in a homogenised state, was determined by an inlet flow-weighted average, shown in Equation 2.

$$\bar{x}_t = \frac{x_{t,l}q_{l,0} + x_{t,r}q_{r,0}}{q_{l,0} + q_{r,0}} \qquad (2)$$

Figures 2 to 5 show the results obtained for the following cases: $\mathrm{Re}_l = 600$ and $\mathrm{Re}_r = 6$, and $\mathrm{Re}_l = 1200$ and $\mathrm{Re}_r = 12$. For both cases, $\bar{x}_t = 0.01$. Figure 2 shows the variation of q^* with the channel number and Figure 3 the variation of a normalised I_s, represented by I_s^*, with the channel number. I_s^* was determined from the normalisation of the time-averaged I_s values in each channel by the average values in the second-row channel. Also, the I_s^* values for channels 1 were not considered since little contact between phases occurs at this point. Figure 4 shows the velocity magnitude and mass fraction of tracer contour maps obtained for each case at a normalised simulation time of $t/\tau = 10$. The scales for Figures 4c and 4d are in logarithmic form. Finally, Figure 5 shows the power spectra obtained for both cases from the computed velocity magnitudes in the x and y-directions for a midpoint in chamber C8 over time. In this Figure, E is the energy curve and λ is a wavenumber calculated as $\lambda = f/\bar{v}_{inj}$, where f is the frequency and $\bar{v}_{inj}$ is the average fluid velocity at the inlets.

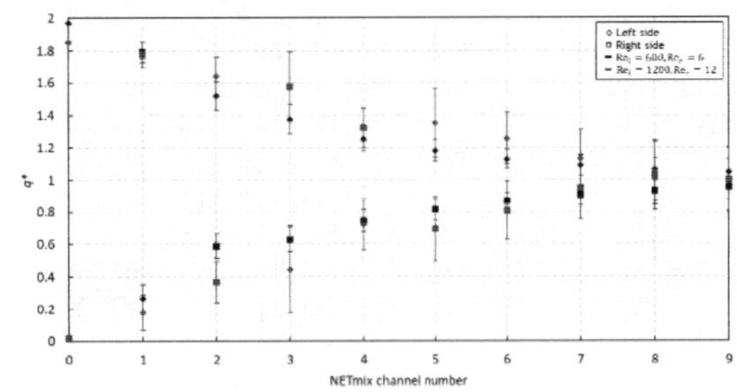

Figure 2. Variation of q^* with the channel number.

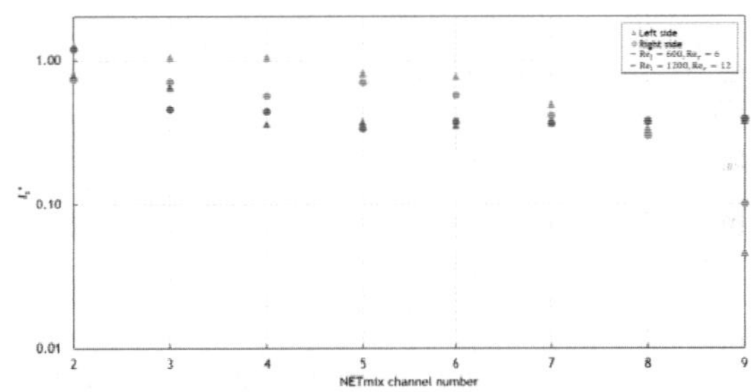

Figure 3. Variation of I_s^* with the channel number.

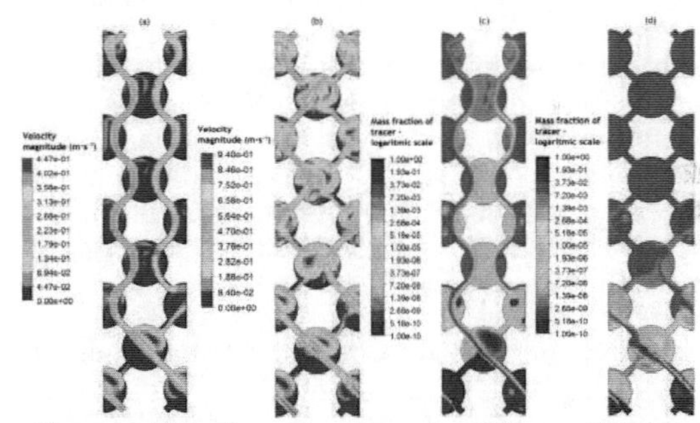

Figure 4. Velocity contour maps for (a) $\mathrm{Re}_l = 600$ and $\mathrm{Re}_r = 6$ and (b) $\mathrm{Re}_l = 1200$ and $\mathrm{Re}_r = 12$, and x_t contour maps for (c) $\mathrm{Re}_l = 600$ and $\mathrm{Re}_r = 6$ and (d) $\mathrm{Re}_l = 1200$ and $\mathrm{Re}_r = 12$ ($t/\tau = 10$).

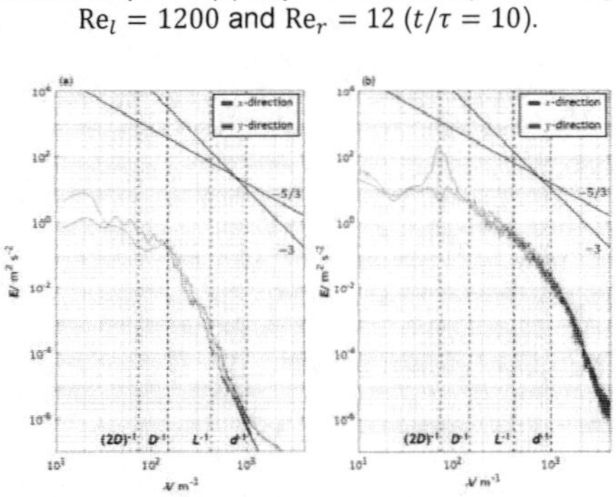

Figure 5. Power spectra for (a) $\mathrm{Re}_l = 600$ and $\mathrm{Re}_r = 6$ and (b) $\mathrm{Re}_l = 1200$ and $\mathrm{Re}_r = 12$.

Figure 2 shows that, in both cases, q^* values in opposite channels along the ExtendedNUB become closer until they reach values close to 1

https://doi.org/10.52202/074122-0078

at the outlets. These results prove that, in fact, this static mixer is able to equally distribute the flows inside its network and attain the desirable condition of similar outlet flowrates for different inlet conditions. It should also be noted that for the case with $\mathrm{Re}_l = 1200$, the difference between both outlet q^* (red curves), $q^*_{l,9} - q^*_{r,9}$, is smaller than the one obtained for the case with $\mathrm{Re}_l = 600$ (black curves). This can be explained by the fact that a higher Re_l increases the turbulence of the flow inside the NETmix, which, in turn, leads to a higher mixing degree of both phases in the chambers and a more uniform flow distribution.

Figure 4a depicts the velocity magnitude contour map for $\mathrm{Re}_l = 600$ and $\mathrm{Re}_r = 6$. The flow along the channels preserves its shape and the flow uniformisation is a gradual process. In contrast, the velocity magnitude contour map obtained for $\mathrm{Re}_l = 1200$ and $\mathrm{Re}_r = 12$ clearly shows a more predominant chaotic behaviour of the flow in the ExtendedNUB, and the formation of more eddies with smaller dimensions in upper chambers. The flow partially loses its shape and this ultimately contributes to the homogenisation of the flowrates in the outlet channels of upper chambers.

Furthermore, regarding mass transfer studies, one can see in Figure 3 that the I_s^* values for the case with $\mathrm{Re}_l = 600$ are higher than the ones obtained for the case with $\mathrm{Re}_l = 1200$ in channels 8 and 9. Higher Re values mean that the fluid has a higher kinetic energy and that the flow dynamics increases, resulting in the formation of more eddies that enhance mixing. Moreover, the I_s^* values obtained for the left and right-hand side channels in both cases are always close and, for the case with $\mathrm{Re}_l = 600$, even coincide in the upper channels. This also attests to the ability of the NETmix to equally distribute the flow along its network.

By comparing Figures 4c and 4d, it is possible to conclude that, in fact, a higher mixing degree is achieved in the case with $\mathrm{Re}_l = 1200$. This is evidenced by the uniform colour distribution seen in chambers C6 and C8 in Figure 4d, which corresponds to a x_t value close to the ideal mass fraction of tracer (0.01). In contrast, there is a gradient of colours along the NETmix in Figure 4c. More precisely, in the upper chambers portions of fluid with a x_t close to 0.01 are perceptible, but these are mingled with portions of fluid with higher concentration of tracer (dark red colour).

Lastly, the power spectrum in Figure 5a shows a slope change from $E \propto \lambda^{-5/3}$ to $E \propto \lambda^{-3}$ at a wavenumber close to D^{-1}. In Figure 5b this change occurs close to d^{-1}. These turning points indicate the respective energy injection scale in 2D turbulence. For $\mathrm{Re}_l = 1200$, the energy injection scale is the channels width. So, in this case, a fully developed dynamic flow can be observed in ExtendedNUB's eighth row, where jets issuing from channels promote mixing by vortex engulfment in the chambers. For $\mathrm{Re}_l = 600$, this flow regime is not attained in the considered length.

Conclusions

The study of the flow distribution in a NETmix operated with asymmetric inlet flowrates showed that this reactor is able to equally distribute the flow in an efficient way for different Re_l and Re_r. The formation of eddies in the chambers promotes intense mixing. The normalised intensity of segregation values along the channels also confirmed a high mixing degree. For the larger Re values, the flow uniformisation and the fully developed vortex dynamics onsets at upstream positions. Finally, the analysis of the velocity fluctuations in the last ExtendedNUB chamber proved that, in fact, a chaotic flow regime can be attained and demonstrated that NETmix reactors are efficient for asymmetric inlet flow conditions. These results show a good potential for the application of the NETmix technology for the production of struvite from the liquid fraction of digestates.

Acknowledgments

This work was financially supported by LA/P/0045/2020 (ALiCE), UIDB/50020/2020 and UIDP/50020/2020 (LSRE-LCM), funded by national funds through FCT/MCTES (PIDDAC).

References

[1] Prasad, R., and Chakraborty, D.: 'Understanding Phosphorus Forms and Their Cycling in the Soil', 2019

[2] Alewell, C., Ringeval, B., Ballabio, C., Robinson, D.A., Panagos, P., and Borrelli, P.: 'Global phosphorus shortage will be aggravated by soil erosion', Nature Communications, 2020, 11, (1), pp. 4546

[3] Di Capua, F., de Sario, S., Ferraro, A., Petrella, A., Race, M., Pirozzi, F., Fratino, U., and Spasiano, D.: 'Phosphorous removal and recovery from urban wastewater: Current practices and new directions', Science of The Total Environment, 2022, 823, pp. 153750

[4] Achilleos, P., Roberts, K.R., and Williams, I.D.: 'Struvite precipitation within wastewater treatment: A problem or a circular economy opportunity?', Heliyon, 2022, 8, (7), pp. e09862

[6] Matos, J., Santos, R.J., Dias, M.M., and Lopes, J.C.B.: 'Mixing in the NETmix Reactor', Frontiers in Chemical Engineering, 2021, 3

[7] Danckwerts, P.V.: 'The definition and measurement of some characteristics of mixtures', Applied Scientific Research, Section A, 1952, 3, (4), pp. 279-296

Evaluation of an OpenFOAM model of an aerated stirred reactor for fermentation modelling and scale-up

Poster.33

PM Pereira[1], *RC Martins*[1,2], *BS Ferreira*[1,2], *FP Bernardo*[1]. *(1) CIEPQPF, Department of Chemical Engineering, University of Coimbra, R. Sílvio Lima, 3030-790 Coimbra, Portugal (2) Biotrend SA, Biocant Park Núcleo 04, Lote 2, 3060-197 Cantanhede, Portugal.*

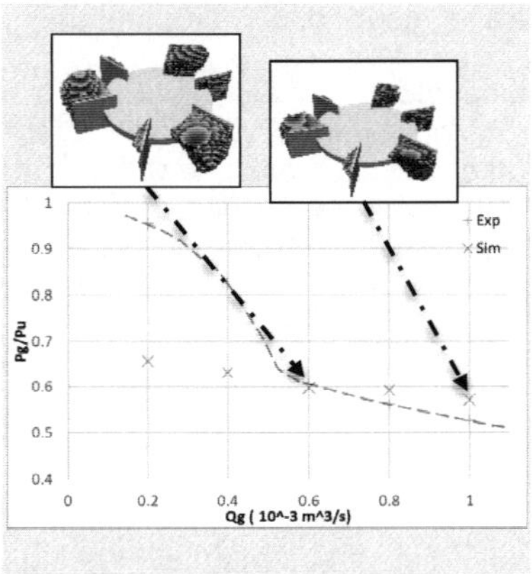

Scale-up of fermentation processes is associated with the formation of gradients of concentration of dissolved substrates and other important solutes. To improve scale-up methods, a model capable of characterising these gradients is necessary For which CFD is an essential tool. In this work, an OpenFOAM Euler-Euler RANS model is compared with experimental data from the literature to establish its validity in the range of operating conditions relevant for industrial fermentation. The model is validated using as main indicator the relative power (P_g/P_u) that is calculated using a custom openFOAM function object. The results show good agreement when the expected regime of operation of the impeller is 3 large cavities and 3 smaller cavities (3-3). The transition to 6 smaller cavities at low gas flow rate is not well predicted. Despite it's limitations this model predicts gassed power with relative accuracy for the 3-3 regime. In this regime it shows promise for future work.

Introduction

The biotechnology industry produces high value-added products like enzymes, biopolymers and biopharmaceuticals. This industry has seen a tendency to grow at higher rate than other process industries and as it expands to higher-volume and lower volume applications the impacts of the scale of operation become crucial aspects that require consideration and optimization. The bioreactor is the core piece of equipment in industrial biotechnology. It provides the environment and aseptic conditions for high performance biomass culture. Despite its ubiquitous presence in the industry, the design and operation of bioreactors remains is the subject of ongoing research, especially the scale-up of industrial fermentation.

The main objectives of a bioreactor's operation are to achieve as homogeneous and controlled as possible conditions in terms of temperature, oxygenation, pH and other parameters to favour the production of the main product of interest while maintaining the viability of the biomass. Multiple designs of bioreactor exist and have been adopted for specialty applications, but the most common and widely used type is the stirred tank reactor (STR).

The scale-up of fermentation processes is an active research topic in industrial biotechnology. In fermentation as in other process/chemical engineering applications the increase in scale introduces inefficiency in mixing [1]. As the scale increases from bench scale to pilot scale, and from this to industrial scale the diminishing efficiency of the mixing introduces spatial gradients, the most important being of substrate and oxygen concentration. These gradients result in undesired and hard to predict biological effects that can lead to lower productivity and selectivity of the process [2].

The existence of spatial gradients in the fermentation tank means that cells are exposed to cyclically fluctuating environmental conditions as they circulate in the fermenter following the prevailing streamlines. If the frequency of these fluctuations is sufficiently low, the cellular apparatus may sense the changing environment and a significant biological response may take place, including changes in metabolism redirecting substrate for undesired metabolic routes that impacts on productivity, yield and product quality [2]. Prediction and avoidance of spatial gradients is thus of paramount importance for process scale-up and optimization.

The most useful tool to obtain a detailed spatial description of gradients in a bioreactor is a model based on computational fluid dynamics (CFD) coupled with transport equations for the most relevant species and kinetic of the biochemical reactions. The CFD models give a non-ideal and spatially discretized picture of the mixing conditions in the bioreactor. The combined model can then be based on totally scale-independent parameters that, if correctly estimated, can be used for *in silico* scale-up experiments.

Models of CFD for aerated stirred tank reactors have become relatively common in the literature [3][4], but CFD models require confirmation with experimental data.

To verify a CFD model, direct data of field values of pressure, velocity or other state variables are not easily available. To substitute direct measurable data, global or macroscopic observations and state variables can be used. Some of the most commonly used state variables are the power expended for the agitation and the mixing time. The observation of the operating regime of the impeller from the type of cavities formed is also a good indication of the model quality [5]. The dimensionless Froude and gas flow numbers are commonly used variables to characterize a stirred tank's mixing characteristics for a given set of operating conditions [6].

The work here presented proposes to verify a CFD model created with the opensource CFD package OpenFOAM®.

https://doi.org/10.52202/074122-0079

Material and Methods

The CFD model here proposed is an Euler-Euler model that represents both the liquid and gaseous phase as interpenetrating continua. Each phase is treated separately with a mass conservation and momentum conservation equation for each phase. The two phases share the same pressure field and exchange momentum with each other. The momentum exchange is modelled by closure relations. The drag is modelled by the Schiller and Naumann model [7] and the virtual mass forces are modelled by a constant coefficient model.

The turbulence model is Reynolds Averaged Navier-Stokes with a particular version of the k-ε suited for Euler-Euler CFD cases called mixture k-ε [8]. This model is designed to describe turbulence in both phases using mass-averaged quantities and only one transport equation for k and another for ε.

The rotation of the impellers is modelled by a multiple reference frame (MRF).

The geometry of the bioreactor here studied was replicated from the description of the equipment used by Warmoeskerken and Smith in the paper **[5]** that provides the experimental data used here to confirm the model. The computer aided design software used was freeCAD, the geometry is presented in figure 1.

The mesh was produced with the openFOAM tools blockMesh and snappyHexMesh. These tools produce a hex dominant mesh, which is the preferred type of mesh topology. The mesh is composed of 517868 elements and the finite volumes at the faces of the impeller blades were refined to reach the an acceptable value of y^+ for the expected velocities in the impeller, the mesh is presented in figure 1.

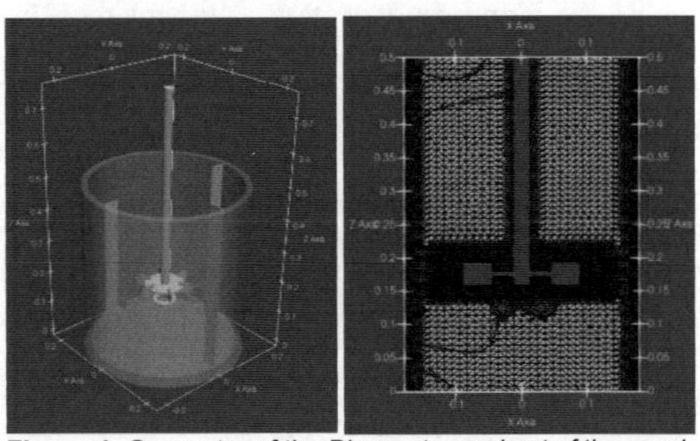

Figure 1. Geometry of the Bioreactor and cut of the mesh.

Predicting the power spent by the impeller was done by a custom openFoam function. This function was created by modifying the already existing forces function object. The original forces function is designed to integrate forces and moment of the forces on surfaces, but this is only implemented for single phase cases. The new function was inspired by the derivation of the mixture k-ε method used in this work [8]. To estimate the viscous forces mass averaged quantities are used. Viscosity, apparent viscosity due to turbulence, velocity field and density were averaged in each finite volume.

The CFD experiments were designed to cover a range of operating conditions relevant for industrial fermentation. To produce data more directly comparable whit the data in the experimental reference [5] the CFD simulations performed keep the rotation speed of the impeller constant at 3 rotations per second. The gas flow rate varied from $2x10^{-4}$ m^3s^{-1} to $1x10^{-3}$ m^3s^{-1} the data points are as seen in figure 2. This gas flow rate corresponds to from 0.2 vvm to 0.9 vvm the higher values being above normal operating conditions.

Results and Discussion

The transient simulations were run until for an interval of 0.25s the power curve showed a net change smaller than 5%. The results of relative power are presented in figure 2.

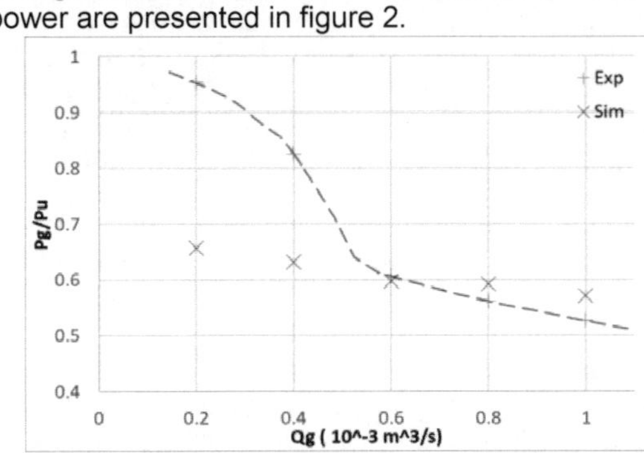

Figure 2. Comparison of relative power consumption by the impeller. The experimental data was extracted from the work of Warmoeskerken and Smith [5].

The simulation data corresponds to simulations done with a bubble size of 3 mm. Simulations were also performed for bubble diameter of 1 mm, but these are not presented because the results were not significantly different.

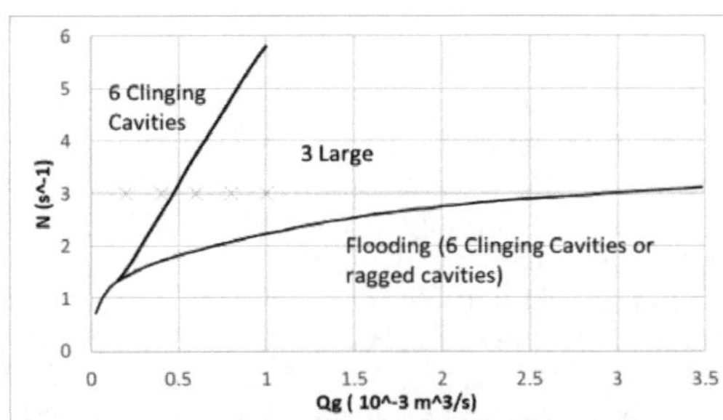

Figure 3. Map of operating regimes of an impeller in an aerated stirred tank. Data was extracted from the work of Warmoeskerken and Smith [5] and adapted from a figure.

The reduction in P_g/P_u with the increase of gas flow rate is predicted by the model. Comparing the simulation results with the experimental data, a lower agreement is obtained at lower gas flow rate. The experimental power curve shows a fast decrease of relative power around 0.0004 m^3/s of gas flow rate [5]. This decrease is the result of a change in the operating regime of the impeller. The impeller changes from operating with six smaller air cavities, one on the trailing side of each impeller blade to the 3-3 regime. The 3-3 regime is characterized by 3 larger cavities interspaced by 3 smaller cavities, on the trailing side of each impeller blade. Figure 3

https://doi.org/10.52202/074122-0079

contains a map of the operating regions of the impeller adapted from the literature [5].

From this map it's observed that the two simulated points of lower gas flow rate are in the 6 small cavities operating regime while the remaining points should be in the 3-3 operating regime.

Defining a gas cavity as a region with gas volumetric fraction of 0.95 or higher the gas cavities can be represented in the post-processing software paraView. A top view of the impeller and the cavities predicted by the simulation are presented in figure 3.

The visual representation of the cavities seems to agree with the trend in the simulated data in figure 2. For the two lower gas flow rates the model is not correctly predicting well the regime of operation of the impeller as, instead of predicting a 6 small cavity regime, a 3-3 regime is predicted. For the three remaining points the model correctly predicts the regime of operation as 3-3 and the gassed power is more accurately predicted.

The work of Maluta et al. with a similar model has reached a similar conclusion that the regime of 6 small cavities shows worse prediction of the relative power. Even if the regime transition is predicted by their work [10].

The correlation of gassed power with the mass transfer coefficient is well documented **[11].** Since the objective of this model is to advance scale-independent modelling of the fermentation process this makes predicting gassed power an important indicator of quality of the model. The good predictive power of the model here presented for tanks with impellers working in the 3-3 regime indicates the suitability of the model for further investigation with gas-liquid mass transfer.

Conclusions

In the work here presented a CFD model was validated using data available in the literature. This CFD model was produced in OpenFoam®, and a new function was also defined to calculate power developed by the impeller in multiphase Euler-Euler cases.

The model showed low sensitivity to the parameter, bubble diameter in the range studied (1-3 mm). The model presented good results for the three higher volumetric gas flow conditions studied (0.0006-0.001 m^3s^{-1}). The two lower gas flow conditions show considerably worse agreement with experimental data. The lower predictive power is possibly associated with a lower ability of the model to describe the cavity formation in the 6 small cavity regime. On the other hand the results seem promising for the 3 large cavities operating regime.

This model was designed for further development of scale independent fermentation models. Future work will aim to further develop scale independent fermentation models and ultimately more rational scale-up methods for fermentation.

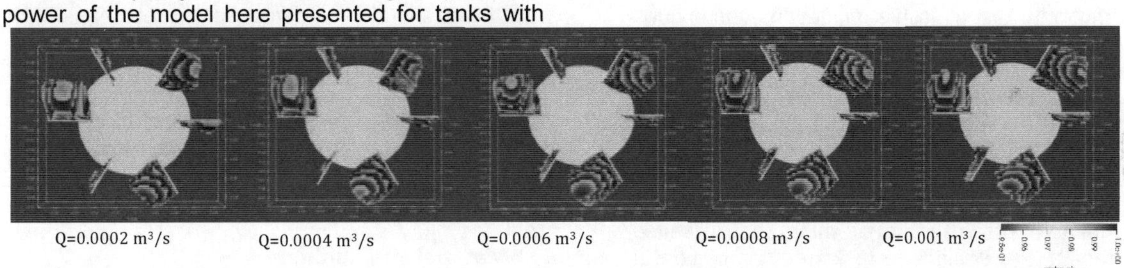

Figure 3. View of the impeller and the modelled air cavities, the cavities are represented in blue to red colour gradations corresponding to the alpha or volumetric fraction of air.

References

[1] MCCONVILLE, Francis X.; KESSLER, Stephen B. Scale-Up of Mixing Processes: A Primer. Chemical Engineering in the Pharmaceutical Industry: Active Pharmaceutical Ingredients, 2019, 241-259.

[2] NADAL-REY, Gisela, et al. Understanding gradients in industrial bioreactors. Biotechnology Advances, 2021, 46: 107660.

[3] SARKAR, Jayati, et al. CFD of mixing of multi-phase flow in a bioreactor using population balance model. Biotechnology Progress, 2016, 32.3: 613-628.

[4] *NADAL-REY, Gisela, et al. Computational fluid dynamics modelling of hydrodynamics, mixing and oxygen transfer in industrial bioreactors with Newtonian broths. Biochemical Engineering Journal, 2022, 177: 108265.*

[5] WARMOESKERKEN, M. M. C. G.; SMITH, John M. Flooding of disc turbines in gas-liquid dispersions: A new description of the phenomenon. Chemical Engineering Science, 1985, 40.11: 2063-2071.

[6] SMITH, J. M. Simple performance correlations for agitated vessels. Fluid Mechanics of Mixing: Modelling, Operations and Experimental Techniques, 1992, 55-63.

[7] SCHILLER, Links. A drag coefficient correlation. Zeit. Ver. Deutsch. Ing., 1933, 77: 318-320.

[8] BEHZADI, A.; ISSA, R. I.; RUSCHE, H. Modelling of dispersed bubble and droplet flow at high phase fractions. Chemical Engineering Science, 2004, 59.4: 759-770.

[9] Handbook of Industrial Mixing, Science and Practice, EL Paul, VA Atiemo-Obeng, SM Kresta (Eds.), John Wiley & Sons, Hoboken, NJ (2003)

[10] MALUTA, Francesco; PAGLIANTI, Alessandro; MONTANTE, Giuseppina. Prediction of gas cavities size and structure and their effect on the power consumption in a gas-liquid stirred tank by means of a two-fluid RANS model. Chemical Engineering Science, 2021, 241: 116677.

[11] Yawalkar, A. A., Heesink, A. B., Versteeg, G. F., & Pangarkar, V. G. (2002). Gas—liquid mass transfer coefficient in stirred tank reactors. The Canadian Journal of Chemical Engineering, 80(5), 840-848.

https://doi.org/10.52202/074122-0079

Role of Solids on Scale Growth on the Wall of a Solid-Liquid Agitated Vessel

Poster.34

***R. Parthasarathy*[1], *S.* *Mohd Khayry*[1], *J. Wu*[3], *N. Ohmura*[2]. *(1) Chemical and Environmental Engineering, School of Engineering, RMIT University, Melbourne, Victoria 3000, Australi, (2) Department of Chemical Science and Engineering, Graduate School of Engineering, Kobe University, 1-1 Rokkodai, Nada, Kobe, 657-8501, Hyogo, Japan, (3) CSIRO Mineral Resources, Clayton, Victoria 3168, Australia*
Email: rajarathinam.parthasarathy@rmit.edu.au

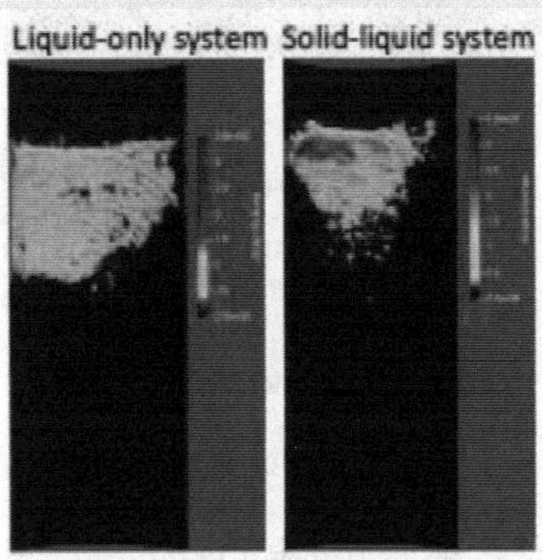

Scale growth pattern on the reactor wall
N = 500 rpm, 1% solids, A310 impeller

Scaling, an unwanted deposit formed on the inner walls of chemical reactors, is undesirable because it reduces reactor capacity, wastes energy, and increases maintenance costs. This work studies the role of suspended inert solids in scale growth in a cylindrical solid-liquid agitated reactor equipped with an axial flow impeller (A310). Scale growth experiments were conducted using four different solids concentrations. Just off-bottom suspension impeller speeds (N_{js}) and fixed impeller speeds were used. The scale weight formed on the reactor wall was lower for the solid-liquid system than the liquid-only (zero-solids) system for all solids concentrations. The scale weight decreased with increasing solids concentration. It also decreased with an increase in the degree of solids suspension. The scale reduction in the solid-liquid system is attributed to the near-wall shear and attrition due to particle collisions on the reactor wall.

Introduction

Scaling is an unwanted deposit formed on the inner walls of chemical reactors. Larger scale growth can lead to serious problems like capacity reduction, wasteful energy consumption, higher maintenance costs and production losses. Commonly used scale-mitigation techniques include injecting anti-scalants, using removable protective layers, and employing a fluid mechanics-based approach. Anti-scalants are specific to the chemical reaction involved; therefore, empirical studies are required for every chemical reaction to determine the type and amount of anti-scalants to be added. The removable protective layer might have limited durability and could lead to the contamination of reactor contents. The scale mitigation technique that has the potential to be used widely without many limitations is based on fluid flow within the reactor. The manipulation of the direction and intensity of the fluid flow could change the shear force near the reactor wall and influence scale formation and growth.

The fluid velocity, a function of variables like baffle type, impeller type and speed, directly influences the fluid shear force acting on the wall. Studies on the role of fluid shear on scale reduction in agitated reactors are rare in the literature. Using experimental and numerical studies, Davoody et al. [1-3] showed that increasing impeller speed reduces the average scale thickness on an agitated reactor wall due to increased near-wall shear stress. They also showed that scale reduction is greater in an unbaffled reactor than in a baffled reactor. The effects of other variables like impeller and baffle types on scale reduction are yet to be understood. Also, the role of inert solids in scale reduction is not well known yet. This work aims to study the role of inert solids on the weight of scale formed on the wall of an agitated reactor using an axial flow impeller (A310). To realistically study the effects of solids on scale growth, they need to be suspended off the tank bottom by running the impeller at an appropriate speed, known as just-off-bottom suspended-impeller speed (N_{js}). This work employed N_{js} for suspending solids of different concentrations and determined the scale weight on the reactor wall. Additionally, the influence of the degree of suspension on scale weight was studied by operating the impeller at a constant speed while varying the solids concentration.

Material and Methods

A dismantlable cylindrical stainless-steel tank of nine parts (four wall segments, four baffles, and a circular base) was used in scale growth experiments. The scale was grown using a chemical reaction involving calcium hydroxide, sodium carbonate and magnesium chloride leading to the formation of the magnesium hydroxide scale. The inert solids used were spherical glass beads with a mean diameter of ~250 µm. All experiments were conducted in the cylindrical reactor of 10 cm diameter at 80°C for 45 minutes under baffled conditions. The reaction liquid was drained at the end of the experiment, and the reactor was disassembled. One of the wall segments was then used in scale weight measurement. The scale weight was determined by measuring the weight of the wall segment before and after the scale growth. Experiments were conducted using solids concentrations of 1, 2, 3 and 4 vol% at their respective Njs values of 500, 570, 610 and 640 rpm, respectively. Experiments were also conducted at the above speeds but varying the solids concentration at

https://doi.org/10.52202/074122-0080

each speed to vary the degree of solids suspension (cloud height).

Results and Discussion

The experimental results of the scale weight in solid-liquid and liquid-only (zero-solids) systems are shown in Figure 1 for four solids concentrations. The scale weight in the solid-liquid system decreases continuously with increasing solids concentration. Compared to the scale weight for 1 vol% solids, the scale weights for 2, 3 and 4 vol% solids were 39.2, 72.5 and 94.1% lower. The scale weight obtained for the liquid-only system (zero-solids) also decreases continuously with increasing impeller speed. The reduction in scale weights in solid-liquid systems compared to zero-solids systems are 22.7, 24.4, 30 and 40% for 1, 2, 3 and 4 vol% solids, respectively.

The above results indicate that the presence of solids significantly reduces the scale weight. The reduction could be not only due to the collision of particles with the scale layer but also to the increase in near-wall liquid-flow-induced shear stress on the scale with an increase in N_{js} as solids concentration increases. However, in the experiments mentioned above, the 'cloud height' of the suspended solids did not reach the liquid surface, which suggests that the contribution to the scale reduction by colliding solids may be lower.

To understand the relationship between the degree of suspension (or cloud height) on scale growth, scale weight results obtained at constant impeller speeds can be used. Each constant impeller speed will equal N_{js} for only one solids concentration. When the impeller operates at N_{js}, the solids are suspended off the tank bottom. If the impeller speed is below N_{js}, only a portion of the solids will be suspended off the bottom, while the remainder will either remain stationary or move along the bottom. If the impeller speed exceeds N_{js}, the cloud height will increase, potentially reaching the liquid surface at very high speeds.

The scale weight results for each solid concentration at 500, 540, 610 and 640 rpm are shown in Figure 2. The graph also shows the results for runs with zero solids at these speeds. The scale weight values decrease with increasing impeller speed regardless of the solids concentration used. The scale weight values are the highest for 500 rpm and lowest for 640 rpm. Also, the scale weight at each speed is the highest for the zero-solids system. It is lower for the solid-liquid system and decreases with an increase in solids concentration, reaches a minimum value and then increases. The scale weight trend for different impeller speeds indicates that the extent of scale formation on the reactor wall is influenced by the near-wall shear stress and the degree of solids suspension, both determined by the impeller speed and solids concentrations used.

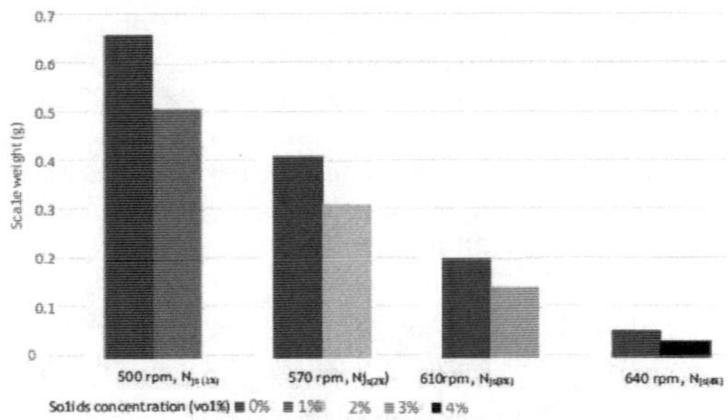

Figure 1. Experimental values of scale weight obtained in solid-liquid and liquid-only (zero-solids) systems, A310 impeller, Baffled reactor.

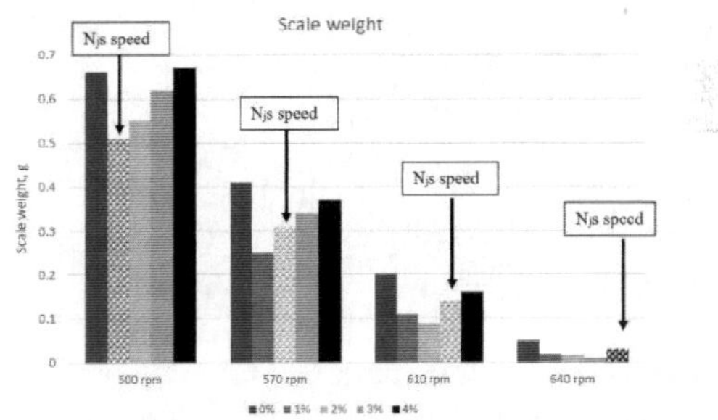

Figure 2. Experimental values of scale weight obtained in solid-liquid systems for various solids concentration at constant speeds, A310 impeller, Baffled reactor.

Conclusions

The weight of the scale formed on the wall of a solid-liquid agitated reactor depends on the solids concentration. Increasing solids concentration decreases the scale weight. In addition, the scale weight is influenced by the degree of solid suspension (cloud height), which is a function of the impeller speed and solids concentration. When the cloud height is low, the scale weight is higher and vice-versa. Therefore, developing an optimized approach for mitigating scale growth on the reactor wall should consider both the solids concentration and impeller speed.

References

[1]. Davoody, M., L. J. W. Graham, J. Wu, I. Youn, A. A. Abdul Raman, & R. Parthasarathy, *56*, (2017) 14582.

[2]. Davoody, M., L.J.W. Graham, J. Wu, P.J. Witt, M. Srinivasan, & R. Parthasarathy, 146, (2019), 11.

[3]. Davoody, M., G. Lane, L.J.W. Graham, J. Wu, S. Madapusi, & R. Parthasarathy, 64, (2018). 3912.

https://doi.org/10.52202/074122-0080

A Novel Mixture Refining System using the NETmix Technology. Proof-of-Concept: Extraction from Green Coffee Beans

Poster.35

Cláudia F. Almeida[1,2], *Simão Mateus*[3], *Luís Esteves*[3], *Hugo Moreira*[4], *Yaidelin A. Manrique*[1,2], *José Carlos B. Lopes*[1,2,5], *Madalena M. Dias*[1,2]. *(1) LSRE-LCM – Laboratory of Separation and Reaction Engineering – Laboratory of Catalysis and Materials, Faculty of Engineering, University of Porto, Rua Dr. Roberto Frias, 4200-465 Porto, Portugal. (2) ALiCE – Associate Laboratory in Chemical Engineering, Faculty of Engineering, University of Porto, Rua Dr. Roberto Frias, 4200-465 Porto, Portugal. (3) Instituto de Ciência e Inovação em Engenharia Mecânica e Engenharia Industrial (INEGI), Campus da FEUP, Rua Dr. Roberto Frias 400, 4200-465 Porto, Portugal. (4) Paralab SA, Rua Dr. Joaquim Manuel Costa 946B, 4420-437 Valbom, Portugal. (5) CoLAB Net4CO2 – Network for a Sustainable CO2 Economy, Rua Dr. Júlio de Matos 828-882, 4200-355, Porto, Portugal.* *ccalmeida@fe.up.pt.*

In this work, a novel continuous process for refining mixtures was designed. The system (shown in the graphical illustration) withstands pressures up to 100 bar and temperatures of 40 °C, thus allowing extractions from natural raw materials with supercritical or liquid carbon dioxide. The carbon dioxide stream is then mixed with another solvent (e.g., water) to refine the resulting extract, which is a complex mixture of various compounds. The NETmix technology, a static, continuous mixing device, was used to promote an efficient mass transfer between phases. Solvent streams are separated due to density differences in a device optimised for this purpose, and the respective extracts are collected for analysis.

Liquid and supercritical carbon dioxide extractions from green coffee beans were chosen as a proof-of-concept case study, with a particular focus on the fractionation of chlorogenic acid and caffeine.

Introduction

Supercritical fluid extraction (SFE) has become a prominent substitute for traditional solvent extraction, fractionation, and isolation of active ingredients, mainly propelled by growing stringent legislation concerning residual solvent levels in food products [1]. This extraction technique has several advantages, including the maintenance of the extracted fractions' properties, such as flavour and aroma; products not contaminated with solvents; and the selective and fractional extraction of compounds, which can be tuned by controlling the solvent power via pressure and temperature adjustments [2, 3]. Carbon dioxide is the most commonly used solvent accounting for over 90 % of all analytical extractions [4]. This substance's appeal rests on the moderate conditions of its critical point (31.1 ºC and 74 bar), making it appropriate for extracting temperature-labile compounds [2-4]. Moreover, it is non-flammable, non-toxic, available at high purity at a relatively low cost, and is easily removed from the extracts upon decompression to atmospheric conditions [3]. CO_2 is adequate for dissolving lipophilic compounds due to its nil dipole moment; however, its lack of polarity makes it unsuitable for extracting polar substances [4]. The typical workaround involves adding a polar modifier or co-solvent (e.g., water, methanol or ethanol) to dissolve polar compounds [3, 4]. Consequently, the resulting extract is often a complex mixture requiring a follow-up liquid-liquid extraction (LLE) for further refinement of the targeted compound. LLE consists of at least a mixing and a separation step, which explores the different solubility of solutes between immiscible liquids [5, 6]. Common drawbacks include the high retention times to obtain a good mixture between phases and using solvents with expensive recovery costs, representing up to 97 % of the operation costs [6, 7].

In this work, a novel continuous system for refining mixtures was designed and tested. The NETmix technology [8] was used for mixing, and the process solvents were carbon dioxide and water, thus addressing the above-exposed weaknesses. The NETmix consists of a static mixer formed by a network of mixing chambers interconnected by transport channels [8, 9].

Material and Methods

Figure 1 shows the simplified process flow diagram of the NETmix refining system, where components such as valves and pressure, temperature and level sensors are not represented. CO_2 is cooled (E-1) before the pump's inlet (E-2) to avoid cavitation. This stream is then heated (E-3) to the process's desired temperature and fed to a tank (E-4) containing the natural raw material, where the extraction will take place. Water, stored in tank E-7, is pressurised through pump E-9 and heated in the heat exchanger E-11. The streams' mixture occurs in the NETmix (E-12), which limits mass transfer resistances, thus promoting product refinement. The outlet stream is forwarded to separator E-13, whose geometry was optimised for phase separation due to fluid density differences. The aqueous phase is collected in tank E-17, and the CO_2-phase extract precipitates in tank E-14.

This work considered extractions from green

https://doi.org/10.52202/074122-0081

coffee beans with liquid/supercritical carbon dioxide and subsequent extract refining with water. Recovered extracts were analysed in high-performance liquid chromatography (HPLC) with phosphoric acid (0.05 %) and acetonitrile.

Results and Discussion

The first step was to define the system's operating requirements. Temperature and pressure ranges were set to 20-40 °C and 75-100 bar, respectively. These conditions are similar to those used in the separation zone of typical SFE systems. The LLE mixing step is promoted in a NETmix device. Computational fluid dynamic (CFD) simulations, later supported by tracer experiments, proved that mixing is achieved above a critical channel's Reynolds number, Re_{crit}, when the flow in the mixing chambers evolves to a self-sustained oscillatory laminar flow regime [9]. Figure 2 displays the Reynolds number in the NETmix channels as a function of temperature, pressure and the mixture's volumetric composition. Simulations were performed in the Aspen Plus V10 commercial software by Aspen Technology, Inc., using the NIST Reference Fluid Thermodynamic and Transport Properties database, i.e., the REFPROP property method. A chamber width of 0.5 mm and depth of 1 mm resulted in a NETmix hydraulic diameter of 0.67 mm, and the total flow rate was fixed at 0.2 L·min^{-1}. Detailed analysis of Figure 2 limits the upper carbon dioxide concentration to 50 % v/v due to the higher Reynolds number for greater CO_2 concentrations. For mixtures with 25 and 30 % v/v of CO_2, calculations near carbon dioxide's critical point are imprecise due to the model's higher errors for this region.

The total flow rate (0.2 L·min^{-1}) was defined to promote a high enough residence time in the separator, thus allowing phase segregation. The volume of the CO_2's reservoir (E-4) was chosen to enable the dissolution of about 15 mg of solute per tank fill. Furthermore, CFD simulations were performed to understand the flow dynamics inside the separator (E-13) as well as to optimise this equipment's geometry. The multiphase volume of fluid (VOF) model was applied to simulate the injection of a mixture of 25 % v/v of supercritical carbon dioxide (at 100 bar and 40 °C) and water into the separator. Maximum water concentration at the carbon dioxide's outlet, and vice-versa, were estimated to not exceed 0.2 % v/v considering a constant pressure difference of 3900 Pa between the separator's outlets. However, the interface's level is not constant, invalidating this assumption. Thus, a new simulation was performed with varying pressure differences between outlets from 3690 to 4150 Pa, and the results indicated a maximum of 6 % v/v of carbon dioxide in the water outlet and nearly 10 % v/v of water in the CO_2 outlet.

The NETmix refining system, shown in the graphical illustration, is currently undergoing tests to extract chlorogenic acid (CGA) and caffeine from green coffee beans with liquid/supercritical carbon dioxide. The resulting extract is refined with a water stream, as previously reported in the literature [10]. Machmudah et al. reported 100 % purity of caffeine in the supercritical carbon dioxide stream and both caffeine and CGA in the water.

Soxhlet extractions with methanol displayed higher yields for the ground coffee bean than for the whole bean (9.4 versus 0.6 % w/w). Thus, ground raw material is used. Effects of temperature, pressure, and mixture composition on the recovery and isolation of CGA and caffeine are being evaluated.

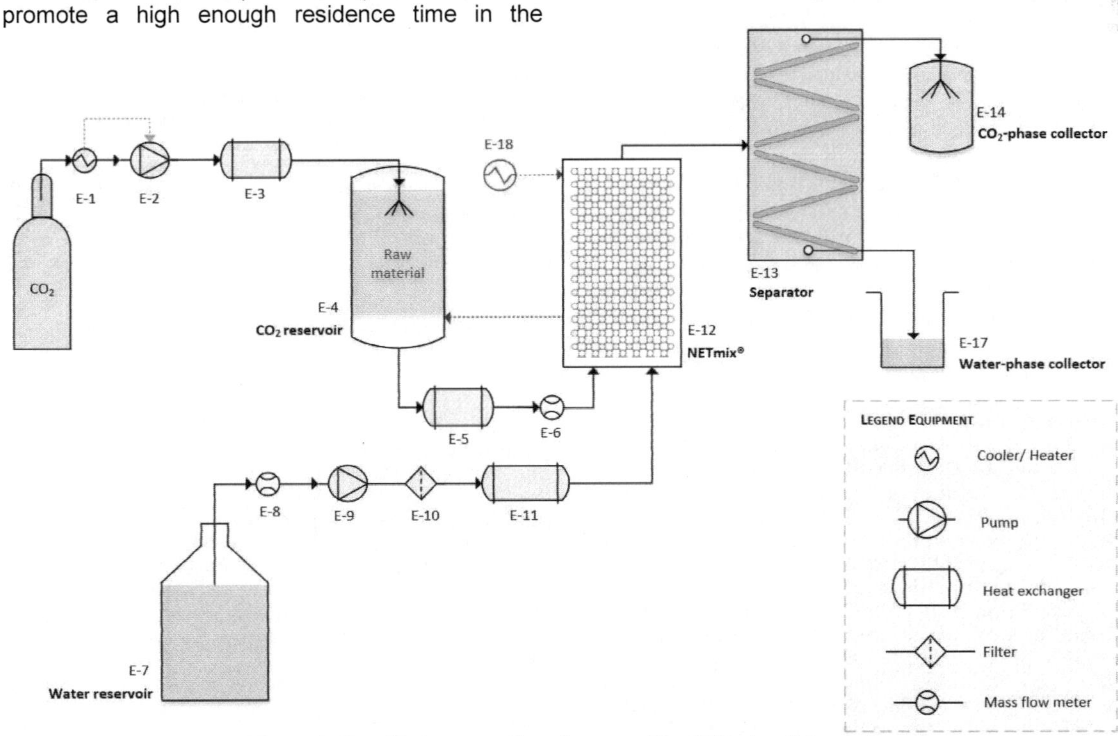

Figure 1. Simplified process flow diagram of the NETmix refining system.

https://doi.org/10.52202/074122-0081

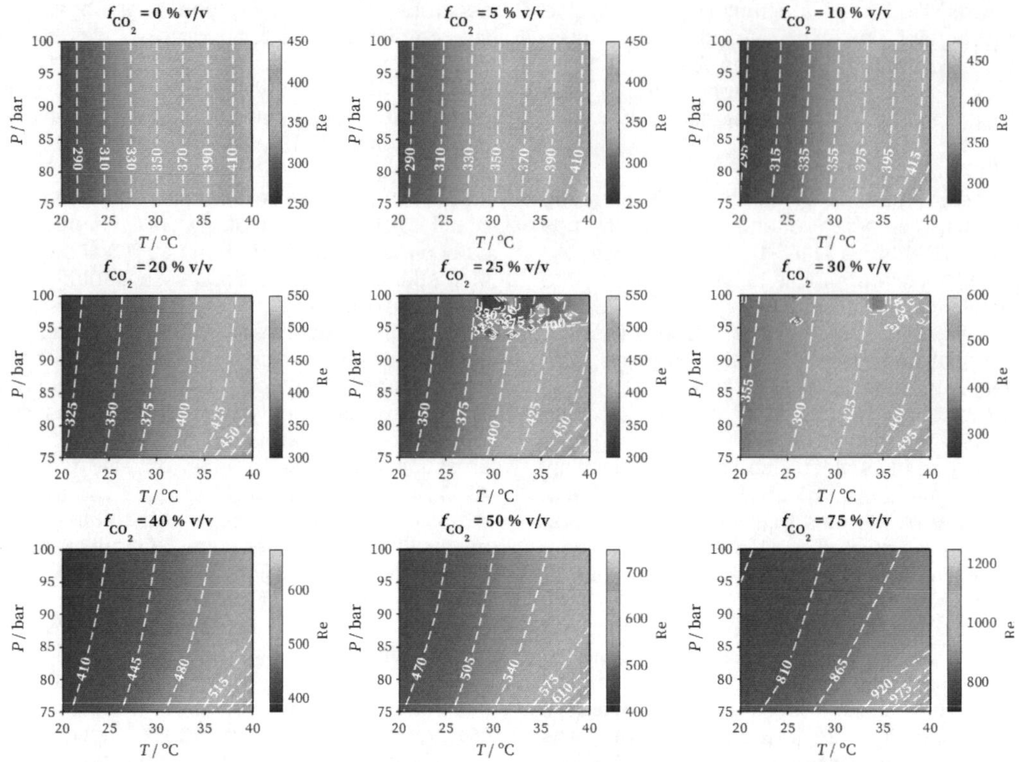

Figure 2. Reynolds number in the NETmix channels as a function of temperature, pressure and mixture's volumetric composition. Carbon dioxide properties were calculated in the Aspen Plus software using the REFPROP property method.

Conclusions

In this work, a novel continuous refining system was designed. Crucial steps included defining the system's operating conditions and elaborating the process flow diagram of each process step to verify the valves' correct position as well as safety issues. Furthermore, CFD simulations were performed to optimise the separator's geometry. The Reynolds numbers at the NETmix's channels were calculated for different conditions of pressure, temperature, and mixture composition.

Preliminary proof-of-concept tests are currently being carried out, considering pressure, temperature, and mixture composition adjustments to extract and isolate caffeine and CGA from green coffee beans. The long-term goal of this work consists of replacing the separation zone of typical SFE systems for the NETmix refining unit, thus improving the extracts' purity.

Acknowledgements

This work was financially supported by: LA/P/0045/2020 (ALiCE), and UIDB/50020/2020 and UIDP/50020/2020 (LSRE-LCM), funded by national funds through FCT/MCTES (PIDDAC); by Project Mobilizador Valor Natural (NORTE-01-0247-FEDER-024479), supported by Norte Portugal Regional Operational Programme (NORTE 2020), under the PORTUGAL 2020 Partnership Agreement, through the European Regional Development Fund (ERDF); Cláudia Almeida acknowledges her PhD scholarship by FCT (ref. 2020.04470.BD). The authors also express their gratitude to the *Instituto de Ciência e Inovação em Engenharia Mecânica e Engenharia Industrial* (INEGI) and Paralab teams, which aided the design and assembly of the NETmix refining system.

References

[1] *Supercritical Fluid Extraction of Nutraceuticals and Bioactive Compounds*. Jose L. Martínez (Eds.), Boca Raton (FL), CRC Press, 2008.

[2] Juliana M. Prado, Renata Vardanega, Isabel C. N. Debien, Maria A. Meireles, Lia N. Gerschenson, Halagur B. Sowbhagya, Smain Chemat in *Food Waste Recovery*, Charis M. Galanakis (Eds.), Academic Press, 2015, 6.

[3] M. K. L. Bicking, in *Encyclopedia of Separation Science*, Ian D. Wilson (Eds.), Academic Press, 2000.

[4] Sudipta Saha, Ashok K. Singh, Amit K. Keshari, Vinit Raj, Amit Rai, Siddhartha Maity in *Ingredients Extraction by Physicochemical Methods in Food*, Alexandru Mihai Grumezescu, Alina Maria Holban (Eds.), London (UK), Academic Press, 2017, 2.

[5] A. Vansteene, J.-P. Jasmin, L. Al Amhad, S.Cavadias, G. Core, C. Mariet, *Proc. of Congrès Français de Mécanique, Lille,* France, 2017, 23.

[6] A. Wojik, T. Haderer, R. Marr, M. Siebenhofer, S. Martens, *Proc. of American Institute of Chemical Engineers Annual Meeting, Austin TX,* USA, 2004.

[7] Gabriella Lestari, Alinaghi Salari, Milad Abolhasani, Eugenia Kumacheva, *Lab on a Chip*, 16 (2016), 2710.

[8] José Carlos B. Lopes, Paulo E. Laranjeira, Madalena M. Dias, António A. Martins, Network Mixer and Related Mixing Process (WO 2005/077508), 2005.

[9] Paulo E. Laranjeira, António A. Martins, Maria Isabel Nunes, José Carlos B. Lopes, Madalena M. Dias, *American Institute of Chemical Engineers Journal*, 4 (2011) 1020.

[10] Siti Machmudah, Kiwa Kitada, Mitsuru Sasaki, Motonobu Goto, Jun Munemasa, Masahiro Yamagata, *Industrial & Engineering Chemistry Research*, 50 (2011) 2227.

CFD Radiation Models for Annular Photoreactors Simulation

Poster.36

Isabel S.O. Barbosa[1,2], *Ricardo J. Santos*[1,2], *Madalena M. Dias*[1,2], *Joaquim L. Faria*[1,2], *Cláudia G. Silva*[1,2]
(1)Laboratory of Separation and Reaction Engineering-Laboratory of Catalysis and Materials (LSRE-LCM), Rua Dr. Roberto Frias, 4200-465 Porto, Portugal;(2) ALiCE Associate Laboratory in Chemical Engineering, Faculty of Engineering, University of Porto, Rua Dr. Roberto Frias, 4200-465 Porto, Portugal, e-mail: isob@fe.up.pt

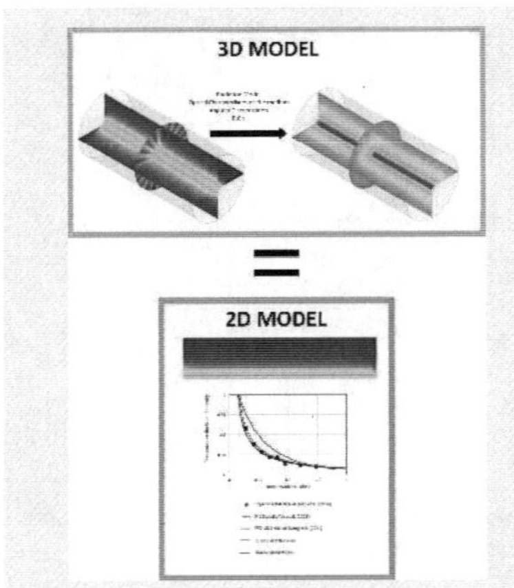

CFD simulations referring to three models, P-1, discrete ordinates (DO), and Monte Carlo (MC) models, for simulating radiation distribution in annular photoreactors are addressed.
This work offers a tutorial for radiation simulation inside an annular reactor. The cases are a three-dimensional (3D) annular photoreactor and a two-dimensional (2D) rectangular enclosure with black walls, and the results are compared with the literature.
The DO model was selected to solve the radiation transfer equation (RTE) as the most reliable model to fit the radiation distribution inside the reactor. The setting-up of mesh, angular discretization, and boundary conditions for CFD simulation of photocatalytic reactors are addressed. When correctly set, the CFD models show an excellent agreement with the literature results.

Introduction

In recent years, the use of photo-assisted technologies has received increasing attention mostly due to the possibility of using solar light, a renewable and economical energy source, to activate several processes, namely water and wastewater treatment, chemical synthesis, and energy production. Radiation distribution is an indispensable parameter in the simulation of photochemical and photocatalytic systems [1]. The light intensity distribution inside the reactors is known by solving RTE (Radiative Transport Equation), and it is dependent on parameters such as type and position of the radiation source, reactor geometry, radiation wavelength, the optical properties of the medium, optical specifications of reactor walls, and operating conditions (inflow and outflow rate, the catalyst type, and chemicals concentrations) [2]. The RTE is an integrodifferential equation based on the conservation principle applied to a ray along a path through a medium and can be expressed as:

$$\frac{dI(\vec{r},\vec{s})}{ds} + (\alpha + \sigma_s)I(\vec{r},\vec{s}) = \alpha n^2 \frac{\sigma T^4}{\pi} + \frac{\sigma_s}{4\pi}\int_0^{4\pi} I(\vec{r},\vec{s}')\phi(\vec{s}\cdot\vec{s}')d\Omega' \quad (1)$$

where $\vec{r}$ is the position vector, $\vec{s}$ is the direction vector, $\vec{s}'$ is the scattering direction vector, s (m) is the path length, α ($\mathrm{m^{-1}}$) is the absorption coefficient, n is the refractive index, σ_s ($\mathrm{m^{-1}}$) is the scattering coefficient, σ ($5.669 \times 10^{-8}\ \frac{\mathrm{W}}{\mathrm{m^2K^4}}$) is Stefan-Boltzmann constant, I ($\frac{\mathrm{W}}{\mathrm{m^2{\cdot}sr}}$) is the radiation intensity which depends on position ($\vec{r}$) and direction ($\vec{s}$), T (K) is the local temperature, ϕ is the phase function and Ω' (sr) is the solid angle. $(\alpha + \sigma_s)$ is the optical thickness or opacity of the medium. An exact analytical solution of the RTE is only possible for one-dimension ideal situations after simplifying assumptions such as uniform radiative properties of the medium and homogeneous boundary conditions.

Numerical simulation has gradually become a powerful technique to study radiation transport in absorbing, emitting and scattering media. Thus, CFD (Computational Fluid Dynamics) became a valuable tool for simulation, enabling an integrated analysis of reactors through simultaneous modelling of hydrodynamics, species mass transport, reaction kinetics and photon flux distribution (RTE) [3].

Several methods have been developed to solve the RTE, such as P-1 Model, Discrete Ordinates Model (DO), which are approximations to the solution of the RTE, and Monte Carlo Model (MC), which involves a statistical modelling. The development of computers and CFD software, including radiation models, enabled the complete simulation of photocatalytic systems. A major challenge in modelling photocatalytic reactors, apart from issues related to catalyst activity, is a poor catalyst illumination efficiency, being the rate of reaction in an immobilized system related to the incident radiation.

In this work, radiation modelling in photoreactors is proposed to evaluate the light intensity distribution in an annular photocatalytic reactor and compare models' reliability and accuracy. Experimental data, P-1 and DO models were compared to solve the RTE in a three-dimensional annular geometry and a 2D enclosure.

Material and Methods

The modelling of radiation in a photocatalytic system enables the determination of the radiation field, which affects the rate of photochemical

https://doi.org/10.52202/074122-0082

reactions. This work describes the implementation of radiation models in CFD codes. The simulations were performed using the CFD ANSYS Fluent R2020 ® commercial code. The case study reactor has an annular geometry with 645 mm of length, 15 mm of external radius and 2.25 mm of internal radius (Figure 1). This reactor has been used in several experimental and numerical works [2,4,5]. In the base case, a three-dimensional annular photoreactor and a two-dimensional rectangular enclosure (simplification of a 3D geometry) with black walls were considered as showcases. A single UV lamp concentric was considered as the radiation source.

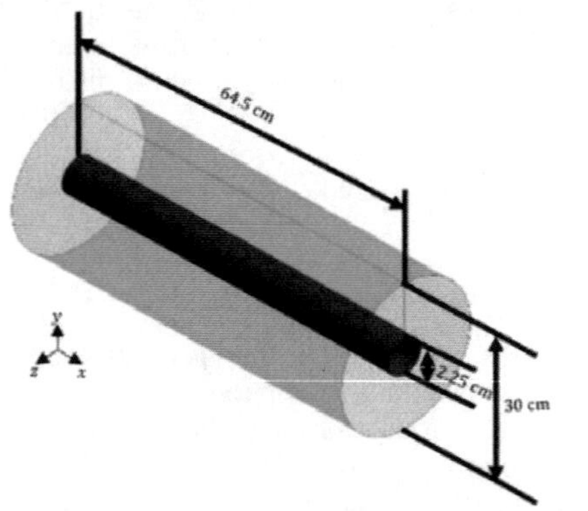

Figure 1. Single-lamp axial annular geometry of the photoreactor.

DO Model using the finite volume method (FVM) was chosen to simulate the RTE. This is a promising method for modelling radiative transfer due to its extensive application and high accuracy. The steady-state solution method was used to solve the radiation and energy equations. Angular discretization and absorption coefficient were investigated to determine their influence in the radiation field. Finally, a comparison between simulation results and previous works [2,4,5] was carried out to validate the results obtained with 2D Axisymmetric and 3D Models.

Results and Discussion

The refinement of the grid in the lamp wall influences the radiation distribution. Only significantly refined meshes around the lamp sleave are able to describe the radiation intensity in cylindrical geometries accurately.

The results of angular discretization show that for the 2D case, the angular discretization does not have a role in the distribution of radiation results, because radial homogeneity is already implicit when simulating only a slice of the reactor. For the 3D case, angular discretization has a significant impact on distribution of light. Figure 2 shows the radiation intensity map for two angular discretization values.

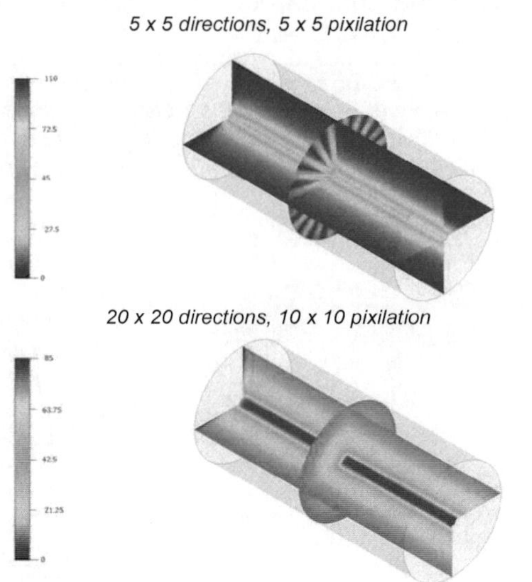

Figure 2. Radiation profiles of the non-absorption case for different angular discretization of 3D model.

The visual effect of the discretization is striking even at the lamp surface with the occurrence of stripes of larger intensity. This heterogeneity is propagated into the medium, as seen in Figure 2. For the 3D case, an increase in angular discretization enables a more uniform light distribution, namely in the lamp wall.

From the results of the optimized cases, it was concluded that the 2D model describes the same radiation profile behaviour as the 3D model. For the study of absorption coefficients (Figure 3), only 2D simulation was performed since, for the non-absorption case, 2D and 3D results are very similar.

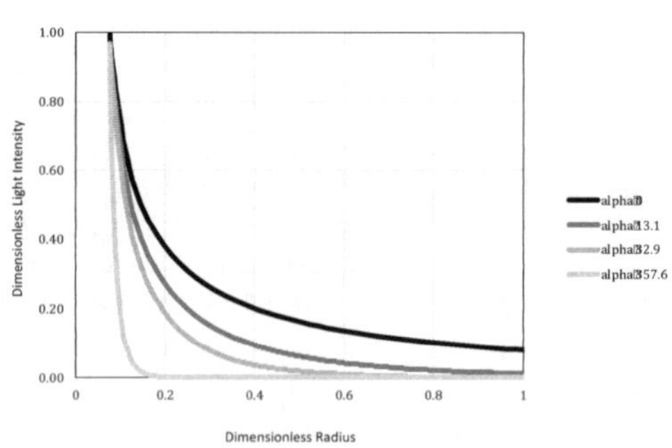

Figure 3. Radiation distribution for 2D axisymmetric model.

The profiles presented in Figure 3. show that the light intensity decreases rather sharply with the increase in the absorption coefficient. For the case with zero absorption coefficient, the decrease in incident radiation is due to the increase of the photonic flow area. When the absorption coefficient is equal to 357.6 m^{-1}, 90% of the incident light is absorbed in the initial ≈1.5 mm away from the lamp. This shows that the exposure of the catalyst to radiation when there are heterogeneities of absorption in the medium will depend on the flow patterns and degree of mixing within the reactor in the photocatalytic systems. If the flow is very segregated, promoting large gradients in the light absorption throughout the reactor, the illumination of the catalyst in the outer wall will be very heterogeneous.

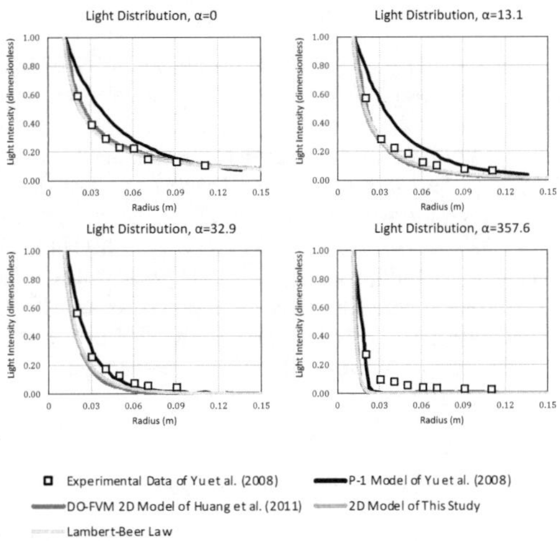

Figure 4. Comparison of models of the relative incident radiation for several absorption coefficients.

From Figure 4, it was concluded that the developed models to simulate radiation distribution inside an annular reactor described the same behaviour reported in the literature. Also, the method to solve RTE by Lambert-Beer Law proposed by Abu-Ghararah, Ziad H. (1997) presents the same curve shape for all absorption coefficients. Thus, this work presents a reliable CFD method to simulate the radiation distribution inside an annular reactor.

Conclusions

The Discrete Ordinate Model is the most reliable model for predicting the radiation distribution inside a cylindrical reactor. Using this model, a base case analysis and sequential steps of simulation radiation inside an annular reactor were described. In order to understand the reliability of the results, they were compared with the ones reported in the literature and with the Lambert-Beer Law equation describing the exact behaviour of the radiation profile. Thus, this work represents a tutorial for radiation simulation inside an annular reactor.

Acknowledgements

This work was financially supported by LA/P/0045/2020 (ALiCE), UIDB/50020/2020 and UIDP/50020/2020 (LSRE-LCM), and the project SuN2Fuel 2022.04682.PTDC - funded by national funds through FCT/MCTES (PIDDAC). I.S.O. Barbosa acknowledges her FCT grant UI/BD/151092/2021.

References

[1] Barbosa, I.S.O, Santos, R.J., Dias M.M., Faria J.L., Gomes, C.G., *Chemical Engineering & Technology 2023*

[2] Huang, Q. S.; Liu, T. Z.; Yang, J.; Yao, L. S.; Gao, L. L., *Chemical Engineering Science 2011*, 66 (17), 3930-3940.

[3] Duran, J. E.; Taghipour, F.; Mohseni, M., *J Photoch Photobio A 2010*, 215 (1), 81-89.

[4] Yu, B., B. Deng, and C.N. Kim, *Chemical Engineering Science 2008*, 63(22); 5552-5558.

[5] Quan, Y., S.O. Pehkonen, and M.B. Ray, *Industrial & Engineering Chemistry Research 2004*, 43(4), 948-955.

https://doi.org/10.52202/074122-0082

CFD – Aided Design of Primary Settlers

Poster.37

M. Carolina Duarte[1], *M.S. C.A Brito*[1,2,3], *A.M. Karpinska*[4], *R.J. Santos**[1,3] *(1) LSRE-LSM, Faculty of Engineering, University of Porto, Rua Dr. Roberto Frias, 4200-465 Porto, Portugal, *rsantos@fe.up.pt. (2) LEPABE - Laboratory for Process Engineering, Environment, Biotechnology and Energy, Rua Dr. Roberto Frias, 4400-465, Porto, Portugal. (3) ALiCE, Faculty of Engineering, University of Porto, Rua Dr. Roberto Frias, 4200-465 Porto, Portugal. (4) Southern Water Services LTd.*

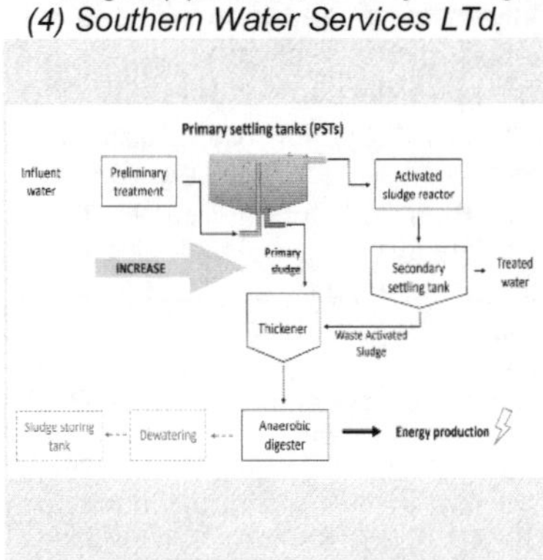

This study aims to improve circular Primary Settling Tanks (PST) design in Wastewater Treatment Plants (WWTP) by addressing limitations of existing guidelines that do not align with recent environmental goals. Advanced computational methods are used to enhance PST efficiency. Two-dimensional Computational Fluid Dynamics simulations investigate circular PST hydrodynamics using data from an English South Coast WWTP. The influence of five inlet designs on tank velocities, turbulent viscosity, and flowrate streamlines were examined. Two influent velocities were tested to assess the effect of inlet variations on tank performance. Additionally, DPM simulations were performed to analyse the percentage of different particle sizes retained within the tank.

Introduction

Adaptation to climate change and achieving Net-Zero Energy and Net-Zero Carbon targets are driving initiatives in the wastewater sector. Converting Wastewater Treatment Plant (WWTP) facilities into energy-neutral and carbon-neutral facilities involves enhancing energy efficiency, utilizing renewable energy sources, and reducing fugitive emissions. Anaerobic Digestion (AD) is a crucial process for generating renewable energy from sludge, which helps offset the reliance on grid power primarily generated from fossil fuel [1].

One of the established methods to increase the production of renewable energy from biogas (methane) is to increase gas yield in AD. This increase relies on the optimum ratio of the sludges produced in a WWTP that is fed to AD: primary sludge and secondary sludge.

Optimizing Primary Sedimentation Tank (PST) operations to enhance primary sludge production and improve the primary-to-secondary sludge ratio relies on various parameters. Geometrical configurations, including depth, inlet, and outlet regions, play a key role in this optimization. The design of inlet, settling, and outlet zones significantly influences sedimentation performance. Optimizing the flow conditions facilitates sediment settling at the bottom of PSTs and the promotion of a turbulent mixing region at the inlet.

The project of WWTPs relies on traditional design processes that used alone fall short in flexibility to accommodate recent environmental and sustainability goals [2,3].

Computational Fluid Dynamics (CFD) has significantly improved PST design by simplifying empirical research, reducing costs and time. CFD modelling is particularly suitable for studying PSTs due to the different settling types involved (discrete, flocculent, hindered, and compression) [4].

The first CFD studies on PSTs date back to the 1970s. With the advancement of computational resources, CFD simulations of PSTs have become more efficient and commercially available. However, the application of CFD in the wastewater sector was, until recently, still limited due to the lack of guidance for modelling.

More recently, it was published the first CFD for Wastewater technical report detailing best modelling practice for particular unit treatment processes [5].

PSTs play a crucial role in nutrient removal and energy consumption reduction in WWTPs. Advanced design tools, such as CFD simulations, are necessary to achieve these goals. This work focuses on using CFD models to optimize primary sludge production and the primary-to-secondary sludge ratio in circular PSTs, aiming to shift WWTPs' energy balance towards neutrality.

Material and Methods

The study focuses on a circular PST with a simplified axisymmetric domain. The 2D surface, shown in Figure 1 represents half of the settler's cross section, which rotates to create the complete circular settler.

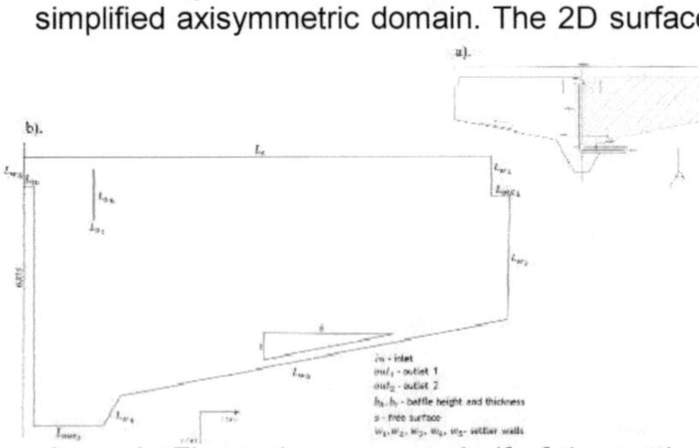

Figure 1. a) circular PST cross section; b) Axisymmetric 2D CFD geometric domain.

The PST's hydrodynamics were simulated using ANSYS Fluent 2022 R1. The flow simulations are based on the conservation laws of physics, i. e.

https://doi.org/10.52202/074122-0083

the mass conservation principle and the momentum conservation (Newton's Second Law). These principles are applied through governing equations (Ansys 2022).

Due to the flow hydrodynamics, the RNG (Renormalization Group theory) k-ε model was the turbulent model used. For turbulent flow, the 2D axisymmetric Navier-Stokes equations were averaged using the decomposition of the velocity components in an average, $\overline{v_i}$, and a fluctuating component, $\overline{v_i}$, which translate into: $v_i = \overline{v_i} + v'_j$.

CFD simulations were set at steady state, i.e., the transient terms are zero and water was set as working fluid. For these simulations, it was used the Coupled scheme for the pressure-velocity coupling; Second order for the pressure and Second order upwind for both turbulent kinetic energy and turbulent dissipation rate. The convergence criteria of the simulations were the residuals as10^{-4}.

The simulations were performed according to the models described using a mesh with 252 584 elements, and a size of 0.0012 m. Different inlet designs were studied in this work. The dimensions and designs of five different inlets are shown in Figure 2. In this work, two velocities at the inlet were tested $v_1 = 2.2\ \mathrm{m\ s^{-1}}$ and $v_2 = 1.0\ \mathrm{m\ s^{-1}}$, being these the higher and mean flow velocities given by the English WWTP. In total 10 tests were carried as shown in Table 1.

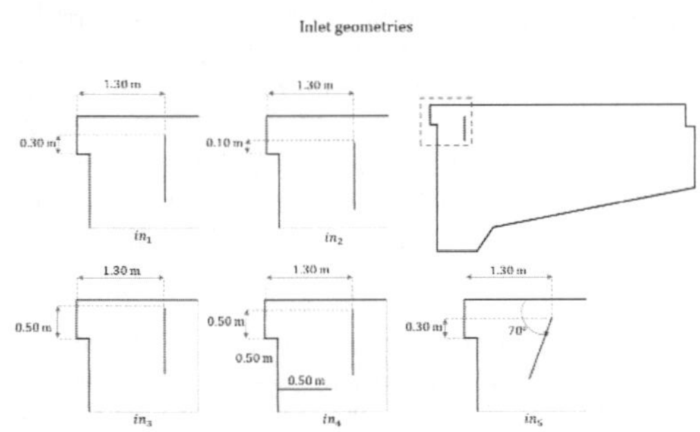

Figure 2. Inlet designs and measurements for CFD simulations.

Table 1. Planification and identification of simulations.

	Influent Velocity $v_c\ [\mathrm{m\ s^{-1}}]$	
$\mathrm{in_j}$	$v_1 = 2.2$	$v_1 = 2.2$
$\mathrm{in_1}$	$\mathrm{in_1\upsilon_1}$	$\mathrm{in_1\upsilon_2}$
$\mathrm{in_2}$	$\mathrm{in_2\upsilon_1}$	$\mathrm{in_2\upsilon_2}$
$\mathrm{in_3}$	$\mathrm{in_3\upsilon_1}$	$\mathrm{in_3\upsilon_2}$
$\mathrm{in_4}$	$\mathrm{in_4\upsilon_1}$	$\mathrm{in_4\upsilon_2}$
$\mathrm{in_5}$	$\mathrm{in_5\upsilon_1}$	$\mathrm{in_5\upsilon_2}$

For the second part of this work, six simulations were carried out with a Discrete Phase Model, DPM, to simulate the flow of six different sized discrete particles in the PST. An Euler-Lagrange approach was implemented for the multiphase simulations. In this case, water is the fluid phase and discrete particles correspond to the dispersed phase. The disperse phase is solved by tracking a large number of particles through the flow field previously calculated. In this case, the flow field used for this study was the one calculated for the conditions $\mathrm{in_1\upsilon_1}$.

The transient state was set in CFD simulations, and particle-particle interactions were neglected. The particle trajectory was computed individually at adaptative time-steps, with a minimum of 0.001 s and a maximum of 1 s.

Results and Discussion

The streamlines shown in Figure 3, obtained for the 10 simulations described in Table 1 show that the main features of the flow in the PST are set from the clearance above the ring baffle. A jet issues from the gap above the baffle, creating a shortcut to the outlet. This jet induces two recirculation zones (vortices). These vortices are unstable and make the flow very sensible to boundary conditions and numerical parameters, as reported from the mesh studies.

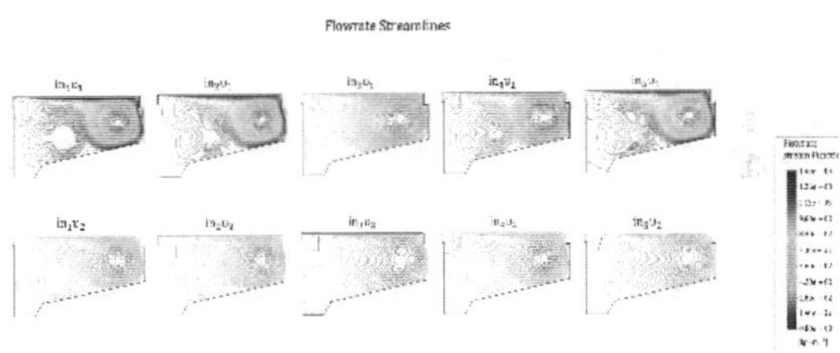

Figure 3. Streamlines for simulations $in_j v_c$.

In simulations in $\mathrm{in_1\upsilon_1}$ and $\mathrm{in_1\upsilon_2}$, the jet issuing over the ring baffle clearly promotes the flow division into the two vortices mentioned above. The left vortex is centred at a lower region and does not contact directly with the inlet jet. The vortex on the right covers most of the length of the inlet jet, presents a well-defined roundish circulation and is closer to the outlet zone.

The same flow patterns are observed for the inlet design with the larger gap above the ring baffle, simulations $\mathrm{in_2\upsilon_1}$ and $\mathrm{in_2\upsilon_2}$. For the smaller gap ($\mathrm{in_3\upsilon_1}$ and $\mathrm{in_3\upsilon_2}$), the flow rate jet issued above the ring baffle is higher, pushing the left vortex to a lower position and making the overall flow dominated by the right vortex. The flow pattern alteration for these conditions shows further the unstable nature of the vortices.

For conditions $\mathrm{in_4\upsilon_1}$ and $\mathrm{in_4\upsilon_2}$, a horizontal baffle (disk baffle) is placed below the ring baffle. Its presence promotes the appearance of a smaller third vortex, that interplays with the left vortex, formed in the inlet region. Although the gap is the same as for $\mathrm{in_3\upsilon_1}$ and $\mathrm{in_3\upsilon_2}$, the overall recirculation patterns of the large vortices are similar for the conditions in $\mathrm{in_1\upsilon_1}$, $\mathrm{in_1\upsilon_2}$, $\mathrm{in_2\upsilon_1}$ and $\mathrm{in_2\upsilon_2}$.

The last inlet design configuration is a tilted baffle. Condition $\mathrm{in_5\upsilon_1}$ exhibits the same flow patterns as simulations as $\mathrm{in_1\upsilon_1}$ and $\mathrm{in_1\upsilon_2}$. But for the lower velocity, condition $\mathrm{in_5\upsilon_2}$, the flow patterns were the same as in $\mathrm{in_3\upsilon_1}$ and $\mathrm{in_3\upsilon_2}$ with the left vortex placed closer to the bottom of the PST. The change in flow patterns shows that the recirculation in the region close to the inlet and its

https://doi.org/10.52202/074122-0083

interaction with the left vortex can be used to set the overall flow patterns for this condition.

The flow features are further analysed from turbulent viscosity profiles, drawn from a cross-section of the 2D PST's geometric domain (line 1 in Figure 4), at r = 5.0 m. The turbulent viscosity profiles are shown in Figure 5.

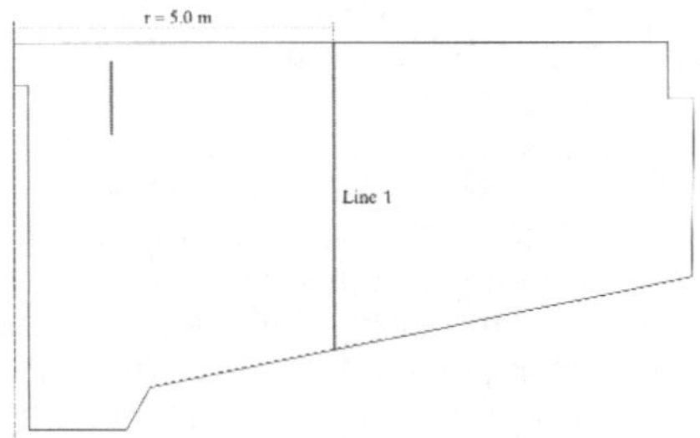

Figure 4. Position of the cross-section of the geometric domain – line 1.

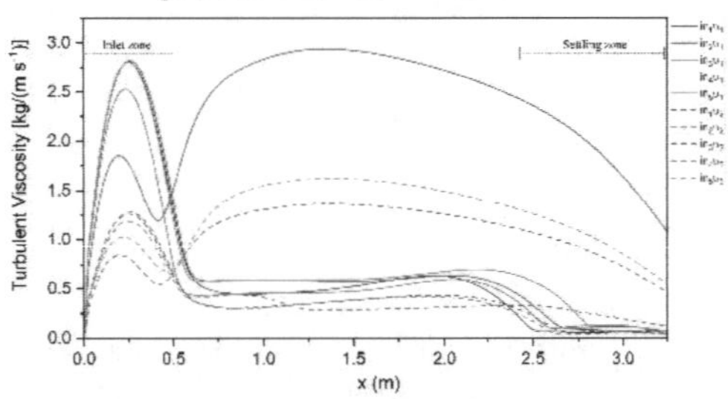

Figure 5. Turbulent Viscosity at line 1.

For a better performance of a PST, as previously stated, the inlet zone design, especially the baffle, should provide a region of turbulence, to enhance particle flocculation with or without mixing a chemical agent. Analysing the Turbulent viscosity shown in Figure 5, it is possible to identify that, for both velocities, the inlet design that allows for a more turbulent flow is in_2. Contrary to this, it is required a reduction of turbulence in the lower region of the PST, to allow a better settling of the particles. Figure 5 shows that, with the exception of designs in_3 and in_5, all designs present a low turbulence flow near the bottom of the PST, thus allowing for a better settling of particles. Overall, the design that presents the worst performance, for both velocities is in_3 and in_5 presents a bad performance for the highest velocity, υ_1.

The DPM simulation results were analysed through the percentage of particles removed, which represent the percentage of particles that settled inside the PST. These results are presented in Table 2.

As shown in Table 2, the removal's percentage increases with the increase of the particle size. It reaches the value 100 % for particle's diameter of $\phi_i > 1.0 \times 10^{-4}$ m. For larger sediments, the settling is faster, and so the removal's percentage is higher.

Table 2. DPM simulations plan, working conditions and removal percentage.

ϕ_i	**Particle diameter** [m]	**Removal [%]**
ϕ_1	1.0×10^{-5}	6.2
ϕ_2	2.0×10^{-5}	10.0
ϕ_3	4.0×10^{-5}	87.0
ϕ_4	1.0×10^{-4}	100.0
ϕ_5	2.0×10^{-4}	100.0
ϕ_6	1.0×10^{-3}	100.0

Conclusions

Different inlet and baffle designs were studied using steady state CFD simulations. The defined parameters for efficiency analysis were the Turbulent Viscosity and the inlet velocity. Results show that PST with the baffle's position further from the free surface for both inlet velocities, 2.2 m s^{-1} and 1.0 m s^{-1} (conditions $\mathrm{in}_2\upsilon_1$ and $\mathrm{in}_2\upsilon_2$) is the one that presents the best efficiency according to the hydrodynamics.

In what regards the DPM simulations, the results of the simulations performed for the six different particle sizes allow to conclude that for diameters ≥1.0 ×10−4 m, 100 % of particles settle.

CFD models developed in this work enable a better understanding of the influence of velocity and geometric variations in the hydrodynamics of a circular primary settling. This work will contribute to further know-how of the design concept and optimization of PSTs in WWTPs, guaranteeing the achievement of higher sedimentation efficiencies.

Acknowledgments
This work was financially supported by LA/P/0045/2020 (ALiCE), UIDB/50020/2020 and UIDP/50020/2020 (LSRE-LCM), funded by national funds through FCT/MCTES (PIDDAC).

References
[1] International Energy Agency, INE. 2021. "Net Zero by 2050 - A Roadmap for the Global energy sector." In.
[2] Metcalf, and Eddy. 2014. WASTEWATER ENGINEERING: TREATMENT AND RESOURCE RECOVERY.
[3] Hendricks, David. 2016. Fundamentals of water treatment unit processes: physical, chemical, and biological (Crc Press).
[4] Samstag , R W, J J Ducoste , A Griborio , and Ingmar Batstone Nopens , DJ Wicks, JD Saunders , S Wicklein , EA Kenny , G Laurent , J. 2016. 'CFD for wastewater treatment: an overview', Water Science and Technology, 74: 549-63.
[5] Julien Laurent, Randal Samstag, Jim Wicks, Ingmar Nopens. 2022. CFD Modelling for Wastewater Treatment Processes: IWA Working Group on Computational Fluid Dynamics. IWA Publishing

https://doi.org/10.52202/074122-0083

An experimental study in a new technology for continuous crystallisation: O-NETmix

Poster.38

***A. F. Freitas*[1,2,3*]**, *Y. A. Manrique*[1,3], *J. C. B. Lopes*[1,3,4], *R. J. Santos*[1,3], *A. Ferreira*[2,3], *M. M. Dias*[1,3]. *(1) LSRE-LCM – Laboratory of Separation and Reaction Engineering – Laboratory of Catalysis and Materials, Faculdade de Engenharia da Universidade do Porto, Rua Doutor. Roberto Frias, 4200-465 Porto, Portuga. (2) LEPABE – Laboratory of Process Engineering, Environment, Biotechnology and Energy, Faculdade de Engenharia da Universidade do Porto, Rua Doutor. Roberto Frias, 4200-465 Porto, Portugal. (3) ALiCE – Associate Laboratory in Chemical Engineering, Faculdade de Engenharia da Universidade do Porto, Rua Doutor. Roberto Frias, 4200-465 Porto, Portugal. (4) Colab NET4CO2 – A Network for a Sustainable CO2 Economy, Rua Júlio de Matos, 828-882, 4200-355 Porto, Portugal. *affreitas@fe.up.pt*

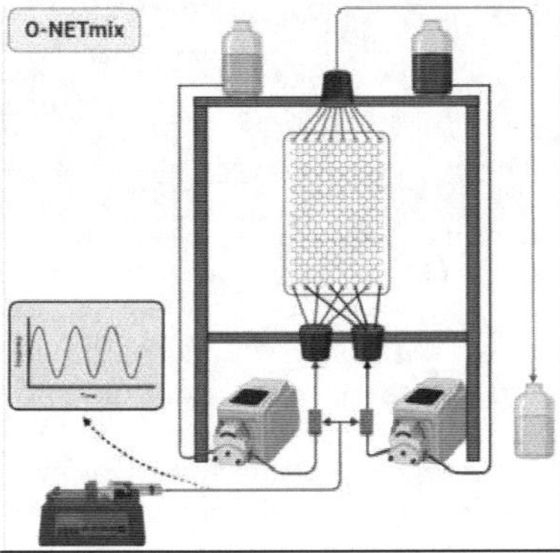

Growing awareness of physical and chemical processes and the creation of new methods and technologies that increased effectiveness and precision influenced the history of crystallisation processes. The oscillatory NETmix (O-NETmix), a new technique developed for continuous crystallisation, is studied in this work. The best characteristics of the oscillatory flow crystalliser (OFC), which can offer high residence times necessary for crystal formation, are combined with the NETmix structure for controlled and efficient mixing in this device. Experiments were conducted with an indigo carmine solution in a NETmix reactor with and without a coupled oscillation system.

Introduction

Crystallisation is one of the oldest and most commonly used techniques in various industrial sectors, namely in the chemical, food and pharmaceutical industries, and consists of separation and purification of products. Combining the formation and purification of particles in a single process makes crystallisation an attractive isolation step in manufacturing [1].

In recent years there have been interesting developments in crystallisation, such as high-throughput, microfluidic and continuous crystallisation. With advances in automation and robotics, high-throughput crystallisation has become increasingly popular. This allows for the rapid parallel screening of many different crystallisation conditions, enabling researchers to identify optimal conditions for crystallising a given molecule [2]. Microfluidic crystallisation has become increasingly popular in recent years due to its ability to produce high-quality crystals on a small scale. This has become a valuable tool for protein crystallography and other applications where small amounts of high-quality crystals are needed [3]. Continuous crystallisation is an approach that enables the production of crystals in a continuous flow. This can be useful for large-scale production and for controlling the size and shape of the crystals produced [4].

Oscillatory flow crystallisation has been used in the pharmaceutical, chemical, and food industries. New advances in flow control, measurement, and modelling have further improved the efficiency and effectiveness of oscillatory flow crystallisation, making it an important tool for producing high-quality crystals.

An oscillatory flow crystalliser works by creating an oscillating flow pattern within the crystallisation solution. This flow pattern promotes the growth of uniform crystals and helps to improve the separation of the crystals from the mother liquor. The oscillatory flow is generated by an external source, such as a pump or an acoustic driver, that creates a periodic flow variation. This flow variation leads to a complex flow field in the crystallisation solution, which promotes the reactants' mixing and transport, improving the crystal growth's kinetics. In addition to improving mixing and transport, the oscillatory flow helps to prevent the formation of large, irregular crystals by continuously stirring and homogenising the solution. This leads to the formation of smaller and more uniform crystals. The specific design of an oscillatory flow crystalliser would depend on the desired outcome and the characteristics of the crystallisation solution [5].

Continuous oscillatory crystallisers combine the features of continuous crystallisation with the benefits of oscillatory flow, resulting in improved crystal quality and increased process efficiency. In this crystalliser, the reactants are fed into the reactor continuously and subjected to an oscillatory flow, which enhances mixing and causes the crystals to nucleate and grow [6]. The crystals are then separated from the mother liquor and continuously removed from the reactor.

The use of oscillatory flow in continuous crystallisation offers several benefits, including: improved crystal quality - oscillatory flow enhances mixing and promotes the growth of high-quality crystals with well-defined morphologies and narrow size distributions; increased efficiency - continuous oscillatory crystallisation offers improved control over the crystallisation process and can result in increased efficiency and productivity; lower capital and

https://doi.org/10.52202/074122-0084

operating costs - continuous oscillatory crystallisation can be more cost-effective than batch crystallisation due to the reduced need for batch-to-batch cleaning and the increased process efficiency.

Continuous oscillatory crystallisers offer a promising solution for crystal production, potentially achieving higher yields and improved product quality compared to traditional crystallisation methods.

The study of the residence time distribution (RTD) is commonly used to describe the hydrodynamic behaviour of a system over time. It can help understand the dynamics of mixing, chemical reactions, and transport. Residence time is the distribution of times that fluid elements spends in the system before leaving it [7]. Tracer experiments can be performed to find the residence time, and the cumulative distribution function (CDF) used to describe the system's behaviour.

The outlet concentration of the tracer is measured as a function of time, t, which gives the cumulative distribution function, $F(t)$ [8]:

$$F(t) = \frac{C_s(t)}{C_f} \qquad (1.1)$$

where $C_s(t)$ is the concentration of the tracer over time, and C_f is the tracer concentration introduced in the reactor to perform the tracer experiment.

The RTD function, $E(t)$, is given by equation (1.2) and describes quantitively how long the fluid elements stayed inside the reactor. [8].

$$E(t) = \frac{d[F(t)]}{d(t)} = \frac{d}{dt}\left[\frac{C_s(t)}{C_f}\right] \qquad (1.2)$$

A normalised RTD $E(\theta)$ is frequently used instead of the function $E(t)$ [9]., since it normalizes the effect of different passage times, τ. If the normalized time is defined as:

$$\theta = \frac{t}{\tau} \qquad (1.3)$$

a dimensionless function $E(\theta)$ can be defined as:

$$E(\theta) = \tau E(t) \qquad (1.4)$$

and plotted as a function of θ The quantity θ represent the number of reactor volumes of fluid based on entrance conditions that have flowed through the reactor in time t.

Material and Methods

For the tracer experiments, water and an aqueous solution of indigo carmine were used as working fluids. Indigo carmine does not alter the properties of the water (e.g. viscosity and surface tension) but acquires a blue colour even when dissolved at low concentrations. A solution with an 8.8 x 10^{-5} M concentration of indigo carmine was used and will be referred to from this point forward as the tracer.

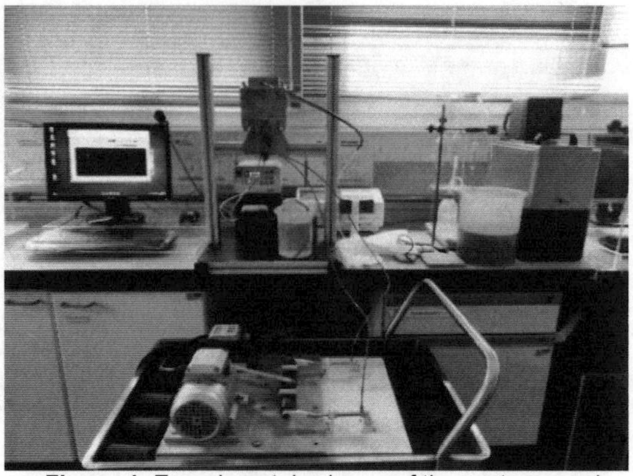

Figure 1. Experimental scheme of the system used for tracer experiences in MicroNETmix with oscillation.

The oscillation of the inlet flow streams in NETmix leads to more flexible control and operation of this device. For this purpose, an experimental set-up was assembled, with a flow oscillation system coupled to the structure of an existing MicroNETmix reactor.

For the tracer experiment with and without oscillation in the MicroNETmix reactor, a Jenway 6300 spectrophotometer with the RPA 2021 software was used (Figure 1). Different Reynolds numbers and different frequencies of oscillation were tested. For Reynolds number were 75 and 100 and the flow oscillation frequencies 6 Hz and 20 Hz. Initially, the whole system was fed with water through a peristaltic pump (Imastec® MCP process), and then the tracer solution, an indigo carmine solution, was passed.

Results and Discussion

The residence time distribution in Figure 2 show that the O-NETmix reactor can be considered as a dispersive flow plug flow reactor [9], similar to the theoretical curves previously reported [10]. The curves are identical for all Reynolds and frequencies of oscillation.

Synthesis of calcium phosphate preliminary results carried out with and without oscillation are shown in Table 1. The average diameter was obtained through Laser diffraction (Coulter LS 230) and the calcium to phosphate ratio was determined by Energy Dispersive Spectroscopy (EDS). From experiments O3 and O4, respectively without and with oscillation, it appears that oscillation decreases the average diameter of the particles. However, experiments O4 and O5 indicate that an increase in frequency increases the average diameter of the particle. The results of EDS are qualitative and intended to verify only the existence of the calcium and phosphorus elements in the material analysed. X-ray diffraction analyses are being carried out to categorise the calcium phosphates produced.

https://doi.org/10.52202/074122-0084

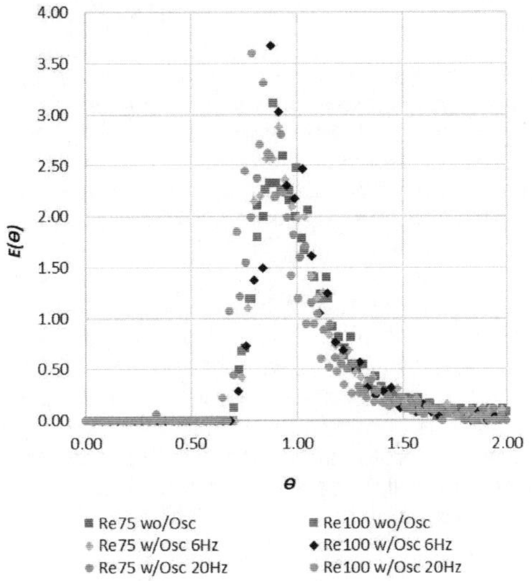

Figure 2. Residence time distribution function for tracer experiments in MicroNETmix with and without oscillation (6 Hz and 20 Hz) at Reynolds numbers 75 and 100.

Table 1. Summary of the synthesis of calcium phosphate with and without oscillatory system and main results.

ID	Oscillation	Average diameter (μm)	Ratio Ca/P
O3	-	10.3	1.52
O4	6 Hz	9.0	1.43
O5	20 Hz	12.4	1.48

Conclusions

The residence time distribution represents the range and distribution of residence times for the fluid elements within a system. The oscillation did not produce significant impact on the RTD, so macromixing is kept.

All experiments carried out in the reactor, with or without the oscillation system coupled, show curves very similar to a laminar flow profile, so the reactor is considered ideal.

The first studies in producing calcium phosphate were conducted in the experimental set-up. There was an impact of oscillation on the calcium phosphate particles' diameter and composition. Thus, although macromixing was the same, the oscillation has an impact on micromixing. Further tests are being conducted in this experimental set-up to draw a complete picture of the impact of flow oscillation on continuous crystallisation.

Acknowledgments

This work was financially supported by LA/P/0045/2020 (ALiCE), UIDB/50020/2020 and UIDP/50020/2020 (LSRE-LCM), and UIDB/00511/2020 - UIDP/00511/2020 (LEPABE) funded by national funds through FCT/MCTES (PIDDAC), and the doctoral scholarship 2020.05252.BD from FCT.

References

[1] Handbook of industrial crystallization. A. S. Myerson, D. Erdemir; A. Y. Lee, (3 ed), Cambridge, Cambridge University, 2019.

[2] B.G. Sumpter, K. Hong, R.K. Vasudevan, I. Ivanov, R. Advincula, Carbon trends, (2023).

[3] J. Ferreira, F. Castro, F. Rocha, S. Kuhn, Chemical Engineering Science, 191 (2018) 232-244.

[4] P.Cruz, F. Rocha, A. Ferreira, Crystengcomm, 18 (2016) 9113-9121.

[5] Crystallization Technology Handbook, A. Mersmann, (2 ed), New York, Marcel Dekker, Inc., 2001.

[6] T. McGlone, N.E.B. Briggs, C.A. Clark, C.J. Brown, J. Sefcik, A.J. Florence, Organic Process Research & Development, 19 (2015) 1186-1202.

[7] Y. Gao, F.J. Muzzio, M.G. Ierapetritou, Powder Technology 228 (2012) 416-423.

[8] Models for Flow Systems and Chemical Reactors, C.Y. Wen, L.T. Fan, New York, Dekker, 1975.

[9] P.V. Danckwerts, Chemical Engineering Science, 2 (1953) 867-944.

[10] P.E. Laranjeira, A.A. Martins, J.C.B. Lopes, M.M. Dias, New York, AIChE Journal, 55(2009) 2226-2243.

https://doi.org/10.52202/074122-0084

Development of Lab-Scale Formulation: Reactive extrusion process for Industrial Mixing Conditions

Poster.39

Elham Delvar[1,2], *Margarida S.C.A. Brito*[1,2], *Arantzazu Santamaria-Echart*[3], *Maria-Filomena Barreiro*[3], *Cláudia G. Silva*[1,2], *Ricardo J. Santos*[1,2]. *(1) ALiCE - Associate Laboratory in Chemical Engineering, Faculty of Engineering, University of Porto, Rua Dr. Roberto Frias, 4200-465 Porto, Portugal, email. (2) Laboratory of Separation and Reaction Engineering – Laboratory of Catalysis and Materials (LSRE-LCM), Faculty of Engineering, University of Porto, Rua Dr. Roberto Frias, 4200-465 Porto, Portugal. (3) Laboratório Para a Sustentabilidade e Tecnologia em Regiões de Montanha, Instituto Politécnico de Bragança,Campus de Santa Apolónia, 5300-253 Bragança, Portugal. rsantos@fe.up.pt*

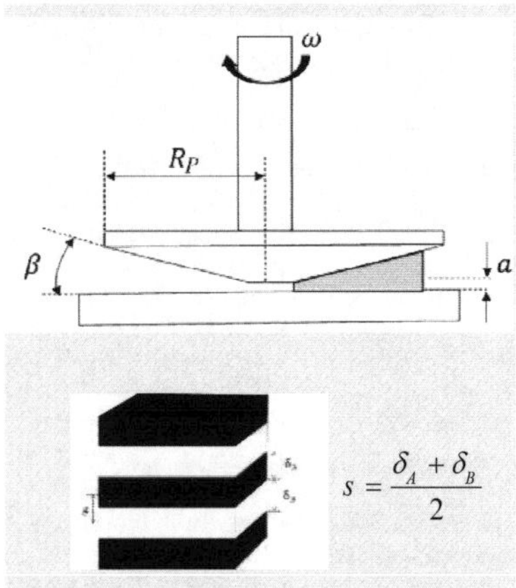

This work aims to introduce a product and process development method for biodegradable polyurethanes (BIOTPUs) for footwear applications, which is an alternative to synthetic polymers. An experimental protocol to assess the reactive polymerization will be developed from CFD simulations and linked to the conditions in an industrial reactive extruder. A protocol connecting the process conditions for BIOTPUs at a laboratory scale using small quantities of raw materials to industrial extrusion processes will be developed.

Introduction

Among the extensive range of polymers employed in footwear, polyurethanes are the most utilized because they have various versatile properties, such as resistance to degradation, durability, strength, and lightness. Biobased polyols, isocyanates, and chain extenders are the most common components in the TPUs formulation. Reactive extrusion (REX) is one of the processing technologies that is being assessed for the large-scale incorporation of biobased TPUs in the footwear industry.

REX is a significant production and processing method with several applications in polymers' processing, modification or functionalization, and recycling. Designing extruders for REX systems requires data on the rheology of raw materials, the reactive mixture, and the kinematics of polymerization. REX enables mixing control, reaction time, and temperature for reactive mixtures of melted polymers with high molecular weights. This geometry is particularly suitable for the mixing and transport of highly viscous materials. The dynamic nature of reacting melts poses a challenge for extruders. When a reactive mixture is highly viscous, mixing phenomena take longer time, causing steeper gradients in temperature and chemical compositions, which impacts polymerization and the final properties of the polymer.

The development of product formulations for REX industrial processes typically occurs at the laboratory scale. However, replicating the intricate mixing conditions found in industrial settings poses a significant challenge. In this study, we assess a rotational rheometric for the fast laboratory-scale formulation development of reactive polymerizations by mimicking industrial mixing conditions. The interfacial area generation and dynamics of mixing scales reduction in the rheometric flow are assessed and faced with those in screw extruders. The rheometric method offers notable advantages, including reduced raw material requirements, shorter testing times, and the evolution of viscosity build-up, which enables the kinetic process (rheokinematic method).

A rheokinematic methodology to a case study involving the reactive extrusion process was applied to demonstrate its effectiveness. By matching the evolution of mixing length scales observed in rotational rheometers to those found in extruder screws, the gap between laboratory processing conditions using minimal raw materials and high-throughput continuous flow reactors can be successfully bridge. This work focuses on studying a rotational device that creates a laminar flow, which holds great significance for the manufacturing industry. The flexibility of processing conditions in the rheometer enables the assessment of a wide range of equipment designs and operational conditions.

Material and Methods

The scale-up of HTMD from the lab-scale to the plant

Low-temperature physical crosslinking biobased thermoplastic polyurethane (TPU) formulations are set as the basis for developing an HTMD – High Throughput Material Development method. So, viscosity is set to 1 Pa•s, density to 1 kg m^{-3}, and the process time window in the scale of one minute, which are typical parameters for reactive polymerization of TPU.

The development of the HTMD method is based

https://doi.org/10.52202/074122-0085

on CFD simulations of the rheometric flow devices and of the REX. The flow Reynolds numbers are low (≤1), and creep flow conditions, for the rheometric and extruder flows, which enables steady-state hydrodynamic simulations. The mixing is simulated from the dynamic simulation of the transport of a tracer over the steady flow field.

CFD model

The geometry of the single screw extruder used in this project consists of one screw that is rotating inside one cylinder. The screw walls are set to angular velocity ω, and the tangential velocity at each position of the extruder wall is $v = \omega R_{in}$, where R_{in} is the distance of the extruder surface to the rotation axis. The cylinder enclosing the extruder screw is stationary.

The single screw has a positive displacement, meaning that the screw transfers a particular product volume according to the extruder's speed and pitch. The rotation of the extruder pushes the material from the inlet to the outlet.

Mixing of Similar Fluids

Ottino et al. (1978) defined a lamellar mixing model, which considers that mixing occurs by stretching and folding in a laminar flow, giving rise to a lamellar structure of two fluids [1]. The thickness of the lamellae ($s(t)$) is continuously thinned by the stretching mechanisms under the laminar flow regime.

In the initial stage of the simulation, for both the plate-plate rheometer and the screw extruder, two similar fluids are positioned side by side. As the system progresses through a series of rotations, the interface separating the two fluids undergoes stretching, eventually forming a distinct lamellar structure. Interestingly, these findings align perfectly with a notable experiment conducted by Mohr et al. in 1957, which examined the stretching of a dye within a laminar flow, mirroring the behavior depicted in Figure 1. Through the utilization of such visualizations, a comprehensive understanding of the mixing process can be achieved. Specifically, this approach enables the characterization of key factors such as the generation of the interfacial area and the measurement of striation thickness. The latter refers to the distance that exists between the two segregated fluids [2,3]. By carefully analyzing these parameters, valuable insights into the intricate dynamics of mixing were gained, enabling the assess the mixing conditions in the lab device and the industrial REX are the same.

Results and Discussion

Kinematics of deformation

The interfacial area and striation thickness in rheometric flows are quantified by the stretching of a material element, as illustrated in Table 1. The evolution of interfacial area is described by $\alpha'(\mathrm{t})$, and the striation thickness is given by $s(t)$. The interfacial area and the striation thickness are quantified for the two most common rotational rheology topologies: plate-plate and cone-plate. The complete development of the equations for the rheometric flows is found in Brito et al. [5].

In a flow with a constant field of deformation, the interfacial area generation and striation thickness evolution between two fluids is given by $\alpha'(t) = \sqrt{1+\dot{\gamma}(t)t}$, and $s(t) = s_0/\sqrt{1+\dot{\gamma}(t)t}$, respectively. For the two rheometric flows, plate-plate and cone-plate, the evolution of $\alpha'(t)$ and $s(t)$ for two fluids set side by side in a plate is given in Table 1.

For a cone-plate rheometer, the shear rate field is constant and depends on the cone angle, β, and the angular velocity of the cone, ω. For this flow, $\dot{\gamma} = \omega/\beta$.

For the screw extruder in each section, there is an eccentric circle rotating with angular velocity ω_{ext}. The screw circle has a radius R_{screw} and rotates inside a cylindrical runner with a radius R_{ext}. The screw circle rotates eccentrically over the runner axis, with a distance r to this axis.

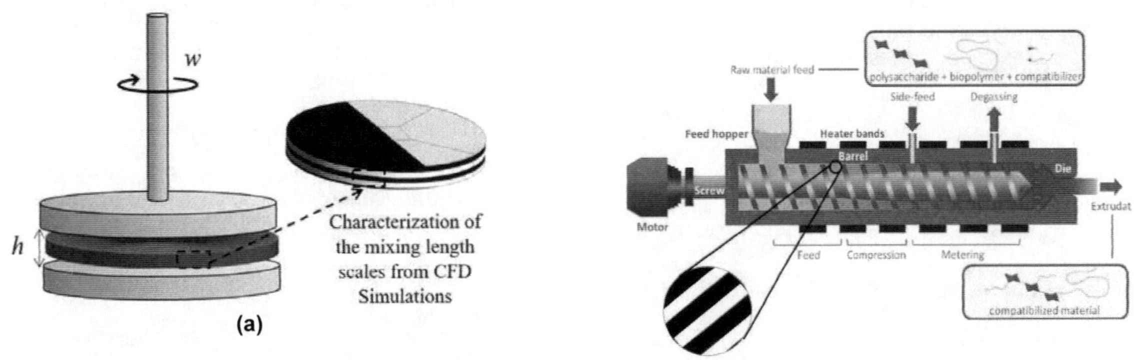

(a)

Figure 1. Schematic of implantation of a rheometric rotational device VS Industrial reactive extruder. The extruder is reprinted from [4] under Creative Commons license.

 https://doi.org/10.52202/074122-0085

Table 1. Mixing scales length obtained in topology

Topology	Contour Maps of phases	Interfacial Area	Striation Thickness
Plate-plate		$\alpha'(t) = \sqrt{1 + \left(\frac{\omega R_p t}{h_p}\right)^2}$	$s(t) = \frac{s_0}{\sqrt{1 + \left(\frac{\omega R_p t}{h_p}\right)^2}}$
Cone-plate		$\alpha'(t) = \sqrt{1 + \left(\frac{\omega t}{\beta}\right)^2}$	$s(t) = \frac{s_0}{\sqrt{1 + \left(\frac{\omega t}{\beta}\right)^2}}$
Screw-extruder		$\alpha'(t) = \sqrt{1 + \left(\frac{\omega_{ext} R_{ext} t}{R_{ext} - R_{screw}}\right)^2}$	$s(t) = \frac{s_0}{\sqrt{1 + \left(\frac{\omega_{ext} R_{ext} t}{R_{ext} - R_{screw}}\right)^2}}$

The maximum and minimum shear rates in each section are, respectively,

$$\dot{\gamma}_{max} = \frac{\omega_{ext} R_{ext}}{R_{ext} - (R_{screw} + r/2)},$$

and

$$\dot{\gamma}_{min} = \frac{\omega_{ext} R_{ext}}{R_{ext} - (R_{screw} - r/2)},$$

An average value is considered for all the domain as,

$$\dot{\gamma} = \frac{\omega_{ext} R_{ext}}{R_{ext} - R_{screw}}.$$

Mixing in the Screw extruder

Figure 2 shows the pathlines in the screw extruder obtained from a 3 CFD simulation. The pathlines show a flow that resembles the Taylor-Couette flow between concentric rotating. Although, in this case, there is an axial velocity component due to screw geometry. Also in Figure 2 is a tracer mass fraction map that shows the creation of an interfacial area in a binary mixture in this flow.

The mixing mechanisms shown in Figure 2 are based on the shear rate field that was well-approximated in the assumptions leading to the screw-extruder equations for α' and $s(t)$ in Table 1.

Figure 2. (left) Pathlines in the extruder and (right) mass fraction maps of a binary mixture.

Product development parameters

The parameters for the rheokinetic formulation are determined by matching the interfacial area generation and striation thickness thinning in the extruder and in the rheometric flows, in this case the cone-plate or plate-plate with two fluids set side-by-side. The rheometric flows are steady but in each location the mixing scales evolve with time, while in the extruder the evolution occurs in space along the axial direction. The relation between time and the position in the extruder, z, is given by

$$t = z/v_z$$

For the cone-plate

$$\omega = \frac{\omega_{ext} R_{ext} \beta}{R_{ext} - R_{screw}}$$

For the plate-plate

$$\omega = \frac{\omega_{ext} R_{ext} h_p}{(R_{ext} - R_{screw}) R_p}$$

where h_p is the gap between the top and bottom plate, R_p is the radius of the plates, and β is the angle between the cone and the plate.

Conclusions

In this study, we assessed a rotational rheometric device for developing polymeric materials in reactive extrusion, matching the evolution of mixing scales in fluids with similar viscosity. CFD results show a similar evolution of homogeneous lamellar structures in the extruder and the rheometric flows.

As a result, the rheometric rotational devices are viable methods for HTDM, enabling them to accurately replicate the mixing scales observed in pilot machines at a laboratory scale.

Acknowledgments

This work was financially supported by LA/P/0045/2020 (ALiCE), UIDB/50020/2020, UIDP/50020/2020 (LSRE-LCM), LA/P/0007/2021 (LA SusTEC), UIDB/00690/2020 and UIDP/00690/2020 (CIMO), funded by national funds through FCT/MCTES (PIDDAC); BioShoes4All (PRR - 000011), supported by PRR - Plano de Recuperação e Resiliência - Bioeconomia.

References

[1] Ottino, J.M., Ranz, W.E., Macosko, C.W.. Chemical Engineering Science. 34 (1978). 877-890.
[2] Mohr, W.D., Saxton, R.L., Jepson, C.H.. Industrial & Engineering Chemistry Research. 49 (1957). 1857-1862.
[3] Ranz, W.E. AIChE Journal. 25 (1979). 41-47.
[4] B. Imré, L. Garcia, D. Puglia, F. Vilaplana, *Carbohydrate Polymers*. 209 (2019). 20.
[5] Brito, M.S.C.A, Matos, J., Torres, M.V.L., Fonte, C.P., Dias, M.M., Lopes, J.C.L., Santos, R.J. *AIChE Journal*, 68 (2022). 4.

An Experimental Study on the Influence of Additive Volume on Mixing Performance in Turbulent Stirred Vessels

Poster.40

S S Mirfasihi[1], *W Basu*[1], *S Sarrafi*[1], *P Martin*[1], *A Kowalski*[2], *C P Fonte*[1], *A Keshmiri*[1]. *(1) The University of Manchester, Oxford Road, Manchester M13 9PL, United Kingdom, sorosh.mirfasihi@manchester.ac.uk, (2) Unilever R&D Port Sunlight Laboratory, Bebington CH63 3JW, United Kingdom*

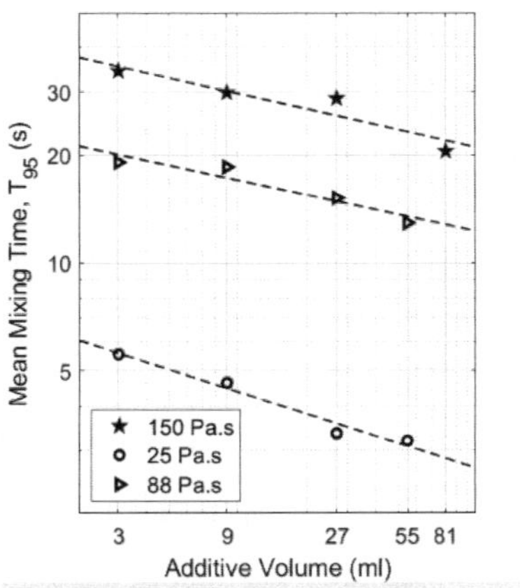

The mixing of highly viscous liquids in stirred reactors is a routine practice in process-based industries such as pharmaceutical, petrochemical, and Fast-Moving Consumer Goods (FMCG). Despite the importance of accurate mixing performance prediction in achieving high-quality production, the dependence of the mixing dynamics on the parameters involved such as the volume and characteristics of the additive liquid, especially in the additive-controlled regime (Re*<100), remains poorly understood. This article presents the preliminary results from an experimental study on the influence of volume fraction on dimensionless mixing time for a Newtonian additive mixed in a turbulently stirred benchtop vessel. The dynamics of the mixing process are characterized using electrical resistance measurements performed via an Electrical Resistance Tomography (ERT) instrument. We describe the methodology of the study including details of the experimental setup, design-of-experiments and the algorithms used for arriving at dimensionless mixing time estimates, present the trends observed in mixing time as a result of varying volume fraction, and propose an interim hypothesis that seeks to explain the observed trends.

Introduction

The mixing of highly viscous additives into a less viscous medium has been the subject of various studies in the past. Given a fully turbulent regime within the vessel, and a relatively low viscosity contrast between the two liquids, the dimensionless mixing time can be calculated using Grenville's correlation [1]. However, as the viscosity of the added liquid increases, this correlation fails to predict mixing times with sufficient accuracy. As part of the CAFE4DM [2] research initiative, we are currently performing a multifactorial study to assess the effect of various parameters such as additive viscosity, impeller speed, volume of additive etc. on mixing time so as to develop a model capable of mixing time prediction with improved accuracy. Having presented the results from other phases of this study corresponding to additive viscosity and impeller speed variation before [3], the authors would like to take this opportunity to focus on the impact of additive volume on mixing time. While a small number of past studies [4, 5] have evidenced a variation in mixing time with the volume of additive being mixed, a rigorous experimental characterization of the effect across a wide range of relevant mixing conditions such as additive viscosity, density etc. is not known from the literature and no analytical models are known to describe this dependence either. In the following sections, we present the methodology of the study including details of the experimental setup and the algorithms used for arriving at mixing time estimates, the preliminary results from the study and a consideration of the next steps to be taken for an improved understanding of the effect of additive volume on mixing time.

Material and Methods

The experimental setup used in this study consists of a standard baffled cylindrical vessel (diameter, T = 0.15 m) with a centrally mounted impeller. The data acquisition setup for real-time detection of fluid properties comprises of an ITS v5r ERT (Electrical Resistance Tomography) instrument [6] and a host computer. The instrument is controlled from the host computer over a USB connection and acquires adjacent pairwise trans-impedance measurements at a rate of 312 frames per second across the 4 active planes (shown in green in Figure 1) by applying ~0.5 mA current at 9.6 kHz. Figure 1 below contains a diagram of the key elements of this setup.

Figure 1. Experimental setup

The vessel is a standard baffled cylindrical tank with diameter, T = 0.15 m, filled with tap water to a volume, Vb = 2.65 litre. The stirrer, a 6-blade Rushton turbine with diameter, D = T/3 = 0.05 m, is positioned at the centre and at a height of H = T/3 = 0.05 m from the bottom surface and was driven using a Heidolph Hei-TORQUE Ultimate 100 digital stirring motor. The impeller speed was

https://doi.org/10.52202/074122-0086

set to N = 1000 rpm = 16.7 rps to achieve a fully turbulent regime with a Reynolds number, Re ~ 4×10^4. With tap water as the bulk liquid, 3 additives with viscosities, μ_a = 25, 88 & 150 Pa.s were used, each prepared by mixing glucose syrup and salt in water for altering the additive viscosity and electrical conductivity values respectively. For each run of the experiment, a predefined volume, V_a, of an additive was delivered from above, approximately midway between the stirrer shaft and a baffle, using truncated plastic syringes. To assess the impact of additive volume on mixing time, 4 values of V_a = 3 ml, 9 ml, 27 ml, and 55 (or 81) ml were used, and each experiment was repeated 15 times to help overcome the influence of experimental variation on the observed trends.

In order to determine the mixing time in each experiment, trans-impedance measurements obtained using the ERT setup were processed to obtain dynamic estimates of the concentration distribution throughout the vessel. Next, the convergence rate of the concentration distribution to homogeneity was characterized using the mixing index, σ_{RMS}, as per the following equation [1]:

$$\sigma_{RMS}(t) = \sqrt{\frac{1}{N_e}\sum_{k=1}^{N_e}\left(\frac{c_{t,k}-c_{0,k}}{c_{\infty,k}-c_{0,k}} - 1\right)^2} \qquad (1)$$

where,

N_e = number of elements,

$c_{t,k}$ = k^{th} measurement/concentration at time 't',

$c_{0,k}$ = k^{th} measurement/concentration prior to the additive being introduced, and

$c_{\infty,k}$ = k^{th} measurement/concentration after mixing has progressed to completion.

Finally, the 95% mixing time, T_{95}, was determined as the time taken for $log_{10}(\sigma_{RMS}^2)$ to decrease from the instant of additive introduction and settle below $log_{10}(1-0.95)^2 = log_{10}(0.05)^2 = -2.6$, as described in [1].

Results and Discussion

Figure 2 shows the box plots of the observed mixing times (T_{95}) against V_a for each additive. In each case, we observe a distinct downward trend i.e. a reduction in mixing time with additive volume. While the mixing time values show significant variance and thus overlapping quartiles, the existence of a trend is particularly evident from a comparison of the cases corresponding to the minimum and maximum additive volumes – the median mixing time for the highest volume is well below the first quartile for the lowest volume in each case.

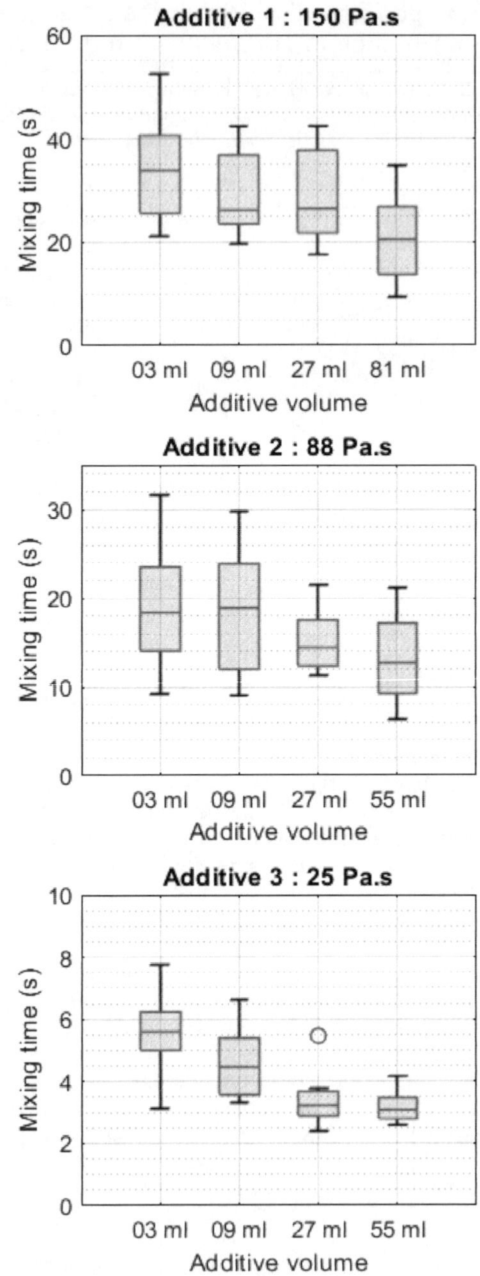

Figure 2. Mixing time trend as a result of variation in additive volume

While it may be counterintuitive for mixing to progress faster when the amount of additive being mixed is increased, this may be explained by the distribution of shear rates over the volume of the vessel. The shear stresses are highest near the impeller and reduce as we get closer to the vessel wall. It is likelier for a larger pulse of additive to enter a high shear zone and disintegrate than a smaller pulse which may remain largely intact and away from such zones. Videos of the mixing process appear to confirm that especially for viscous additives, a 3 ml pulse tends to travel around the vessel while slowly dissolving whereas

https://doi.org/10.52202/074122-0086

a 27 ml pulse of the same additive begins to rend apart immediately upon entering the vessel. Also, the injection of higher volume fractions of the viscous material decreased the additive's residence time in the low-pressure zones which are formed behind the baffles, resulting in an increase in the overall mixing effectiveness measured by dimensionless mixing time.

Conclusions

A negative correlation was observed between dimensionless mixing time and additive volume in a set of mixing experiments featuring Newtonian additives of vastly different viscosities. Further experimental investigation is necessary to validate this novel finding by ruling out the possible influence of confounding factors. Upon validation, an explanation of the observed phenomenon may then be pursued via computational and analytical means.

Acknowledgments
The authors would like to acknowledge Unilever and the CAFE4DM consortium for funding and technical input.

References

[1] R. K. Grenville and A. W. Nienow, "Blending of Miscible Liquids," Handb. Ind. Mix., pp. 507–542, Jan. 2004, doi:10.1002/0471451452.CH9.

[2] "CAFE4DM - The University of Manchester." https://www.cafe4dm.manchester.ac.uk/ (accessed Oct. 18, 2022).

[3] S. S. Mirfasihi, W. Basu, T. L. Rodgers, C. P. Fonte, A. Keshmiri, P. Martin and A. J. Kowalski, "An experimental study using Electrical Resistance Tomography on the influence of additive viscosity on mixing time in turbulently stirred vessels," 26th International Congress of Chemical and Process Engineering 2022. https://2022.chisa.cz/

[4] A. B. Pandit, P. R. Gogate, and V. Y. Dindore, "Effect of tracer properties (volume, density and viscosity) on mixing time in mechanically agitated contactors," *10th Eur. Conf. Mix.*, pp. 385–393, Jan. 2000, doi: 10.1016/B978-044450476-0/50049-2.

[5] T. L. Rodgers, M. Cooke, F. R. Siperstein, and A. Kowalski, "Mixing and dissolution times for a cowles disk agitator in large-scale emulsion preparation," *Ind. Eng. Chem. Res.*, vol. 48, no. 14, pp. 6859–6868, 2009, doi: 10.1021/ie900286s.

[6] "v5r (Electrical Resistance Tomography) | ITS Ltd." https://www.itoms.com/products/v5r-electrical-resistance-tomography/ (accessed Oct. 11, 2022).

https://doi.org/10.52202/074122-0086

AUTHOR INDEX